Springer Atmospheric Sciences

More information about this series at http://www.springer.com/series/10176

David A. Randall · J. Srinivasan
Ravi S. Nanjundiah · Parthasarathi Mukhopadhyay
Editors

Current Trends in the Representation of Physical Processes in Weather and Climate Models

Springer

Editors
David A. Randall
Department of Atmospheric Science
Colorado State University
Fort Collins, CO, USA

Parthasarathi Mukhopadhyay
Indian Institute of Tropical Meteorology
Pune, Maharashtra, India

J. Srinivasan
Centre for Atmospheric and Oceanic Sciences
Indian Institute of Science
Bengaluru, Karnataka, India

Ravi S. Nanjundiah
Indian Institute of Tropical Meteorology
Pune, Maharashtra, India

and

Centre for Atmospheric and Oceanic Sciences
Indian Institute of Science
Bengaluru, Karnataka, India

ISSN 2194-5217 ISSN 2194-5225 (electronic)
Springer Atmospheric Sciences
ISBN 978-981-13-3395-8 ISBN 978-981-13-3396-5 (eBook)
https://doi.org/10.1007/978-981-13-3396-5

Library of Congress Control Number: 2018962118

This Springer imprint is published by the registered company Springer Nature Singapore Pte Ltd.
The registered company address is: 152 Beach Road, #21-01/04 Gateway East, Singapore 189721, Singapore

Dev Raj Sikka
(former Director, Indian Institute of Tropical Meteorology, Pune)
(1932–2017)

Foreword

An international workshop on 'Representation of Physical Processes in Weather and Climate Models' at IITM, Pune, India, on 13–16 February 2017 was an event of great significance for many reasons. First of all, there is an urgent national need for improving monsoon rainfall forecasts, as monsoon rainfall plays a big role in determining the country's agricultural production and hence also its gross national product and the state of its economy. There are other related reasons too: droughts, storms and cyclones can cause much destruction, to both human and material. Secondly, the recent availability of much higher computing power for weather forecasting enables Indian scientists now to set their goals higher and expect that the quality of forecasts will have a significant improvement in the near future. Thirdly, there is a great deal of research going on elsewhere in the world that is potentially of immediate relevance to the monsoons, and a detailed discussion of these advances, in India and abroad, can be most useful in identifying newer lines of attack that hold promise.

From these and other points of view, INTROSPECT 2017 appears to have been a very successful meeting. I was unfortunately unable to be present at the meeting, but the availability of material that one can examine on the net has been a great help. The final recommendations made during the panel discussion on 16 February 2017 provide a very useful collection of comments and suggestions made at the meeting by the participants, including specific proposals for future action. These have dealt with a variety of issues, including what kinds of data are currently available and what else needs to be measured or acquired, what new facilities would be required and how one may proceed about it, what frequency of observations is best and other related matters.

The value of collaboration, both within India and with foreign agencies and scientists, was rightly emphasized at the workshop. An annual meet in India would be useful for assessing any improvements in model fidelity since the previous meet. Global inter-comparisons would be valuable, and model outputs should be available, especially to the academic community, to analyse the implications of inter-comparison results. The emphasis on these issues shows how important they are for further progress.

An interesting point that came up during the panel discussion showed concern that, for the limited number of modelling experts India has, perhaps we are using too many modelling and data assimilation tools. Dr. Bechtold felt that two models should really be enough (GFS and UKMO), along with their data assimilation systems. A related question was about whether India should have its own model. Prof. Kinter pointed out that there is no need to start from scratch and it would be better to improve some other model known to be well documented and validated.

At the end, there were a set of unanimous recommendations. One of these was for the formation of a Web-based consortium that would keep MoES institutions and interested academic groups in touch with each other, exchange ideas, have dialogues, etc. An annual workshop in which all model developers in India could get together for discussions and exchange of views is another recommendation. The title for the present meeting, namely INTROSPECT 2017, already implies that more such meetings are intended; it was suggested that a two-to-three-year period between successive meetings might be about right. An advisory committee involving global experts on weather and climate modelling can play a useful role in improving the efforts in India.

It is clear that the 2017 workshop was most effective and successful; it can be a landmark in the history of Indian meteorology if the recommendations at the workshop are all followed. I must congratulate MoES, IITM and all the others, particularly those from abroad, who made the meeting so interesting and useful. It has the potential to trigger systematic and substantial progress in Indian efforts to improve monsoon forecasting in the not too distant future.

Bengaluru, India
February 2018

Roddam Narasimha
Jawaharlal Nehru Centre for
Advanced Scientific Research

Preface

An International Workshop on "Representation of Physical Processes in Weather and Climate Models (INTROSPECT)" was held during 13–16 February 2017 at the Indian Institute of Tropical Meteorology in Pune, India. The workshop focused on the significant advancements made in the field of numerical weather prediction models and the challenges ahead. It brought together eminent scientists from across the globe to discuss the plausible future developments in the representation of unresolved physical processes in high-resolution models. A large number of masters and doctoral students and early career scientists attended the lectures and participated in the tutorial sessions conducted by the senior experts. All the talks and presentations were live-streamed and later uploaded in the website (https://www.tropmet.res.in/introspect/) and IITM youtube. The workshop ended with a consensus recommendation on the representation of physical processes in numerical models which can serve as guidelines for future developments in parameterization of physics in high-resolution weather prediction models. Most of the presentations in INTROSPECT are compiled in this book. The editors thank the authors of each chapter for their excellent summaries of their talks. We hope that this book will be an important reference for students and researchers in the coming decade.

Fort Collins, USA — Prof. David A. Randall
Bengaluru, India — Prof. J. Srinivasan
Pune, India — Prof. Ravi S. Nanjundiah
Pune, India — Dr. Parthasarathi Mukhopadhyay
26 January 2019

Acknowledgements

The convener and organizer of INTROSPECT 2017 gratefully acknowledge the encouragement, guidance and advice of Dr. M. Rajeevan, Secretary, Ministry of Earth Sciences (MoES), Government of India, in organizing the workshop. We are grateful to Director, IITM, Pune, for all the advice and support for organizing the workshop. MoES, Government of India, is thankfully acknowledged for the financial support and all other logistic clearances. We express our gratitude to Dr. Roddam Narasimha for kindly agreeing to write the Foreword of the book. Our special thanks to Dr. R. Krishnan for constant encouragement and support to organize the workshop. We are grateful to all the speakers and authors of the chapters for their valuable contribution. We express our thanks and gratitude to the editors of the book.

Contents

About the Editors

Prof. David A. Randall received his Ph.D. in atmospheric sciences from the University of California, Los Angeles, in 1976. He served as Faculty Member in the Department of Meteorology, Massachusetts Institute of Technology, from 1976 to 1979. Subsequently, he served as Meteorologist at the Global Modeling and Simulation Branch, NASA Goddard Space Flight Center (1979–1988). Since 1988, he has been a member of the faculty in the Department of Atmospheric Science, Colorado State University, where he is currently University Distinguished Professor. He has received prestigious awards including the Jule Charney Award from the American Meteorological Society, 2014; Coordinating Lead Author for the Fourth Assessment of the Intergovernmental Panel on Climate Change, which shared the 2007 Nobel Peace Prize; Fellow, American Geophysical Union, 2002; NASA's Medal for Distinguished Public Service, 2006; the Meisinger Award from the American Meteorological Society, 1994; NASA's Medal for Exceptional Scientific Achievement, 1988; and Fellow, American Meteorological Society, 1986. He has published more than 230 peer-reviewed research papers and 2 books.

Prof. J. Srinivasan is currently Distinguished Scientist at Divecha Centre for Climate Change, Indian Institute of Science, Bengaluru. He received his Ph.D. from Stanford University and then worked as a member of faculty in the Department of Mechanical Engineering, IIT Kanpur, until 1982, when he joined the Department of Mechanical Engineering, Indian Institute of Science, Bengaluru. He worked as NRC Senior Resident Research Associate for NASA in Langley, Virginia, USA, from 1993 to 1995. He subsequently served as Chairman of the Centre for Atmospheric and Oceanic Sciences (1996–2005) and as Chairman of the Mechanical Sciences Division, Indian Institute of Science, Bengaluru (2005–2009). He established the Divecha Centre for Climate Change, Indian Institute of Science, Bengaluru, in 2009 and was its Chairman until 2016. His research interests include monsoon models, monsoon variability, simple climate models, satellite meteorology and renewable energy. He served as Lead Author on the IPCC's Fourth and Second Assessment Reports and as Review Editor on its Third Assessment Report. He is Fellow of Indian Academy of Engineering, Indian National Science Academy, Indian Academy of Sciences and Indian Meteorological Society and has published 140 research papers in peer-reviewed journals.

Ravi S. Nanjundiah received his Doctor of Philosophy in atmospheric sciences from the Indian Institute of Science, Bengaluru, India, in 1992. He worked as Postdoctoral Fellow in the Computer Hardware, Advanced Mathematics and Model Physics (CHAMMP for studying global change) project at the Mathematics and Computer Science Division of Argonne National Laboratory from 1993 to 1995. He has been Faculty Member of the Centre for Atmospheric and Oceanic Sciences, Indian Institute of Science, Bengaluru, since 1995. He served as Chairman, Centre for Atmospheric and Oceanic Sciences, from 2013 to 2017. Since March 2017, he has been Director of the Indian Institute of Tropical Meteorology, Pune, India. He served as Associate Editor for the Journal of Earth System Sciences (published jointly by the Indian Academy of Sciences, Bengaluru, and Springer) from 2008 to 2014. He was awarded the Sir C. V. Raman Young Scientist Award for the year 2000 in the field of earth sciences and the NVIDIA Innovation Award for the year 2013. He has published over 75 papers in peer-reviewed journals.

Dr. Parthasarathi Mukhopadhyay is currently working as Senior Scientist at IITM, Pune, India. He received his Ph.D. from Savitribai Phule Pune University in 2005 and has since worked extensively in the fields of numerical modelling of mesoscale systems, thunderstorms and tropical cyclones. He has particularly contributed to the development of cloud and convective processes in numerical models. He was awarded the Silver Jubilee Award from IITM, Pune, for the best research paper in 2002 and 2010 and has received the Ministry of Earth Science's Certificate of Merit for outstanding contributions in Atmospheric Science and Technology in 2015. As Adjunct Professor in the Department of Atmospheric and Space Sciences, Savitribai Phule Pune University, he has published 46 papers in peer-reviewed journals.

Model Systematic Errors in the Annual Cycle of Monsoon: Inferences from Process-Based Diagnostics

H. Annamalai

Abstract Forecasting monsoon rainfall using dynamical climate models has met with little success, partly due to models' inability to represent the monsoon precipitation annual cycle accurately. Here, we review and examine the nature and dynamical causes of their biases. We discuss the coupled nature of the monsoon annual cycle from observations and then present errors in multi-model-mean, climatological fields of ocean–atmosphere variables determined from CMIP5. We argue that in CMIP-era models, there is a spatial redistribution in the organization of convection, and precipitation biases are longitudinally oriented with "wet-west" and "dry-east" over the tropical Indian Ocean, with wet (dry) biases prominent over the climatological dry (wet) regions. Irrespective of resolutions and varied physical parameterizations employed in CMIP-era models, the robustness in the biases across the suite of models suggests that multiple processes and their interactions lead to these persistent errors. We review recent literature that addressed the source(s) of model errors and indicate the importance of examining both atmospheric and oceanic fast processes. After discussing the unique nature of observed convection peak during May over the western Indian Ocean, we demonstrate through idealized experiments, how errors in the representation of ocean–atmosphere feedbacks along the equatorial Indian Ocean impact monsoon precipitation errors. We apply process-based diagnostics to identify the relative role of moist and radiative processes and show how systematic errors in certain parameterizations could anchor model biases in precipitation. Despite devoted efforts by the model development teams, persistence of model errors leads us to ask: are there fundamental limits to realistically simulating the monsoon annual cycle? Can a concerted observational and modeling effort enhance models' fidelity in simulating the monsoon? We summarize the pertinent issues on modeling, and limitations on observations to constrain model physics, and stress the need for coordinated activities across diagnostics, modeling, and observational personnel.

Keywords Systematic errors · Processes representation · Fundamental limits

---

H. Annamalai (✉)
Department of Oceanography, International Pacific Research Center, IPRC/SOEST, University of Hawaii, 1680 East West Road, Honolulu, HI 96822, USA
e-mail: hanna@hawaii.edu

D. A. Randall et al. (eds.), *Current Trends in the Representation of Physical Processes in Weather and Climate Models*, Springer Atmospheric Sciences,
https://doi.org/10.1007/978-981-13-3396-5_1

1 Introduction

a. Background

In South and Southeast Asia, farming and agriculture-related markets employ about two-thirds of the workforce, and therefore, the seasonal-mean monsoon precipitation during boreal summer and winter seasons exerts significant impacts on socioeconomic conditions in the region, particularly with long-lasting imprints on small-scale farmers during years of consecutive drought-like conditions. As such on timescales of days-to-seasons, demands for accurate prediction of spatial distribution of precipitation are increasing. However, skill in spatial average (not to mention spatial distribution) of monsoon precipitation prediction by dynamical climate models remains low (Del Sole and Shukla 2002). One attribution to this low skill is model errors or biases in simulating monsoon annual cycle (Sperber et al. 2013; Annamalai et al. 2017). Identifying the source(s) of model errors and suggesting pathways for model improvements have been very demanding and challenging.

In the past few decades, concerted research from observations, sensitivity experiments with modeling and theory has demonstrated that monsoon results from complex (and yet unknown) interactions among the ocean, atmosphere, and land components of the climate system. Despite focused efforts by model development teams, realistic simulation of the monsoon annual cycle, particularly precipitation characteristics, has met with slow progress. Compared to observations, Fig. 1 summarizes the multi-model-mean (MMM) errors or biases in precipitation (ΔP) over the Asian-Australian monsoon region throughout the annual cycle. Here, biases or errors are defined by differences between the MMM fields from a suite of Coupled Model Intercomparison Project Phase 5 (CMIP5) models (Taylor et al. 2012) and observations. The spatial patterns as well as amplitude of these errors have been persisting for over a decade (Sperber et al. 2013) suggesting that model limitations in multiple processes and their interactions lead to these systematic errors (Annamalai et al. 2017). Briefly, wet bias is noted over the southern tropical Indian Ocean with a local maximum centered at 60°E, and during the course of the annual cycle, the pattern is meandering between 15°S and 10°N. During boreal summer, the structure of the wet bias is prominent all along the low-level climatological cross-equatorial flow path (Annamalai et al. 2017) with the bias core situated over the western Arabian Sea. As regards dry bias, prominent patterns include: (a) the monsoon trough extending from the Indian subcontinent into the central-northern Bay of Bengal during boreal summer, (b) the eastern equatorial Indian Ocean (EEIO) during boreal fall and winter, and (c) parts of Bay of Bengal during all seasons. Compared to observations, modeled wet bias is persistent over the near-equatorial western-southern Indian Ocean, a region that is climatically "dry", and the dry bias is prominent in climatically "wet" spots. In summary, there is a spatial redistribution in convection over the tropical Indian Ocean, and can we identify model processes that cause it?

Identifying the cause(s) for model biases, however, is difficult to assess because intricate interactions among ocean, atmosphere, and land components are recognized to anchor the monsoon; as a result, misrepresentation of a process in one component

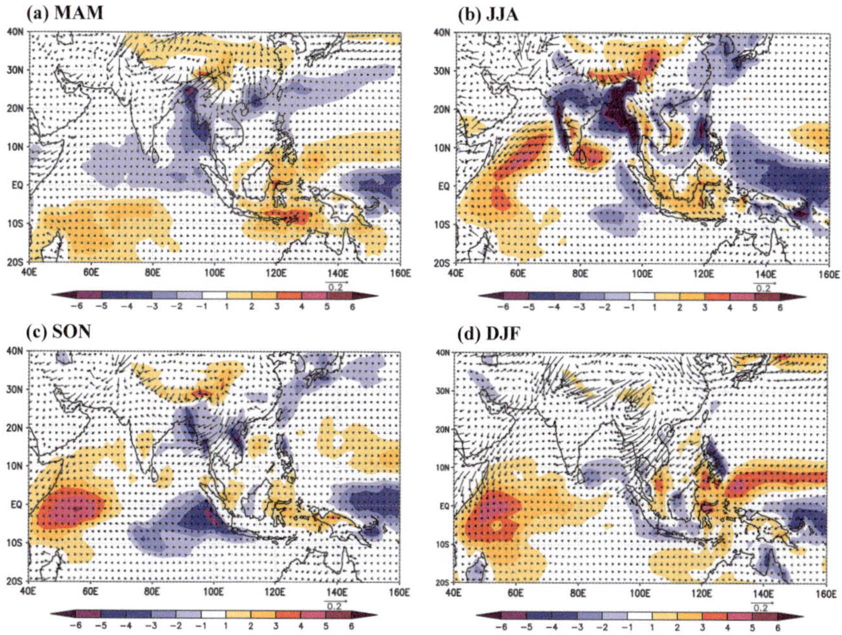

Fig. 1 Seasonal mean climatology difference between CMIP5 multi-model-mean (MMM) and observations of precipitation (shaded, mm/day) and wind stress (N/m^2): **a** March–May; **b** June–August; **c** September–November; and **d** December–February. Reference vector is also given

can lead to errors in other components. Further, in a coupled system, errors generated in one season can amplify due to differing background condition and also persist throughout the annual cycle (Sect. 3). An indicator of the difficulty of the problem is that such model errors have persisted for the last decade, despite considerable efforts to eliminate them.

b. Present study
In this chapter, first we present and discuss the coupled nature of the tropical Indian Ocean-monsoon climate systems, and highlight the role of large-scale air–sea interactions in shaping monsoon precipitation. Second, we discuss the systematic model biases in ocean–atmosphere variables pointing out how errors in one model component could cascade into another. Third, we review recent publications that identify the possible source(s) of model errors, and emphasize the need for realistically representing the coupled processes along the equatorial Indian Ocean (EIO), with particular focus on the oceanic processes during intermonsoon (April–May and October–November) seasons. Fourth, we apply process-based diagnostics specifically vertically integrated moisture and moist-static energy (MSE) budgets to CMIP5 MMM solutions to identify the representation of adiabatic (advection) and diabatic (fluxes of radiation, sensible, and latent heat) sources in priming column MSE. We examine if the error sources are seasonally invariant and ascertain if ΔP are entirely due

to ΔSST or not. Finally, we close the chapter by stressing the need for sustained observations of dynamics and thermodynamics of atmosphere and ocean systems to constrain model physics for improved representation of processes that are expected to improve simulation of the monsoon annual cycle.

2 Coupled Nature of the Monsoon—Tropical Indian Ocean Climate Systems

In this section, after summarizing the salient observational aspects of the monsoon annual cycle (Sect. 2.1), we identify coupled ocean–atmosphere processes that anchor the precipitation annual cycle over the tropical Indian Ocean (Sect. 2.2).

2.1 Monsoon Annual Cycle

To a first order, monsoon annual cycle can be viewed as "primarily driven by the seasonal displacement of the Intertropical Convergence Zone (ITCZ), which is anchored by the north-south migration of the Indo-Pacific warm pool (regions where SST is >28 °C)." Figure 2a, b plot observed climatology of SST (contours) and precipitation (color shading) during boreal summer and winter. "During winter, the ITCZ is 'zonally elongated' and resides around the equatorial latitudes while it is 'diagonally oriented' during summer, stretching from central India to the tropical western Pacific. During both seasons, there is intense rainfall (>8 mm/day) only in regions where SST is high (>28 °C), suggesting the latter is a necessary condition for the former (e.g., Graham and Barnett 1987)." The SST/precipitation relationship, however, is not one-to-one, indicating that other factors such as moisture availability in the atmosphere, tropospheric stability (Raymond 2000) and cloud-radiative feedbacks (Stephens et al. 2008) collectively determine column water vapor (CWV) that subsequently impact rainfall intensity (Bretherton et al. 2004; Bretherton 2007). *Do the CMIP-era models realistically represent the processes that determine CWV?* The difficulty lies in the fact that processes involved in priming CWV, amongst others, are moisture–convection, and cloud–radiation feedbacks that are a consequence of convection themselves. Furthermore, orography–flow interactions also contribute to rainfall intensity. More importantly, there are regional differences in the SST/precipitation relationship between the tropical Indian Ocean and west Pacific. *Do climate models capture these regional aspects and represent the diagonally oriented ITCZ during boreal summer?*

Annamalai et al. (2017) note that "during winter (summer), despite southward (northward) displacement of the thermal equator, SST over the Bay of Bengal (Equatorial Indian Ocean) remains warm with values around 28 °C, suggesting the role of oceanic processes in its maintenance. Important aspects of these oceanic pro-

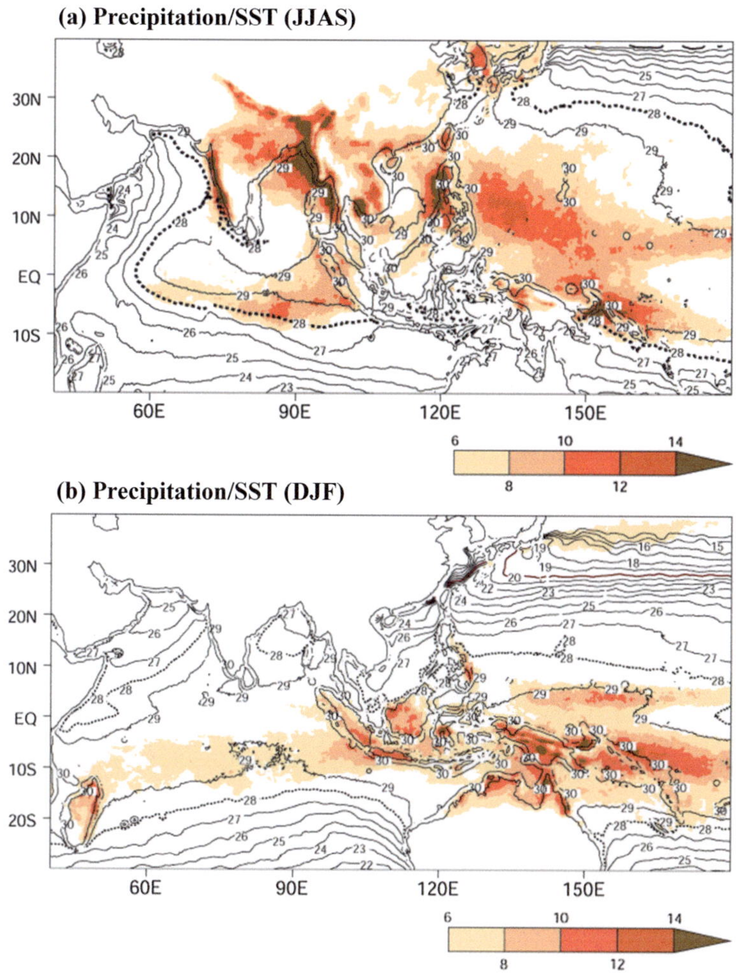

Fig. 2 **a** Boreal summer (June–September) climatology of precipitation (shaded; mm/day) and SST (contours; °C) constructed from TRMM/TMI products (1998–2015), and **b** same as (**a**) but for the boreal winter (December–February) season

cesses, such as upper-ocean stratification and equatorial eastward flowing Wyrtki Jets (WJs)," and their impacts on SST evolution are discussed next.

2.2 Ocean–Atmosphere Interactions in Shaping Monsoon Precipitation

In the northern Indian Ocean, there is a remarkable east–west asymmetry in SST and precipitation during the monsoon (Fig. 2a), which to a large extent results from ocean processes. Just prior to the monsoon, SST is the warmest of all the tropical oceans throughout the northern Indian Ocean (Joseph 1990). During the monsoon, there is an intense upwelling of cold water along the Somali and Omani coasts driven by the cross-equatorial low-level (Findlater) jet, and SST drops to about 23–24 °C. Subsequently, horizontal advection by ocean currents, in conjunction with evaporative cooling, cools SST in the central Arabian Sea (McCreary et al. 1993) and these processes limit the westward extension of the warm pool. During summer, SST cooling in conjunction with descent forced by Bay of Bengal convection weakens the rainfall over the Arabian Sea. During winter, lack of upwelling along the Somali and Omani coasts leads to SST being warmer there (Fig. 2b) than during the summer. In Sect. 5.1, we will show how a weakened monsoon circulation in CMIP5 models could be attributed to the modeled wet bias over the western Arabian Sea (Fig. 1b) during summer.

In contrast, SST over the Bay of Bengal remains high because upwelling along the east coast of India is weak or absent (McCreary et al. 1993), and rainfall is much stronger there resulting in upper-ocean salinity and temperature stratifications (Shenoi et al. 2002; Seo et al. 2009). During winter, cold and dry northeasterly monsoon winds are directed away from the Asian continent and due to excessive surface cooling, SST drops to <27 °C over the northern Arabian Sea but remains warm (~28 °C) over the Bay of Bengal (Fig. 2b). The near-surface stratification is impacted locally by the surface freshwater flux (precipitation and river runoff) and remotely through advection. The freshwater thins the surface mixed layer and generates a barrier layer that prevents the entrainment of cold subsurface waters (Lukas and Lindstorm 1991). Furthermore, solar radiation can penetrate the mixed layer to warm the barrier layer, thereby generating temperature inversions (Sengupta and Ravichandran 2001). The maintenance of time-mean SST during winter and spring anchors the ensuing summer monsoon. Observing and modeling details of the thin mixed-layer, upper-ocean stratifications in temperature and salinity, and their collective impact on SST evolution remains a grand challenge (Annamalai et al. 2018).

Quoting from Annamalai et al. (2017), "The equatorial Indian Ocean (EIO) differs from the other equatorial oceans in that it lacks trade winds, owing to the strong atmospheric convection over the eastern EIO and Maritime Continent (Fig. 2a, b). As a consequence, the EIO experiences semiannual westerly winds during the intermonsoon periods when the ITCZ crosses the equator, with a stress magnitude of about 0.4 N/m^2. They force the eastward-flowing WJs (Wyrtki 1973), which attain velocities of the order of 80 cm/s near the surface (Han et al. 1999; Fig. 5a). The WJs carry mass and heat from the western to the eastern EIO and are instrumental in maintaining the warm pool over the eastern EIO (Rao and Sivakumar 2000), so that even during summer warm and wet conditions prevail over the eastern EIO (Fig. 2a).

The WJs are predominantly forced by the near-equatorial winds but modeling studies suggest that salinity-induced barrier layer increases the jet speed by trapping the wind momentum in a thinner mixed layer (Han et al. 1999; Masson et al. 2003). Note that over the eastern EIO, precipitation occurs all year round (Fig. 2). A few points of interest are: (i) during the annual cycle, Bjerknes' feedback occurs along the EIO in which WJs are an important component (Annamalai et al. 2017) and (ii) WJs maintain eastern EIO warm pool that promotes local precipitation that in turn creates a salinity-induced barrier layer, an important process to increase WJ speed itself." We will show that misrepresentation of EIO processes could lead to systematic biases in ocean–atmosphere variables over the tropical Indian Ocean including the dry bias over South Asia during boreal summer (Sects. 4 and 5).

In contrast to the seasonal reversal in the winds over the northern Indian Ocean, the southern Indian Ocean experiences southeasterly trades throughout the annual cycle, and the wind stress curl between the south easterlies in the southern Indian Ocean and westerlies along the EIO implies an upwelling zone between 15°S and 5°S (McCreary et al. 1993) resulting in a shallow thermocline over the southwest Indian Ocean (SWIO; Xie et al. 2002), termed "thermocline ridge". Quoting Xie et al. (2002), "In the tropical oceans, wind-induced upwelling combined with a shallow thermocline results in a minimum in climatological SST and reduced precipitation, features noticeable along the thermocline ridge (Fig. 2a, b)." Nagura et al. (2013) showed that errors in wind stress ($\Delta\tau$; Fig. 1) are the primary candidates for the biases in modeled thermocline ($\Delta D20$), a finding discussed in Sect. 3. Interannual variations in the monsoon onset date are linked to variations in boreal spring SST over the ridge region through their impacts on the poleward migration of the ITCZ (Annamalai et al. 2005), particularly during years after the peak phase of El Niño when fluctuations in thermocline depth and SST are strongly coupled (Xie et al. 2002).

Atmospheric convection, through its effect on the vertical distribution of diabatic heating, influences surface winds that determine oceanic mixed-layer and thermocline processes, and subsequently SST. While high-mean SST through its impact on surface fluxes and CWV is a necessary condition for the occurrence of monsoon convection, in regions of high-mean precipitation such as Bay of Bengal and EEIO, freshwater-forced upper-ocean stratification impacts, perhaps determines SST (Shenoi et al. 2002; Seo et al. 2009). In summary, large-scale air–sea interactions play an active role in the monsoon precipitation annual cycle.

3 Persistent Model Errors over the Asian-Australian Monsoon Region

Here, we begin with a discussion of biases in seasonal-mean ΔP and $\Delta\tau$ over the broader Asian-Australian monsoon (Sect. 3.1) followed by discussions of biases in ocean variables (Sect. 3.2). Then, we interpret the biases in the EIO coupled system (Sect. 3.3).

3.1 Precipitation and Wind Stress Errors

As mentioned in Sect. 1a, in CMIP-era models, there is a spatial redistribution in the organization of convection and precipitation: biases are aligned "wet-west and dry-east" over the tropical Indian Ocean latitudes (15°S–20°N). The wet bias is strongest during fall and winter seasons while the dry bias peaks during summer. Over the annual cycle, the modeled wind stress (also winds in the lower troposphere extending from 1000 to 700 hPa—not shown) shows easterly bias over the northern Indian Ocean extending into parts of South China Sea, with stronger intensity during boreal summer (Fig. 1b) and winter (Fig. 1d) seasons. Easterly bias prevails along the near-equatorial central-eastern Indian Ocean throughout the year with a peak during boreal fall season (Annamalai et al. 2017; Fig. 7a). Over the southern Indian Ocean, modeled southeast trades are also weaker throughout the annual cycle.

In a broader view, the wet-west is dynamically connected with the dry-east by wind biases. For example, the near-equatorial easterly $\Delta\tau$ bias is interpreted as a Kelvin-wave response to positive ΔP over the western EIO (Fig. 1c), and the northeasterly bias associated with anticyclonic circulation features over the northern Indian Ocean is interpreted as a Rossby wave response to negative ΔP over the Bay of Bengal (Fig. 1b). Indeed, irrespective of resolutions and physical parameterizations employed, in almost all CMIP5 models, ΔP and $\Delta\tau$ are robust. It should be mentioned here that dry ΔP noted in boreal summer over the Bay of Bengal (Sperber et al. 2013) is not unique to summer but prevails during all seasons with varying strength (Fig. 1).

Throughout the year, modeled rainfall is stronger over the marginal seas of the Maritime Continent region (10°S–10°S; 100°–140°E), and wet ΔP maximizes in spring with signatures of westerly $\Delta\tau$ (Fig. 1a). During fall and winter seasons, dry ΔP prevails over the island chain of Borneo and Java. Dry ΔP is noted off Philippines for most of the annual cycle. In summary, "wet" bias over the marginal seas and "dry" bias over land are prominent features over the Maritime Continent.

Over the tropical western Pacific longitudes 140°–160°E a latitudinally oriented ΔP pattern, "wet-north and dry-south", is evident. In all seasons, modeled East Asian monsoon front extending from the Korean peninsula to Japan is weaker with dry bias peaking during spring (Fig. 1a). During winter season ΔP is weakly wet over northern Australia (Fig. 1d) but in all seasons and centered on 140°E northerly $\Delta\tau$ is noted. While the dynamical linkage between ΔP and $\Delta\tau$ biases are consistent over the tropical Indian Ocean, for reasons unclear, such linkages are not evident over the tropical western Pacific and Maritime Continent regions.

3.2 SST and Thermocline Depth Errors

While there are both wet and dry ΔP patterns (Fig. 1), throughout the domain (with few exceptions) modeled errors in SST (ΔSST, Fig. 3) and depth of the thermocline ($\Delta D20$, Fig. 4) are cold and deep, respectively. In regions where variations in ther-

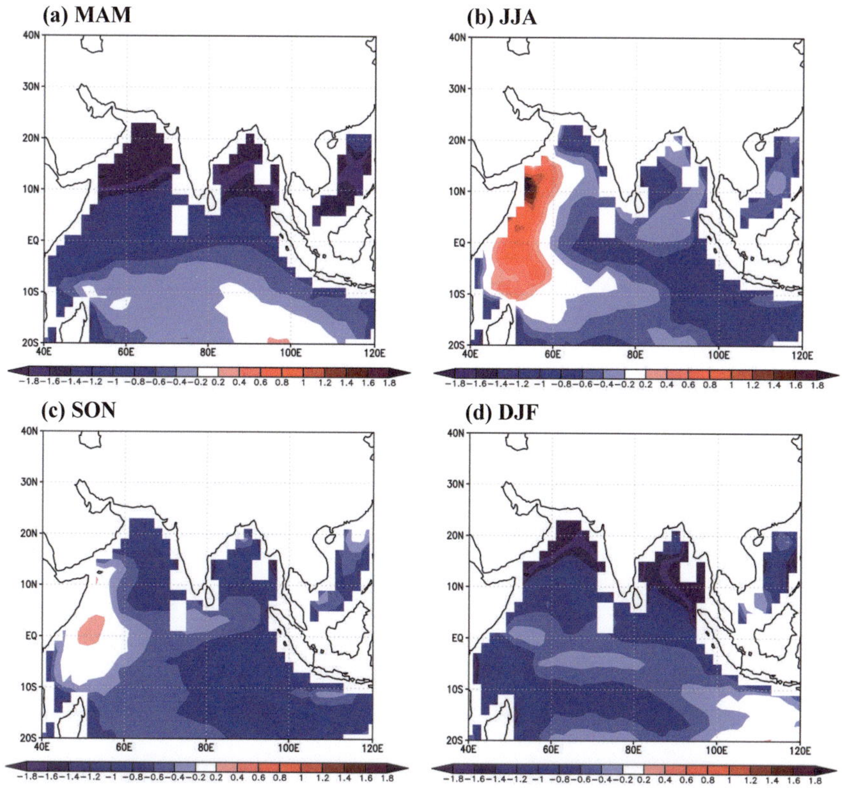

Fig. 3 Seasonal mean climatology difference between CMIP5 multi-model-mean (MMM) and observations of SST (°C): **a** March–May; **b** June–August; **c** September–November; and **d** December–February

mocline depth are known to impact SST, biases in them are physically consistent. They include warm ΔSST and deeper ΔD20 over the western EIO during fall, and along the Somali and Omani coasts during summer, and shallow ΔD20 and cold ΔSST over the EEIO during fall, and weakly positive or near-normal ΔSST along the thermocline ridge over the SWIO.

In CMIP-era models, persistence of northern Indian Ocean ΔSST, particularly those over the Arabian Sea is attributed to model biases in the simulation of winter monsoon circulation (Fig. 1d; Levine et al. 2013; Sandeep and Ajayamohan 2014) while the western Indian Ocean ΔSST during summer and fall seasons is attributed to weakened equatorial WJs during intermonsoon seasons and summer monsoon circulation (Annamalai et al. 2017), issues further discussed in Sect. 3.3. For the cold ΔSST over the Arabian Sea, our interpretation is that the wet ΔP over the near-equatorial southern Indian Ocean during boreal winter (Fig. 1d) forces a stronger local Hadley-type circulation with descent over the subtropical latitudes whose return low-level flow advect cold continental air, leading to excessive surface cooling over

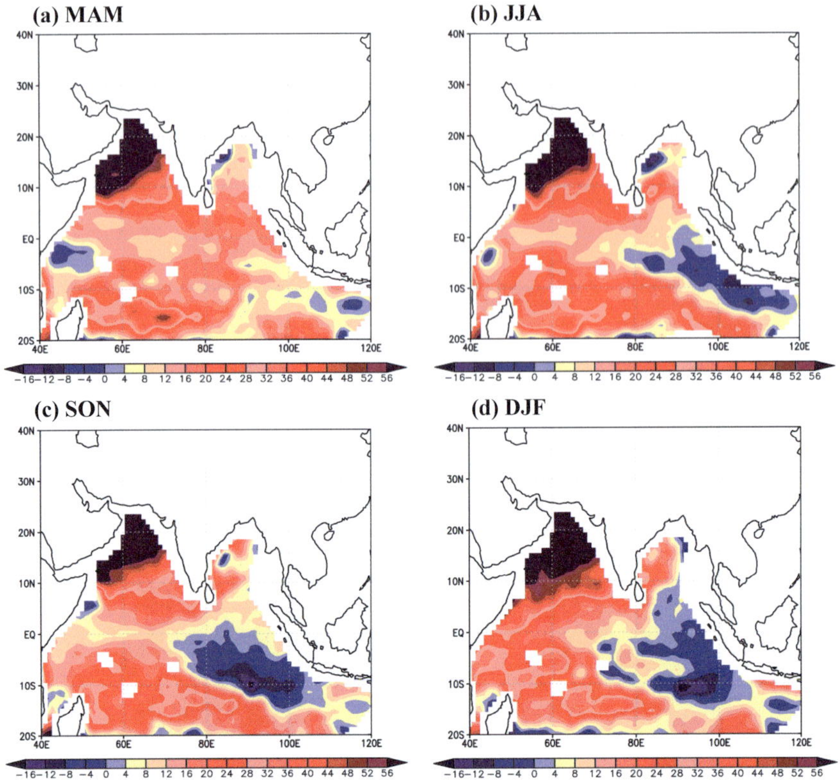

Fig. 4 Seasonal mean climatology difference between CMIP5 multi-model-mean (MMM) and observations of thermocline depth (m) as measured by depth of the 20 °C isotherm: **a** March–May; **b** June–August; **c** September–November; and **d** December–February

the northern Arabian Sea. The source of moist processes that determine wet ΔP is discussed in Sect. 5.2.

Due to persistent easterly bias ($\Delta\tau$) in the EIO and weakened southeast trades leading to a weak wind stress curl and Ekman pumping velocity, ΔD20 in CMIP-era models are deeper by 20–30 m (Fig. 4) over thermocline ridge region (Tozuka et al. 2010; Nagura et al. 2013). In the models, however, the largest ΔD20 occurs over the northern Arabian Sea (Fig. 4), and an examination of MMM mixed-layer depth (see Fig. 1 in Nagura et al. 2018) shows that largest biases (ΔMLT) occur in the northern Arabian Sea. Nagura et al. (2018) examined the processes that account for ΔD20 there. Briefly, their process-based study revealed that in most models ΔMLT influences ΔD20 and variations in ΔMLT are strongly linked to biases in the density stratification (jump) across the bottom of the mixed layer than to surface cooling biases. The density jump is in turn determined primarily by sea-surface-salinity biases (ΔSSS) that are advected into the northern Arabians Sea by the west India coastal current (see their Fig. 12), and the source of ΔSSS is the rainfall deficit associated with the models' weak, summer monsoon (Fig. 1b). Ultimately, then, ΔD20 is linked to deficit of monsoon rainfall over the Bay of Bengal. It suffices to mention here that largest ΔD20 over the northern Arabian Sea is due to nonlocal processes further highlighting the intrinsic difficulties in identifying source of model errors.

3.3 Misrepresentation of the Equatorial WJs and Coupled Processes Along the EIO

WJs transport mass and heat from the western to eastern EIO. Given that the time it takes for an equatorial Kelvin wave to travel the Indian Ocean basin is only a few days, we can expect that errors in wind stress ($\Delta\tau$) will quickly imprint on the simulated WJs. In the CMIP5 models, a direct consequence of the $\Delta\tau$ bias (Fig. 7a) is that both the spring and fall WJs, as measured by the depth-integrated (0–100 m) zonal current at 0°N, 85°E, are weaker by 50–60% compared to observations (Annamalai et al. 2017), and this weakness is apparent everywhere along the equator (compare Fig. 5b and 5a), with biases in the amplitude, phasing, and duration of the WJs; moreover, the summertime westward flow is erroneously strong. The weak WJs and stronger westward flow result in shallow (deep) thermocline biases in the eastern (western) EIO, features that persist for most of the annual cycle (Fig. 4). The errors in WJs and the resultant ΔD20 lead to ΔSST (Fig. 3), and to ΔP locally (Sect. 5) and subsequently onto $\Delta\tau$ along the EIO and the cycle continues leading to erroneous Bjerknes' feedback in the EIO.

4 Possible Sources for Monsoon Precipitation Errors

As mentioned above, it is recognized that complex interactions among the ocean, atmosphere, and land components of the climate system anchor the monsoon annual cycle. In this scenario, the errors in ocean–atmosphere variables are not mutually independent, and errors in one variable (e.g., $\Delta\tau$) can cascade into others (e.g., ΔD20), and errors that develop in one season can persist into the next seasons (e.g., boreal winter Δ SST persisting into the following spring and summer seasons over the Bay of Bengal), and errors are nonlocal in space (e.g., ΔP in Bay of Bengal impacting ΔD20 over the northern Arabian Sea). How do we then reconcile this problem and identify sources of errors, and improve the representation of processes and their interactions in models? Next, we review results from recent studies that attempted to isolate source(s) of model errors.

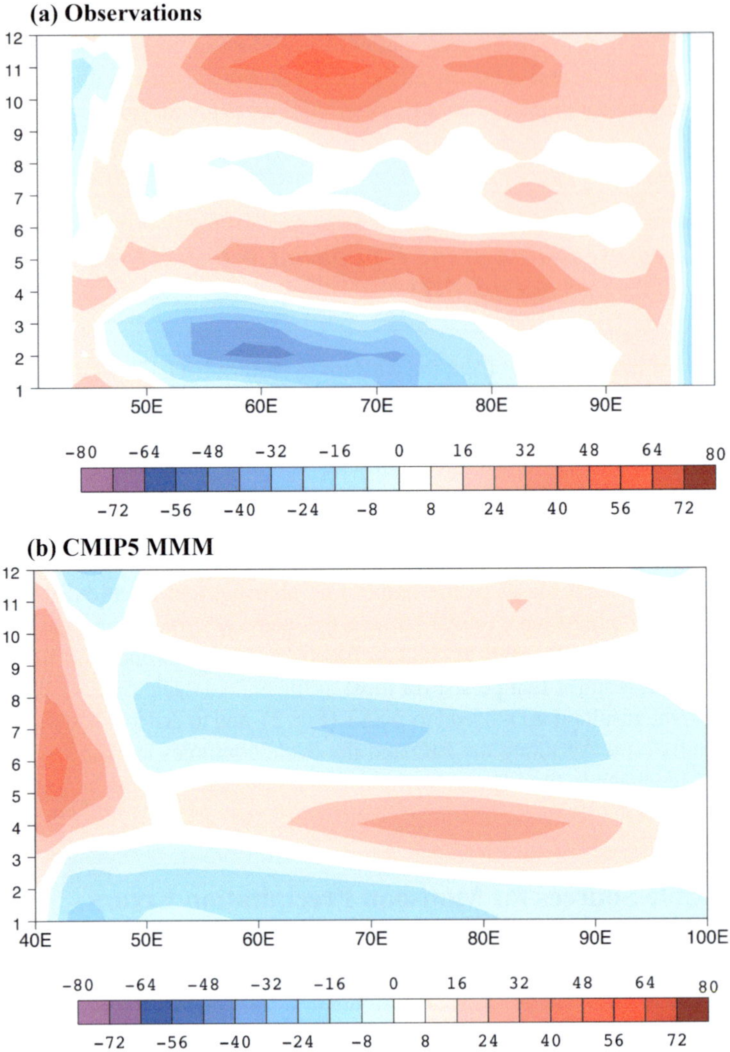

Fig. 5 Monthly evolution (ordinate) of the equatorial (3°S–3°N) zonal current (cm/s) from **a** observations and **b** CMIP5 MMM. The observations used here come from OSCAR product. Adopted from Annamalai et al. (2017)

4.1 Atmospheric Processes

Martin et al. (2010) and Ma et al. (2014) invoked initialized forecast approach as in Numerical Weather Prediction (NWP) to study the development of model errors in the Unified Model and in a suite of models, respectively. Initialized on 01 May, both studies noted a rapid growth of ΔP over the western EIO in the first 5 days of the simulation. The authors suggest that "it is a direct impact of parameterizations and not due to a nonlinear feedback process operating on longer time-scales." Note that NWP forecasts are initialized with "observational estimates" (initial states generated by data assimilation system) and therefore in this experimental setup, biases in SST and large-scale circulation are minimized, paving ways to assess the growth of model errors due to limitations in parameterizations (Klinker and Sardesmukh 1992; Phillips et al. 2004; Rodwell and Palmer 2007). The approach is a promising one.

Bush et al. (2015) incorporated changes to entrainment/detrainment values in the convection scheme of the Unified Model and performed initialized forecasts. While enhanced entrainment effectively reduced ΔP over the western EIO, modeled precipitation showed unrealistic wet bias over the tropical western Pacific. Zhao et al. (2018) reported that the systematic biases over the monsoon region persist in the three generations of the Geophysical Fluid Dynamics Laboratory (GFDL) model versions (from AM2.0 to the latest version AM4.0). In a series of experiments with changes to entrainment rates in AM4.0, particularly incorporating lateral mixing in the deep convective plume to a linear function of free tropospheric column relative humidity, Zhao et al. (2018) note that simulated precipitation increases over the tropical western Pacific for increased entrainment rate. A similar sensitivity to enhanced entrainment in Global Forecast System (GFS) model shows increased precipitation over the near-EIO and decreased precipitation over India (Mapes, 2017 personal communication), further aggravating model systematic errors. One inference is that Asian monsoon is comprised of multiple regional heat sources and given their close proximity to the equator, perturbations to one of the precipitation centers are quickly communicated to others by equatorial waves (Annamalai and Sperber 2005).

In climate models, why do ΔP persist over the western EIO? Let us consider the following scenario. Climatologically, the western EIO receives much less rainfall (dotted line in Fig. 6a) compared to the eastern EIO throughout the annual cycle (thick line in Fig. 6a). An examination of vertical velocity profile (not shown) suggests weak descent (strong ascent) over western (eastern) EIO for the most part of the annual cycle. Therefore, the western EIO is climatically a "dry" region and higher amount of MSE is required to trigger convection and maintain precipitation there. Climatologically during May, observed SST peaks around 30 °C (Fig. 6b) and promotes local evaporation (not shown), and a weak cross-equatorial flow develops (Fig. 6c) and transports moisture from the southern Indian Ocean. These processes lead to the accumulation of CWV in the western EIO. As a result, a local maximum in precipitation (~4 mm/day) is observed (Fig. 6a). Immediately thereafter in June, for reasons given in Sect. 2.2, SST drops to ~26 °C and rainfall decreases subsequently. It is during this "time window" of the annual cycle that the model errors begin to emerge, say wet bias, warm SST over the western EIO, and easterly $\Delta\tau$ in the EIO (Annamalai et al. 2017). Then, due to nonlinear atmospheric feedbacks, ΔP amplify, and the coupled nature of the monsoon-Indian Ocean climate systems anchors its persistence.

Hanf and Annamalai (2018) diagnosed a series of NCAR_CAM4/5 model solutions and reanalysis products. Their budget analyses revealed that after the "burst" of precipitation during late May–early June over the western EIO, accumulation of dry

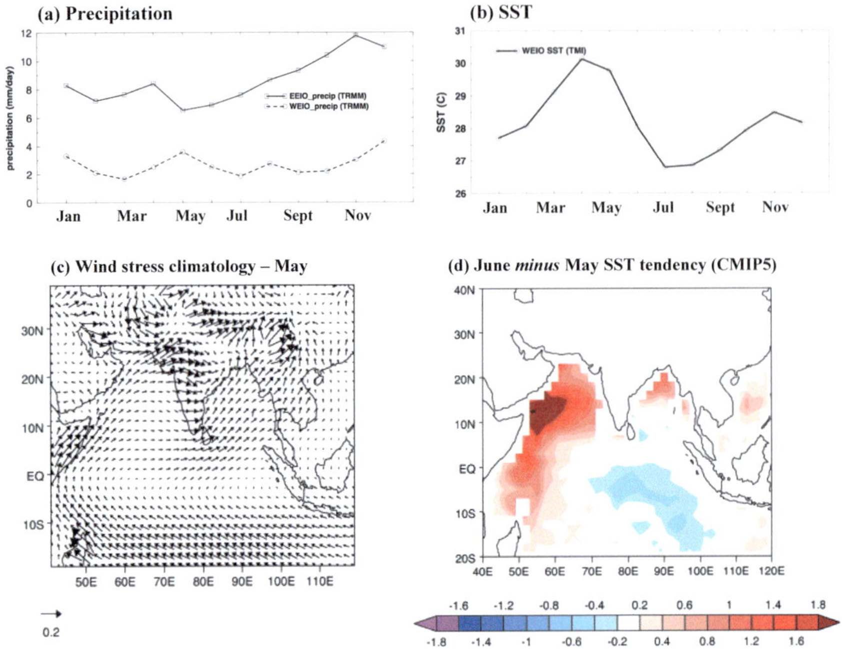

Fig. 6 **a** Observed monthly precipitation (mm/day) climatology averaged over eastern EIO (5°S–5°N; 90–105°E) and western EIO (5°S–5°N; 50–60°E); **b** observed monthly SST (°C) climatology averaged over western EIO; **c** observed wind stress (N/m^2) climatology during May; and **d** June *minus* May SST (°C) tendency bias in CMIP5 MMM. Adopted from Annamalai et al. (2017)

and cold air in the lower troposphere by horizontal advection is the dominant term in anchoring the demise of convection in mid-June. Encouraged by this, Hanf and Annamalai (2018) performed sensitivity experiments by increasing entrainment rates in convection scheme employed in NCAR-CAM4. Their solutions showed a reduction of convective precipitation over the western EIO and an increase of large-scale precipitation along the monsoon trough. In contrast to earlier studies on entrainment sensitivity, their results showed a reduction of model precipitation biases over South Asia.

In cumulus convection schemes, entrainment and detrainment coefficients have notable imprints on cloud properties and precipitation partitioning (Neale et al. 2008; Martin et al. 2010; Zhao et al. 2018), and models whose convection schemes are most responsive to free tropospheric moisture performs better (Lin et al. 2012; Neale et al. 2008). Details of entrainment role on vertical processes and their interactions on moisture–convection and cloud–radiation feedbacks, and precipitation partitioning are discussed in Hanf and Annamalai (2018).

4.2 Model Resolution and Orography

Few studies examined the sensitivity of modeled monsoon precipitation to higher horizontal resolutions. Goswami et al. (2015) and Abhik et al. (2017) performed integrations with CFSv2 coupled model with two horizontal resolutions, viz., ~100 km and ~38 km. In both solutions, ΔP over the tropics and particularly over the monsoon region remain identical, implying insensitivity of ΔP to resolution, at least in this model configuration.

Boss and Hurley (2013) hypothesized that in regions of sharp MSE gradient such as the Hindu Kush (along 73°E) if the height of the model orography is not realistically resolved, then horizontal advection of low MSE air will result in a dry bias over India. In CMIP3/5 models, the average height of orography along 73°E is around 3 km compared to ~5 km in observations. To validate their hypothesis, Boos and Hurley (2013) truncated model orography height in NCAR_CCSM4 to ~1.0 km and their sensitivity run developed a dry bias over India. While encouraging, solutions do not capture the corresponding biases over the western EIO and Bay of Bengal as well as reducing model orography height to ~1.0 km deserves attention.

4.3 Coupled Processes

Annamalai et al. (2017) adopted a different approach. In CMIP5 models, their analysis indicates that an easterly wind stress bias ($\Delta\tau$) along the EIO begins during April–May and peaks during November (Fig. 7a); the severity of the $\Delta\tau$ is that the WJs, eastward-flowing equatorial currents during the intermonsoon seasons (April–May and October–November), are almost eliminated (Fig. 5b). An erroneous, east-to-west SST gradient (warm west and cold east) develops in June. An examination of SST tendency errors (June *minus* May) in CMIP5 MMM clearly shows a warming (cooling) tendency over western (eastern) EIO and the emergence of SST gradient along the EIO (Fig. 6d). The structure of the model errors indicates that they arise from Bjerknes feedback in the EIO. In CMIP5 models, reversed gradient in precipitation along the EIO, and its persistence result from errors in the coupled processes.

To test their hypothesis that the ΔP over South Asia is due to weakened WJs leading to a too strong Bjerknes feedback, Annamalai et al. (2017) forced a coupled model, namely, Coupled model for Earth Simulator (CFES) developed at JAMSTEC, Japan with the easterly $\Delta\tau$ (Fig. 7a) noted along the EIO. In one of the sensitivity solutions where the easterly $\Delta\tau$ is introduced only during intermonsoon seasons (April–May; October–November), the sensitivity run develops westward equatorial currents throughout the year, suppressing the WJs in comparison to the control run (Fig. 7b). The anomalous forcing also excites oceanic Rossby and Kelvin waves that tend to deepen (shallow) the thermocline depth (Fig. 7c) in the western EIO (eastern EIO). In response, warm (cool) SST and wet (dry) conditions develop in each region

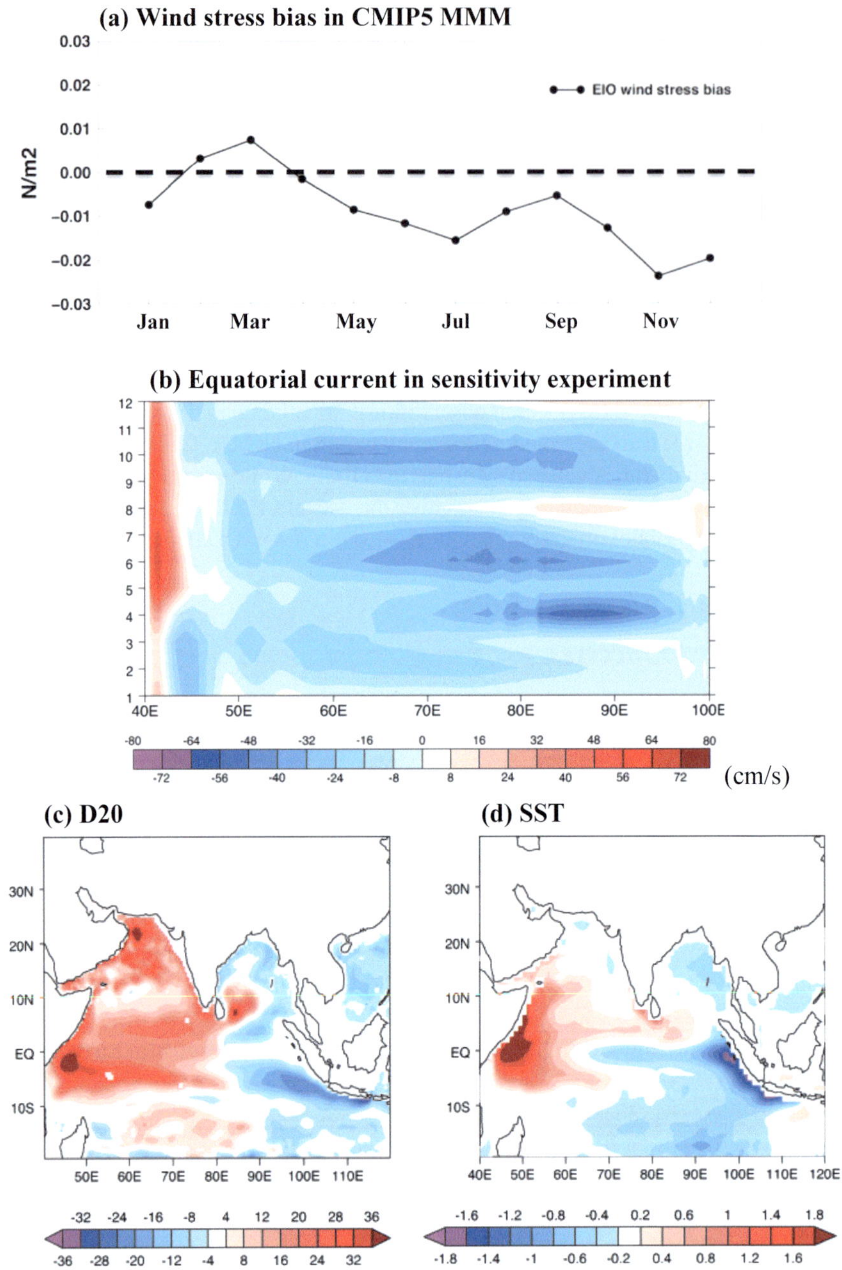

Fig. 7 **a** Equatorial Indian Ocean (3°S–3°N; 40°–100°E) monthly mean wind stress climatology difference between CMIP5 MMM and ERA-INT; **b** monthly evolution (ordinate) of the equatorial (3°S–3°N) zonal current (cm/s) from sensitivity experiment; **c** seasonal (June-September) mean differences in thermocline (m) between CFES_EXP and CFES_CTL; and **d** same as (**c**) but for SST (C). Adopted from Annamalai et al. (2017)

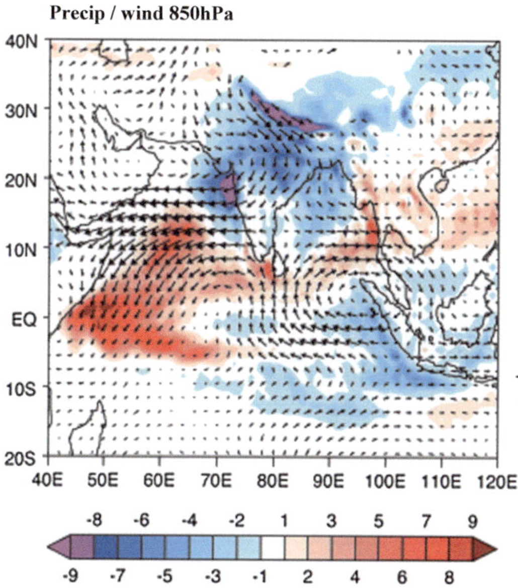

Fig. 8 Seasonal (June–September) mean differences in precipitation (shaded, mm/day) and wind 850 hPa (m/s) between CFES_EXP and CFES_CTL. The reference wind vector is also shown. Adopted from Annamalai et al. (2017)

(Fig. 7d). Additionally, the simulated cold SST bias along the equatorial central Indian Ocean closely resembles errors in the CMIP5 models (Fig. 3b). Strikingly, the spatial pattern in ΔP and low-level $\Delta\tau$ (Fig. 8) look identical to the CMIP5 MMM biases (Fig. 1b). Since errors in the sensitivity experiment are driven only from the equator, it follows that the South Asian precipitation errors are as well. What processes account for this causal linkage?

5 Moisture and MSE Budget Analyses

In this section, we begin with biases in specific humidity (Δq) during summer, and introduce budget equations (Sect. 5.1). Then, we discuss biases in atmospheric processes in regions where ΔP and ΔSST are collocated, and in others, they are not and provide interpretations on how EIO processes impact ΔP over South Asia (Sect. 5.2). Specifically, during summer ΔP and ΔSST (Figs. 1b and 3b) are collocated over the western EIO, termed REG1 (10°S–5°N, 40°–60°E) and they are not over the Arabian Sea, termed REG2 (5°–15°N, 60°–85°E). What are the leading processes that determine ΔP, and are they seasonally and regionally invariant? If they are, can we improve their representation in models? We address these issues in Sect. 5.3.

5.1 *Biases in Specific Humidity and Budget Equations*

To motivate our budget analyses, Fig. 9 shows biases in specific humidity (Δq) for the planetary boundary layer (PBL; Fig. 9a) and free troposphere (Fig. 9b) during boreal summer. Vertical integrations for the PBL extend from surface to 850 hPa and for the free troposphere from 700 to 400 hPa. Here, PBL (free troposphere) moisture is associated with triggering of convection (intensity of convection). Figure 9a shows that over REG1 (REG2), Δq for the PBL is wet (dry). In contrast, in the free troposphere, a prominent wet bias is seen over REG2 (Fig. 9b). Throughout the troposphere, dry bias exists poleward of 10°N with a maximum along the monsoon trough. Keeping in mind that in deep tropics Δq determines bias in precipitable water we proceed with budget diagnostics to unravel the sources of model errors to ΔP.

The moisture budget relates precipitation to moisture convergence and heating. However, one limitation of moisture budget is that convergence-related variables are largely a result of the convection itself rather than a cause (Neelin and Held 1987; Raymond et al. 2009), particularly over the Indo-Pacific warm pool where SST gradients are nearly absent. The MSE budget deals with moisture and temperature processes that are critical to understand the interaction between small-scale convection and large-scale circulation (Back and Bretherton 2006); moreover, it is useful for determining the thermodynamic processes that anchor moisture divergence (Su and Neelin 2002).

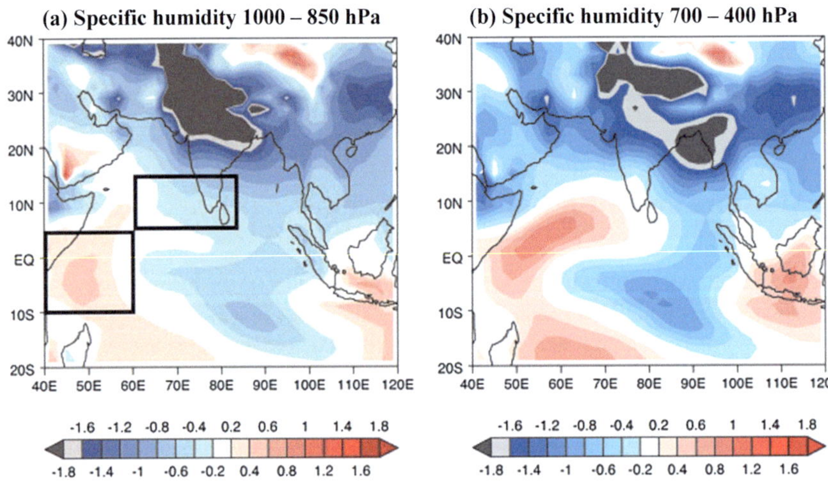

Fig. 9 Seasonal (June–September) mean specific humidity (g/kg) climatology difference between CMIP5 MMM and ERA-INT: **a** integrated from 1000 to 850 hPa and **b** integrated from 700 to 400 hPa. The boxed areas illustrate two different regions, namely, REG1 (10°S–5°N; 40–60°E) and REG2 (5–15°N; 60–85°E), used for estimating atmospheric budget analyses. Adopted from Annamalai et al. (2017)

The vertically integrated moisture budget is

$$P = -\left\langle \omega \frac{\partial q}{\partial p} \right\rangle + LH - \langle \boldsymbol{V} \cdot \nabla q \rangle, \quad (1)$$

where P is precipitation; $\langle \boldsymbol{V} \cdot \nabla q \rangle$ and $\langle \omega q_p \rangle$ are horizontal and vertical advection (or horizontal divergence) of moisture, respectively; and LH is the latent heat flux at the surface. In these quantities, $\boldsymbol{V}$ is the horizontal velocity vector, ∇ is the gradient operator, ω is vertical pressure velocity, q is the specific humidity, and angle brackets designate vertical integration. Using (1), we obtain moisture budget terms for both ECMWF Interim Reanalysis (ERA-Interim) and MMM fields; their difference, labeled by a prime, measures the model error of each term. Here, all the terms are expressed in energy units (W/m^2).

MSE (h) is defined by

$$h = C_P T + gz + Lq, \quad (2)$$

where C_p is the specific heat at constant pressure, T is temperature, g is the gravitational acceleration, z is geopotential height, L is latent heat of vaporization at 0 °C, and q is specific humidity. The vertically integrated MSE budget is then given by

$$\left\langle \frac{\partial h}{\partial t} \right\rangle = -\langle \boldsymbol{V} \cdot \nabla h \rangle - \left\langle \omega \frac{\partial h}{\partial p} \right\rangle + LH + SH + \langle LW \rangle + \langle SW \rangle + R. \quad (3)$$

In (3), SH is sensible heat flux at the surface, $\langle LW \rangle$ and $\langle SW \rangle$, net column longwave and shortwave heating rates, and R is a residual. The terms $\langle \boldsymbol{V} \cdot \nabla h \rangle$ and $\langle \omega h_p \rangle$ represent horizontal and vertical advection of MSE, respectively, and the former can be split into horizontal advection of moisture $L\langle \boldsymbol{V} \cdot \nabla q \rangle$ and temperature $C_p\langle V \cdot \nabla T \rangle$. The sum, $\langle F_{rad} \rangle = \langle LW \rangle + \langle SW \rangle$, is the net radiative flux. As for the moisture budget terms, we evaluate the terms in (3) using both ERA-Interim and MMM fields and indicate their differences by a prime.

In our analysis, we assume that (3) is quasi-steady, that is, $\langle h_t \rangle \approx 0$, an approximation that is valid at seasonal timescales. In that case, (3) represents a balance between the net flux of h into the column and moist processes that determine convective instability (Annamalai 2010). Rearranging the terms in Eq. 3 leads to

$$-\left\langle \omega \frac{\partial h}{\partial p} \right\rangle = -\langle \boldsymbol{V} \cdot \nabla q \rangle - \langle \boldsymbol{V} \cdot \nabla T \rangle + LH + SH + \langle F_{rad} \rangle + R. \quad (4)$$

In the deep tropics, the "weak temperature gradient" (WTG) approximation is valid, and hence the contribution from $\langle \boldsymbol{V} \cdot \nabla T \rangle$ in (4) is small (Sobel et al. 2001). In tropical, moist convective regions, then, q determines h and precipitation is directly proportional to $\langle \omega h_p \rangle$. As such, the MSE budget provides insights into how processes such as $\langle \boldsymbol{V} \cdot \nabla h \rangle$, LH, SH, $\langle LW \rangle$, and $\langle SW \rangle$ force convection (Raymond et al. 2009). For the budget terms, vertical integrations are from 1000 to 100 hPa, and the

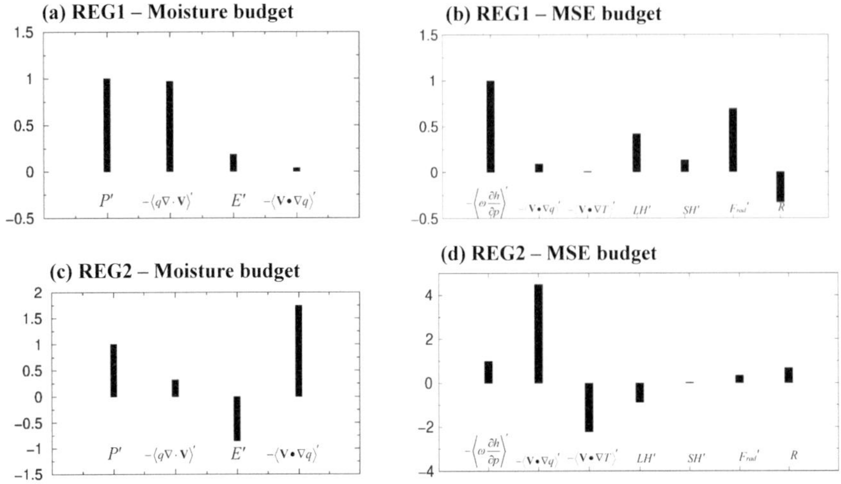

Fig. 10 **a** Various terms of the moisture budget averaged over REG1 (10°S–5°N, 40°–60°E); **b** moist-static energy budget terms averaged over REG1; **c** same as (**a**) but averaged over REG2 (5°–15°N, 60°–85°E); and **d** same as (**b**) but averaged over REG2. The budgets are estimated separately for the ERA-INT, and individually for each of the CMIP5 models and then MMM budget terms are calculated. Shown here are the differences (CMIP5 MMM minus ERA-INT). Note that in the moisture (MSE) budget, all the terms are scaled by precipitation ($\langle \omega h_p \rangle$). For MSE budgets, the residual term (R) is also shown. Adopted from Annamalai et al. (2017)

terms averaged over REG1 (REG2) are shown in the top (bottom) panels in Fig. 10. Note that in the moisture (MSE) budget, all the terms are scaled by precipitation ($\langle \omega h_p \rangle$). In estimating the budgets, note the vertical integrals through the depth of the troposphere are mass integrals (Neelin and Su 2005), and therefore, all the terms are expressed in energy W/m^2. More details on the derivation of moisture and MSE budgets are available in Yanai et al. (1973) and Neelin and Held (1987).

5.2 *Budget Diagnostics During Boreal Summer*

Over REG1, the moisture budget terms (Fig. 10a) indicate that the precipitation error P' (or ΔP) is largely determined by moisture convergence, $-\langle \omega q_p \rangle'$, with a lesser contribution from evaporation $LH'E'$. Both factors point toward SST errors as determining P' (Figs. 1b and 3b), the moisture-convergence error caused by the error in the equatorial SST gradient (Lindzen and Nigam 1987; Back and Bretherton 2009) and evaporation error due to local warm SST bias. The MSE terms (Fig. 10b) indicate that errors in convective forcing, measured by $\langle \omega h_p \rangle'$, are opposed by errors in the horizontal advection of cold/dry air $\langle \mathbf{V} \cdot \nabla h \rangle'$, with most of the positive contributions from diabatic terms ($\langle F_{rad} \rangle'$ and LH') that favors convective instability. The errors in F'_{rad} arise from enhanced cloud cover trap more LW. This cloud–radiation feedback process enhances radiative heating that in turn increases $\langle\omega h_p\rangle'$. We conclude that in REG1 SST biases cause errors in atmospheric moist processes, which in turn lead to errors in PBL-specific humidity (Fig. 9a).

In REG2, ΔP are not collocated with Δ SST and, consistent with this property, contributions to ΔP from errors in moisture convergence, $-\langle\omega q_p\rangle'$, and evaporation LH' are either weak or opposite in sign (Fig. 10c). The reduced evaporation is attributable to the cold SST bias (Fig. 3b) and to the weakened low-level monsoon westerlies (Fig. 1b). Instead, the moisture budget indicates that positive rainfall errors are primarily due to the horizontal advection term $-\langle \boldsymbol{V} \cdot \nabla q\rangle'$, a conclusion supported by the dominance of this term to favor convective instability in the MSE budget there (Fig. 10d). Note that similar to REG1, in REG2 too, errors in $\langle F_{rad}\rangle'$ favor convective instability. In both regions, is the intensity of ΔP influenced by errors in F'_{rad}?

Given the excess rainfall in REG1, the moisture flux into REG2 and beyond into South Asia is reduced. In REG2, what are the moist processes that promote the positive ΔP or how does $-\langle \boldsymbol{V} \cdot \nabla q\rangle'$ dominate as the principal source of convective forcing? It happens because of the decreased precipitation over South Asia (Fig. 1b). In response to this forcing, an atmospheric Rossby wave is generated that radiates westward into the Arabian Sea. It is associated with northeasterly winds (Fig. 1b), which advect climatological moisture out of South Asia into REG2, causing precipitation there. An examination of various components of the term $-\langle \boldsymbol{V} \cdot \nabla q\rangle'$ confirms the dominance of anomalous wind advecting climatological moisture (not shown).

Let us now focus on understanding how EIO processes impact ΔP over South Asia (Fig. 8). We conjecture that it results from a feedback loop between rainfall over South Asia and REG2. Because of the positive rainfall errors over REG1, moisture transport into South Asia is decreased thereby reducing rainfall there (Annamalai et al. 2005). As noted above, this South Asian forcing generates northeasterly wind errors that extend to the Arabian Sea (REG2) as a Rossby wave response. There, they weaken the moisture flux out of REG2 and, hence, enhance local precipitation. *Once organized, positive rainfall errors in REG2 further reduce moisture transport to South Asia, which in turn weakens South Asian rainfall even more,* and the cycle continues. Budget results for the equatorial wind stress bias sensitivity solutions confirm results presented in Fig. 10, and hence not shown. Details are in Annamalai et al. (2017).

5.3 Budget Diagnostics During Other Seasons

Here, to understand if the error sources are seasonally invariant, we examine budgets over the western-southern Indian Ocean in other seasons when wet bias is prevalent. The moisture budgets for both fall and winter seasons (Fig. 11a, b) confirm that ΔP is largely balanced by errors in moisture convergence, $-\langle\omega q_p\rangle'$ and that is supported by errors in low-level winds (Fig. 1c, d). Therefore in CMIP5 models, does this property imply ΔSST is the primary initiator for ΔP? To be precise, is there a

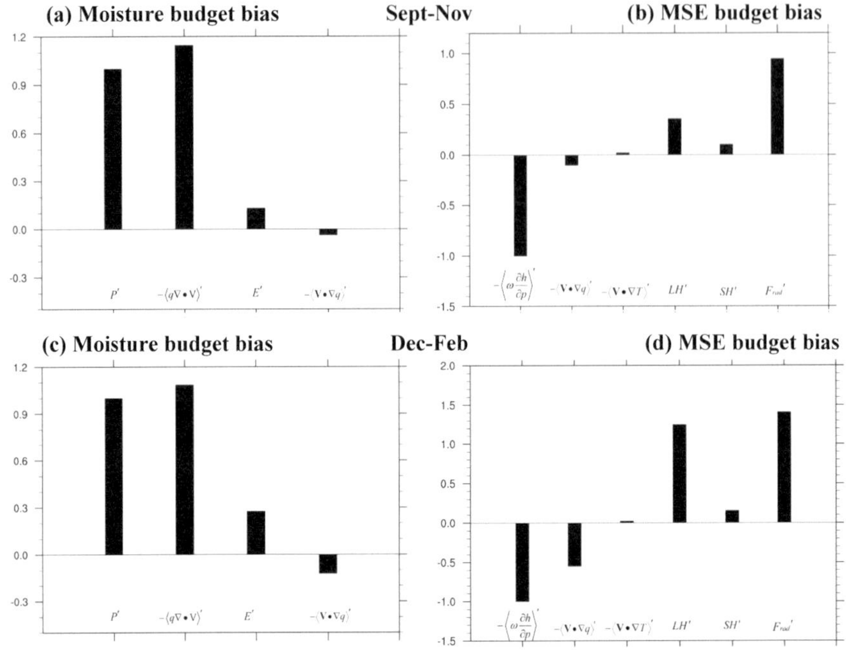

Fig. 11 **a** Various terms of the moisture budget averaged over the wet bias region (12°S–5°N, 40°–70°E) during boreal fall season (September–November); **b** same as (**a**) but for moist-static energy budget terms; **c** same as (**a**) but averaged over the wet bias region (20°S–8°N, 48°–70°E) during boreal winter season (December–February); and **d** same as (**c**) but for moist-static energy budget terms. The budgets are estimated separately for the ERA-Interim, and individually for each of the CMIP5 models and then MMM budget terms are calculated. Shown here are the differences (CMIP5 MMM minus ERA-INT). Note that in the moisture (MSE) budget, all the terms are scaled by precipitation ($\langle \omega h_p \rangle$)

clear error in SST gradient either longitudinally or latitudinally to trigger the low-level convergence? An examination of ΔSST maps (Fig. 3c, d) supports the notion for errors in SST gradient along the EIO in boreal fall (Fig. 3c) to drive easterly bias (Fig. 1c), and a north–south (cold–warm) gradient in winter (Fig. 3d) to drive northeasterly bias (Fig. 1d). Note that unlike in summer, northeasterly bias in winter over the northern Indian Ocean is not forced by ΔP over the Bay of Bengal.

The MSE budgets (Fig. 11c, d) for both seasons indicate that diabatic terms (LH' and $\langle F_{rad} \rangle'$) are dominant factors in convective forcing $\langle \omega h_p \rangle'$. While the adiabatic terms are sinks of MSE in the column, the diabatic terms are sources of MSE, and therefore, any errors in estimating their strength by limitations in model parameterizations will lead to feedback between diabatic and adiabatic terms resulting in large ΔP. For example during winter (Fig. 11d), errors in LH' and $\langle F_{rad} \rangle'$ make equal contributions suggesting model improvements are needed in their formulation or parameterization. Furthermore, estimated MSE divergence $\langle \omega h_p \rangle'$ crucially depends on the model's ability to represent vertical velocity accurately, which in turn depends on cloud–convection–radiation interactions. An obvious limitation is that computation of the MSE budget from reanalysis data has substantial errors and uncertainties (Back and Bretherton 2006), and therefore, a residual component arises. On the other hand, the leading processes identified are expected to be robust and reproducible (Neelin and Su 2005).

In summary, in CMIP models wet precipitation bias is not always associated with similar signatures in SST over the tropical Indian Ocean. The budget diagnostics performed here reveal that larger contributors for the precipitation biases come from errors in diabatic terms (e.g., F_{rad}). The cloud-radiative feedback processes are known to be a dominant single uncertainty factor in climate models (Randall et al. 2007). Given that they are a consequence of convection, small errors in convective parameterization (e.g., entrainment/detrainment rates) are amplified by cloud schemes interacting with radiation parameterization, resulting in precipitation biases through nonlinear feedbacks. Thus, a concentrated effort on modeling the interactive convection–cloud–radiation physics, and a continuous assessment through process-based diagnostics during model development should be given top priority.

6 Discussion and Recommendations

Despite dedicated efforts by the modeling community to improve processes representation, the persistence nature of model precipitation errors over the Asian-Australian monsoon region, leads us to ask: are there fundamental limits to realistically simulating the monsoon? Will a concerted observational and modeling effort enhance models' fidelity in simulating the monsoon? Given the model errors are persisting for about two to three decades, one can say that modeling the monsoons is a daunting task in climate science. While advancement in supercomputing provides a leap to run models at very horizontal and vertical resolutions to better represent small-scale processes (e.g., convective scales), the processes that determine monsoon are interactive and intertwined and will limit our ability to model the monsoons. On the modeling front, what are the efforts needed to improve the simulation of monsoon annual cycle, particularly the heterogeneous regional distribution of precipitation structure and amplitude? For that to be successful, do we have a clear understanding of how dynamic and thermodynamic processes, flow–orography interactions, and ocean–atmosphere interactions determine the precipitation structure and strength? Are there enough observations for a comprehensive understanding and to constrain model physics? Here, we summarize the pertinent issues on modeling and limitations on observations, and suggest recommendations for improving our understanding and modeling the monsoons.

In the CMIP-era database, irrespective of horizontal and vertical resolutions, and a spectrum of varied physical parameterizations employed, the structure of the precipitation errors is robust across all models. Such a behavior across models suggests that the errors may arise due to multiple processes and their interactions, and/or certain processes are missing in all models. The fact that "fast atmospheric" processes have been identified as leading source of model errors (Martin et al. 2010; Ma et al. 2014), improvements to cloud-cumulus schemes, and cloud–radiation feedbacks stand out as top priorities. However, no detailed process-based diagnostics at a level close enough to parameterization schemes has been performed yet to identify the limitations in the representation of individual processes and their interactions. For example, do the errors emerge during convective triggering and/or development, lack of interactions with environmental humidity or lack of communication between convection–cloud schemes (large-scale precipitation) or response of radiation to convection? Are the errors due to communication deficiencies between shallow and deep convection schemes or due to limitations in charging and discharging of column MSE? We believe that such a thorough examination is expected to first identify the initial source in the chain of error sources, and then to provide pathways for improving representations of processes.

Could the precipitation errors in the CMIP5 models arise from errors due to coupled rather than fast atmospheric processes, a view supported by the results of Annamalai et al. (2017)? To check this idea, Fig. 12 compares precipitation and SST climatologies in the AFES (Fig. 12a) and CFES (Fig. 12b) control simulations. In both runs the atmospheric model is the same, the two solutions differing in that AFES is forced with daily-varying, observed SST for the period 1979–2010. The simulated precipitation in AFES is biased with higher precipitation (>4–8 mm/day) over REG1 (~west of 65°E), a bias similar to that in the CMIP5 models and in other atmosphere-only models (Ma et al. 2014). In contrast, this erroneous feature is clearly rectified by the coupling in the CFES solution: While the structure of precipitation along the EIO is present in the AFES and CFES solutions, its amplitude distribution is more realistic in CFES, implying the need for the correct representation of EIO coupled processes. Similar results are noticeable in MIROC5 coupled (Sperber et al. 2013) versus uncoupled solutions (Ma et al. 2014). Motivated by this idea, Hanf and Annamalai (2018) diagnosed coupled and uncoupled simulations of NCAR_CESM and concluded that realistic representation of regional air–sea interactions along the WEIO is important to an improved simulation of precipitation and reduce systematic errors over the Arabian Sea. In this scenario, errors in either atmospheric or oceanic variables could lead to errors in coupled processes due to feedbacks. In other words, even if the fast atmospheric processes are realistically represented initially, errors in the representation of near-surface wind could cascade into surface fluxes, ocean currents and thermocline properties with subsequent effects on SST, and onto atmospheric convection. Stand-alone ocean modeling studies forced by atmospheric winds and near-surface fluxes show diversified SST and thermocline simulations (e.g., McCreary et al. 1993). In CMIP5-era coupled models, do we need to do systematic process-based diagnostics to assess the representation of oceanic processes such as thermocline properties, and mixed-layer physics, particularly in regions such as Bay of Bengal where mixed-layer depths are shallow? In regions of high-mean precipitation such as the eastern EIO and Bay of Bengal, do the models represent the upper-ocean stratifications in temperature and salinity realistically, and how errors in their representation lead to errors in SST?

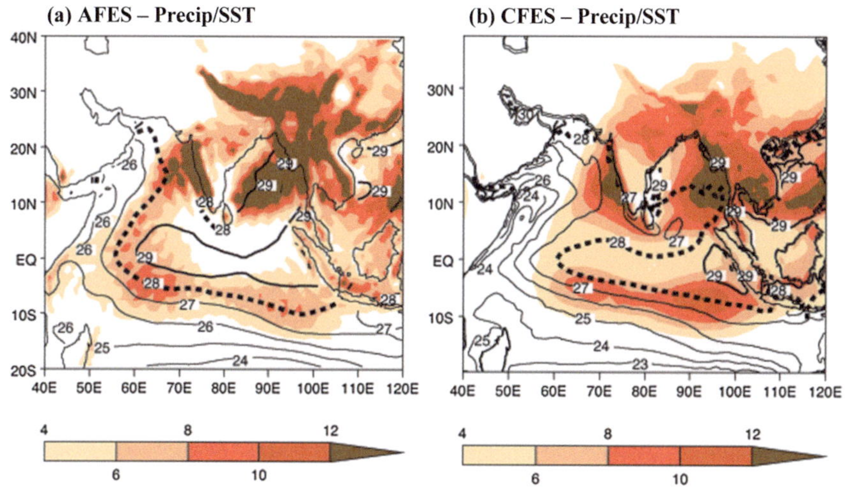

Fig. 12 The seasonal (June–September) mean precipitation (shaded, mm/day) and SST (contour, °C) climatology from control solutions for: **a** AFES and **b** CFES. Note that AFES is forced with climatologically varying daily-observed SST. Adopted from Annamalai et al. (2017)

Or, are the model errors due to our lack of observational knowledge on the interactive dynamic, and thermodynamic processes, and flow-orography interactions in determining the heterogeneous spatial distribution of precipitation, and realistically represent them in models? In that regards, we have a fair understanding of the importance of large-scale forcing (e.g., orography) and boundary forcing (e.g., SST, snow cover) on the large-scale tropical flow patterns (Hoskins and Rodwell 1995; Charney and Shukla 1981), but roles of thermodynamic and oceanic process in shaping the mean precipitation and its spectrum of variability are just beginning to unfold. For a realistic representation of these processes, sustained observations are needed to constrain the model physics. Observing and modeling the details of the thin mixed-layer and the upper-ocean stratifications in temperature and salinity particularly over regions of high-mean precipitation, and their collective impact on SST evolution remains a challenge (Annamalai et al. 2018). *We stress here and conclude that dedicated and sustained observational efforts are absolutely needed for improving the representation of leading physical processes in numerical models.*

One inference is that Asian monsoon is comprised of multiple regional heat sources and given their close proximity to the equator perturbations to one of the precipitation centers are quickly communicated to others by the equatorial wave response that subsequently impact the precipitation characteristics of those centers (Annamalai and Sperber 2005; Annamalai 2010). What are the leading processes that determine these regional precipitation centers, and do we have adequate observations to address this and related questions? Given the true complexity of the monsoon-Indian Ocean climate systems, coordinated activities across disciplines, and among diagnostics, modeling and observational personnel are needed.

Acknowledgements This research work is funded by the National Science Foundation (NSF) under grant 1460742. The author also acknowledges the financial support provided by the Indian Monsoon Mission and JAMSTEC. Jan Hafner is thanked for his assistance with diagnostics.

References

Abhik, S., R.P.M. Krishna, M. Mahakur, M. Ganai, P. Mukhopadhyay, and J. Dudhia. 2017. Revised cloud processes to improve the mean and intraseasonal variability of Indian summer monsoon in climate forecast system: Part 1. *Journal of Advances in Modeling Earth Systems* 9: 1–28. https://doi.org/10.1002/2016ms000819.

Annamalai, H. 2010. Moist dynamical linkage between the equatorial Indian Ocean and the South Asian monsoon trough. *Journal of Atmospheric Sciences*, *67*, 589–610.

Annamalai, H., and K.R. Sperber. 2005. Regional heat sources and the active and break phases of boreal summer intraseasonal (30–50 day) variability. *Journal of the Atmospheric Sciences* 62: 2726–2748.

Annamalai, H., P. Liu, and S.P. Xie. 2005. Southwest Indian Ocean SST variability: Its local effect and remote influence on Asian Monsoons. *Journal of Climate* 18: 4150–4167.

Annamalai, H., B. Taguchi, J.P. McCreary, M. Nagura, and T. Miyama. 2017. Systematic errors in south Asian monsoon simulation: Importance of Equatorial Indian Ocean processes. *Journal of Climate* 30 (19): 8159–8178. https://doi.org/10.1175/jcli-d-16-0573.1.

Annamalai, H., M. Nagura, and K. Richards. 2018. Indian Ocean and South-Asian monsoon: Upper-ocean processes relevant to monsoon annual cycle. A chapter in IndOOS Review, Jakarta, March 2018.

Back, L.E., and C.S. Bretherton. 2006. Geographic variability in the export of moist static energy and vertical motion profiles in the tropical Pacific. *Geophysical Research Letters* 33: L17810.

Back, L.E., and C.S. Bretherton. 2009. On the relationship between SST gradients, boundary layer winds, and convergence over the tropical oceans. *Journal of Climate* 22: 4182–4196.

Boos, W.R., and J.V. Hurle. 2013. Thermodynamic bias in the multi-model mean boreal summer monsoon. *Journal of Climate* 26: 2279–2287.

Bretherton, C. 2007. Challenges in numerical modeling of tropical circulations. In *The global circulation of the atmosphere*, ed. T. Schneider and A.H. Sobel, 302–330. Princeton University Press.

Bretherton, C.S., M.E. Peters, and L.E. Back. 2004. Relationships between water vapor path and precipitation over the tropical oceans. *Journal of Climate* 17: 1517–1528.

Bush, S.J., et al. 2015. The effect of increased convective entrainment on Asian monsoon biases in the Met UM general circulation model. *Quarterly Journal of the Royal Meteorological Society* 141: 311–326.

Charney, J.G., and J. Shukla. 1981. Monsoon predictability. In *Monsoon dynamics*, ed. Lighthill and R.P. Pearce, 99–109. Cambridge Press.

Del Sole, T., and J. Shukla. 2002. Linear prediction of Indian monsoon rainfall. *Journal of Climate* 15: 3645–3658.

Goswami, B.B., R.P.M. Krishna, P. Mukhopadhyay, M. Khairoutdinov, and B.N. Goswami. 2015. Simulation of the Indian summer monsoon in the superparameterized Climate Forecast System version 2: Preliminary results. *Journal of Climate* 28 (22): 8988–9012. https://doi.org/10.1175/837JCLI-D-14-00607.1.

Graham, N.E., and T.P. Barnett. 1987. Sea surface temperature, surface wind divergence, and convection over the tropical oceans. *Science* 238: 657–659.

Han, W., J.P. McCreary, D.L.T. Anderson, and A.J. Mariano. 1999. Dynamics of the eastern surface jets in the equatorial Indian Ocean. *Journal of Physical Oceanography* 29: 2191–2209.

Hanf, F., and H. Annamalai. 2018. Assessing monsoon biases in a climate model: Sensitivity to entrainment. *Journal of Climate* (revised).

Hoskins, B.J., and M.J. Rodwell. 1995. A model of the Asian summer monsoon. Part I: The global scale. *Journal of the Atmospheric Sciences* 52: 1329–1340.

Joseph, P.V. 1990. Warm pool over the Indian Ocean and monsoon onset. *Tropical Ocean-Atmosphere Newsletter* (Winter): 1–5.

Klinker, E., and P.D. Sardesmukh. 1992. The diagnosis of mechanical dissipation in the atmosphere from large-scale balance requirements. *Journal of the Atmospheric Sciences* 49: 608–627.

Levine, R.C., A.G. Turner, D. Marathayil, and G.M. Martin. 2013. The role of northern Arabian Sea surface temperature biases in CMIP5 model simulations and future projections of Indian summer monsoon rainfall. *Climate Dynamics* 41: 155–172.

Lin, et al. 2012. TWP-ICE global atmospheric model intercomparison: Convection responsiveness and resolution impact. *Journal of Geophysical Research* 117: D09111. https://doi.org/10.1029/2011jd017018.

Lindzen, R.S., and S. Nigam. 1987. On the role of sea surface temperature gradients in forcing low-level winds and convergence in the tropics. *Journal of the Atmospheric Sciences* 44: 2418–2436.

Lukas, R., and E.J. Lindstorm. 1991. The mixed layer of the western equatorial Pacific Ocean. *Journal Geophysical Research* 96: 3343–3357. https://doi.org/10.1029/90jc01951.

Ma, H.-Y., et al. 2014. On the correspondence between mean forecast errors and climate errors in CMIP5 models. *Journal of Climate* 27: 1781–1798.

Martin, G.M., S.F. Milton, C.A. Senior, M.E. Brooks, S. Ineson, T. Reichler, and J. Kim. 2010. Analysis and reduction of systematic errors through a seamless approach to modeling weather and climate. *Journal of Climate* 23: 5933–5957.

Masson, S., C. Menkes, P. Delecluse, and J.-P. Boulanger. 2003. Impacts of salinity on the eastern Indian Ocean during the termination of the fall Wyrtki Jet. *Journal of Geophysical Research* 108: 3067. (https://doi.org/10.1029/2001jc000833).

McCreary, J.P., P.K. Kundu, and R. Molinari. 1993. A numerical investigation of dynamics, thermodynamics and mixed-layer processes in the Indian Ocean. *Progress in Oceanography* 31: 181–244.

Nagura, M., et al. 2013. Longitudinal biases in the Seychelles Dome simulated by 35 ocean-atmosphere coupled general circulation models. *Journal of Geophysical Research: Oceans* 118. https://doi.org/10.1029/2012jc008352.

Nagura, M., J.P. McCreary, and H. Annamalai. 2018. Origins of coupled model biases in the Arabian Sea climatological state. *Journal of Climate* 31 (5): 2005–2029. https://doi.org/10.1175/JCLI-D-17-0417.1.

Neale, R.B., J.H. Richter, and M. Jochum. 2008. The impact of convection on ENSO: From a delayed oscillator to a series of events. *Journal of Climate* 21: 5904–5924.

Neelin, J.D., and I.M. Held. 1987. Modeling tropical convergence based on the moist static energy budget. *Monthly Weather Review* 115: 3–12.

Neelin, J.D., and H. Su. 2005. Moist teleconnection mechanisms for the tropical South American and Atlantic sector. *Journal of Climate* 18: 3928–3950. https://doi.org/10.1175/jcli3517.1.

Phillips, T.J., and Coauthors. 2004. Evaluating parameterizations in general circulation models: Climate simulation meets weather prediction. *Bulletin of the American Meteorological Society* 85: 1903–1915.

Randall et al. (2007). The physical science basis. In *Contribution of working group I to the fourth assessment report of the intergovernmental panel on climate change*, eds. S. Solomon, D. Qin, M. Manning, Z. Chen, M.C. Marquis, K.B. Averyt, M.Tignor, and H.L. Miller, p. 996. Cambridge, United Kingdom, and New York, NY, USA: Cambridge University Press.

Rao, R.R., and R. Sivakumar. 2000. Seasonal variability of near-surface thermal structure and heat budget of the mixed layer of the tropical Indian Ocean from a new global ocean temperature climatology. *Journal of Geophysical Research* 105: 985–1015.

Raymond, D.J. 2000. Thermodynamic control of tropical rainfall. *Quarterly Journal of the Royal Meteorological Society* 126: 889–898.

Raymond, D.J., S.L. Sessions, A.H. Sobel, and Z. Fuchs. 2009. The mechanics of gross moist stability. *JAMES* 1: 1–20.
Rodwell, M.J., and T.N. Palmer. 2007. Using numerical weather prediction to assess climate models. *Quarterly Journal of the Royal Meteorological Society* 133: 129–146.
Sandeep, S., and R.S. Ajayamohan. 2014. Origin of cold bias over the Arabian Sea in climate models. *Scientific Reports* 4: 6403.
Seo, H., S.P. Xie, R. Murtugudde, M. Jochum, and A.J. Miller. 2009. Seasonal effects of Indian Ocean freshwater forcing in a regional coupled model. *Journal of Climate* 22: 6577–6596.
Sengupta, D., and M. Ravichandran. 2001. Oscillations of Bay of Bengal sea surface temperature during the 1998 summer monsoon. *Geophysical Research Letters* 28: 2033–2036.
Shenoi, S.S.C., D. Shankar, and S.R. Shetye. 2002. Differences in heat budgets of the near surface Arabian Sea and Bay of Bengal: Implications for the summer monsoon. *Journal of Geophysical Research* 107 (C6): 3052. https://doi.org/10.1029/2000jc000679.
Sobel, A., J. Nilsson, and L.M. Polvani. 2001. The weak temperature gradient approximation and balanced tropical moisture waves. *Journal of the Atmospheric Sciences* 58: 3650–3665.
Sperber, K.R., et al. 2013. The Asian monsoon: An intercomparison of CMIP3vs CMIP5 simulations of the late 20th century. *Climate Dynamics* 41 (9–10): 2711–2744. https://doi.org/10.1007/s00382-012-1607-6.
Stephens, G.L., S.V. Heever, and L. Pakula. 2008. Radiative-convective feedbacks in idealized states of radiative-convective equilibrium. *Journal of the Atmospheric Sciences* 65 (3989): 3916.
Su, H., and J.D. Neelin. 2002. Teleconnection mechanisms for tropical Pacific descent anomalies during El Niño. *Journal of the Atmospheric Sciences* 59: 2694–2712.
Taylor, K.E., R.J. Stouffer, and G.A. Meehl. 2012. An overview of CMIP5 and the experiment design. *Bulletin of the American Meteorological Society* 93: 485–498.
Tozuka, T., T. Yokoi, and T. Yamagata. 2010. A modeling study of interannual variations of the Seychelles Dome. *Journal Geophysical Research* 115: C04005. https://doi.org/10.1029/2009jc005547.
Xie, S.P., H. Annamalai, F. Schott, and J.P. McCreary. 2002. Structure and mechanisms of South Indian Ocean climate variability. *Journal of Climate* 15: 864–878.
Yanai, M., S. Esbensen, and J.H. Chu. 1973. Determination of bulk properties of tropical cloud clusters from large-scale heat and moisture budgets. *Journal of the Atmospheric Sciences* 60: 611–627.
Wyrtki, K. 1973. An equatorial jet in the Indian Ocean. *Science* 181: 262–264.
Zhao, et al. 2018. *The GFDL global atmosphere and land Model AM4.0/LM4.0: 2. Model description, sensitivity studies, tuning strategies. Journal of Advances in Modeling Earth Systems* 10: 735–769. https://doi.org/10.1002/2017MS001209.

Challenges in Tropical Numerical Weather Prediction at ECMWF

Peter Bechtold

Abstract We describe the challenges in the coming decade in global numerical weather prediction and in the tropics in particular. The ECMWF forecasting system is our benchmark. These challenges comprise four main areas of developments: making optimal use of the available observational data to obtain the best analysis, advanced ensemble methods to predict the uncertainties in the analyses and forecasts, model developments to better represent shallow and deep convection and associated circulations and finally necessary advances in computational efficiency called scalability.

Keywords Numerical weather prediction · Tropical convection
Observation feedback

The current strategy of the European Centre for Medium-Range Weather Forecast aims at having by 2025, a fully coupled ensemble-based atmosphere and ocean prediction system at horizontal resolution of 5 km and with at least 137 and 75 vertical levels in the atmosphere and ocean, respectively. This ambitious goal poses numerous challenges in terms of model development, high-performance computing and the use of observational data.

Surely, we are still at the beginning of an important development period, and it turns out that the challenges in model development, data assimilation and computational efficiency are strongly interlinked. However, foremost, we want to improve the model, which means to improve on the accuracy of the predicted weather parameters, notably winds, 2-m temperature, precipitation and clouds and to further extend the predictability limit of global scale circulations and their anomalies. This requires a careful analysis of the systematic model errors, a better understanding of the influence of the tropics on the global weather and tools to estimate the flow-dependent uncertainties of the forecasts.

The challenges and potentials in tropical predictability are best appreciated by reflecting on some of the major achievements. In the following, we focus in the

P. Bechtold (✉)
ECMWF, Shinfield Park, Reading RG29AX, UK
e-mail: peter.bechtold@ecmwf.int

D. A. Randall et al. (eds.), *Current Trends in the Representation of Physical Processes in Weather and Climate Models*, Springer Atmospheric Sciences,
https://doi.org/10.1007/978-981-13-3396-5_2

frame of the ensemble system on challenges in predicting tropical convection and tropical circulations, while making best use of the available observing system. Of particular focus with respect to systematic model errors and model resolution will be the role of the convection and microphysical parametrizations.

1 Evolution of the Observing System

Progress in numerical weather prediction is strongly linked to the availability of satellite data and the advances in data assimilation methods. It is the introduction of the four-dimensional variational data assimilation system at ECMWF end 1997 (Rabier et al. 2000) which allowed to make full use of satellite radiances, which are now the most important data source over the oceans. Figure 1 displays the number of satellite data products actively assimilated at ECMWF since 1996 when the three-dimensional variational data assimilation system was first introduced. The number of satellite observations assimilated in each assimilation cycle (each analysis) has increased from roughly 10 million in 1996 to 80 million in 2017. We are now eagerly awaiting the arrival of fully processed data form the ADM Aeolus mission which should by the end of 2018 provide for the first time extensive wind profiles for the tropical regions.

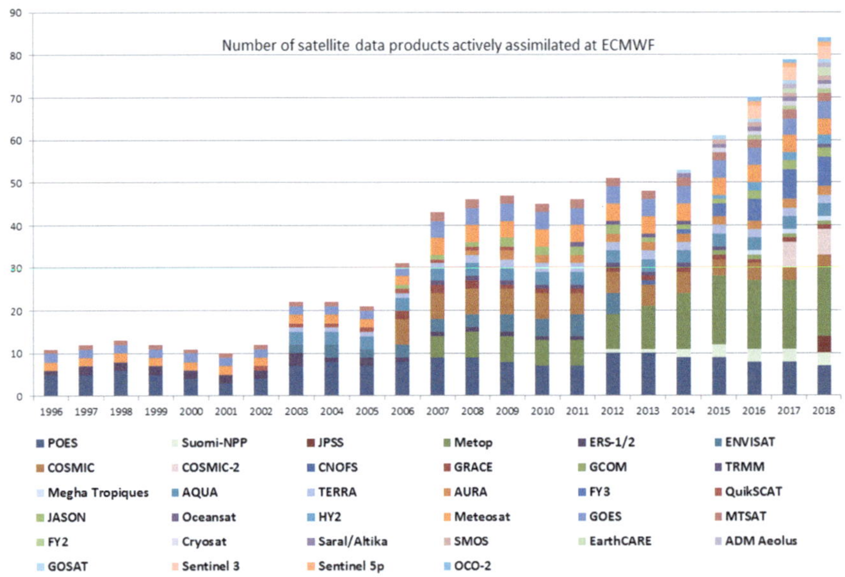

Fig. 1 Number of satellite data products actively assimilated at ECMWF since 1996, including prospect for 2017–2018. The legend lists the different satellite platforms/missions

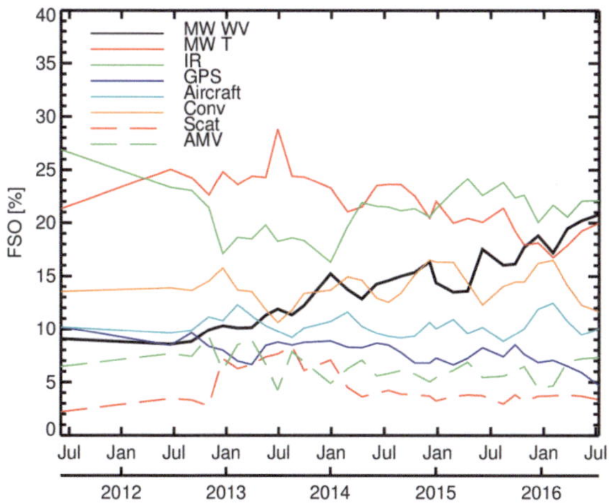

Fig. 2 Impact on the quality of the forecasts by the different observing systems: microwave (MW) water vapour plus cloud and temperature data, infrared (IR) temperature data, stratospheric temperatures derived from the global positioning system (GPS), aircraft, conventional observations (radiosondes, ships, buoys and synoptic data), near-surface winds over oceans from scatterometer (Scat) and atmospheric motion vectors (AMV) derived from geostationary satellite cloud and moisture tracking

Figure 2 displayed in more detail about the impact on the forecast quality of the different observing systems for the period 2011–2017. We distinguish between conventional observations (e.g. radiosondes, buoys), aircraft data, scatterometer measuring near-surface winds over the oceans, atmospheric motion vectors derived from satellite tracking of cloud and humidity features, temperature data in the stratosphere from GPS (using the bending angles of the signal) and satellite data from infrared sounders and microwave temperature as well as microwave water vapour, cloud and rain measurements. It is only since 2010 that microwave data (brightness temperatures) under all-sky conditions, i.e. in clear sky, cloudy and precipitating scenes, is directly assimilated in the four-dimensional variational system (Bauer et al. 2010). It is evident from Fig. 2 that temperature data from infrared sounders and microwave has the largest impact on the forecasts. However, there is a strong rise in the impact of the microwave all-sky humidity data, which by the end of 2017 will be the most beneficial data source.

2 Representing Model Uncertainty

Ensemble forecasts, comprising multiple individual forecasts (50 in the current ECMWF system) are used to measure the uncertainty or confidence in the underlying deterministic forecast. The uncertainty arises from errors in the initial conditions

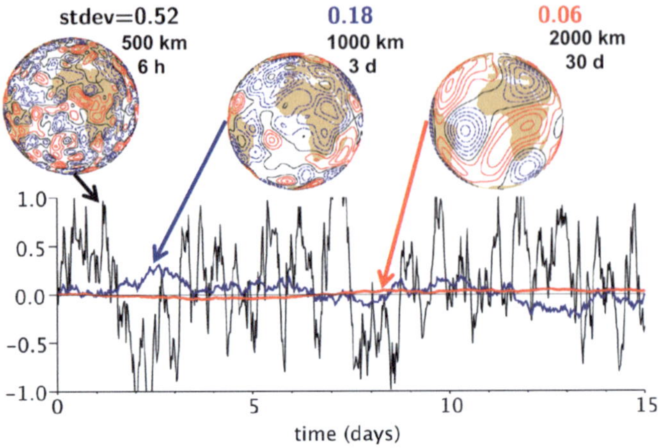

Fig. 3 The three global scale patterns with characteristic standard deviations and temporal and spatial correlation scales as used in the SPPT to generate the random 2D field ξ. The SPP only uses the first pattern

and the model itself (Leutbecher and Palmer 2008). Here, we will only discuss the representation of errors in the model physics.

In the current operational model, the physics uncertainty is represented by the stochastically perturbed total tendency scheme (SPTT) where the total physics tendencies of the prognostic model variables q, T, u and v are multiplied by a factor 1 + ξ where ξ is a 2D random field that is obtained through a first-order autoregressive process in spectral space. The three-scale pattern with different temporal and spatial correlation scales and standard deviations that have been implemented is shown in Fig. 3. The dominant pattern is the 'synoptic' pattern with spatial scales of 500 km and temporal scales of 6 h.

However, we think that the SPPT scheme cannot represent the uncertainties in the physics at the process level. For example, it cannot modify the shape of the heating profile and cannot add perturbations in regions where the tendency is zero. Therefore, we have developed an alternative approach named the stochastic perturbed parameter (SPP) approach where a number of carefully chosen parameters of each physical process (scheme) are perturbed (Ollinaho et al. 2017). The perturbed parameters for the vertical diffusion and subgrid orography, the large-scale cloud scheme, the radiation and the convection scheme are listed and illustrated in Fig. 4. All parameters are sampled from a positive definite log-normal distribution, with the nominal parameter value being multiplied by a factor ξ, with exception of the convective momentum transport which is sampled from a Gaussian distribution. However, parameters are given specific standard deviations. Once sampled, these parameters evolve with a synoptic-scale pattern as pictured in the center of Fig. 3. The SPP scheme also has its limitations. We might not have included all necessary parameters and one needs to keep in mind that changing a parameter is not as powerful as changing the underlying

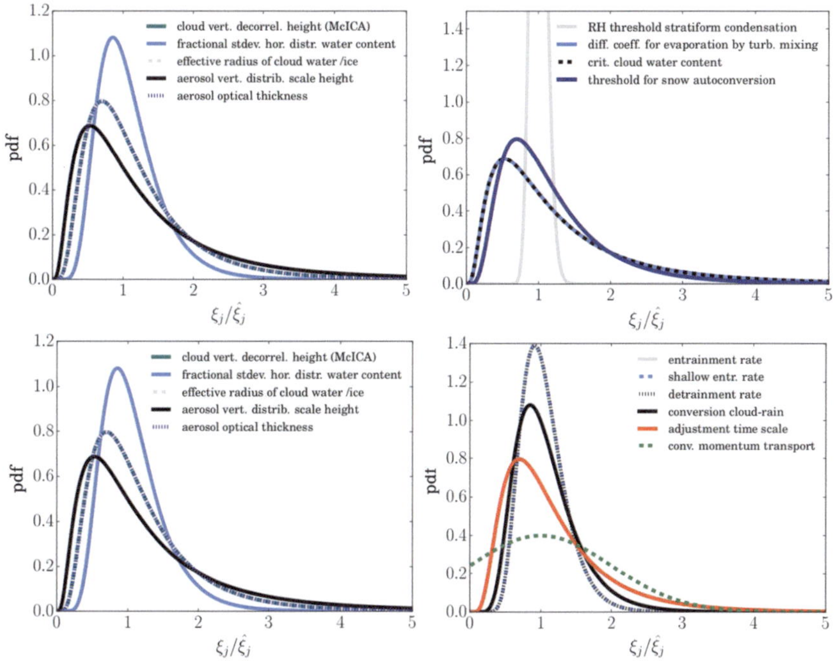

Fig. 4 List of perturbed parameters and their distributions as a function of the normalised value of the random variable ξ for the turbulent diffusion and subgrid orography, the large-scale cloud scheme and the radiation scheme and the convection. All parameters, except the convective momentum transport are sampled form a log-normal distribution, but have different standard deviations

equation or process. This was also the reason why we perturbed the whole convective momentum transport as a 'parameter'. As discussed in Ollinaho et al. (2017), the SPP scheme is still experimental and the challenge is to develop it further so that it can become operational and become also part of the ensemble data assimilation system. In the remainder of this article, we will make further use of the SPP for the convection parameters only, to estimate the uncertainty in the forecasts of convective precipitation and convective momentum balance.

3 Tropical Convectively Coupled Waves

The modulation of tropical rainfall on synoptic to intra-seasonal scales is mainly associated with the propagation of convectively coupled waves. The accurate representation of the structure and propagation of these waves is therefore of highest priority in tropical numerical weather prediction. The tropical convectively coupled waves are readily detected by the satellite observing system including infrared imagers/sounders as well as microwave sounding channels from the Special Sensor

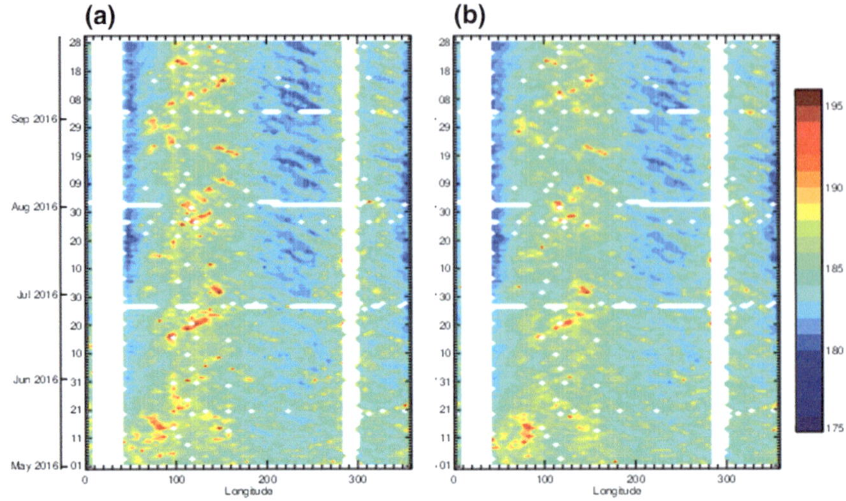

Fig. 5 Hovmöller diagrams of 10 GHz brightness temperatures over the equatorial belt during May–September 2016, **a** as obtained from SAPHIR and **b** as predicted by the daily IFS forecasts which are concatenated in time

Microwave Imager/Sounder (SSMIS) and the Sondeur Atmosphérique du Profile d'Humidité Intertropicale par Radiométrie (SAPHIR). Note that these instruments may contain hundreds of channels with major sensitivities to either water vapour, droplets, rain or frozen particles.

Figure 5 displayed Hovmöller diagrams of 10 GHz brightness temperatures over the equatorial belt as observed by SAPHIR and as predicted by the IFS short-range forecasts which have been concatenated in time. The 10 GHz channel is sensitive to frozen particles and therefore these microwave brightness temperatures are representative for the upper troposphere and also meaningful over land areas, with convection being characterized by high brightness temperatures. Strong convective activity is apparent in Fig. 5 over the Indian Ocean and West Pacific in both the observations and the forecasts, however the convective anomalies are somewhat larger in the observations. The convective activity is associated mainly with eastward propagating Kelvin waves and westward moving Rossby waves. The Madden Julian Oscillation (MJO) is also apparent as a roughly 30–35-day oscillation of larger clusters with enhanced Rossby and Kelvin wave activity.

While we can conclude from Fig. 5 that the analysis and short-range forecasts reasonably represent the convective activity in space and time, the assimilation of satellite data that is sensitive to cloud and rain is far from obvious. Ideally, the model must provide detailed and accurate information on the microphysics (size distribution and phase of droplets/crystals and precipitating particles) in order to accurately compute brightness temperature via the advanced radiative transfer model RTTOV (fast Radiative Transfer for TOVS). Here, we are at our second challenge. In global numerical weather prediction models, most tropical precipitation is produced by a mass flux convection parametrisation employing often rather simplified microphysics. In the IFS convection scheme,[1] the convective cloud model and the microphysics are diagnostic and volume mean concentrations of rain and falling frozen particles as required by a microwave or radar observation operator (the procedure to convert from model variables into observables) are not available, but only the liquid and solid precipitation fluxes at every vertical level. Clearly, we would like to have detailed information on convection and microphysics as it is available from convective-scale models (though these bring other challenges, see Sect. 6). Fortunately, it can be shown (Geer et al. 2009; Kessler 1969) that reasonable estimates of the volume mean convective rain and snow content can be derived from the precipitation fluxes using

$$q_{r,s} = 10^{-3}\rho^{-1}\left(a_{r,s}F_{r,s}\right)^{b_{r,s}} \tag{1}$$

where the subscripts r and s denote rain and snow, respectively, q is the specific mass mixing ratio (kg kg^{-1}), F the precipitation flux (kg m^{-2} s^{-1}), ρ the air density (kg m^{-3}) and a and b are constants accounting for the different fall speeds of rain and frozen particles. Mean in-cloud convective rain and snow water contents can then be estimated by dividing by the convective fraction which can be reasonably assumed to be constant below the melting levels with values O (5%).

After this small excursion on some challenges in assimilating satellite data in convective areas, we want to illustrate the predictability and structure of the convectively coupled waves in longer range forecasts. Herman et al. (2016) have extracted Kelvin wave signals from multi-year integrations with the IFS with the aid of spectrally filtering the outgoing longwave radiation, followed by a regression of other

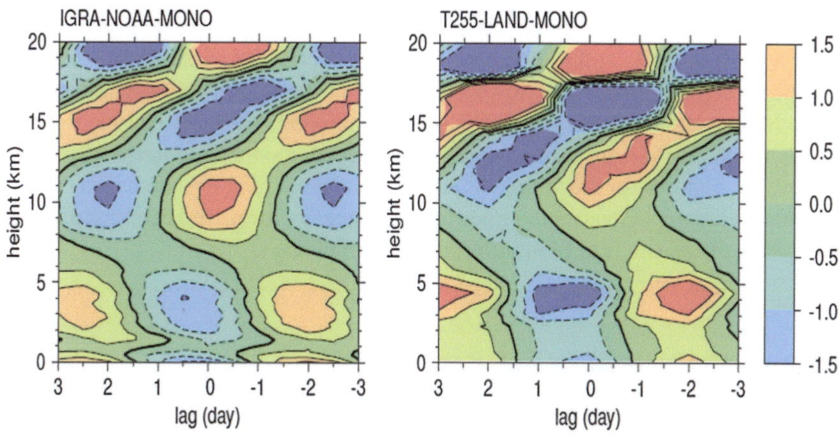

Fig. 6 The temperature anomaly (K) of the eastward-moving Kelvin wave as obtained from the IGRA West Pacific sounding network and as extracted from the IFS long integrations at the radiosondes locations

[1] https://www.ecmwf.int/en/elibrary/17117-part-iv-physical-processes.

variables onto the spectrally filtered field. They showed that the Kelvin wave signal is in agreement with linear theory and the observations. The vertical structure of Kelvin waves as a function of the lag, with 0 denoting minimum outgoing longwave radiation and maximum convective activity, from radiosonde data in the West Pacific and the IFS is displayed in Fig. 6. Both the observations and the model show the characteristic baroclinic structure with an amplitude of up to 2 K and different tilts in the troposphere and the stratosphere. Getting this structure right is absolutely crucial for wave propagation and amplification as it means that at lag 0, the convective heating in the upper troposphere coincides with the warm phase of the wave, therefore generating available potential energy and converting it into kinetic energy. As discussed in Hirons et al. (2013), models with insufficient convective entrainment or mixing, in general, do not get the moisture structure and temporal correlations between the convective heating and phase of the wave right and therefore are unable to reproduce and maintain realistic tropical variability.

The predictability of these large-scale wave structures with the IFS has continuously increased since 2006 due to many factors, more realistic ocean coupling, evolution of the ensemble system and stochastic physics, the microphysics, but particularly with the revisions to the convection scheme including a realistic strong entrainment (Bechtold et al. 2008). This is documented in Fig. 7a for the predictability of the MJO where the time axis denotes the model version operational during that year. Using a correlation value of 0.6 as a predictability limit, the average predictability of the MJO has increased from about 15 days in 2002 to about 28 days in 2016. Note that, the predictability of the MJO is largely insensitive to the atmospheric model's horizontal resolution. For a more detailed discussion of the thermodynamic characteristics of the MJO, the reader is referred to Benedict and Randall (2007), the predictability of the MJO and its teleconnections with the northern hemispheric circulation is discussed in Vitart and Molteni (2010). The MJO predictability by the IFS is still significantly higher than that by other centres issuing sub-seasonal ensemble forecasts (Fig. 7b). However, we expect that the potential predictability of the MJO is around one cycle, i.e. 45 days. Therefore, we think that there is still a large potential for improvements in the MJO prediction and consequently in the monthly predictions for the middle latitudes. Also, modelling experience has shown that relatively small changes to either the latent and/or radiative heating profiles or the surface fluxes can lead to improvements of a few days.

4 Diurnal Cycle of Convection

The accurate representation of the diurnal cycle of convection is another important challenge in terms of data assimilation, prediction of mesoscale convective systems and even seasonal predictions. Indeed, it has been shown in numerous studies (e.g. Zhang et al. 2016; Birch et al. 2014) that the diurnal cycle of convection is strongly interlinked with the intensity and northward extension of the West African monsoon

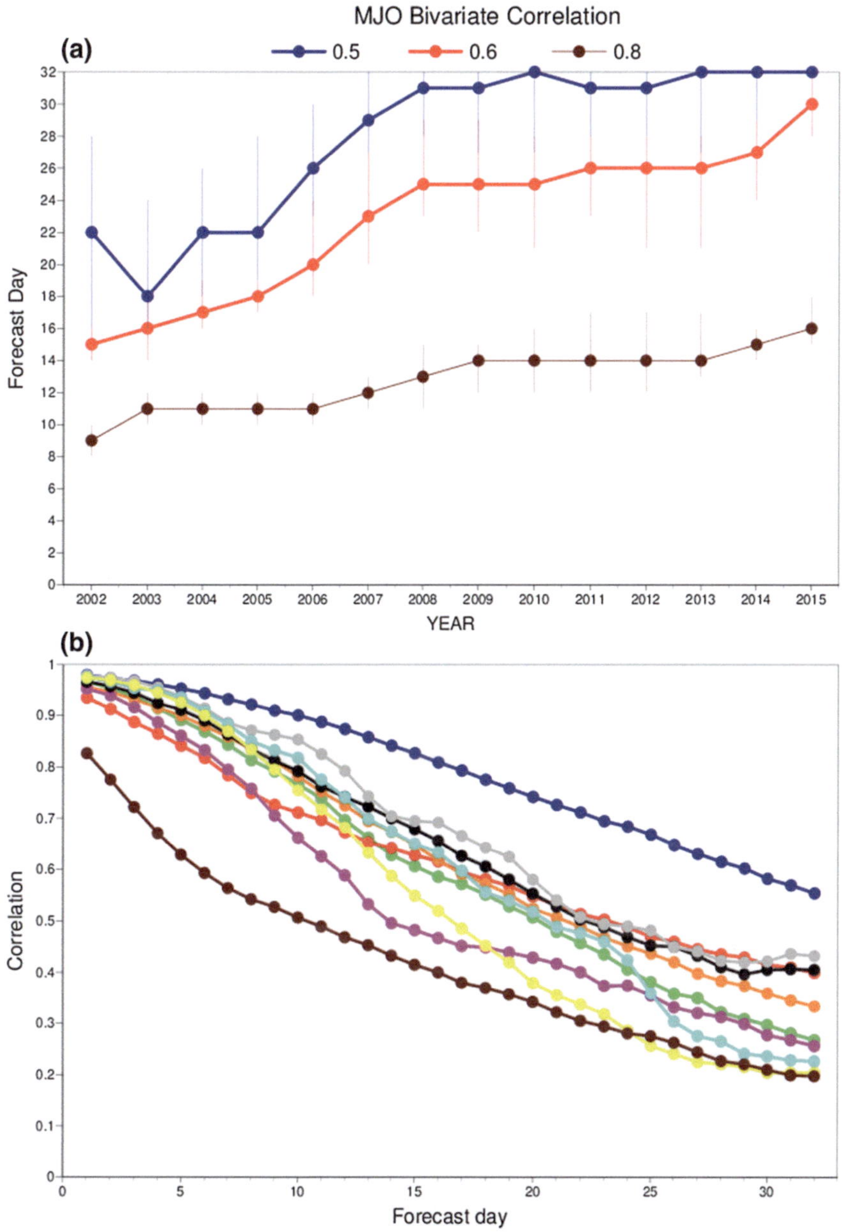

Fig. 7 **a** Evolution of the predictability of the MJO in the IFS coupled system as measured by a bivariate correlation coefficient including winds and outgoing longwave radiation. The year denotes the specific IFS cycle as all reforecasts have been run for the same reference period and with the same initial conditions from the European Interim Reanalysis (ERAI). **b** Same as figure (**a**), but showing the MJO bivariate correlation coefficient as a function of forecast lead time. Colours denote forecasts form different operational Centres issuing seasonal to sub-seasonal forecasts, the IFS is denoted by the dark blue line

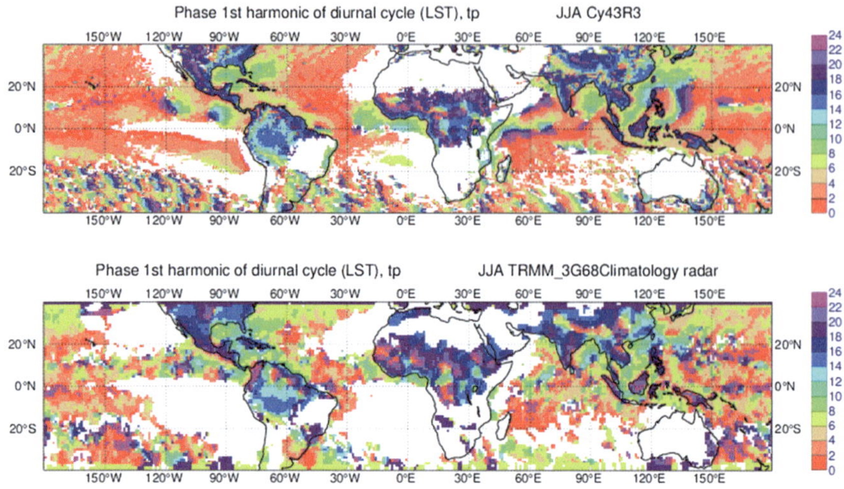

Fig. 8 Phase (LST) of the diurnal cycle of precipitation as obtained as the first harmonics from an ensemble of 1-year coupled integrations during 2000–2004 and a 10-year rainfall climatology from the TRMM radar

and also affects the propagation of the MJO through the maritime continent (Hagos et al. 2016).

The diurnal cycle of rainfall over the tropical belt during boreal summer has been extracted in Fig. 8 from a 10-year climatology of the Tropical Rainfall Measuring Mission (TRMM) radar and from an ensemble of coupled 1-year integrations with the latest operational version of the IFS. The observations (Fig. 8b) show maximum rainfall over water during the early morning hours, while over land maximum rainfall typically occurs during the late afternoon and early evening hours. The model (Fig. 8a) is able to reproduce the land–sea contrast and most regional variations as apparent over the continental United States and India. As discussed in Bechtold et al. (2014), the main contributors for a realistic representation of the diurnal cycle of convection in large-scale models are a mass flux convection scheme with a strongly entraining updraught plume and a convective closure that takes into account both the rapidly evolving surface forcing and the slowly evolving storage of conditional instability in the free troposphere.

However, when looking more into the detail, some local but important systematic model errors become apparent. These errors have been statistically evaluated with the aid of short-range ensemble forecasts using 25 ensemble members with perturbations to the convective parameters (Sect. 2, Fig. 4d). All forecasts start at 00 UTC. Figure 9 shows the domain mean rainfall rate (mm day^{-1}) as a function of time (UTC) over the southern Great Planes of the United States as observed by the NEXRAD radar plus rain gauge network (black solid line) and two versions of the IFS ensemble (operational, yellow and experimental, blue). The spread in the ensembles is denoted by 25th and 75th percentile. The model reproduces the increase in precipitation

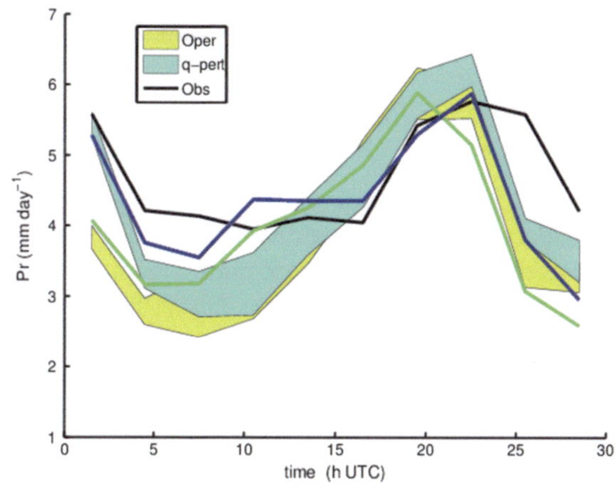

Fig. 9 Domain mean rainfall over the central Great Planes of the US during summer 2015 as observed by the NEXRAD radar + rain gauge product (thick black line) and from two sets of ensemble prediction with 25 members with the operational model version (yellow) and a version with additional moisture perturbations in the convection (sky blue). Shadings denote the 25th and 75th percentile of the distribution and the solid blue and green lines the unperturbed forecasts. LST = UTC-6 h

during the afternoon with peak rain rates during late afternoon, but while significant precipitation persists in the observations during the first half of the night, the model rainfall drops off too quickly. In addition, the spread in model precipitation during the first half of the night is small, meaning the error is systematic and there is not enough variability in the occurrence of night-time mesoscale convective systems. The experimental model version with revised convective moisture perturbations improves on the results in the very short range but is not able to maintain the signal over several days. Similar results have been obtained for the Sahel region, which like the southern Great Planes is marked by a low-level jet and the propagation of mesoscale convective systems that are generated upstream during daytime. While there is a substantial amount of studies on the propagation of organized convective systems during the night, and theories like convective outflow by downdraughts, lifting by bores and dynamical interactions with the low-level jet have been brought forward (e.g. Laing et al. 2012; Blake et al. 2017), important challenges in both understanding and modelling of these systems remain.

5 The Tropical and Subtropical Boundary Layer

5.1 Cloud and Radiation Biases

Significant model biases in terms of low clouds and radiation persist in the tropical and subtropical boundary layers. These are related to the formulation of the turbulent and/or shallow convective mixing and microphysics. Figure 10a displays the annual mean difference in the net shortwave radiative flux at the top of the atmosphere between an ensemble of 1-year coupled integrations with the latest operational cycle of the IFS and the NASA CERES-EBAF product (Loeb et al. 2009). The subtropical anticyclones with predominantly shallow cumulus (Cu) clouds appear as regions that are too reflective (negative bias, the sign convection is negative upward). In contrast, the stratocumulus (Sc) regions off the West Coasts of the continents, as well as the Southern hemisphere storm tracks appear as regions with a positive radiation bias, indicating a too transparent atmosphere with too much downwelling shortwave radiation at the surface. As discussed in Li et al. (2013), these characteristic radiation biases are shared by most general circulation models participating in the CMIP5 5th Coupled Model Intercomparison Project. The systematic radiative flux biases could be linked to either an insufficient representation of the low cloud cover and/or errors in the optical thickness, i.e. the liquid or ice water path of the models.

For the IFS, it can be demonstrated that the radiation biases in the tropical and subtropical trade wind and Sc regions are primarily due to errors in the liquid water path and not the cloud cover. In Fig. 11, we have plotted the column integrated liquid water path along a section of the East Pacific from Los Angeles to Hawaii during the MAGIC's campaign (Lewis 2016; Kalmus et al. 2014), and compare the results of short-range model forecasts to a composite of satellite retrievals from MODIS on-board Aqua and Terra and from microwave radiometer measurements taken on-board a commercial container ship. The retrievals agree in that they indicate a significant underestimation of the model's liquid water path in the Sc regions, consistent with not reflective enough clouds, and a modest overestimation in the shallow Cu regions, consistent with too reflective clouds. The reasons for the large errors in the Sc off the West coasts of the continents are yet to be fully understood, but preliminary experimentation suggests that it is linked to the formulation of mixing and condensation in the boundary-layer, as well as to errors in the vertical advection of moisture in the semi-Lagrangian advection scheme.

Finally, it can also be shown that the shortwave radiation biases in the Southern hemispheric storm tracks are neither primarily due to errors in the cloud cover or the total condensate path, but due to errors in the phase of the cloud condensate (Forbes et al. 2016). Indeed, liquid water clouds or clouds with a liquid water layer near their top are more reflective than ice clouds and the former occurs in the cold air outbreak regions with shallow and congestus convection (Hu et al. 2010). We have therefore revised both the mixed-phase microphysics in the convection scheme, computing only saturation with respect to the liquid phase for shallow convection and the detailed warm phase microphysics of the stratiform cloud scheme.

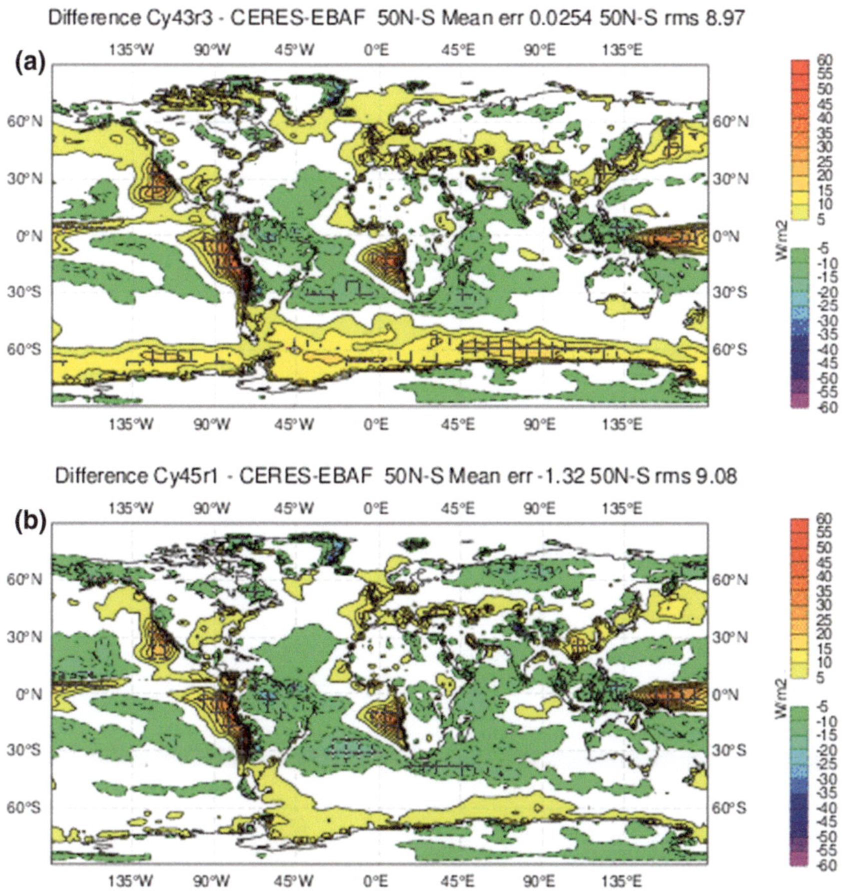

Fig. 10 **a** Climatological annual mean difference in net shortwave radiation at the top of the atmosphere (W m^{-2}) between an ensemble of 1-year atmosphere–ocean coupled model integrations with Cy43r3 of the IFS (operational 2017) for 2000–2004 and the NASA CERES-EBAF product. Significant differences are highlighted by hatches. Negative values indicate areas that are too reflective and positive values areas that are too transparent with too much downwelling shortwave radiation at the surface. **b** Same as figure (**a**), but for model Cy45r1 (to be implemented in 2018) with revised convective mixed-phase microphysics and revised warm phase microphysics in the stratiform cloud scheme

The first results are promising (Fig. 10b) and in that, the shortwave biases over the Southern and Northern hemispheric storm tracks have been largely removed. The mixed-phase revisions to the microphysics have been introduced operationally in 2018, but further work and evaluations in data assimilation are necessary to fully address the radiation/cloud errors in the cold-air outbreak regions.

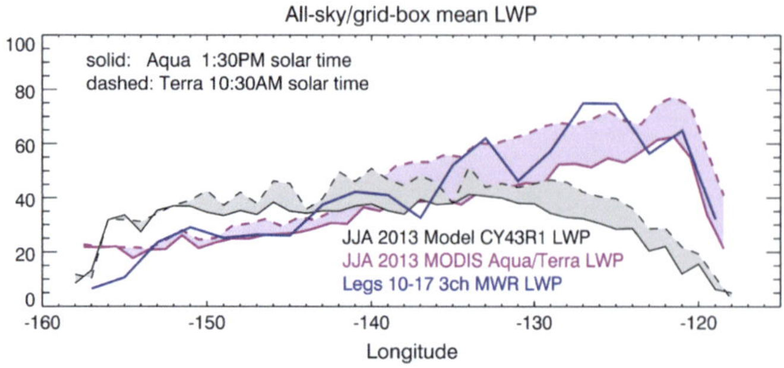

Fig. 11 The column integrated liquid water path along a section of the East Pacific as obtained from the IFS short-range forecasts (black) and as obtained by satellite retrievals with MODIS Aqua/Terra (pink) and microwave radiometer measurements on-board ship (blue). The shading denotes the measurement uncertainty related to the diurnal cycle as Aqua and Terra observe at different solar times

5.2 *Winds and Convective Momentum Transport*

Accurate analysis and forecasts of tropical winds are vital for accurate tropical and extratropical predictions on all time scales. This is a difficult task as tropical motions

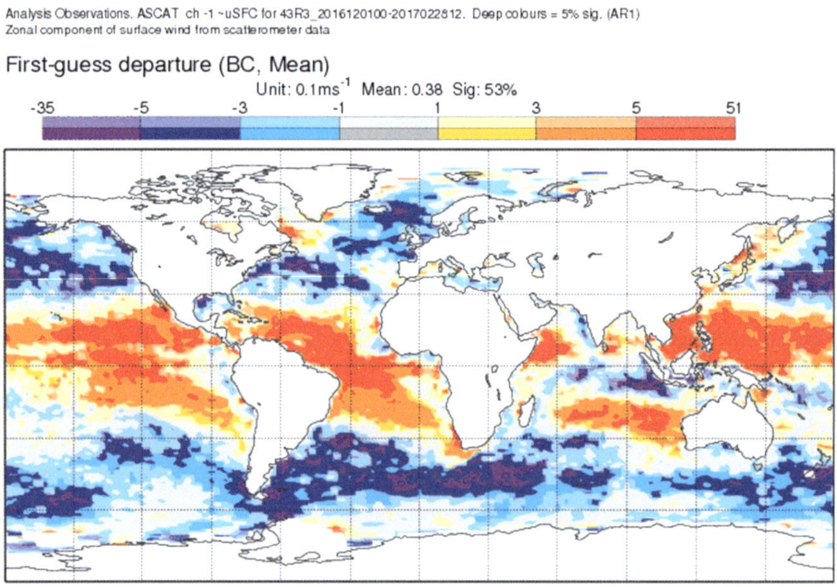

Fig. 12 Mean difference (first-guess departure) between the zonal wind speed as derived from ASCAT scatterometer and the short-range forecasts from a full analysis cycle for December 2016 to February 2017

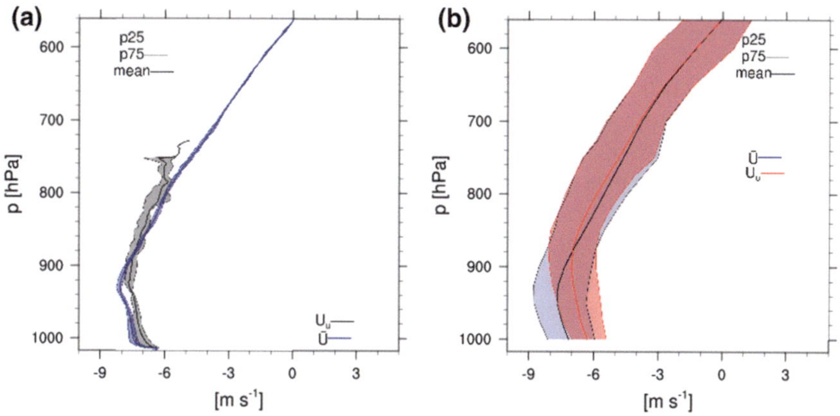

Fig. 13 Mean zonal wind component and mean updraught zonal wind speed during 16–18 December 2004 averaged over a 50 × 50 km domain centred near Barbados island in the subtropical Atlantic: **a** as obtained from a large domain Large Eddy Simulation, and **b** as extracted from 25-member ensemble forecasts with the IFS using parameter perturbations

have a strong unbalanced component (Žagar et al. 2005) as geostrophic balance vanishes near the equator and as the motions are largely heating driven with additionally a strong contribution from the transport of momentum in convective draughts. In addition, direct wind observations that could help to constrain the analysis are relatively sparse over the tropical oceans. As listed in Figs. 1 and 2, the principal wind data is for surface winds from scatterometer and to a lesser extent from aircraft data at flight level, while atmospheric motion vectors (AMVs) from geostationary satellite have a good global coverage but are derived products with a relatively large error.

As a reliable estimate of the near-surface wind errors of the IFS, we have plotted in Fig. 12 for the period December 2016 to February 2017 the mean difference between the zonal wind components are derived from scatterometer data and the IFS short-range forecasts. These differences are obtained as part of the variational analysis and are called first-guess departures. Figure 12 indicates an easterly wind bias of 0.3–0.5 m s^{-1} of the forecasts in the tropical and subtropical regions. Obviously, the model also shows a compensating westerly wind bias in the extratropical storm tracks as the global angular momentum is conserved. Interestingly, the tropical low-level easterly wind bias is shared by many CMIP5 models (D. Kim and M.-S. Ahn, personal communication). At present, we do not have an explanation of this, but we have been looking at the formulation of the turbulent surface drag, the turbulent heat fluxes and the convective momentum transport in the trade wind region.

In order to evaluate the parametrized convective momentum transport by shallow convection, we took advantage of Large Eddy Simulation (LES) data of organized shallow convection during the Rain in Cumulus over the Ocean (RICO) experiment. The data has 25 m horizontal resolution and covers a 50 × 50 km domain centred at 18°N/62°W near Barbados Island in the subtropical Atlantic; the period is 16–28 December 2004. For the same period, we ran the IFS ensemble system at 16 km resolution with convective parameter perturbations only (see Fig. 4d) and extracted from the global data momentum diagnostics over the LES domain. A detailed description of the analysis method and further results are given in Schlemmer et al. (2017). In Fig. 13, we only present the domain mean zonal momentum and the zonal momentum in the convective updraughts. In both, the LES and the IFS, the updraught zonal wind speed is smaller than that in the environment below 900 h Pa, decelerating down the flow, but the updraught zonal momentum exceeds the environmental momentum in the upper part of the trade wind layer. It can be shown that the average effect of the shallow convective momentum transport is a reduction of the environmental wind shear, therefore the momentum transport is downgradient. We could further demonstrate that the convective parameter perturbations reasonably sample the wind uncertainties. The main error source in the representation of the convective momentum transport could be related to the specification of the updraught momentum at cloud base, and to a lesser extent to the representation of the across-cloud pressure gradient. The systematic easterly low-level wind bias could not, however, be related to uncertainties in the convective momentum transport.

6 Deep Convection at Higher Resolution

As mentioned in the introduction, our goal is to have by 2025 a global ensemble system at 5 km horizontal resolution. An important challenge we, therefore, need to cover is the formulation of the deep convective parametrization in the resolution range between 10 and 1 km where deep convective motions gradually become more resolved by the model dynamics. There is extensive literature on this subject (see e.g. Arakawa and Wu 2013; Gerard 2015) and the basic questions are how to formulate the subgrid convective fluxes and the convective closure at these resolutions. Denoting by ω the vertical velocity (Pa s^{-1}), the subgrid organized convective flux of any variable ϕ can be written as

$$\overline{\omega'\Phi'} = \overline{\omega\Phi} - \bar{\omega}\bar{\Phi} = \sigma(1-\sigma)\left(\bar{\omega}^c - \bar{\omega}^e\right)\left(\bar{\Phi}^c - \bar{\Phi}^e\right) \tag{2}$$

where σ denotes the updraught fractional area and superscripts c and e denote updraught and environmental values, respectively. In the IFS convection, parametrization and most other convection schemes in operational models, σ is supposed to be small, which simplifies (2) to

$$\overline{\omega'\Phi'} \approx \sigma\bar{\omega}^c\left(\bar{\Phi}^c - \bar{\Phi}\right) = -g^{-1}M^c\left(\bar{\Phi}^c - \bar{\Phi}\right) \tag{3}$$

where M^c is the convective mass flux, g is gravity and where the environmental value of ϕ has been replaced by the predicted grid-scale value. Obviously, there is a problem with (3) in the limit $\sigma \to 1$ when convection is fully resolved; while (2) converges to zero, (3) does not. However, using (2) requires the knowledge of σ which is difficult to determine accurately as no theoretical closure exists for this parameter,

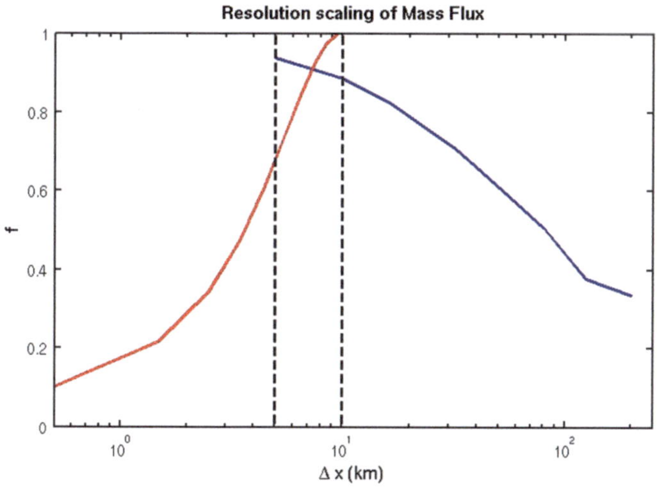

Fig. 14 Horizontal resolution dependent mass flux scaling factor f, representing two regimes, $\Delta x < 8$ km (red) and $\Delta x > 8$ km (blue). The vertical dashed lines denote the 5 and 10 km resolutions

though the authors above have tried to circumvent the problem. In contrast, we and others (e.g. Grell and Freitas 2014; Kwon and Hong 2017) opted for a pragmatic approach, where we keep (3) but multiply the fluxes by an empirical scaling factor *f*. This method also has the advantage that it does not introduce resolution dependent convective profiles and therefore potential model biases, as in (3), the fluxes are computed with respect to the grid-mean value and not with respect to a convective environment. The empirical function *f* has been evaluated in collaboration with the German Weather Service who employed the non-hydrostatic ICON model in the resolution range resolutions $\Delta x < 10$ km. The ICON model makes use of the same convection parametrization as the IFS. A set of experiments with variable horizontal resolutions from 1 to 10 km has been run over Europe and *f* has been evaluated as to give best model performance in terms of rainfall against radar and routine temperature and wind scores. The horizontal resolution dependency of the mass flux scaling factor *f* is displayed in Fig. 14 for two regimes, $\Delta x < 8$ km, representing the $(1-\delta)$ factor and for $\Delta x > 8$ km. In the former regime, the mass flux at 1 km is about 10% of its nominal value at 10 km. The regime $\Delta x > 8$ km represents the decreasing mass fluxes with lower resolution, as the large-scale forcing, i.e. the horizontal divergence decreases with decreasing resolution. This effect must be explicitly taken into account in an equilibrium closure for the convective available potential energy CAPE (Bechtold et al. 2014). Due to computational constraints, we have not yet extensively evaluated the scaling in Fig. 14 for the high-resolution regime, but have only done a few preliminary case studies at a global resolution of 5 km.

Figure 15 provides an illustration of the effect of the deep convective parametrization at 5 km resolution (spectral truncation TCo1999), where we have run short-range forecasts with the deep convection scheme turned off, and with the operational deep convection using the scaling as in Fig. 14. The 24-h rainfall estimated for 18 July 2015 over Central Africa from the NOAA-FEWS product (Fig. 15a) reveals three large convective areas associated to African easterly waves with maximum rainfall of 50–75 mm. While the model run without the deep convection parametrization produces a very noisy precipitation field and typically overestimates convective rainfall by a factor of two, the simulation with scaled convective fluxes reasonably reproduces the main convective features, though overestimates the areas with light rain and underestimates the maxima. Note that, the errors are larger over Western Africa, where rainfall typically occurs later during the night.

These 5 km global simulations are still experimental but clearly illustrate the need for a performant convection parametrization at these resolutions and in particular in the tropical regions.

7 Conclusions

We have presented some of the important challenges in global numerical weather prediction. These challenges were put into the context of the ECMWF strategy which aims to have by 2025, an ensemble based atmosphere–ocean coupled earth system model at a horizontal resolution of 5 km. As discussed, the largest potential for improving predictability is in the tropical regions. This stems from the predictability of the dominant large-scale tropical waves which can be estimated to be of O (20–30 days) for the Kelvin waves and up to 45 days for the MJO. While the model realistically represents the vertical structure of these waves, model errors exist due to uncertainties in the convective and stratiform heating profiles. Furthermore, it can also be shown that the initial errors are large, especially in the upper tropical troposphere, as the analysis is insufficiently constrained by wind observations in the tropics and the ensemble data assimilation system has difficulties to estimate the full uncertainties related to the tropical circulations (Žagar 2017). We are currently developing a physically based method to take into account the uncertainties inherent to the model physical parametrizations, called stochastically perturbed parameter perturbations (SPP). We hope that future versions of the ensemble forecasts, including the ensemble data assimilation will benefit from these developments. We also have large expectations in the ADM AEOLUS wind lidar mission that it is finally scheduled for 2018–2023, and should help to constrain the tropical wind analysis and considerably improve the forecasts.

We have also elaborated on the importance of accurate mixed-phase microphysics and its consistency across the model physics, i.e. boundary layer scheme, cloud scheme and convection scheme. Indeed, accurate mixed-phase microphysics is particularly important in making full use of the data assimilation of advanced microwave radiances in all-sky conditions in the tropics, and in reducing the shortwave radiation biases in the storm tracks that are related to convective cold air outbreaks.

The model's representation of deep and shallow convection has also been evaluated in terms of precipitation, cloudiness and winds. Concerning deep convection,

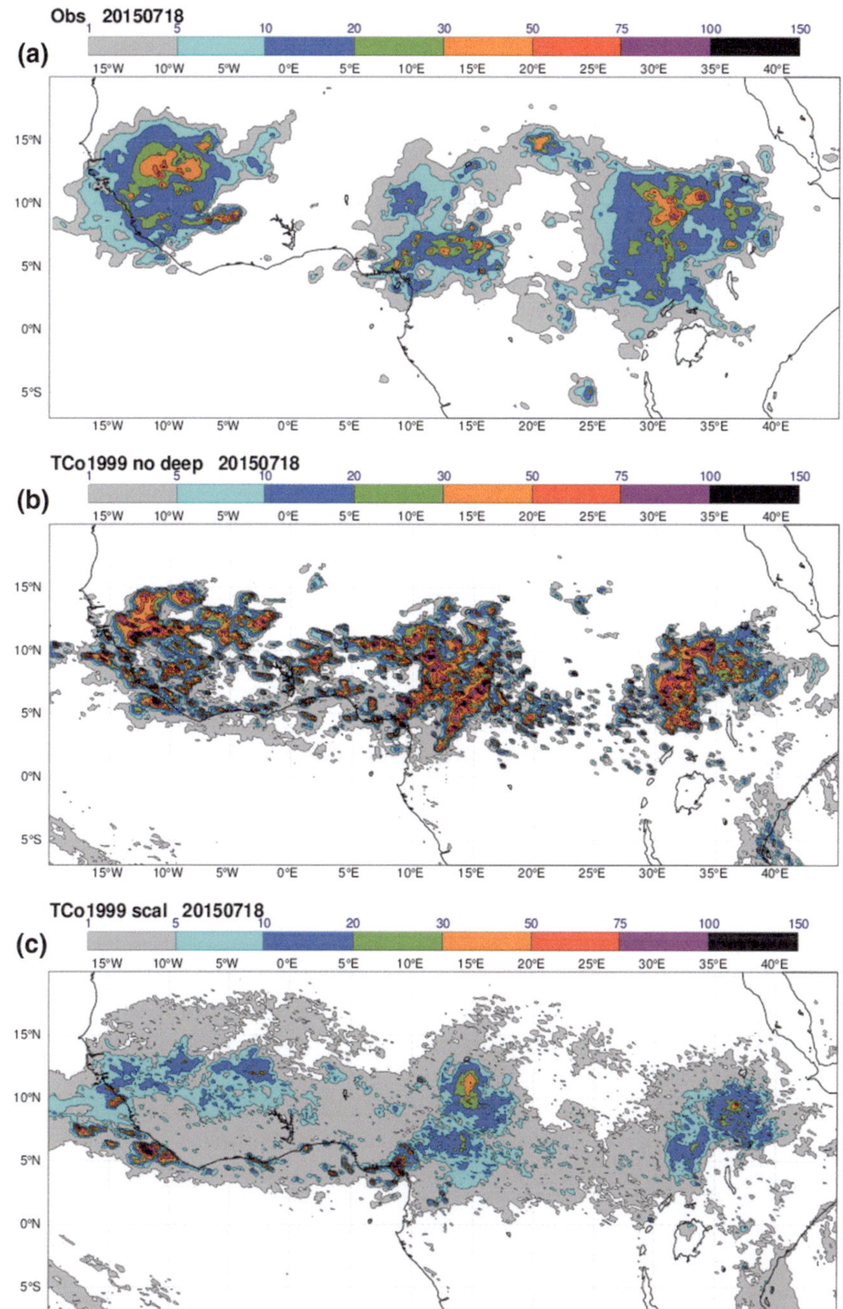

Fig. 15 24-hour rainfall totals for 18 July 2015 over Central Africa as **a** from the NASA-FEWS satellite product and from spectral truncation TCo1999 (5 km) IFS simulations **b** without and **c** with the deep convection parametrization using the flux scaling as in Fig. 14

the main remaining systematic model error is in the underestimation of night-time convection, in particular, in the Sahel region and the Southern Great Planes of the US which are characterized by propagating mesoscale convective schemes that are dynamically forced by a low-level jet. We do not yet have a solution for this problem, though preliminary experimentation has shown a large sensitivity to the evaporation and advective elevated moisture perturbations in the convection parametrization and resolved microphysics.

The model also has characteristic biases in the subtropical Sc-to-Cu transition region off the West coasts of the continents. With the aid of satellite and in situ observations it could be demonstrated that the liquid water content in the Cu region is overestimated by 20–30% while it is strongly underestimated in the Sc regions. The latter appeared to be sensitive to both the formulation of vertical mixing and subgrid condensation in the physics and to truncation errors in the semi-Lagrangian vertical advection of moisture. The trade wind flow in the model's short-range forecast has an easterly wind bias of roughly 0.5 m s^{-1} against scatterometer data. An easterly low-level wind bias is also present in most CMIP5 models. Though we could not yet identify the root cause of this bias, we were able with the aid of large domain LES data to validate the shallow convective momentum transport which turned out to be predominantly downgradient, and this independently of the organization of shallow convection.

Finally, we presented the empirical scaling that is applied to the deep convective fluxes in the 10–1 km resolution range in order to ramp down the contribution from the parametrized fluxes to the total mass overturning. This formulation provides a smooth transition and more importantly, avoids strong grid-scale overturning with unrealistic divergent outflow. We do not yet apply any other modification to the deep convection scheme in the high-resolution limit. However, we consistently couple the convective cloud to the resolved microphysics through the detrainment of liquid and solid condensate as well as rain and snow. It is clear that in the resolution limit of 1–2 km the model should also be able to reproduce organized convection such as squall lines and the characteristic outflow. However, as experienced by regional model systems (e.g. Hans Wang, Singapore Met Service, M. Köhler, D. Klocke and B. Stevens using the ICON model of German Weather Service, personal communications) even at these grid spacings, it is difficult to reproduce the correct scale and frequency of occurrence of these features.

We could only present a selection of challenges related to advances in earth system modelling in the tropics. There are of course many other physical challenges such as the resolution of the ocean model and problems in coupling all fluxes with the atmospheric fluxes. Other notable challenges are the modelling of aerosols and their influence on the tropospheric radiation budget, and the accurate simulation of the stratospheric circulation which is strongly driven by the ozone radiative effect and friction from convectively generated gravity waves and large-scale inertia–gravity waves.

Last but not least, there is the computing challenge. As projected in Bauer et al. (2015), the tenfold increase in the number of observations, until 2025 will not pose any significant computational problems. However, how to deal in real-time operational forecasts with the expected 50 fold increase in computing time when going from a 16 km ensemble to a 5 km ensemble or the 1000 fold increase in computing time and 100 fold increase in storage when going to a 1 km ensemble with even more prognostic equations is an open question. It will certainly require a clever use of the then existing computing resources, e.g. by trading ensemble size against ensemble resolution and precision.

References

Arakawa, A., and C.-M. Wu. 2013. A unified representation of deep moist convection in numerical modelling of the atmosphere. Part I. *Journal of the Atmospheric Sciences* 70: 1977–1992.

Bauer, P., A. Thorpe, and G. Brunet. 2015. The quiet revolution of numerical weather prediction. *Nature* 525: 47–55.

Bauer, P., A.J. Geer, P. Lopez, and D. Salmond. 2010. Direct 4D-Var assimilation of all-sky radiances: Part I. Implementation. *Quarterly Journal of the Royal Meteorological Society* 136: 1868–1885.

Bechtold, P., M. Köhler, T. Jung, M. Leutbecher, M. Rodwell, F. Vitart, and G. Balsamo. 2008. Advances in predicting atmospheric variability with the ECMWF model, 2008: From synoptic to decadal time-scales. *Quarterly Journal of the Royal Meteorological Society* 134: 1337–1351.

Bechtold, P., N. Semane, P. Lopez, J.-P. Chaboureau, A. Beljaars, and N. Bormann. 2014. Representing equilibrium and non-equilibrium convection in large-scale models. *Journal of the Atmospheric Sciences* 71: 734–753.

Benedict, J.J., and D.A. Randall. 2007. Observed characteristics of the MJO relative to maximum rainfall. *Journal of the Atmospheric Sciences* 64: 2332–2354.

Birch, C.E., D.J. Parker, J.H. Marsham, D. Copsey, and L. Garcia-Carrera. 2014. A seamless assessment of the role of convection in the water cycle of the West African monsoon. *Journal Geophysical Research* 119: 2890–2912.

Blake, B.J., D.B. Parsons, K.R. Haghi, and S.G. Castleberry. 2017. The structure, evolution and dynamics of a nocturnal convective system simulated using the WRF-ARW model. *Monthly Weather Review* (early online release).

Forbes, R., A. Geer, K. Lonitz, and M. Ahlgrimm. 2016. Reducing systematic errors in cold-air outbreaks. *ECMWF Newsletter* 146: 17–22.

Geer, A.J., P. Bauer, and C.W. O'Dell. 2009. A revised cloud overlap scheme for fast microwave radiative transfer in rain and cloud. *Journal of Applied Meteorology and Climatology* 48: 2257–2270.

Gerard, L. 2015. Bulk mass-flux perturbation formulation for a unified approach of deep convection at high resolution. *Monthly Weather Review* 143: 4038–4063.

Grell, G.A., and S.R. Freitas. 2014. A scale and aerosol aware stochastic convective parameterization for weather and air quality modelling. *Atmospheric Chemistry and Physics* 14: 5233–5250.

Hagos, S.M., C. Zhang, Z. Feng, C.D. Burleyson, C. De Mott, B. Kerns, J.J. Benedict, and M.N. Martini. 2016. The impact of the diurnal cycle on the propagation of Madden-Julian oscillation convection across the maritime continent. *Journal of Advances in Modeling Earth Systems* 8: 1552–1564. https://doi.org/10.1002/2016MS000725.

Herman, M.J., Ž. Fuchs, D. Raymond, and P. Bechtold. 2016. Convectively coupled Kelvin waves: From linear theory to global models. *Journal of the Atmospheric Sciences* 73: 407–428.

Hirons, L.C., P. Inness, F. Vitart, and P. Bechtold. 2013. Understanding advances in the simulation of intraseasonal variability in the ECMWF model. Part II: The application of process-based diagnostics. *Quarterly Journal of the Royal Meteorological Society* 139: 1427–1444. https://doi.org/10.1002/qj.2059.

Hu, Y., S. Rodier, K.-M. Xu, W. Sun, J. Huang, B. Lin, P. Zhai, and D. Josset. 2010. Occurrence, liquid water content and fraction of supercooled water clouds from combined CALIOP/IIR/MODIS measurements. *Journal of Geophysical Research* 115 (D4): D00H34.

Kalmus, P., M. Lebsock, and J. Teixeira. 2014. Observational boundary layer energy and water budgets of the stratocumulus-to-cumulus transition. *Journal of Climate* 27: 9155–9170.

Kessler, E. 1969. On the distribution and continuity of water substance in atmospheric circulation. In *Meteorological monographs*, vol. 10. Boston, MA: American Meteorological Society.

Kwon, Y.C., and S.-Y. Hong. 2017. A mass flux parameterization scheme across gray-zone resolutions. *Monthly Weather Review* 145: 583–598.

Laing, A.G., S.B. Trier, and C.A. Davis. 2012. Numerical simulation of episodes of organized convection in tropical northern Africa. *Monthly Weather Review* 140: 2874–2885.

Leutbecher, M., and T. Palmer. 2008. Ensemble forecasting. *Journal of Computational Physics* 227 (7): 3515–3539.

Lewis, E.R. 2016. Marine ARM GPCI investigation of clouds (MAGIC) field campaign report. In *DOE ARM climate research facility*, ed. R. Stafford. DOE/SC-ARM-16-057.

Li, J.-L.F., D.E. Walliser, G. Stephens, S. Lee, T. L'Ecuyer, S. Kato, N. Loeb, and H.-Y. Ma. 2013. Characterizing and understanding radiation budget biases in CMIP3/CMIP5 GCMs, contemporary GCM, and reanalysis. *Journal Geophysical Research* 118: 8166–8184. https://doi.org/10.1002/jgrd.50378.

Loeb, N.G., B.A. Wielicki, D.R. Doelling, G.L. Smith, D.F. Keyes, S. Kato, N. Manalo-Smith, and T. Wong. 2009. Toward optimal closure of the Earth's top-of-atmosphere radiation budget. *Journal of Climate* 22: 748–766.

Ollinaho, P., S.-J. Lock, M. Leutbecher, P. Bechtold, A. Beljaars, A. Bozzo, R. Forbes, T. Haiden, R. Hogan, and I. Sandu. 2017. Towards process-level representation of model uncertainties: Stochastically perturbed parametrisations in the ECMWF ensemble. *Quarterly Journal of the Royal Meteorological Society* 143: 408–422.

Rabier, F., H. Järvinen, E. Klinker, J.-F. Mahfouf, and A. Simmons. 2000. The ECMWF operational implementation of four-dimensional variational assimilation. I: Experimental results with simplified physics. *Quarterly Journal of the Royal Meteorological Society* 126: 1143–1170.

Schlemmer, L., P. Bechtold, I. Sandu, and M. Ahlgrimm. 2017. Uncertainties related to the representation of momentum transport in shallow convection. *Journal of Advances in Modeling Earth Systems*.

Vitart, F., and F. Molteni. 2010. Simulation of the MJO and its teleconnections in the ECMWF forecast system. *Quarterly Journal of the Royal Meteorological Society* 136: 842–855.

Žagar, N. 2017. A global perspective of the limits of prediction skill of NWP models. *Tellus A* (early online release). https://doi.org/10.1080/16000870.2017.1317573.

Žagar, N., E. Andersson, and M. Fisher. 2005. Balanced tropical data assimilation based on a study of equatorial waves in ECMWF short-range forecast errors. *Quarterly Journal Royal Meteorological Society* 131: 987–1011.

Zhang, G., K.H. Cook, and E.K. Vizy. 2016. The diurnal cycle of warm season rainfall over West Africa. Part II: Convection-permitting simulations. *Journal of Climate* 29: 8439–8454.

Convection Initiation in Climate Models Using the Heated Condensation Framework: A Review

Rodrigo J. Bombardi, Ahmed B. Tawfik, Lawrence Marx, Paul A. Dirmeyer and James L. Kinter III

Abstract This chapter presents a review of the Heated Condensation Framework (HCF) theory and applications. The HCF offers an alternative approach to methods of parameterizing convection based on parcel theory. Formulated to take into account the role of atmospheric mixing within the boundary layer, the HCF uses profiles of temperature and humidity to quantify how conditioned the atmosphere is to moist free convection. The initiation of convection is evaluated based on both the availability of large-scale convective instability and local surface heating. Therefore, the HCF can be applied as a trigger function for convective parameterizations. When compared to conventional convective trigger criteria, the HCF triggers returns less false positives, and when implemented into the CFSv2 model, the HCF trigger improves the representation of the Indian monsoon and tropical cyclone intensity. From a climate perspective, applying the HCF trigger to the Community Earth System Model reduces convective overactivity in the model and improves the frequency of intense precipitation events. The use of the HCF as a convective trigger is still under active investigation, and strategies for including the effects of remote dynamical forcings and model sub-grid triggering are being explored.

Keywords Convective trigger function · CFSv2 · CESM · Convection initiation
Convective inhibition

R. J. Bombardi (✉) · L. Marx · P. A. Dirmeyer · J. L. Kinter III
Department of Atmospheric, Oceanic, and Earth Sciences, Center for Ocean-Land-Atmosphere Studies, College of Science, George Mason University, 4400 University Drive, Fairfax, VA 22033, USA
e-mail: rjbombardi@gmail.com

A. B. Tawfik
National Center for Atmospheric Research, Boulder, USA

D. A. Randall et al. (eds.), *Current Trends in the Representation of Physical Processes in Weather and Climate Models*, Springer Atmospheric Sciences,
https://doi.org/10.1007/978-981-13-3396-5_3

1 Introduction

1.1 Convective Triggering: Background and Gaps in Knowledge

There have been several strategies for representing convective initiation in models. First, it should be recognized that convective initiation in models has a slightly different meaning than that which occurs in the nature. In nature, convective initiation broadly refers to the transition from dry convection to moist convection and, consequently, the formation of clouds. Convective initiation in models refers to the criterion or set of criteria that must be met to activate a convective parameterization. These two definitions may, of course, converge if a model is capable of resolving individual clouds or groups of clouds. However, Earth Systems Models (ESMs) and General Circulation Models (GCMs) operate on resolutions from tens to hundreds of kilometers; much greater than the cloud-resolving scale of hundreds of meters. There have been several strategies employed for activating convection in models, and most, if not all, use a parcel-based approach that begins from the standpoint of a mechanically lifted parcel of air that then encounters some degree of entrainment as it is lifted. For example, the Arakawa–Schubert convection scheme is activated when a parcel is lifted from the maximum level of moisture static energy below 700 hPa and reaches the Level of Free Convection (LFC) less than 150 hPa from its launch level (e.g., Pan and Wu 1995; Hong and Pan 1998; Lee et al. 2008; Han and Pan 2011). The Zhang-McFarlene scheme (Zhang and McFarlene 1995) employs a less sophisticated method only requiring that Convective Available Potential Energy (CAPE) be greater than 70 J kg^{-1} for the convection scheme to be activated. More exhaustive descriptions of various convective triggering mechanisms can be found in Suhas and Zhang (2014).

Note that in both examples, the triggering mechanism relies on lifting a parcel of air from a given level; not coincidently these lifted air parcels often originate from within the boundary layer where the destabilizing effects of surface heating are most pronounced. Additionally, note that, the two examples do not include convective inhibition in their triggering criteria. This has led to two primary flaws. First, convective precipitation in models is phase-locked with the diurnal solar cycle with a noon peak. This diurnal phase locking has been shown to be caused by an oversensitivity to surface energy fluxes. Take the Zhang-McFarlene triggering mechanism as an example, where CAPE must be greater than 70 J kg^{-1} to activate the convection scheme. How often is this condition met during the summer in the midlatitudes? According to soundings from the Integrated Global Radiosonde Archive (IGRA; Durre et al. 2006), approximately 78% of boreal summer days in the midlatitudes have CAPE greater than 70 J kg^{-1}. This clearly suggests that without some representation of inhibition modeled convection is likely to be overactive and coincide with the time of peak buoyancy generation (i.e., around solar noon).

The second flaw is the lack of balance between large-scale and local forcing. Large-scale here refers to synoptically forced ascent or instability production such as would occur along a frontal boundary, and local refers to the buoyancy produced by surface heating and moistening. The need to account for these two forcing factors is necessary because convection is often triggered due to the combined effects of local and large-scale forcings. CAPE-based triggers, such as in the Zhang-McFarlene scheme, have attempted to incorporate some large-scale effects by replacing CAPE with a time rate of change in CAPE (referred to as dCAPE in the literature; Xie and Zhang 2000; Zhang 2002) that quantifies how efficiently CAPE is generated in the free troposphere by large-scale advection. The convection scheme is activated when dCAPE is positive (i.e., CAPE is being generated faster than it is being consumed) and the relative humidity of the lifted parcel at the LFC must be greater than some threshold, usually RH > 80%. Using dCAPE has been shown to improve model representation of the Madden–Julian Oscillation (MJO) and the Intertropical Convergence Zone (ITCZ) (Zhang 2002; Song and Zhang 2009).

1.2 Where the Heated Condensation Framework Fits In

Despite the many advances and approaches used to address these two flaws, they all operate within the same parcel-based paradigm. This is not to suggest that parcel-based approaches are incorrect or inappropriate but to say that common issues and limitations will arise when applied to triggering mechanisms. Specifically, parcel theory begins from the starting point of mechanical lifting, i.e., a forcing that generates lift and displaces air parcels. One of the common issues that need to be addressed is how much mixing a parcel experiences when it is lifted. This has led to many sensitivity studies exploring various entrainment rates along the lifted parcel's trajectory and demonstrating the sensitivity of convective precipitation to the entrainment rate (e.g., Wang et al. 2007; Bechtold et al. 2008; Neale et al. 2008; Chikira 2010; Oueslati and Bellon 2013). The heated condensation framework (HCF; Tawfik and Dirmeyer 2014; Tawfik et al. 2015a, b) does not approach convective initiation from a lifted parcel approach but rather from the perspective of a gradually growing boundary layer. The development of the HCF was motivated by recognizing the boundary layer properties of the dry-to-moist convective transition in weakly forced regimes. In particular, this convective transition occurs when boundary layer mixing produces relatively constant vertical profiles of potential temperature and specific humidity, and as a consequence a relative humidity profile with a maximum near the boundary layer top. If there is sufficient moisture available, saturation occurs at the boundary layer top and moist convection occurs.

By approaching convective initiation from the perspective of a well-mixed boundary layer rather than using mechanically lifted parcels, we expect the HCF to have its own set of advantages and limitations. The objective of this chapter is to present the HCF theory, its implementation in an operational forecast model and a climate model, and describe future directions of the HCF development within the context of improving modeled convection. Section 2 presents the HCF theory and formulation. The validation of the HCF theory using observations is explored in Sect. 3. Section 4 discusses the application of the HCF for forecasts using the National Centers for Environmental Prediction (NCEP) Climate Forecast System version 2 (CFSv2; Saha et al. 2014a, b). Climate applications of the HCF trigger are explored in Sect. 5. Future advances of the HCF are discussed in Sect. 6 and Sect. 7 contains the final remarks.

2 The Heated Condensation Frameworks: Theory and Formulation

As mentioned before, the HCF evaluates convective initiation from the perspective of a gradually growing boundary layer using vertical profiles of temperature and specific humidity. The formulation of the HCF begins from the basic observation that moist convection triggered by surface heating is a slow process. Specifically, cloud-topped Planetary Boundary Layers (PBLs) develop gradually, more closely resembling a local buoyancy forced regime rather than some synoptic scale mechanical forcing that would follow a lifted parcel approach. The HCF was developed in an effort to represent this buoyancy driven forcing and allow for the influence of local surface heating on convective initiation to be returned within the context of the synoptic scale instabilities.

The HCF contains a set of diagnostic variables that serve to quantify some aspect of convective initiation. Only vertical profiles of temperature and humidity are required to return these variables. We use standard thermodynamic diagrams (i.e., skew-T log-P charts) to illustrate the construction of the HCF variables in Fig. 1. The first step is to mimic a gradually growing PBL by increasing the lowest level temperature by some increment (Fig. 1a). Next, we find where this new temperature intersects the environmental temperature profile when following the dry adiabat. This level of intersection defines a new hypothetical boundary layer referred to as the potential mixed layer or PML. The humidity profile is then mixed from the top of the PML to the surface resulting in a constant mixing ratio value throughout the PML (Fig. 1b). After mixing is performed, we check whether saturation has occurred at the top of the PML. If not, then we return to the first step and increase the lowest level temperature by another increment and repeat the humidity mixing step until saturation occurs. Once saturation occurs at the top of the PML, several HCF variables are determined. The first variable of interest is the buoyant mixing potential temperature (θ_{BM}), which is also referred to as the convective threshold (Fig. 1c). The θ_{BM} is equal to the lowest level temperature that was needed for saturation to occur, and similarly, the height of the PML is referred to as the Buoyant Condensation Level (BCL) once saturation is achieved. Another important variable is the potential temperature deficit (θ_{def}), which is the sum of all the temperature increments needed to achieve condensation and is a measure of convective inhibition.

As θ_{def} represents inhibition, moist convection is activated when θ_{def} reaches zero (i.e., no additional heating is required to achieve saturation at the top of the PBL).

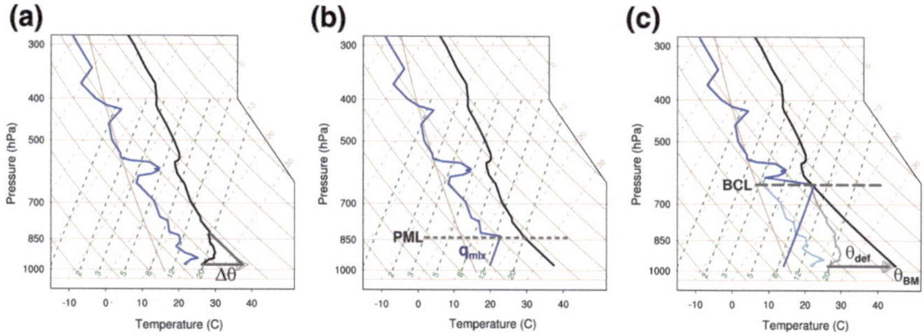

Fig. 1 Thermodynamic profiles (e.g., skew-T log-P diagrams) of temperature (black) and dew point temperature (blue) illustrating the steps for calculating the Buoyant Condensation Level (BCL), buoyant mixing temperature (θ_{BM}), and mixed layer specific humidity (q_{mix}). Dashed green lines represent constant mixing ratio lines; dashed tan lines represent isotherms; and solid tan lines are dry adiabats. **a** The first step where θ_{2m} is perturbed by some increment, $\Delta\theta$. **b** The height, the potential mixed level (PML), where the perturbed surface parcel ($\theta_{2m} + \Delta\theta$) is neutrally buoyant and the humidity profile is mixed from the PML to the surface. **c** The θ_{2m} is perturbed until saturation occurs at the PML, and the level is identified as the BCL. The total potential temperature increment is θ_{def}, which is necessary to reach θ_{BM} from the initial θ_{2m}. Faint blue and gray lines in (**c**) refer to the initial profile shown in (**a**) as a reference (Fig. 1 from Tawfik and Dirmeyer 2014)

This strict condition is, of course, a tunable parameter in models that can either be more stringent or relaxed depending on its interaction with other model components. One advantage of the HCF approach is that the triggering criterion does not depend on the user selecting a specific parcel for ascent to determine inhibition. Specifically, the BCL height, and as a consequence, the value of θ_{def}, will only change if the temperature and humidity profiles evolve and are not particularly sensitive to changes at a single level. In contrast, parcel-based approaches must arbitrarily select one or more representative parcels to lift, making the returned values of CAPE and CIN sensitive to the state of the selected parcel as well as to any entrainment assumptions. By not requiring parcel selection within the HCF, we remove two potential user biases: the criterion for choosing a parcel and the mixing that the parcel encounters upon ascent. Another advantage is more model design related; because the HCF was designed with the idea of separating the surface forced contribution from the atmospheric background state, it fits well with the current compartmentalized design of ESMs and forecast models. This model design advantage will be discussed in more detail in Sect. 6.

Nevertheless, the HCF does contain a disadvantage when compared to traditional parcel-based convective triggers. The primary shortcoming is that the HCF does not explicitly account for mechanical lifting and thus can underestimate convective initiation due to phenomena such as convective outflow boundaries and strong frontal boundaries. Incorporating mechanical lift in the HCF is an avenue of active research and will be discussed further in Sect. 6.

Detailed and thorough description of the calculation of BCL and θ_{def} can be found in Tawfik and Dirmeyer (2014), Tawfik et al. (2015a), and Bombardi et al. (2016). Computer code for calculating the HCF variables can be accessed at https://github.com/abtawfik/coupling-metrics/blob/master/heated_condensation/static/hcfcalc.f90. The next section presents some results comparing the Lifted Condensation Level (LCL) to the BCL followed by validating the accuracy of the HCF theory in representing convective triggering using observational data.

3 Validating the HCF Theory

The θ_{def} diagnostic variable has been shown to accurately capture convective triggering due to maritime land–sea breeze circulation (Tawfik and Dirmeyer 2014), as well as responding to low-level jet and mesoscale convection systems over the United States Central Plains (Tawfik and Dirmeyer 2014; Tawfik et al. 2015a). Here, we expand on how the HCF behaves during the emergence of a low-level jet over the United States Central Plains resulting in the development of mesoscale convective systems, and compare the HCF to the traditional parcel-based approach. In particular, we will compare the BCL to the LCL during an Intensive Observational Period (IOP) from the International H2O Project field campaign during June 6–13, 2002 over the Atmospheric Radiation Measurement Central Facility located in Lamont, Oklahoma (Weckwerth et al. 2004; Fig. 2). During this IOP, radiosondes were launched every 3 h, resolving the diurnal cycle. The radiosondes were also collocated with the observed cloud fraction as a function of height from the Active Remote Sensing of Clouds (ARSCL) value-added product. The ARSCL provides hourly measurements at 45-m vertical resolution enabling us to identify instances of convective initiation.

Before comparing the BCL to the LCL, it is helpful to recall how each method identifies convective initiation. For the LCL, convection is initiated when the LCL is below the PBL height; however, some studies have shown that this is a necessary but not sufficient condition for initiation (Santanello et al. 2011; Gentine et al. 2013). The BCL defines convection initiation when the PBL intersects the BCL. Figure 2 shows the large diurnal variability of the LCL height, which was calculated using 2-m temperature and humidity. The large diurnal swings make interpretation of the LCL difficult when comparing midday to nighttime conditions. This is in contrast to the BCL height, which is relatively stable throughout the diurnal cycle and only changes when the atmospheric background state is significantly modified. A sharp transition occurred on June 7, corresponding to a large change in synoptic conditions, namely the appearance of a low-level jet transporting mid-level moisture to the ARM-SGP site (Fig. 2). We see that on the first day (June 6) the BCL is well above an attainable height for the PBL and in the proceeding days, we only get three intersections (marked by the red arrows in Fig. 2) despite having substantial shallow low-level cloud cover (color contours). Upon closer inspection using NEXRAD radar and GOES observations (not shown here), we found that most observed cloud covers was advected into the ARM-SGP observation site, and that locally initiated convection clearly occurred only on June 9 and 12 (the last two red arrows) as diag-

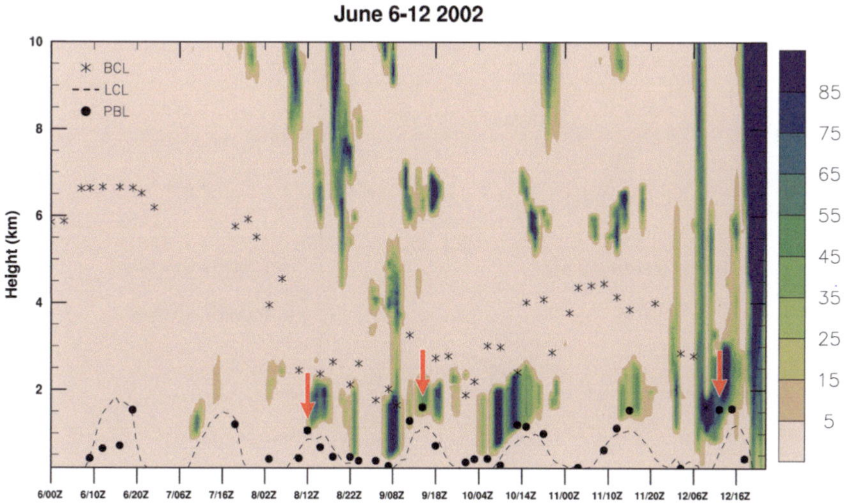

Fig. 2 Comparison of the observed BCL, LCL, and PBL heights with the corresponding cloud fraction during the IOP June 2002 period at the ARM-SGP central site. Red arrows indicate convection triggered locally according to the HCF

nosed by the HCF. The event identified by the HCF on June 8 was not clearly a local initiation event, however. If a triggering criterion using an LCL deficit were applied (e.g., trigger when the LCL is below the PBL) we would have returned several false positives for local initiation, such as on the evening of June 6 (2000Z), and most of the day on June 10–11. While not definitive and only for one site, this analysis shows promise in applying the HCF trigger more broadly in ESMs and forecast models to avoid false alarm convective initiation events that may be due to nonlocal, advected cloud cover.

The HCF trigger applied in the Simplified Arakawa-Schubert (SAS; Han and Pan 2011) deep convection scheme was evaluated using radiosonde data from the Dynamics of the Madden Julian Oscillation project (DYNAMO; Ciesielski et al. 2014). Figure 3 shows a stand-alone comparison between the HCF trigger and the original SAS trigger (mentioned in the Introduction). A range of precipitation values from the Tropical Rainfall Measuring Mission (TRMM; Huffman et al. 2007) was used as a proxy for atmospheric convection, similar to the methodology used by Suhas and Zhang (2014). Skill scores were calculated considering a 2 by 2 contingency table and comprising all 28 locations listed in Fig. 3d. The HCF trigger outperforms the original criterion in SAS as shown by the Heidke Skill Score (Fig. 3a), the Equitable Threat Score (Fig. 3b), and the bias (Fig. 3c). The original SAS trigger criterion initiates convection too frequently (Fig. 3d), which is a well-known problem in SAS and in most convective parameterizations (e.g., Dai 2006; Suhas and Zhang 2014). The impacts of the HCF trigger on SAS are further explored in the next section.

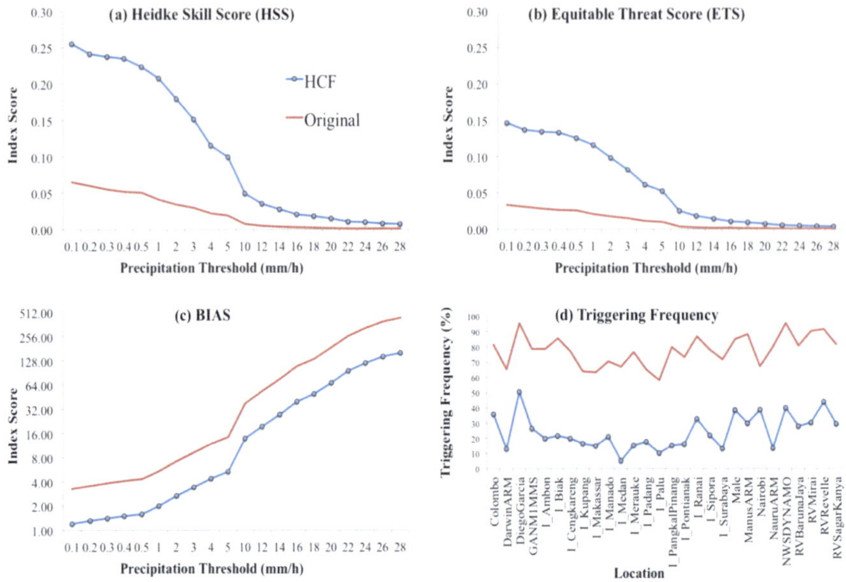

Fig. 3 Comparison between the HCF trigger and the original triggering criterion shown by **a** the Heidke Skill Score; **b** the Equitable Threat Score; **c** the Bias; and **d** the triggering frequency for each sounding release location (Fig. 2 from Bombardi et al. 2016)

4 The HCF Convective Trigger for Forecast Applications: The CFSv2 Model

Two separate studies were carried out to evaluate the impact of using the HCF as a convective triggering criterion in the CFSv2 (Bombardi et al. 2015, 2016). The first study (Bombardi et al. 2015) used the operational version of the CFSv2 (Pan and Wu 1995; Hong and Pan 1998) and the triggering criterion for the deep convection parameterization was relaxed. In other words, if either the original criterion or the HCF criterion were met, the deep convection parameterization was allowed to initiate (thus the triggering was relaxed). The second study (Bombardi et al. 2016) used an experimental version of the CFSv2, with revised parameterizations for shallow and deep convection (Han and Pan 2011), and stricter triggering criterion than in Bombardi et al. (2015). In the latter study, the convective parameterizations (both shallow and deep) were only allowed to initiate if the HCF triggering criterion was met (thus the triggering criterion was stricter). Table 1 shows a summary of the experiments carried out using the CFSv2. The HCF subroutines implemented into the CFSv2 can be accessed at https://github.com/rjbombardi/HCF_CFSv2.

A common aspect of both studies was the attempt to improve the representation of Indian summer monsoon rainfall in the CFSv2. The operational version of the CFSv2 (CTRL) has well-known negative precipitation biases over the Indian subcontinent and positive precipitation biases over the tropical Indian and Western Pacific Oceans

Table 1 Description of CFSv2 experiments published in Bombardi et al. (2015, 2016)

Study	Experiment name	Deep convection scheme	Shallow convection scheme	Convective trigger criterion	Triggering frequency
Bombardi et al. (2015)	CTRL	Operational	Operational	Original	Original
	HCF	Operational	Operational	Original + HCF applied to deep convection	Increased (relaxed)
Bombardi et al. (2016)	CTRL2	Revised SAS (Han & Pan)	SAS-based shallow cumulus (Han & Pan)	Original	Original
	HCFv2	Revised SAS (Han & Pan)	SAS-based shallow cumulus (Han & Pan)	HCFv2 applied to both deep and shallow convection	Decreased (restricted)

(Fig. 4a, b; Zhu and Shukla 2013; Goswami et al. 2014; Silva et al. 2014; Saha et al. 2014b; Bombardi et al. 2015, 2016), even at higher spatial resolution (Abhik et al. 2016). The convective parameterization highly overestimates the incidence of light rainfall and underestimates the incidence of moderate rainfall (Bombardi et al. 2015; Ganai et al. 2015).

When the triggering criterion in the CFSv2 convective parameterization was relaxed (HCF, Bombardi et al. 2015), there was a small but significant increase in the Indian summer rainfall, reducing the original bias in the model (Fig. 4d). Bombardi et al. (2015) showed that this improvement in the representation of Indian summer rainfall was in part due to the fact that by relaxing the triggering criterion, the convective parameterization was invoked more often. The increase in the convection frequency resulted in an increase in precipitation, which reduced the precipitation bias on seasonal timescales.

The precipitation biases are much reduced in the experimental version of the CFSv2 (CTRL2), with revised parameterizations (Han and Pan 2011; Ganai et al. 2015), over both land and ocean in comparison to CTRL (Fig. 4b, c). When the triggering criterion in CTRL2 was made more restrictive (HCFv2; Bombardi et al. 2016), the simulated summer rainfall over land was slightly improved (Fig. 4e). In this case, the changes in seasonal precipitation cannot be attributed to changes in the triggering frequency of the convective parameterizations. The HCF trigger changed the timing of convection in the CFSv2, which led to a completely different simulation from CTRL2.

Both relaxed and stricter triggering criteria resulted in improvement of the representation of the onset date of the rainy season over India (Fig. 5). The onset date of the rainy season is calculated using only daily precipitation data. The methodology for

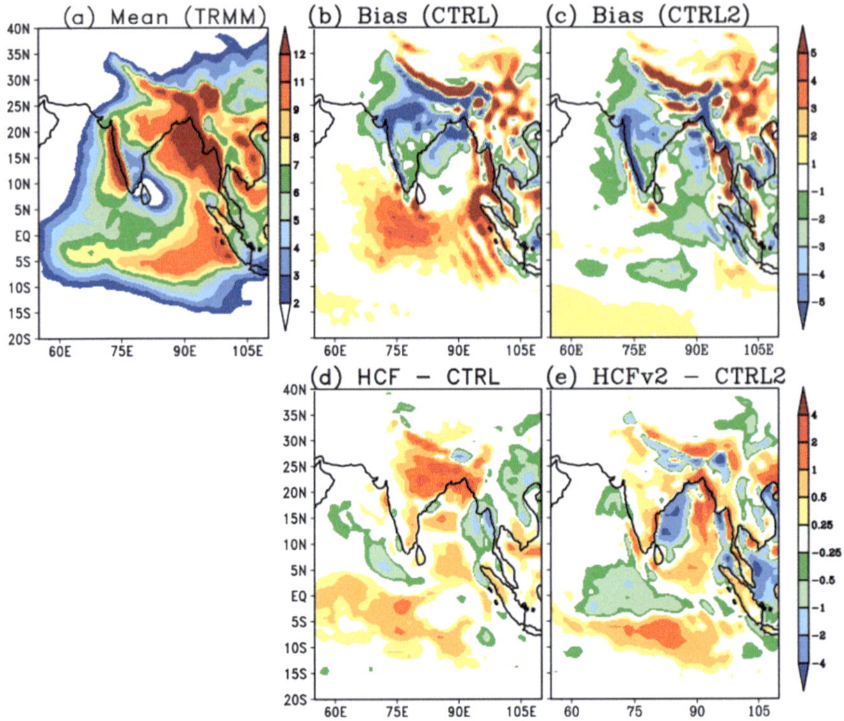

Fig. 4 **a** Mean JJAS precipitation from TRMM; JJAS precipitation bias for **b** CTRL ad **c** CTRL2; **d** mean JJAS precipitation difference between HCF and CTRL and **e** mean JJAS precipitation difference between HCFv2 and CTRL2

calculating the onset of the rainy season is explained in detail in Bombardi and Carvalho (2009) and further expanded in Bombardi et al. (2017). The observations show that the rainy season onset date (in Day of Year—DOY) occurs between mid-May (DOY 135) and early June (DOY 160) over central India (Fig. 5a), consistent with previous studies (e.g., Joseph et al. 1994; Fasullo and Webster 2002). CTRL initiates the rainy season over India too late in comparison to observations (Fig. 5b). When the triggering criterion is relaxed, the onset of the rainy season is better represented by the CFSv2 (Fig. 5c). This happens because a relaxed triggering criterion allows the model to initiate convection more often, increasing the amount of precipitation and, therefore, reducing the existing dry precipitation bias (Bombardi et al. 2015).

Because the precipitation biases are much reduced in CTRL2 in comparison to CTRL (Fig. 4) that is also reflected in the CFSv2 representation of the rainy season (Fig. 6b, d). However, the onset of the rainy season still happens later than observed in CTRL2 (Fig. 6a, d). When the triggering criterion was made stricter, the CFSv2 represented the onset of the rainy season over India earlier than observed (Fig. 6a, e). Unfortunately, the simulations did not provide enough information to identify the

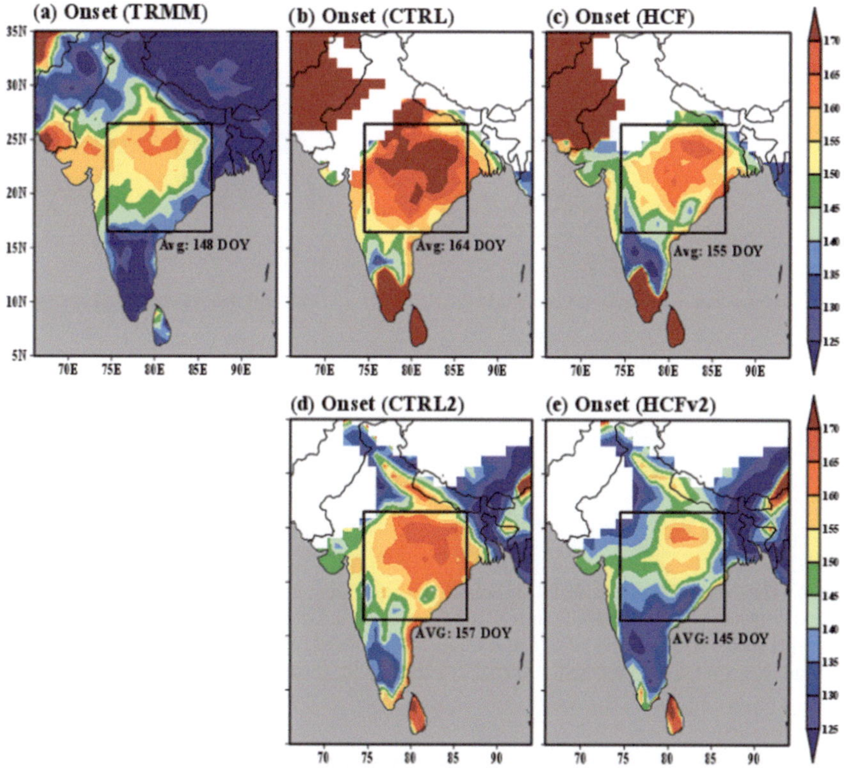

Fig. 5 Mean onset date [DOY] for **a** TRMM, **b** CTRL experiment, **c** HCF experiment, **d** CTRL2 experiment, and **e** HCFv2 experiment. The figure shows the spatial average of onset dates for central India (land region inside the box)

mechanisms whereby the HCF trigger altered the onset of the rainy season (Bombardi et al. 2016).

Bombardi et al. (2016) also investigated the impact of the HCF trigger on the diurnal cycle of convection over the U.S. and on tropical cyclones over the Atlantic Ocean. While CTRL2 simulates a maximum of convection during nighttime over the Central U.S., the maximum precipitation simulated by HCFv2 fell during the afternoon. That is, the HCF trigger improved the representation of late afternoon and early evening convection, while nighttime deep convection was better represented by the control experiment.

Tropical cyclones were simulated with higher resolution (T382, ~38 km) in comparison to the other experiments described in this section. The HCF trigger had no clear impact on tropical cyclone tracks. However, there was a significant improvement in hurricane intensity when the HCF trigger was used. The authors found that the original convective trigger in the CFSv2 initiated convection too frequently and, as a consequence, inhibited the intensification of tropical cyclones. When the trig-

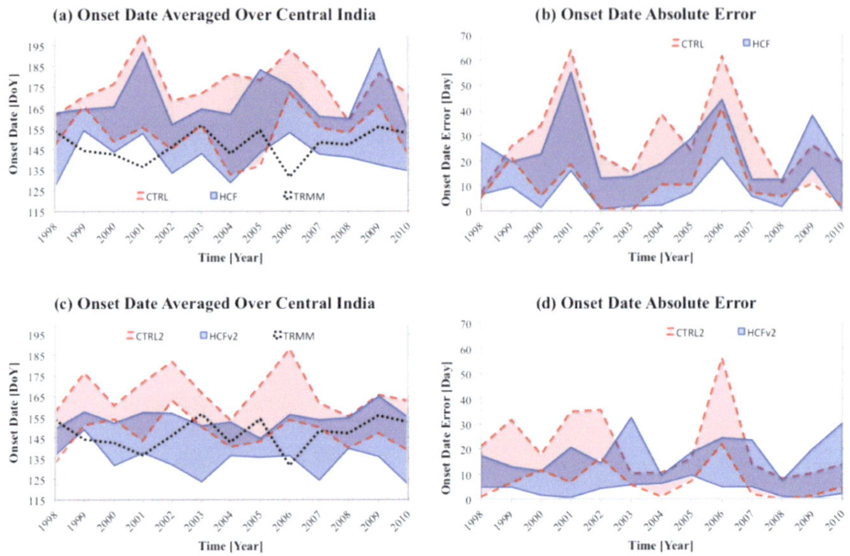

Fig. 6 Time series of observed (TRMM) and simulated rainy season onset date over Central India for experiments **a** CTRL and HCF, and **c** CTRL2 and HCFv2. Time series of onset date absolute error for experiments **b** CTRL and HCF, and **d** CTRL2 and HCFv2. The envelopes show the maximum and minimum values among the four hindcast members for each year

gering mechanism was made stricter, the atmosphere was allowed to become more unstable, more moisture was injected into the mid-troposphere, increasing cloud cover and releasing more latent heat, which intensified tropical cyclones (Bombardi et al. 2016).

5 The HCF Convective Trigger for Climate Applications: The Community Earth System Model

In addition to the forecast studies discussed above, the HCF was also implemented as an additional triggering criterion within the deep convection parameterization of the Community Earth System Model (CESM; Neale et al. 2012) to explore its impacts on climate-relevant precipitation characteristics (Tawfik et al. 2017). In particular, the mean precipitation, activation frequency of the deep convective parameterization, and the precipitation intensity distribution were examined. The initial motivation for this study arose from the seemingly arbitrary convective triggering condition in the Zhang-McFarlane deep convective parameterization (Zhang and McFarlane 1995; hereafter referred to as the ZM scheme).

Within the ZM scheme, deep convection is activated when CAPE is greater than 70 J kg^{-1}. As mentioned in the introduction this condition is easily met during mid-

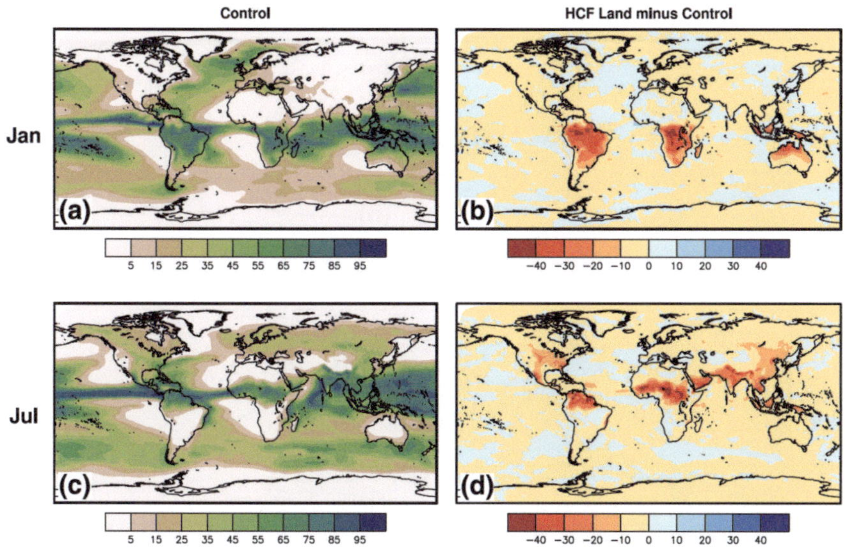

Fig. 7 Percent of hours the deep convection scheme is called for all 25 months for (left) the Control and (right) the difference between the Control and the HCF Land simulation during (top) January and (bottom) July (Fig. 5 from Tawfik et al. 2017)

latitude spring and summer resulting in very frequent activation of the ZM scheme in the CESM. Specifically, we can see from Fig. 7 that the ZM scheme, when using the default CAPE triggering criterion, is active more than 75% of hours over the tropics and 50% of hours over the subtropics and mid-latitude summer. That means there is deep convection at least 12 h of each day during each hemisphere's convectively active season. A consequence of ZM overactivity is convection occurring too early in the day, typically beginning several hours after sunrise with a maximum near midday (Fig. 8).

To alleviate the overactivity and early onset problems we implemented the HCF trigger in addition to the default trigger: convection is triggered when θ_{def} is less than or equal to zero and CAPE is greater than 70 J kg^{-1}. It should also be noted that the implementation in Tawfik et al. (2017) is slightly different from that described in the previous section. In the study by Tawfik et al. (2017), sub-grid land surface information from the Community Land Model (CLM; Oleson et al. 2013) is used to calculate the convective triggering deficit, θ_{def}. The advantage of this approach is that the sub-grid elements within the CLM communicate directly with the assumed sub-grid representation of the ZM scheme in the atmospheric model component (the Community Atmosphere Model, CAM; Neale et al. 2012). Because the assumed departure from convective quasi-equilibrium is more commonly found over land rather than the ocean (Zhang 2002; Donner and Phillips 2003), the HCF trigger is only applied over land, where the simulation is referred to as HCF Land. The details

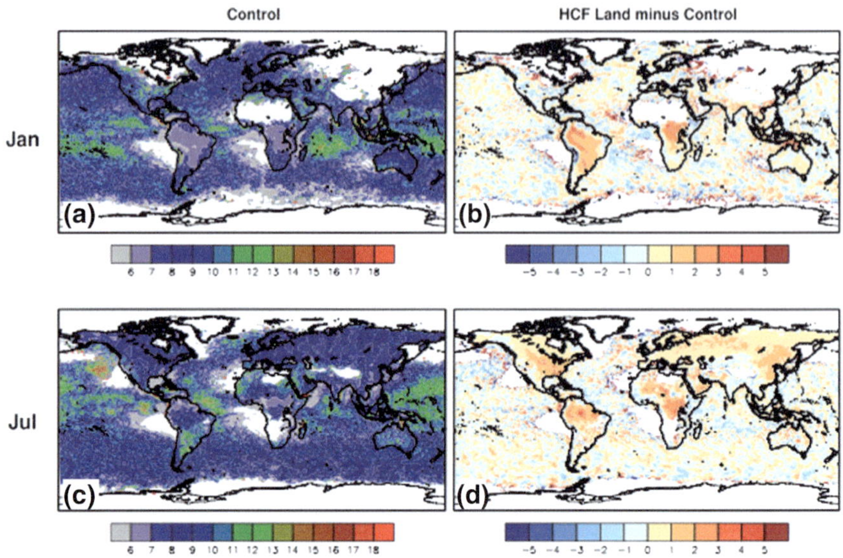

Fig. 8 (left) The average time of convective initiation (local hour) for the control and (right) the difference [hours] between the control and the HCF land simulation for (top) January and (bottom) July in Local Solar Time (LST). Gray shading indicates morning and red shading indicates late afternoon for the individual model configurations (Fig. 2 from Tawfik et al. 2017)

of the sub-grid implementation and results can be found in Tawfik et al. (2017), but here we will describe some of the most relevant results.

When applying the HCF trigger, we can see that overactivity is reduced by 40% in the tropics and 10–20% in the midlatitudes over land (Fig. 7). Similarly, the onset time of convection is delayed by 1–4 h over the same regions (Fig. 8), producing the desired outcome of implementing the HCF triggering. As expected, the reduced ZM activity results in less precipitation falling from the ZM scheme (Fig. 9). The greatest reductions over land can be found over the Amazon during the wet season and throughout the tropics with 50–200 mm less precipitation being derived from the ZM scheme. However, the CESM has many other components that can respond to this reduction in convective activity. For example, Zarzycki et al. (2014) and Duffy et al (2003) have shown that large-scale precipitation can compensate for reductions in the parameterized convective activity in the CESM. This compensation is evident over the tropics when looking at total precipitation (Fig. 10). It can be seen that despite the broad 50–200 mm reductions in precipitation from the ZM scheme, total precipitation (that is the sum of ZM precipitation and large-scale precipitation) increases by more than 200 mm over some tropical regions, such as the leeside of the Andes mountains. In general, the response of total precipitation to the inclusion of the HCF trigger is mixed spatially, returning pockets of both increased and decreased precipitation. This mixed response further highlights the complex interactions that can occur in

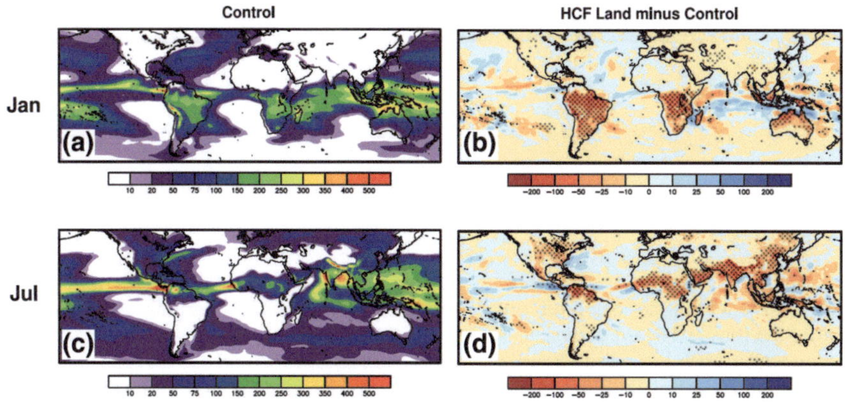

Fig. 9 The average monthly convective precipitation (mm) derived from the ZM scheme for (left) the Control and (right) the difference between the Control and the HCF Land runs for (top) January and (bottom) July. Regions with differences significant at a 95% confidence interval are indicated by hash marks (Fig. 6 from Tawfik et al. 2017)

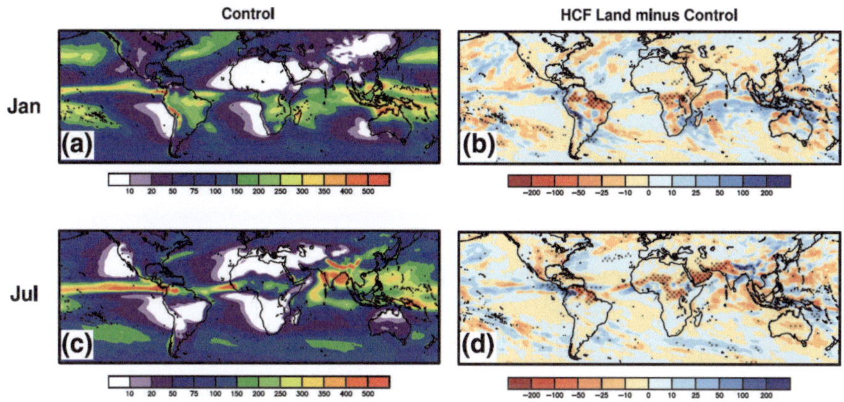

Fig. 10 The average monthly total precipitation (mm) for (left) the Control and (right) the difference between the Control and the HCF Land runs for (top) January and (bottom) July. Regions with differences significant at a 95% confidence interval are indicated by hash marks (Fig. 7 from Tawfik et al. 2017)

models, making net model improvement very unlikely when changing only a single portion of a single parameterization, as was done in Tawfik et al. (2017).

Among the most striking improvements when including the HCF triggering was an increase in the frequency of intense precipitation events (Fig. 11). In particular, there was an order of magnitude increase in precipitation failing at a rate between 10 and 20 mm h^{-1} when using the HCF trigger. While this is still roughly two orders of magnitude lower than the TRMM observations (shown in black in Fig. 11a), including the HCF triggering pushes the frequency distribution in the right direction. There were similar improvements found when evaluating how much precipitation failing at a

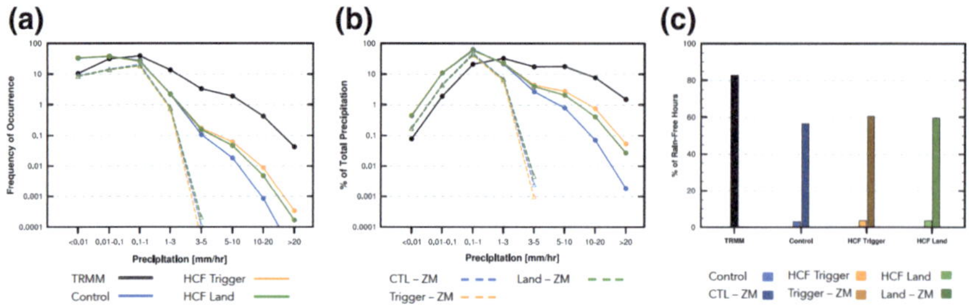

Fig. 11 The **a** frequency and **b** percent contribution of precipitation falling within each intensity bin for the (black) TRMM observations, (blue) Control simulation, (orange) HCF Trigger simulation, (green) HCF Land, and (dashed lines) the contribution only from the ZM deep convection scheme for the Control, HCF Trigger, HCF Land for July. Precipitation bins are in mm h^{-1}. Percent accumulations are for all land and ocean points between 55° S and 55° N. **c** The percent of rain-free hours for (black) TRMM observations, (blue) Control simulation, (orange) HCF Trigger simulation, (green) HCF Land; dark bar shading are for ZM deep convection precipitation and brighter shading is for the total precipitation (Fig. 9 from Tawfik et al. 2017)

given intensity contributes to the total precipitation amount (Fig. 11b). It is important to note that the increased high-intensity precipitation events are exclusively derived from the large-scale precipitation scheme and not the ZM scheme (dashed lines in Fig. 11), again demonstrating the interplay between large-scale and parameterized precipitation in the model.

6 Future Advances of the HCF

As we have discussed, the HCF shows some promise when used as a convective triggering criterion in both forecast and climate applications. However, there are several shortcomings that are being actively developed. For example, the HCF is a diagnostic framework, which does not lend itself to including dynamically forced vertical motion. As a result, the HCF may actually be underestimating triggering events in strong dynamically forced regimes such as along frontal boundaries. To make the HCF more broadly applicable, work is underway to extend the HCF to be more prognostic rather than diagnostic. Specifically, the local time derivative is applied to the θ_{def} value, and upon decomposition of terms, a term explicitly capturing the impact of mean grid cell vertical velocity on convective initiation is present. Because this work is ongoing, the full derivation and subsequent analysis are not presented here.

Another extension of the HCF that is being explored is the concept of sub-grid interactions between the land surface model and the convective parameterization. In the study outlined in Sect. 5, Tawfik et al. (2017) used the sub-grid surface flux information within the Community Land Model (CLM) to directly interact with the Zhang-McFarlane convection scheme. This was done by first calculating the θ_{BM}

value in the Community Atmosphere Model (CAM) and then passing this value to each sub-grid land surface tile within CLM. Triggering is evaluated at each CLM sub-grid tile by comparing the sub-grid 2-m potential temperature (θ_{2m}) against the convective threshold (θ_{BM}). If the value is less than or equal to zero, convective initiation is said to be triggered within the CLM sub-grid tile. The convectively active fraction (σ_{clm}) of the CLM sub-grid is then passed back to CAM and the ZM convection scheme, and triggering occurs if the value is greater than zero.

One implication of this method that is not discussed in Tawfik et al. (2017) is the meaning behind the convectively active land fraction σ_{clm}. Specifically, at the heart of convective parameterizations is the assumption that the grid cell size in a model "must be large enough to contain an ensemble of cumulus clouds but small enough to cover only a fraction of a large-scale disturbance" (Arakawa and Schubert 1974), i.e., $\sigma \ll 1$ (asymptotic assumption). Ongoing work is attempting to use σ_{clm} as derived from the Tawfik et al. (2017) study as a measure of the σ described by Arakawa and Schubert (1974). The σ_{clm} value will be used then to identify regions and instances in the model where the asymptotic assumption no longer holds, and therefore turn off the convective parameterization (i.e., turn off the convection scheme when σ_{clm} is close to or equal to one). In this regard, the HCF trigger would be a dynamically scale-aware trigger that only allows the convective parameterization to operate when one of its primary assumptions holds. This approach may be particularly appealing for nonuniform refined grids that may approach convective-permitting scales over a certain region while maintaining greater than 10 km grid cell sizes outside of the refined region (Zarzycki et al. 2014). Preliminary results show that requiring convection to occupy 75% or less of the grid dramatically reduces the frequency in which convection occurs from 60 to 80% to 10 to 20%. However, it is still unclear whether the σ_{clm}, defined by Tawfik et al. (2017), can be interchangeably used with the σ defined within the classical Arakawa and Schubert (1974) derivation. σ_{clm} communicates information regarding the land surface activation, but does not necessarily quantify the portion of the atmospheric grid cell covered by clouds.

7 Final Remarks

The HCF has been shown to be a useful method to evaluate the initiation of convection from the perspective of a gradually growing boundary layer, using vertical profiles of temperature and specific humidity. It offers an alternative approach to methods based on parcel theory. Most convective parameterizations initiate convection too often. When compared to conventional metrics, the HCF returns a smaller number of false positives for convective initiation.

As a triggering criterion for convective parameterizations in the CFSv2 model, the HCF has shown improvements in the representation of summer Indian monsoon rainfall and the onset date of the rainy season over India. In addition, the HCF trigger results in improvements in the representation of the intensity of hurricanes at a spatial resolution of T382. Some of the improvements in the CFSv2 representation of the

Indian monsoon and tropical cyclones due to the implementation of the HCF trigger can be simply explained by changes in the frequency of the initiation of the convective parameterizations. However, the frequency of the initiation of convection was not the only effect of the HCF trigger in the CFSv2 reforecasts, there was also the effect of changing the timing of convection. As seen in the observational analyses (Sect. 3), the HCF trigger reduces the amount of false positives and, therefore, improves the representation of the timing of convection.

Acknowledgements This study was primarily supported by the National Monsoon Mission, Ministry of Earth Sciences, Government of India. Additional support comes from NSF (AGS-1338427), NOAA (NA14OAR4310160 and NA15NWS4680018), and NASA (NNX14AM19G). The radiosonde data were collected as part of DYNAMO, which was sponsored by NSF, NOAA, ONR, DOE, NASA, JAMSTEC, [Indian and Australian funding agencies]. The involvement of the NSF-sponsored National Center for Atmospheric Research (NCAR) Earth Observing Laboratory (EOL) is acknowledged. The data are archived at the DYNAMO Data Archive Center maintained by NCAR EOL.

References

Abhik, S., P. Mukhopadhyay, R.P.M. Krishna, et al. 2016. Diagnosis of boreal summer intraseasonal oscillation in high resolution NCEP climate forecast system. *Climate Dynamics* 46: 3287–3303. https://doi.org/10.1007/s00382-015-2769-9.

Arakawa, A., and W.H. Schubert. 1974. Interaction of a cumulus cloud ensemble with the large-scale environment, Part I. *Journal of the Atmospheric Sciences* 31: 674–701. https://doi.org/10.1175/1520-0469(1974)031%3c0674:IOACCE%3e2.0.CO;2.

Bechtold, P., M. Köhler, T. Jung, F. Doblas-Reyes, M. Leutbecher, M.J. Rodwell, F. Vitart, and G. Balsamo. 2008. Advances in simulating atmospheric variability with the ECMWF model: From synoptic to decadal time-scales. *Quarterly Journal Royal Meteorological Society* 134 (634): 1337–1351. https://doi.org/10.1002/qj.289.

Bombardi, R.J., and L.M.V. Carvalho. 2009. IPCC global coupled model simulations of the South America monsoon system. *Climate Dynamics* 33: 893–916. https://doi.org/10.1007/s00382-008-0488-1.

Bombardi, R.J., E.K. Schneider, L. Marx, et al. 2015. Improvements in the representation of the Indian summer monsoon in the NCEP climate forecast system version 2. *Climate Dynamics* 45: 2485–2498. https://doi.org/10.1007/s00382-015-2484-6.

Bombardi, R.J., A.B. Tawfik, J.V. Manganello, et al. 2016. The heated condensation framework as a convective trigger in the NCEP climate forecast system version 2. *Journal of Advances in Modeling Earth Systems* 8: 1310–1329. https://doi.org/10.1002/2016MS000668.

Bombardi, R.J., K.V. Pegion, J.L. Kinter, et al. 2017. Sub-seasonal predictability of the onset and demise of the rainy season over monsoonal regions. *Frontiers of Earth Science*. https://doi.org/10.3389/feart.2017.00014.

Chikira, M. 2010. A cumulus parameterization with state-dependent entrainment rate. Part II: Impact on climatology in a general circulation model. *Journal of the Atmospheric Sciences* 67 (7): 2194–2211. https://doi.org/10.1175/2010jas3317.1.

Ciesielski, P.E., H. Yu, R.H. Johnson, et al. 2014. Quality-controlled upper-air sounding dataset for DYNAMO/CINDY/AMIE: Development and corrections. *Journal of Atmospheric and Oceanic Technology* 31: 741–764. https://doi.org/10.1175/JTECH-D-13-00165.1.

Dai, A. 2006. Precipitation characteristics in eighteen coupled climate models. *Journal of Climate* 19: 4605–4630. https://doi.org/10.1175/JCLI3884.1.

Donner, L.J., and V.T. Phillips. 2003. Boundary layer control on convective available potential energy: Implications for cumulus parameterization. *Journal of Geophysical Research Atmospheres* 108 (D22): 4701. https://doi.org/10.1029/2003jd003773.

Duffy, P.B., B. Govindasamy, J.P. Iorio, J. Milovich, K.R. Sperber, K.E. Taylor, M.F. Wehner, and S.L. Thompson. 2003. High-resolution simulations of global climate, Part 1: Present climate. *Climate Dynamics* 21 (5–6): 371–390. https://doi.org/10.1007/s00382-003-0339-z.

Durre, I., R.S. Vose, and D.B. Wuertz. 2006. Overview of the integrated global radiosonde archive. *Journal of Climate* 19 (1): 53–68. https://doi.org/10.1175/JCLI3594.1.

Fasullo, J., and P.J. Webster. 2002. Hydrological signatures relating the Asian summer monsoon and ENSO. *Journal of Climate* 15: 3082–3095. https://doi.org/10.1175/1520-0442(2002)015%3c3082:HSRTAS%3e2.0.CO;2.

Ganai, M., P. Mukhopadhyay, R.P.M. Krishna, and M. Mahakur. 2015. The impact of revised simplified Arakawa-Schubert convection parameterization scheme in CFSv2 on the simulation of the Indian summer monsoon. *Climate Dynamics* 45: 881–902. https://doi.org/10.1007/s00382-014-2320-4.

Gentine, P., A.A.M. Holtslag, F. D'Andrea, and M. Ek. 2013. Surface and atmospheric controls on the onset of moist convection over land. *Journal of Hydrometeorology* 14 (5): 1443–1462. https://doi.org/10.1175/JHM-D-12-0137.1.

Goswami, B.B., M. Deshpande, P. Mukhopadhyay, et al. 2014. Simulation of monsoon intraseasonal variability in NCEP CFSv2 and its role on systematic bias. *Climate Dynamics* 43: 2725–2745. https://doi.org/10.1007/s00382-014-2089-5.

Han, J., and H.-L. Pan. 2011. Revision of convection and vertical diffusion schemes in the NCEP global forecast system. *Weather Forecast* 26: 520–533. https://doi.org/10.1175/WAF-D-10-05038.1.

Hong, S.-Y., and H.-L. Pan. 1998. Convective trigger function for a mass-flux cumulus parameterization scheme. *Monthly Weather Review* 126: 2599–2620.

Huffman, G.J., D.T. Bolvin, E.J. Nelkin, et al. 2007. The TRMM multisatellite precipitation analysis (TMPA): Quasi-global, multiyear, combined-sensor precipitation estimates at fine scales. *Journal of Hydrometeorology* 8: 38–55. https://doi.org/10.1175/JHM560.1.

Joseph, P.V., J.K. Eischeid, and R.J. Pyle. 1994. Interannual variability of the onset of the Indian summer monsoon and its association with atmospheric features, El Niño, and sea surface temperature anomalies. *Journal of Climate* 7: 81–105. https://doi.org/10.1175/1520-0442(1994)007%3c0081:IVOTOO%3e2.0.CO;2.

Lee, M.-I., S.D. Schubert, M.J. Suarez, et al. 2008. Role of convection triggers in the simulation of the diurnal cycle of precipitation over the United States Great Plains in a general circulation model. *Journal Geophysical Research* 113: D02111. https://doi.org/10.1029/2007JD008984.

Neale, R.B., J.H. Richter, and M. Jochum. 2008. The impact of convection on ENSO: From a delayed oscillator to a series of events. *Journal of Climate* 21 (22): 5904–5924. https://doi.org/10.1175/2008JCLI2244.1.

Neale, R.B., et al. 2012. Description of the NCAR community atmosphere model (CAM 5.0). Technical Note NCAR/TN-486 + STR, National Center for Atmospheric Research, 2009–38451.

Oleson, K.W., et al. 2013. Technical description of version 4.5 of the community land model (CLM). NCAR Technical Note, NCAR/TN-503 + STR. https://doi.org/10.5065/d6rr1w7m.

Oueslati, B., and G. Bellon. 2013. Convective entrainment and large-scale organization of tropical precipitation: Sensitivity of the CNRM-CM5 hierarchy of models. *Journal of Climate* 26 (9): 2931–2946. https://doi.org/10.1175/JCLI-D-12-00314.1.

Pan, H.-L., and W.-S. Wu. 1995. Implementing a mass flux convective parameterization package for the NMC medium range forecast model. NMC Off Note 409, 40 pp. http://www.emc.ncep.no.

Saha, S., S. Moorthi, X. Wu, et al. 2014a. The NCEP climate forecast system version 2. *Journal of Climate* 27: 2185–2208. https://doi.org/10.1175/JCLI-D-12-00823.1.

Saha, S.K., S. Pokhrel, H.S. Chaudhari, et al. 2014b. Improved simulation of Indian summer monsoon in latest NCEP climate forecast system free run. *International Journal of Climatology* 34: 1628–1641. https://doi.org/10.1002/joc.3791.

Santanello, J.A., C.D. Peters-Lidard, and S.V. Kumar. 2011. Diagnosing the sensitivity of local land-atmosphere coupling via the soil moisture-boundary layer interaction. *Journal of Hydrometeorology* 12 (5): 766–786. https://doi.org/10.1175/JHM-D-10-05014.1.

Silva, G.A.M., L.M.M. Dutra, R.P. da Rocha, et al. 2014. Preliminary analysis on the global features of the NCEP CFSv2 seasonal hindcasts. *Advances in Meteorology* 2014: 1–21. https://doi.org/10.1155/2014/695067.

Song, X., and G.J. Zhang. 2009. Convection parameterization, Tropical Pacific double ITCZ, and Upper-Ocean biases in the NCAR CCSM3. Part I: Climatology and atmospheric feedback. *Journal of Climate* 22 (16): 4299–4315. https://doi.org/10.1175/2009jcli2642.1.

Suhas, E., and G.J. Zhang. 2014. Evaluation of trigger functions for convective parameterization schemes using observations. *Journal of Climate* 27: 7647–7666. https://doi.org/10.1175/JCLI-D-13-00718.1.

Tawfik, A.B., and P. Dirmeyer. 2014. A process-based framework for quantifying the atmospheric preconditioning of surface-triggered convection. *Geophysical Research Letters* 41: 173–178. https://doi.org/10.1002/2013GL057984.

Tawfik, A.B., P.A. Dirmeyer, and J.A. Santanello. 2015a. The heated condensation framework. Part II: Climatological behavior of convective initiation and land-atmosphere coupling over the conterminous United States. *Journal of Hydrometeorology* 16: 1946–1961. https://doi.org/10.1175/JHM-D-14-0118.1.

Tawfik, A.B., P.A. Dirmeyer, and J.A. Santanello. 2015b. The heated condensation framework. Part I: Description and Southern Great Plains case study. *Journal of Hydrometeorology* 16: 1929–1945. https://doi.org/10.1175/JHM-D-14-0117.1.

Tawfik, A.B., D.M. Lawrence, and P.A. Dirmeyer. 2017. Representing subgrid convective initiation in the community earth system model. *Journal of Advances in Modeling Earth Systems* 9. https://doi.org/10.1002/2016ms000866.

Wang, Y., L. Zhou, and K. Hamilton. 2007. Effect of convective entrainment/detrainment on the simulation of the tropical precipitation diurnal cycle. *Monthly Weather Review* 135 (2): 567–585. https://doi.org/10.1175/MWR3308.1.

Weckwerth, T.M., D.B. Parsons, S.E. Koch, et al. 2004. An overview of the International H2O Project (IHOP_2002) and some preliminary highlights. *Bulletin of the American Meteorological Society* 85: 253–277. https://doi.org/10.1175/BAMS-85-2-253.

Xie, S., and M. Zhang. 2000. Impact of the convection triggering function on single-column model simulations. *Journal of Geophysical Research: Atmospheres* 105 (D11): 14983–14996. https://doi.org/10.1029/2000JD900170.

Zarzycki, C.M., M.N. Levy, C. Jablonowski, J.R. Overfelt, M.A. Taylor, and P.A. Ullrich. 2014. Aquaplanet experiments using CAM's variable-resolution dynamical core. *Journal of Climate* 27 (14): 5481–5503. https://doi.org/10.1175/JCLI-D-14-00004.1.

Zhang, G.J. 2002. Convective quasi-equilibrium in midlatitude continental environment and its effect on convective parameterization. *Journal of Geophysical Research: Atmospheres* 107 (D14): ACL 12-1. https://doi.org/10.1029/2001jd001005.

Zhang, G.J., and N.A. McFarlane. 1995. Sensitivity of climate simulations to the parameterization of cumulus convection in the Canadian climate centre general circulation model. *Atmosphere-Ocean* 33 (3): 407–446. https://doi.org/10.1080/07055900.1995.9649539.

Zhu, J., and J. Shukla. 2013. The role of air-sea coupling in seasonal prediction of Asia-Pacific summer monsoon rainfall. *Journal of Climate* 26: 5689–5697. https://doi.org/10.1175/JCLI-D-13-00190.1.

Cloud Microphysics Across Scales for Weather and Climate

Andrew Gettelman, Hugh Morrison and Greg Thompson

Abstract Cloud microphysics describes the evolution of condensed water in the atmosphere and is critical for weather and climate. This chapter describes the methods used for representing microphysical processes in weather and climate models, from explicit bin schemes used for small-scale simulation up to bulk treatments often used in global models. Of particular importance is how the cloud microphysical treatments are coupled to the rest of the cloud schemes in a numerical model that includes clouds. The key issues include the presentation of sub-grid inhomogeneity in humidity and dynamics. In addition, treatment of cold clouds in a "mixed phase" where liquid and ice may co-exist is important. We discuss current approaches including more comprehensive representations of ice and snow, treatment of rimed ice (graupel or hail), and coupling to unified turbulence schemes. Finally, we discuss possible paths forward for simulating cloud microphysics.

Keywords Clouds · Ice · Microphysics

1 Introduction

Cloud microphysics, the evolution of condensed water in the atmosphere, is critical for weather and climate. Cloud processes span 15 orders of magnitude in the Earth system, from the nanometer (10^{-9} m) scale of individual aerosols that nucleate clouds to the 10^6 m scale of the Earth's general circulation that govern cloud systems (Fig. 1).

Cloud microphysics is defined as the set of processes controlling the evolution of condensed water in the atmosphere. This includes the formation cloud drops (liquid) or ice crystals (solid) on aerosol particles, and the growth of these cloud "hydrometeors" (drops and crystals) by vapor diffusion, collision, and collection among hydrometeors, breakup, and fallout of hydrometeors from gravitational settling. It

A. Gettelman (✉) · H. Morrison · G. Thompson
National Center for Atmospheric Research (NCAR), 3090 Center Green Dr., Boulder, CO 80301, USA
e-mail: andrew@ucar.edu

D. A. Randall et al. (eds.), *Current Trends in the Representation of Physical Processes in Weather and Climate Models*, Springer Atmospheric Sciences,
https://doi.org/10.1007/978-981-13-3396-5_4

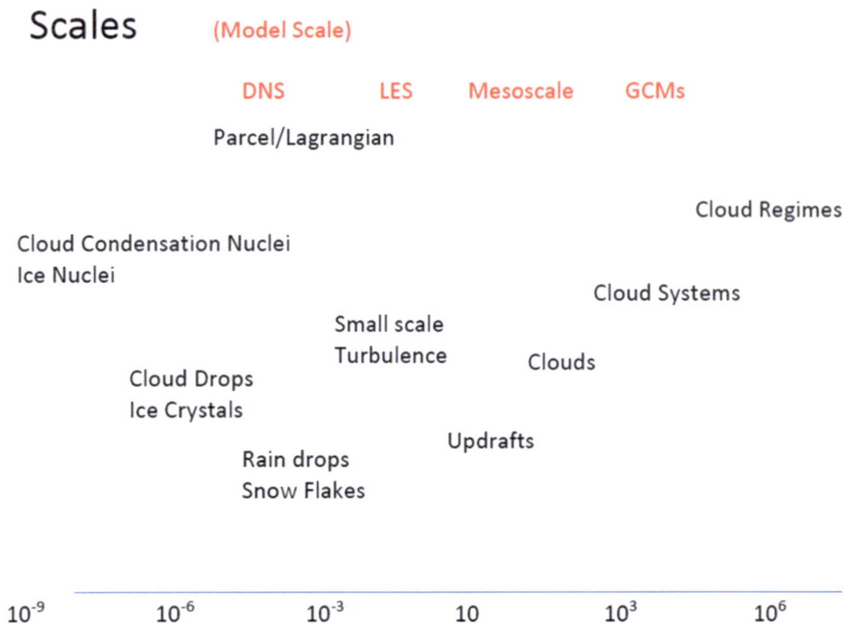

Fig. 1 Schematic of scales: aerosol particles (CCN, ice nuclei) = nanometers, cloud drops = micrometers, raindrops/snow flakes = millimeters, turbulence = meters, updrafts = 10s–100s of meters, cloud scales = kilometers, cloud systems = 10s–100s of kilometers, general circulation driven systems = 1000s of km

also includes processes that change the phase of cloud hydrometeors, i.e., melting and freezing.

Cloud microphysical processes are critical for weather events, including the development and evolution of precipitation. Cloud microphysical processes control the timing, intensity, and phase of precipitation. The freezing process, coupled with cloud dynamics, affects the evolution of hail and graupel. Furthermore, cloud microphysical processes and the evolution of condensate in clouds affect the buoyancy and energetics of storms, through latent heating and the weight of the condensate. This affects the structure and dynamics of clouds, through evaporation influenced downdrafts and cold pools, to buoyant updrafts driven mainly by latent heating from droplet condensation.

In addition, cloud microphysical processes are critical for climate. The radiative effects of clouds are determined by the number and size of liquid droplets, and the number, size, and shape (habit) of ice crystals. The same bulk water mass partitioned in a different number of liquid drops, or a different shape and number of ice crystals will have very different radiative properties. Cloud microphysical processes and precipitation may also alter the persistence of clouds, further altering their radiative effects. The mixed-phase regime, where either liquid or ice can exist (temperatures from 0 to −40 °C, or 273 to 233 K) is particularly important, where the microphysical details of cloud phase will strongly impact the lifetime of clouds and their radiative effects.

This chapter will describe how cloud microphysics is represented in numerical models and how it couples into the rest of the Earth system. Having motivated the discussion of microphysics, we turn our attention to the methods used for representing microphysical processes (Sect. 2), and how such treatments are coupled to the rest of a numerical model that includes clouds (Sect. 3). We will then describe some current issues and new approaches (Sect. 4), and discuss possible paths forward for simulating cloud microphysics in Sect. 5.

2 Methods for Simulating Microphysics

Here, we will conceptually describe what microphysical processes do, and how they work. Various commonly used techniques will be described, as well as a breakdown of key categories of microphysical processes. We will conclude the section with a discussion of scales, and how the model scale helps determine the appropriate level of complexity.

2.1 *What Microphysics Does*

Cloud microphysics governs the transformation and evolution of condensed water in clouds. It is usually considered a local microscale process, as opposed to cloud processes where turbulence and cloud-scale dynamics are important for moving water in coherent motions.

In order to frame the problem, it is useful to understand the basic types of model that contain cloud microphysics. Small-scale models (scales<10s of m) are often idealized, and focused on a series of processes. Box and parcel models represent a zero-dimensional population of cloud drops with a specified environment (e.g., KiD: Shipway and Hill 2012). Eulerian models represent cloud drops in discrete atmospheric volumes in one–three dimensions. These models can have very different scales. The smallest-scale turbulence-resolving models have scales of centimeters up to ~1 m, sometimes called Direct Numerical Simulation (DNS), and these models sometimes use Lagrangian methods for representing cloud particle evolution. Progressively, larger scale models include Large Eddy Simulation (LES) models with scales from about 1 to 100 m, in which the smaller turbulent eddies are sub-grid scale and hence, their effects are parameterized (e.g., Khairoutdinov and Kogan 2000). LES usually represent clouds with a population (or size distribution) of cloud drops and ice crystals in each grid cell as described below. Also using this approach are limited area mesoscale models with scales from about 1 to 10s of km. General Circulation Models (GCMs), global models that have resolutions from 3 to 25 km for weather applications, and from 25 to 200 km for climate applications (e.g., Stier et al. 2005; Neale et al. 2010).

2.2 *Types of Schemes*

We next describe different conceptual methods for representing cloud microphysics. We will focus on schemes used in Eulerian models, which typically represent categories of hydrometeors statistically in a grid box whose size contains a statistically representative sample of hydrometeors. Other possibilities include "Lagrangian" schemes that explicitly track the evolution of individual hydrometeors or representative hydrometeors in a population (the so-called "superdroplet" method, Shima et al. 2009), which have been typically used in parcel and DNS models and more recently LES.

Microphysics schemes typically divide hydrometeors into "categories" or "classes", representing the various types of liquid, low-density ice, and high-density rimed ice (graupel and hail) particles that occur in the atmosphere. The more complex a scheme, the more classes are represented. Early schemes had a single category (e.g., Slingo 1985; Wetherald and Manabe 1988). Later, schemes had categories for liquid and ice (e.g., Ose 1993; Lohmann and Roeckner 1996; Rotstayn et al. 2000). These schemes typically still retained a fixed transition between liquid and ice often as a function of temperature (e.g., Rasch and Kristjansson 1998), which is actually broadly representative of observations (Korolev et al. 2003). Schemes developed for weather prediction models often add additional categories for precipitation water, i.e., rain and snow (four categories, Zhao et al. 1997). When small-scale vertical motions are permitted in mesoscale and cloud models, graupel and/or hail is often explicitly represented (e.g., Lin et al. 1983; Rutledge and Hobbs 1984). Current developments in the field seek to unify categories, especially for the ice phase (cloud ice, snow, and graupel, hail) to avoid ad hoc partitioning of ice between categories and conversion processes between them (e.g., Hashino and Tripoli 2007; Morrison and Milbrandt 2015).

There are several methods for representing the population of cloud particles in a numerical model (not including Lagrangian methods noted above). Figure 2 represents these schematically. The most explicit treatment is a bin scheme, where the number and/or mass of particles in each size or mass "bin" is represented. There are typically different sets of bins for each type of hydrometeor represented (see below). Bin schemes are computationally expensive because they need to estimate the evolution of the number and/or mass in each bin, and because each bin represents a separate physical quantity ("species") to track and advect. Advection, particularly conservative and positive definite or monotonic advection of microphysical variables, is often computational expensive in Eulerian models. However, bin schemes are "direct" in the sense that the microphysical processes that vary across size bins (such as fall speeds) or depend on interactions between size bins (like the collection of cloud drops by rain) can be estimated explicitly. As such, bin schemes are most

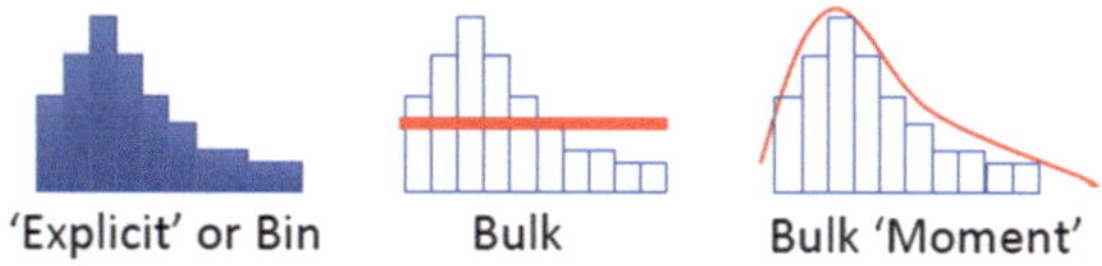

Fig. 2 Types of cloud microphysics schemes

appropriate for small-scale models, usually to simulate weather, for which precipitation evolution is critical, but they are not integrated for a long time because of the computational expense.

The second type of representation is known as a "bulk" representation, where a single value is used for the total bulk mass mixing ratio of each hydrometeor category, representing all sizes (e.g., Lin et al. 1983). Sometimes the particle number is also represented in diagnostically. Such a formulation is computationally efficient (only one species per category, while a bin scheme might have 30 species per category), but by necessity must approximate many processes that are size dependent (fall speeds, collision, etc.). Some single moment schemes for mesoscale models also include representations of the mass of graupel and hail (Hong and Lim 2006; WSM6).

A bulk scheme need not just represent the total mass mixing ratio of a hydrometeor category. Bull schemes now often represent a size distribution with two or more moments, and an assumed functional form (e.g., Fig. 2). If a distribution shape is assumed (often a gamma function) for each class, then the size distribution can vary with the predicted moments. These methods are used typically for mesoscale models over large areas with complex ice processes (Thompson and Eidhammer 2014; Milbrandt et al. 2016) or for long integrations in large-scale (global) models (Lohmann et al. 1999; Morrison and Gettelman 2008). The third moment is proportional to the bulk mass (for liquid drops), and the zeroth moment is the bulk number. Two-moment schemes predicting both the bulk mass and number of one or more categories are common now for mesoscale and even global models. Some schemes introduce a third moment to represent the skewness of the distribution such as through the sixth moment, which is proportional to radar reflectivity (e.g., Milbrandt and Yau 2005).

2.3 *How We Simulate Microphysics*

Microphysics is an important component of the hydrologic cycle. The water lifecycle contains processes that evolve and transform water in the atmosphere. Microphysics focuses on the distribution and evolution of condensed water in the atmosphere. Sources and sinks include condensation and evaporation and precipitation at the surface. Sometimes, microphysics includes the condensation process itself, which is often tightly coupled to cloud and turbulence schemes, and sub-grid variability in a model. We will term this "macrophysics" and discuss it only briefly. A model needs to represent those processes that change the state (the different moments) of each

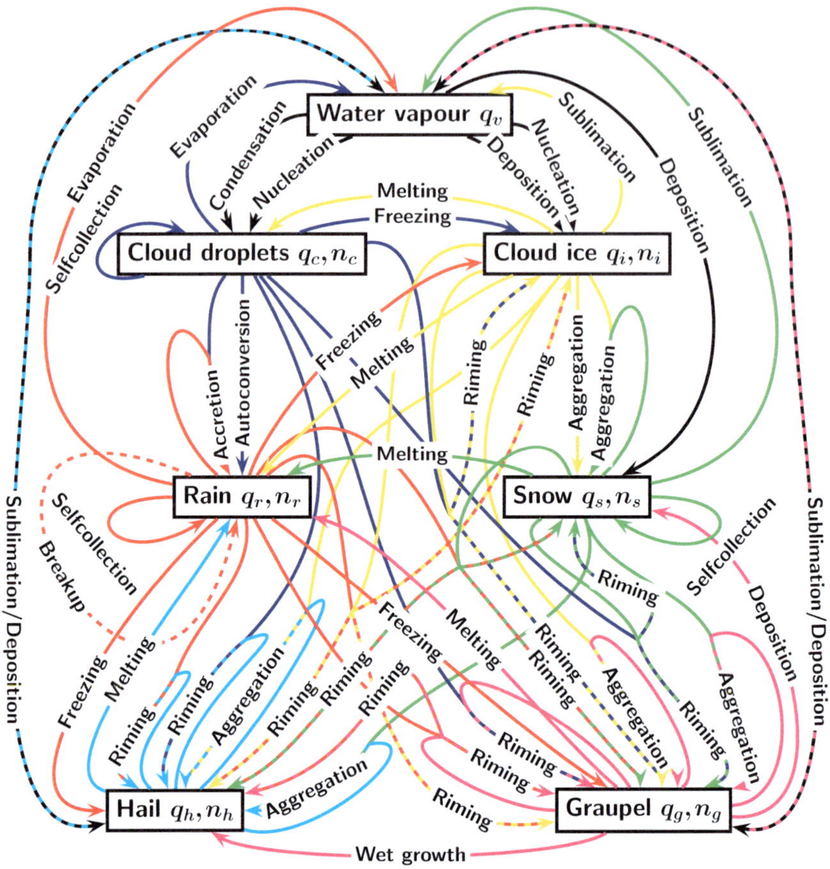

Fig. 3 Schematic diagram of processes in the six-class, 2-moment microphysics scheme of Seifert and Behang (2001). Figure courtesy of Axel Seifert

category or class of hydrometeors. These processes will also change the temperature of the environment if phase changes occur (through latent heating or cooling).

While the list of processes and their interactions vary across schemes, and can rapidly get very complicated (Fig. 3), they fall into several different groups. These include condensation and evaporation, freezing and melting, and precipitation formation by collision-coalescence. The overall goal is to describe the important transformations of water substance in the atmosphere. The most important transformations will, of course, depend on the application and scale. For hazardous weather events, the most important transformations may be related to hail, graupel, and copious supercooled liquid associated with severe storms. For climate, on the other hand, it might be cirrus cloud microphysics that governs radiative transfer from pervasive high ice clouds in the tropics. So, different microphysics schemes will focus on different processes, adding more detail to some or neglecting others entirely.

2.3.1 Nucleation/Activation

Water does not spontaneously condense in the atmosphere. The process of nucleating cloud drops and ice crystals occurs when there are aerosol particles that can activate condensation or freezing. In addition, the freezing process also rarely occurs with pure water, but often with a water solution or due to the presence or collision with a foreign object. As a result, representing the formation stage of cloud droplets and ice crystals is not a simple process. This is usually defined as the nucleation process for cloud drops that occurs when aerosols are "activated" in the presence of supersaturation (over liquid and ice).

Activation and nucleation processes are used for initializing cloud drops in schemes where drop and crystal number is treated. Nucleation is the increase in cloud drop/crystal number. It is often the result of an activation process for aerosols. Activation is the process determining how aerosols initialize cloud drops or ice crystals, and is dependent on the aerosol population, the ambient humidity, and the micro or cloud-scale vertical velocity. The vertical velocity used for activation in large-scale models is often a "sub-grid" vertical velocity that does not match the resolved grid scale velocity. Schemes that treat aerosols and drop number usually contain some version of these processes to determine cloud drop number (e.g., Gettelman et al. 2010; Thompson and Eidhammer 2014). Microphysical schemes that do not predict drop number often have a very simple representation of drop number or size either specified as a constant (Rasch and Kristjansson 1998) or diagnosed from bulk aerosol information (Wilson and Ballard 1999).

Note that while activation of liquid clouds is a rapid process, and depends on the aerosol population and vertical velocity, ice nucleation is much more complex. There are many different pathways for ice formation with either liquid or solid aerosol particles, and with pathways involving the liquid phase, or almost pure ice phase nucleation. These pathways are too complex for discussion here, but the basic pathways are heterogenous freezing onto solid particles (often dust) at relative humidities of ~120–130% over ice (RHi) and heterogenous freezing of solution drops containing sulfate at ~160% RHi. The difference in ambient supersaturation means that the availability of ice nucleating particles (INP) can have important impacts on cloud microphysics, and vastly change cloud properties (e.g. Morrison et al. 2009). There are also pathways for condensation/deposition of liquid onto aerosol and rapid freezing in the mixed-phase regime. Hoose and Möhler (2012) provide a nice review of these heterogeneous ice nucleation processes.

2.3.2 Condensation/Evaporation and Deposition/Sublimation

Condensation is the growth of droplets by incorporation of water vapor into the droplet surface. In the atmosphere it generally occurs at near-equilibrium, meaning that supersaturation with respect to liquid water tends to be small (less than a few %). Deposition of vapor onto ice is more complex as reflected by the wide variety of vapor-grown ice particle shapes, and ice often grows in strongly nonequilibrium conditions in the atmosphere with large supersaturations. The habit, or shape, of ice particles grown by vapor diffusion depends strongly on temperature, with supersaturation also important. For cold clouds below freezing (273 K), supercooled liquid is possible down to nearly −40 °C (233 K), and freezing generally requires an ice nucleation process. The reverse of condensation is the evaporation process which removes liquid drops in sub-saturated regions by conversion to vapor. The analogous process for ice is usually referred to as sublimation.

Most microphysics schemes represent condensation and evaporation of cloud liquid water by the "saturation adjustment" technique, which assumes the relative humidity inside clouds is at equilibrium and any excess vapor is instantaneously condensed as cloud liquid. Conversely, cloud water is evaporated instantaneously if there is any deficit of vapor to bring in-cloud relative humidity back to equilibrium. In contrast to saturation adjustment, more detailed multi-moment schemes and all bin schemes explicitly calculate condensation and evaporation of cloud droplets using the modeled relative humidity field. Most schemes similarly calculate ice deposition and sublimation "explicitly" using the modeled relative humidity field with respect to ice. For parameterizing condensation/deposition and evaporation/sublimation in models, complexities arise because of sub-grid inhomogeneity of relative humidity and condensate. This includes the evaporation of falling precipitation in sub-saturated regions.

Condensation/deposition and evaporation/sublimation are accompanied by latent heating (condensation or deposition) or cooling (evaporation or sublimation). Condensation and evaporation are critical processes for interactions of microphysics with the energy budget, and the resulting heating and cooling terms are important for driving cloud dynamics in many situations, including convective systems, where gust fronts and downdrafts result from evaporative driven cooling.

2.3.3 Collection/Growth and Precipitation

A population of cloud particles (drops and/or crystals) can evolve through collection, which represents collision and coalescence of cloud particles (often referred to as "aggregation" for ice). In bulk schemes, collection can move particles across microphysical categories, for example from cloud liquid to rain. This is generally parameterized as "autoconversion" in bulk schemes, which represents the growth of embryo rain or drizzle drops from collision-coalescence of smaller cloud droplets. The collection of cloud droplets by the existing raindrops in bulk schemes is called "accretion". Autoconversion and accretion in bulk schemes are usually based upon other detailed models with bin microphysics (e.g., Khairoutdinov and Kogan 2000; Seifert and Beheng 2001) or to in situ observations (Wood 2005). Many bulk schemes treat the growth and conversion from cloud ice to snow (from aggregation as well as vapor deposition) in a similar way, but the parameterization of this conversion is fairly ad hoc. In contrast, bin schemes explicitly model the collision-coalescence equation (quasi-stochastic collection equation) rather than treating collection as a conversion process between cloud and precipitation categories. The collection of supercooled liquid by snow through riming can lead to the growth of snow, or potentially conversion to graupel or hail. The details of these conversion processes differ widely among microphysics schemes.

Multi-moment bulk schemes that predict bulk number generally also account for collision-coalescence or aggregation that produces particles that remain within a given hydrometeor class, which is called "self-collection". This process does not affect the bulk mass, and hence is not included in one-moment bulk schemes. Multi-moment bulk schemes and bin schemes also parameterize particle breakup, mainly for raindrops. In multi-moment bulk schemes this is fairly ad hoc, but modeled explicitly using a breakup kernel in bin microphysics schemes. Ice crystals may also breakup and shatter when they collide, but this process is often neglected in models or represented implicitly in bulk schemes by limiting aggregation growth.

2.3.4 Freezing/Melting (Phase Change)

The phase change of cloud drops from liquid to ice (and vice versa) is important for the evolution of clouds. Phase changes lead to latent heating or cooling, affect the physical properties of hydrometeors (especially fall speed), and alter growth rates: ice particles will grow rapidly at the same temperature and humidity where liquid drops may evaporate due to different vapor pressures over liquid and ice. This vapor deposition process is often known as the Bergeron–Findeisen–Wegener process but is also subject to limitations based on the cloud environment (Korolev 2007). The heating and cooling associated with solid–liquid phase changes can have a significant impact on cloud dynamics and evolution, in a similar way to evaporation and condensation, although the enthalpy of fusion is nearly an order of magnitude smaller than the enthalpy of condensation.

In complex clouds with high upward vertical velocities (like thunderstorms), cloud drops may freeze in updrafts, or melt as they sediment, or even re-freeze if the process is repeated. They can also take up (collect) supercooled liquid water by collision, a process called "riming". This can lead to the growth of graupel and hail particles. Under very high riming rates, the liquid water may not freeze entirely due to the associated latent heating, a process called "wet growth" that is associated with hail. Some of the liquid water accumulated on hail during this process may be shed. Models that have high enough resolution to explicitly represent strong vertical motions associated with deep convection often use microphysics schemes that treat graupel and/or hail.

2.3.5 Complexities

There are several important complexities in cloud microphysical processes to highlight. Because it is a metastable regime where hysteresis (the history of cloud particles) matters, the mixed-phase regime is complex and representations of processes in this regime are uncertain (since they are nonequilibrium processes and depend on the history of cloud processes).

In addition, the interaction of cloud drops with aerosols (sometimes called aerosol–cloud interactions) is highly complicated. This encompasses a series of processes illustrating the coupled nature of microphysical processes. Aerosols alter the activation and nucleation process, resulting in different cloud drop and crystal numbers. The different number concentrations and size distributions affect the evolution of cloud drops, potentially altering drop/crystal sedimentation and formation of precipitation, as well as even impacting evaporation and condensation, with feedbacks to the cloud dynamics.

A common problem of representing cloud microphysical processes is a huge range of relevant scales, as noted above. Many of the microphysical processes described previously depend on small-scale particle–particle or particle–environment interactions. While in principle, these are governed by first principles, such as condensation dependent on vapor pressures, humidity and surface tension, practically, there are significant scale mismatches between these processes and the model resolution. A distribution of cloud drops will behave differently than a single drop, due to competition, and due to air motion at all scales from the Kolmogorov scale to the cloud system scale and beyond. This makes formulating microphysical process rates difficult for large-scale models. Often empirical techniques are used to fit large-scale processes (such as autoconversion) to small-scale detailed model treatments that are more complex. For example, the full calculation of stochastic collection with a bin scheme has been used to empirically derive process rates (e.g., Khairoutdinov and Kogan 2000; Kogan 2013). Process rates can also be empirically from observations. The difficulty, of course, is determining the representativeness of the fits.

2.4 *Different Scales and Complexity for Different Types of Models*

As noted above, different levels of microphysics scheme complexity are used for different scales. Models representing shorter timescale "weather" features such as LES and limited area mesoscale models are typically focused on detailed representations of precipitation. They typically represent convective scale updrafts which will generate mixed-phase hydrometeors, so typically will have ice, liquid and a mixed phase. In addition, separate classes for rain and snow are often represented, resulting in five or more classes (liquid, ice, rain, snow, graupel, hail, e.g., Hong and Lim 2006). LES models often use explicit microphysics (bins) when focused on cloud simulation (Khairoutdinov and Kogan 2000). Mesoscale models range widely from simple bulk schemes (Lin et al. 1983; Kessler 1969) to complex single moment schemes (Hong and Lim 2006), to moment based schemes (Morrison et al. 2005; Thompson and Eidhammer 2014), to full bin schemes (Khain et al. 2004). Not all classes need have the same representation (sometimes, a mix of bulk and moments are used).

In global or regional weather models, the goal is to produce a forecast or representation of a particular atmospheric state at a point in time and space. Since much of this is targeted at significant events, processes leading to extreme events are treated. This usually means a focus on the representation of detailed hydrometeors, including graupel and hail. It is critical in weather models to represent the formation and evolution of precipitation and how condensation and evaporation feedback on the environment. These models typically have a detailed treatment of the mixed phase with a representation of graupel and/or hail (e.g., Khain et al. 2015; Thompson and Eidhammer 2014).

Climate models have generally had cruder treatments of cloud microphysics. The goal of a climate model is the representation of the global hydrological cycle and its effect on the global energy budget. The role of precipitation is secondary, and often more as a surface flux of water. The models do not have the horizontal resolution (often ~100 km) to properly represent extremes of precipitation. Furthermore, climate models (and some weather models) usually assume that only a fraction of a grid box need be cloudy. The cloud microphysics is usually assumed to represent "in-cloud" quantities, and the model has to account for out of cloud regions and have energy and mass balance. The goal of a climate model is usually the representation of the energy budget, and as such the radiative effects of clouds are very important.

Both these classes of model struggle with the sub-grid variability of cloud properties at unresolved scales. Most climate models need to treat fractional cloudiness, and many weather models do too. All models are representing the sub-grid distribution of cloud drop and crystal sizes with some approximation (either a set of bins, or a fixed number of fixed functions). Furthermore, we have seen how sub-grid dynamics (vertical velocity) is critical for microphysical process rates. We return to the importance of sub-grid treatments later, but the variation on scales smaller than the grid is important for how microphysics is coupled to other aspects of a complete "cloud" scheme.

3 Microphysics in the Context of Cloud Schemes

Cloud microphysics is usually only one piece of a larger suite of cloud parameterizations used to characterize the evolution of clouds in large-scale models. In an ideal word, microphysics would be the only cloud scheme: given a state of humidity, temperature, and vertical velocity, the microphysics could locally assume uniform properties and estimate the local state of clouds, with only knowledge of a single column to represent fall speeds (if the time step were short enough, even sedimentation could be handled by the large-scale model). A model of clouds would require dynamics, cloud microphysics and radiative transfer, and that would be it. Microphysics would affect dynamics through the heating, cooling, and radiative transfer at the resolved scale.

But in practice, there are other processes at the sub-grid scale, such as turbulent and convective motions that need to be parameterized. The interactions between different "parts" of the cloud scheme are critical.

3.1 Microphysics in a Cloud Scheme

Cloud microphysics treats the local processes responsible for transformations of water, and calculates the speciation of different classes of cloud hydrometeors. Ideally, these species would simply be advected around, heating and cooling from microphysics applied, and the atmospheric dynamics would then send a revised state to cloud microphysics. But because most models have variability and forcing that is below the grid scale, and timesteps that may be long relative to microphysical processes, typically models contain other cloud schemes. Cloud microphysics needs to interact in complex ways with many of these other schemes to completely represent clouds in a weather or climate model.

These other schemes typically include representations of variable condensation, saturation adjustment and fractional cloudiness. They also include representation of convectively unstable motion. Models also typically represent small-scale turbulent motions and associated cloud fields. The interactions between these schemes and the cloud microphysics may be complex.

3.1.1 Macrophysics and Fractional Cloudiness

Many microphysics schemes include large-scale condensation within them. For larger scale models with lower resolution (larger horizontal scale), usually, some partial cloud cover is assumed with a separate scheme. Fractional cloudiness is typically used for models with grid spacings larger than 10–20 km or so. Recently, even finer scale models have been including representations of a clear and cloud part of a model grid box to better resolve variability. Cloud microphysics needs to remain consistent with a large-scale cloud scheme, for example, by making sure that process rates for in-cloud quantities are calculated on those in-cloud quantities only in a part of the grid box. These in-cloud values are then usually converted to a grid average for other computations.

3.1.2 Convective Motions

Small or even large-scale convective motions (due to vertically coherent buoyancy instabilities) usually occur on scales much smaller than a model grid box, and convective motions (both deep convection and shallow convection) are parameterized. Deep and shallow convection are often separate parameterizations in many models. Most models with scales larger than a few km (3–5 km) have at least a deep convection parameterization. Very few of these representations are linked with cloud microphysical schemes consistently. Most convective schemes have very simple cloud microphysics: usually, just a precipitation efficiency or autoconversion term, and most convective schemes track only water mass in cloud, often just diagnostically. There have been a few efforts to embed or link complex microphysical schemes to deep convective parameterizations in large-scale models (Song and Zhang 2011).

3.1.3 Turbulence Schemes

Condensation of clouds in the planetary boundary layer is often handled by a moist turbulence scheme that helps determine whether the boundary layer is stable or unstable. The latter often leads to overturning turbulent motions and clouds in shallow updrafts. Recently, some of these schemes have been linked to vertically coherent convective turbulence, which is natural for shallow convection, and more difficult for deep convection (CLUBB, EDMF). These turbulence schemes can be integrated with cloud microphysical parameterizations: essentially treating the condensation from such a scheme as a cloud macro-physical or saturation adjustment scheme, and the microphysics operating on the condensed water.

3.1.4 Aerosols

Finally, cloud microphysics schemes, particularly for climate applications, are often linked to aerosol models for studies of aerosols on clouds, and studies of anthropogenic effects on clouds. This has become common to understand changes in cloud radiative properties for climate models, and in a more limited way to understand possible effects on precipitation by convective clouds in smaller scale high-resolution models. Aerosols are often advected, and sometimes evolve with their own chemical model (which can also be two moments: mass and number of aerosols). The link to cloud microphysics is usually a "ratchet": aerosol activation is computed based on the model state, and the activated number is used to adjust the prognostic microphysical cloud drop number.

Cloud microphysics also impacts aerosol distributions through wet scavenging, which may be very efficient for particles, and very efficient in high vertical velocity updrafts. The number of aerosols scavenged is a function of solubility, and the formation of precipitation. Care must be taken with advanced cloud microphysics schemes in representing the aerosol budget consistently.

3.2 Scales and Sub-grid Assumptions

As should be obvious from much of the above discussion, one of the big challenges for cloud microphysical parameterization is representing the variation of clouds on the sub-grid scale. In general, when scales are separated by many orders of magnitude (i.e., for cloud microphysics at the 10^{-6} m scale to the cloud scale of 100 m to km), then a statistical approach can be taken (populations of cloud drops/crystals). But when the variations reach near the scales of interest, such as variability in cloud water concentrations within clouds or a cloud field of between 10 m and 1 km or so, this is within an order of magnitude or two of the scale of models at 10 km. Then the scales of variation are too large to represent statistically, and too small to resolve fully. Many microphysical processes fall into this category: if there is variation in cloud properties across a grid box, what value should be used to estimate process rates for microphysical transformations? For example, if there are spatial variations of cloud-scale liquid water in a grid box, how can a single process rate be determined?

There are several new approaches to this problem. One is to use simplified mathematical treatments for either process rates and/or cloud mass and number distributions that can be integrated over the distribution. This works for a few simple processes describable with exponentials (Morrison and Gettelman 2008; Lebsock et al. 2013), but not for most processes.

Another approach is to define distributions of macro-scale properties (like cloud liquid water or drop number) as a distribution of states in a grid box. This is the essence of a size distribution but think of it as a size distribution for clouds. From such a population or distribution, sub-columns can be drawn and the microphysics scheme runs on a series of these sub-columns. This method conceptually allows each sub-column to be treated as uniform properties. Sub-column approaches are commonly used now in model radiation codes where fractional cloudiness needs to be discrete. A similar method of sub-columns could be used to represent variable amounts of liquid water in parts of a grid box. Sub-column methods can also include a convective sub-column representative of the state in a convective parameterization. These approaches hold promise: but may be computationally expensive if several different instances of cloud a microphysical scheme must be run each time step.

3.3 *Discussion/Example*

Here, we will illustrate the concepts of a cloud microphysics scheme by describing an example from the National Center for Atmospheric Research (NCAR) Community Earth System Model. The example is a two-moment, four-class microphysics scheme (Morrison and Gettelman 2008; Gettelman and Morrison 2015), Illustrated schematically in Fig. 4. It is similar in concept to other common two-moment or semi-two-moment schemes (Morrison et al. 2005; Thompson and Eidhammer 2014). The four classes are liquid, ice, rain, and snow. All classes are prognostic, which is common for mesoscale models, but rare for global models. Size distributions are assumed to be gamma functions. It has been assumed that the advection of falling hydrometeors is not important in global models. Gettelman et al. (2015) illustrated, however, that there are important differences in solutions that result from the vertical transport of falling hydrometeors: rain and especially snow do not have to hit the ground in

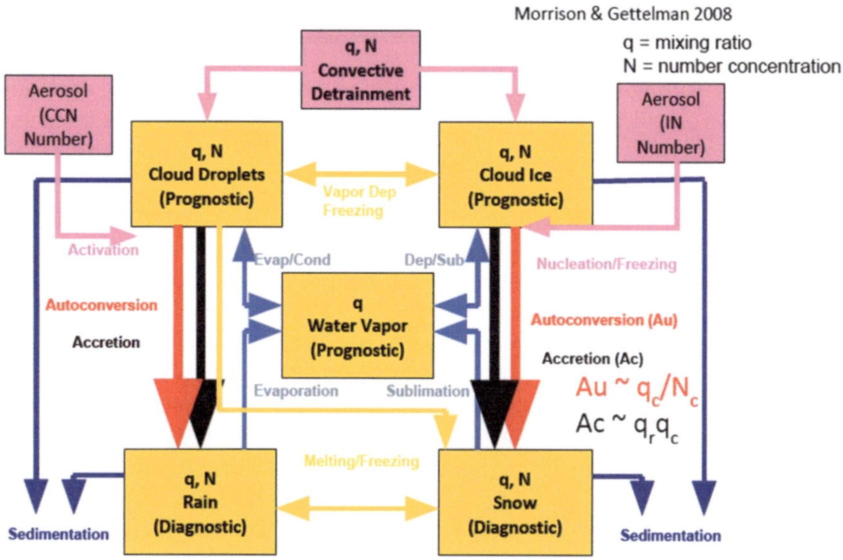

Fig. 4 Schematic of a two-moment, four-class microphysics scheme

one-time step. The Morrison-Gettelman scheme is a microphysics solver that does not contain a large-scale condensation step (this happens before the scheme runs, and condensation is predicted by a moist turbulence scheme). The Morrison-Gettelman scheme predicts drop and crystal number by relying on external ice nucleation and drop activation routines. Most bulk microphysics schemes handle condensation and some aerosol processes internally.

Note that simpler and more complex schemes exist, even for bulk schemes. Simpler schemes would reduce moments for some classes, or even all classes (e.g., Hong and Lim 2006) or make precipitation diagnostic, as in most climate model schemes (e.g., Morrison and Gettelman 2008). More complex schemes might add a third moment to each class to represent skewness, or even reflectivity (Milbrandt and Yau 2005). Some further advances are discussed in Sect. 5: Paths Forward.

Figure 4 illustrates key processes in the Morrison-Gettelman (2008) scheme. As mentioned, condensation, activation of aerosols, and convective detrainment are passed into the scheme. The goal of the scheme is to define the evolution and transformation of condensed water in the atmosphere, and its properties. These properties define the physical evolution (e.g., the drop size defines a fall speed) as well as interactions (cloud drops interacting with aerosols or chemical species, and the absorption and emission of radiation by cloud drops). Condensation is the source of cloud water (including ice), and microphysical processes are responsible for the evolution of water.

The main large-scale loss processes for cloud water forced by the environment is evaporation, which can be large scale or more turbulent. The main microphysical loss process is the formation and evolution of precipitation processes. This process can be described as a stochastic collection process, where growth occurs differentially and then differential fall speeds yield differential collection of smaller drops by larger drops, vastly altering the size distribution. The process can be relatively explicitly defined with a bin microphysics scheme where each bin has a defined fall speed and a collection kernel for all other bins. However, this stochastic collection process is not well defined for bulk size distributions. In bulk models, the transformation to precipitation is usually called an autoconversion process. Once precipitation forms, a collection or Accretion process ensues. These processes are sensitive to the cloud microphysics, and form a regulator on the loss function for cloud water and precipitation. Sedimentation is also an important process that results from both falling precipitation and falling cloud drops and crystals (the latter sediment much more slowly).

Phase transitions are also important in understanding and representing cloud microphysics. The freezing and melting process (like the condensation process itself) results in the uptake (melting) or release (freezing) of latent heat that can be important for driving cloud dynamics. And the character of processes for ice is different: including the evolution of ice crystals and their radiative effects. Furthermore, the evolution of the "mixed" phase between liquid, and ice tends to occur where rapid motions keep hydrometeors out of equilibrium: that is, in extreme storms or convective events. These spawn unique phenomena such as hail and graupel formation, which can be large ice or mixed-phase hydrometeors of centimeter or even many centimeter sizes. Hail and graupel can result in extremely damaging weather events, and are associated with severe storms (including lightning storms and tornadoes). Representing these complex extremes is an important part of many microphysics schemes. Note that typically, these extreme mixed-phase hydrometeors only occur with high updraft velocities, which are often not represented (but can be parameterized) in large-scale models, and are only resolved as scales get down to 100 m or so.

Cloud microphysics schemes generally treat grid scale updrafts and processes. However, they are usually coupled to several "sub-grid" schemes in models. For example, condensation schemes for global models at scales larger than 20 km or so typically will represent condensation in only part of a grid box, and maintain a fractional cloudiness. Generally, clouds are considered as uniform, or as distributions of properties, but clouds can be defined individually by assuming different "sub-columns" of an atmospheric grid box, and running microphysical processes to each sub-column separately before combining them into a grid box average. This requires a method to generate sub-columns, and current experiments couple these columns usually to some standard sub-grid representation of quantities in a grid box (e.g., a pdf distribution whose integral equals the grid box mean value). These methods can help mitigate scale issues by trying to more explicitly treat small-scale variability. For example, because many processes like autoconversion are not linear, the average process rate of several different sub-columns with different water path, drop numbers, and updrafts will generally NOT be the same as the autoconversion rate calculated from the grid box average of these quantities. In addition, as the scale gets smaller, the sources of variance go down, and the sub-columns will look more similar.

Furthermore, cloud microphysics schemes can be tightly coupled to other condensation schemes, such as convective parameterizations, turbulence, or fractional cloudiness. This often happens with distributions or PDFs produced by the schemes. The goal of many models is to have consistent sub-grid variability.

But coupling cloud schemes can be a problem in itself. Parameterizations may interact in unexpected and not complementary ways. For example, microphysics schemes may try to evaporate water, but this may humidify a grid box enough to cause the cloud to form from a fractional cloudiness scheme. One analogy is to think of a set of cloud schemes as a team of sled dogs: each dog is a complex parameterization. The dogs need to be trained on their own, but they also need to be trained together to produce a coherent team. The best sled dog is no good if it is in the wrong position on the team, it can be a detriment rather than a help. One of the difficulties of trying to construct a cloud physics suite for any scale of models is enforcing this consistency among schemes and getting them to work well together. Smaller scale models (i.e., LES or mesoscale models) have less to represent at the grid scale (e.g., they may not have moist convective parameterizations, and just handle convective motions on the grid scale), and may have fewer dogs pulling the sled. Or unified turbulence schemes (boundary layer and convective motions) may also reduce the number of schemes and their complex interactions.

4 Current Issues/Uncertainties

Cloud microphysics in large-scale models has evolved rapidly over the last few decades. Two-moment microphysical schemes are common in mesoscale models (Morrison et al. 2005; Thompson et al. 2008) and advanced global climate models (Morrison and Gettelman 2008; Lohmann et al. 2007). For weather models, the schemes are designed to represent most or all cloud motions, and typically include mixed-phase hydrometeors, while global schemes do not include such treatments, but typically include detailed aerosol models to represent effects of aerosols on cloud drop number and cloud evolution. Nonetheless, critical uncertainties remain in treatments of cloud processes. We detail a few of these uncertainties related to ice, the mixed phase. and convective microphysics. We then discuss numerics and some of the critical interactions with microphysics and other cloud schemes in models.

4.1 Unified Ice

Almost all microphysics schemes, whether bin or bulk, divide hydrometeors into "classes" (liquid, ice, graupel) with their own properties. Almost all these schemes also separate cloud particles that are suspended with slow fall speeds with large falling precipitation particles. Some of this is purely practical: large drops tend to have positive feedbacks on growth (collection, seeder-feeder process) and grow larger and fall faster, leading to bimodal size distributions that separate. For liquid (cloud drops and rain), the spherical nature of drops permits analytic treatments based on detailed models and bulk aerodynamic formulas.

However, for ice, the situation is more complex. Ice crystal growth and sedimentation (fall velocity) are dependent on particle shape and size, and are more of a continuous distribution. The separation of ice into precipitating and suspended ice is therefore artificial. Thus, there are now efforts to attempt to combine ice and snow (frozen precipitation) into a single class with additional properties (for example, mass, number, and a dimensional relationship). This has been implemented in schemes for mesoscale and weather prediction models (Morrison and Milbrandt 2015) and even for global models (Eidhammer et al. 2016). The biggest advantage of these methods is avoiding artificial transitions from suspended ice to falling ice (snow), which have large and abrupt changes in cloud properties like fall speed that are not observed (observations show a more continuous evolution).

4.2 Graupel/Hail

As with unified ice schemes, the addition of mixed-phase hydrometeors as a separate class can be an advantage for representing completely the distribution of cloud drops, but can also create problems of artificial separation between classes. Mixed-phase or rimed hydrometeors (ice with a liquid coating) exist in clouds with updrafts significant enough to lift melting hydrometeors back above the freezing level, sometimes for multiple cycles (hail). Mixed-phase hydrometeors have typically been treated as a separate class in small-scale models that resolve convective scale updrafts. They have not been put into large-scale (global) models for weather or climate forecasting, as these models typically do not have the resolution to resolve significant updrafts. But with increasing resolution of such models, it will soon become necessary to treat these microphysical processes.

Because the evolution of these hydrometeors is critical, and they are interrelated with other hydrometeors and the cloud environment, new treatments to unify the ice phase naturally try to include a rimed fraction as one of the unified ice particle properties to be predicted. Predicting density and rimed fraction can separate hail and graupel effectively. Multiple instances of a unified class have also been explored (Morrison and Milbrandt 2015), enabling the evolution of ice and mixed-phase hydrometeors to be explored.

4.3 Convective Microphysics

Many extreme weather events result from convective clouds. Extreme precipitation and large hail are examples of how cloud microphysics in the presence of convection can result in severe weather events. Representing these events in models is difficult in large-scale models because most extreme precipitation is generated from convective cloud systems and requires representation of convective updraft velocities. In "cloud resolving" or "convective permitting" models with resolutions from 100 m to 1 km, often these updrafts are represented explicitly. In models with horizontal resolutions larger than this, sub-grid motions must be represented. Often, these sub-grid motions are handled within a convective scheme with minimal representation of cloud microphysics. A few recent treatments (Zhang et al. 2005; Song and Zhang 2011) have actually introduced two-moment bulk microphysics in cumulus convection schemes. Song et al. (2012) and Zhang et al. (2005) illustrate that the inclusion of more complex microphysics in convection tends to increase precipitation overall (an improvement), but it is highly dependent on the individual convective scheme. Another approach is to represent convection as part of a distribution of turbulence, and run cloud microphysics consistently on sub-column representations.

4.4 Aerosol–Cloud Interactions

Cloud drops and ice crystals nucleate on aerosol particles, or on solution drops. This explains why clouds over the ocean (with fewer available aerosols) have lower droplet concentrations than over land for example. Thus, changes to the distribution of aerosol particles may change the nucleation of cloud drops. The activation of aerosols to cloud drops and ice crystals and the nucleation of cloud drops is an important control on the number of cloud drops. This process is represented in many models and in cloud microphysical schemes. The resulting impact of changes in cloud drop number then propagates through microphysical processes to eventually impact cloud extent, precipitation and thus cloud lifetime and cloud radiative effects. There are currently a lot of uncertainties, many in the cloud microphysics, in understanding how changes in aerosols thus impact weather and climate. The complexity comes from having to understand how aerosol effects on cloud drop (and crystal) number propagate through cloud microphysical processes, and feedback on the cloud and on the cloud environment.

4.5 Numerics

We have deliberately kept most of this analysis and discussion away from some of the detailed numerics of cloud microphysics schemes. Cloud microphysical processes are generally fast and equilibrium processes. However, some microphysical processes can create numerical issues, such as sedimentation of larger precipitation-class hydrometeors at velocities that will violate Courant–Fredrichs–Levy (CFL) conditions. These processes are then typically sub-stepped in time to prevent this. Constraints on condensation, or condensation timescales are a critical issue. Typically, large-scale models run on timesteps of 10–30 min, sometimes down to 5 min.

But numerically, microphysics schemes may only be convergent for timesteps below 1 min (Gettelman et al. 2015). As long as equilibrium timescales for microphysical processes are fast and the space scale has orders of magnitude separation from the microscale ($O \sim 10^{-6}$ m, or even up to 1 m), then microphysics can be viewed as a small-scale problem, well separated from the resolved scale. However, microphysics does not act in a vacuum and much of the problem comes from interactions between the microphysics and other cloud schemes.

4.6 Interactions with Other Schemes

Interactions between microphysics and other cloud schemes are both a confounding problem and an opportunity. Many typical models contain several cloud microphysics schemes, buried in different representations of large-scale clouds, deep convection, and the boundary layer. These representations may not all be consistent, and that creates problems, when say a bulk deep convective scheme detrains mass into a stratiform cloud scheme that tracks mass and number. Generally one of the greatest difficulties in simulating clouds is the connections between schemes.

Integration of microphysics with sub-grid schemes is a new advance. In Sect. 4.3, we discussed the integration of cloud microphysical processes into the parameterized deep convection in a global model as an example. In this case, cloud microphysics is running on a single instance of a high updraft velocity. More complex treatments exist, however, linking microphysics to other schemes. One recent advance is coupling microphysics to a unified moist turbulence scheme (Bogenschutz et al. 2013, 2018), where the sub-grid state is tracked as a joint distribution of total water potential temperature and vertical velocity (Golaz et al. 2002). This enables cloud microphysics to be run on the means of the distribution, or even on sub-samples drawn from the distribution for a consistent treatment of cloud microphysical processes (Thayer-Calder et al. 2015). Sub-grid samples of cloud properties are often used with current radiative transfer routines in models that have "partial" cloudiness (climate models). Typically with resolutions below 5–10 km, clouds are assumed uniform in a grid box.

5 Path Forward

Cloud microphysical representations in models have come a long way, and are now advanced representations of different microphysical process rates that encapsulate decades of knowledge and observations. Advances in computational performance have enabled increased complexity, and the transfer of knowledge from detailed models to larger scale models of weather and climate. This has enabled attacking new problems, such as mixed-phase clouds, convective motions and interactions of cloud microphysics with aerosols.

Current schemes treat several different "classes" of hydrometeors, either as bulk descriptions of a size distributions with one to three moments, or by representing a discrete size distribution with a series of bins. Recent advances include treating higher order moments of the distributions for bulk schemes, and trying to create more unified treatments of hydrometeors and precipitation, especially for the ice phase. These advances hold promise for increasing fidelity of schemes for large-scale models without increasing computation time, since they can actually have fewer classes of hydrometeors.

In addition, the coupling of cloud microphysics to the condensation and cloud environment in models is critical. Cloud microphysics is a sub-grid process, but needs to be coupled appropriately to other sub-grid processes. Broadly, cloud microphysics needs to be coupled to the sub-grid dynamics, as updrafts determine the evolution of a cloud. This means that turbulence schemes and cloud microphysics need to be coupled at a closer level than in most models. Sub-grid schemes are not just "plug and play", but need to be treated consistently. With advanced turbulence schemes and convective schemes that provide vertical velocity information, cloud microphysical schemes can use this sub-grid information to act differently on different sub-columns within a grid box.

These strategies argue strongly for cloud microphysical schemes that can work across all scales of models. The desire for sub-seasonal prediction on the several week scale, or the need to forecast land–atmosphere interactions for continental storms and air–sea fluxes for tropical cyclones necessitate global coupled models for weather prediction with schemes that conserve mass and energy. High-resolution climate simulations focused on extreme events demand complexity for mixed-phase hydrometeors typically found in smaller scale models. Thus, there is a convergence between the weather scale and the climate scale, necessitating cloud microphysics that can work across scales. The key as noted above is interactions with the cloud dynamics that forces cloud microphysics, and representation of a distribution (or joint distribution) of updrafts and cloud water, if necessary in discrete samples, or if the grid is small enough, the distributions should collapse to a traditional uniform state. Such a configuration requires a complete and efficient microphysics scheme that can handle a range of updraft scales from slow updrafts in thin tropopause cirrus or gentle turbulence in Arctic mixed-phase clouds, all the way to deep convective updrafts in mesoscale convective systems. Such schemes need to be tested in a range of models, hopefully in common frameworks, or ideally unified atmosphere models from weather to climate.

References

Bogenschutz, P.A., A. Gettelman, H. Morrison, V.E. Larson, C. Craig, and D.P. Schanen. 2013. Higher-order turbulence closure and its impact on climate simulation in the community atmosphere model. *Journal of Climate* 26 (23): 9655–9676. https://doi.org/10.1175/JCLI-D-13-00075.1.

Bogenschutz, P.A., A. Gettelman, C. Hannay, V.E. Larson, R.B. Neale, C. Craig, and C.-C. Chen. 2018. The path to CAM6: Coupled simulations with CAM5.4 and CAM5.5. *Geoscientific Model Development* 11 (1): 235–255. https://doi.org/10.5194/gmd-11-235-2018.

Eidhammer, Trude, Hugh Morrison, David Mitchell, Andrew Gettelman, and Ehsan Erfani. 2016. Improvements in global climate model microphysics using a consistent representation of ice particle properties. *Journal of Climate* 30 (2): 609–629. https://doi.org/10.1175/JCLI-D-16-0050.1.

Gettelman, A., and H. Morrison. 2015. Advanced two-moment bulk microphysics for global models. Part I: Off-line tests and comparison with other schemes. *Journal of Climate* 28 (3): 1268–1287. https://doi.org/10.1175/JCLI-D-14-00102.1.

Gettelman, A., X. Liu, S.J. Ghan, H. Morrison, S. Park, A.J. Conley, S.A. Klein, J. Boyle, D.L. Mitchell, and J.-L. F. Li. 2010. Global simulations of ice nucleation and ice supersaturation with an improved cloud scheme in the community atmosphere model. *Journal of Geophysical Research* 115 (D18216). https://doi.org/10.1029/2009JD013797.

Gettelman, A., H. Morrison, S. Santos, P. Bogenschutz, and P.M. Caldwell. 2015. Advanced two-moment bulk microphysics for global models. Part II: Global model solutions and aerosol-cloud interactions. *Journal of Climate* 28 (3): 1288–1307. https://doi.org/10.1175/JCLI-D-14-00103.1.

Golaz, J.-C., V.E. Larson, and W.R. Cotton. 2002. A PDF-based model for boundary layer clouds. Part I: Method and model description. *JAS* 59: 3540–3551.

Hashino, T., and G.J. Tripoli. 2007. The spectral ice habit prediction system (SHIPS). Part I: Model description and simulation of the vapor deposition process. *Journal of the Atmospheric Sciences* 64: 2210–2237.

Hong, Song-You, and Jeong-Ock Jade Lim. 2006. The WRF single-moment 6-class microphysics scheme (WSM6). *Asia-Pacific Journal of Atmospheric Sciences*. http://www.dbpia.co.kr.

Hoose, C., and O. Möhler. 2012. Heterogeneous ice nucleation on atmospheric aerosols: A review of results from laboratory experiments. *Atmospheric Chemistry and Physics* 12 (20): 9817–9854. https://doi.org/10.5194/acp-12-9817-2012.

Kessler, Edwin. 1969. On the distribution and continuity of water substance in atmospheric circulations. In On the Distribution and Continuity of Water Substance in Atmospheric Circulations. Edited by Edwin Kessler, 1–84. Meteorological Monographs. Boston, MA: American Meteorological Society. https://doi.org/10.1007/978-1-935704-36-2_1.

Khain, A., A. Pokrovsky, M. Pinsky, A. Seifert, and V. Phillips. 2004. Simulation of effects of atmospheric aerosols on deep turbulent convective clouds using a spectral microphysics mixed-phase cumulus cloud model. Part I: Model description and possible applications. *Journal of the Atmospheric Sciences* 61 (24): 2963–82. https://doi.org/10.1175/JAS-3350.1.

Khain, A.P., K.D. Beheng, A. Heymsfield, A. Korolev, S.O. Krichak, Z. Levin, M. Pinsky, et al. 2015. Representation of microphysical processes in cloud-resolving models: Spectral (bin) microphysics versus bulk parameterization. *Reviews of Geophysics* 2014RG000468. https://doi.org/10.1002/2014RG000468.

Khairoutdinov, M.F., and Y. Kogan. 2000. A new cloud physics parameterization in a large-Eddy simulation model of marine stratocumulus. *Monthly Weather Review* 128: 229–243.

Kogan, Yefim. 2013. A cumulus cloud microphysics parameterization for cloud-resolving models. *Journal of the Atmospheric Sciences* 70 (5): 1423–1436. https://doi.org/10.1175/JAS-D-12-0183.1.

Korolev, Alexei. 2007. Limitations of the Wegener–Bergeron–Findeisen mechanism in the evolution of mixed-phase clouds. *Journal of the Atmospheric Sciences* 64 (9): 3372–3375. https://doi.org/10.1175/JAS4035.1.

Korolev, Alexei V., George A. Isaac, Stewart G. Cober, J. Walter Strapp, and John Hallett. 2003. Microphysical characterization of mixed-phase clouds. *Quarterly Journal of the Royal Meteorological Society* 129 (587): 39–65. https://doi.org/10.1256/qj.01.204.

Lebsock, Matthew, Hugh Morrison, and Andrew Gettelman. 2013. Microphysical implications of cloud-precipitation covariance derived from satellite remote sensing. *JGR* 118 (12): 6521–6533. https://doi.org/10.1002/jgrd.50347.

Lin, Y.-L., R.D. Farley, and H.D. Orville. 1983. Bulk parameterization of the snow field in a cloud model. *Journal of Applied Meteorology and Climatology* 22: 1065–1092.
Lohmann, U., and E. Roeckner. 1996. Design and performance of a new cloud microphysics scheme developed for the ECHAM general circulation model. *Climate Dynamics* 12: 557–572.
Lohmann, U., J. Feichter, C.C. Chuang, and J.E. Penner. 1999. Prediction of the number of cloud droplets in the ECHAM GCM. *Journal Geophysical Research* 104: 9169–9198.
Lohmann, U., P. Stier, C. Hoose, S. Ferrachat, E. Roeckner, and J. Zhang. 2007. Cloud microphysics and aerosol indirect effects in the global climate Model ECHAM5-HAM. *Atmospheric Chemistry and Physics* 7 (2): 3245–3446.
Milbrandt, J.A., and M.K. Yau. 2005. A multimoment bulk microphysics parameterization. Part I: Analysis of the role of the spectral shape parameter. *Journal of the Atmospheric Sciences* 62 (9): 3051–3064. https://doi.org/10.1175/JAS3534.1.
Milbrandt, J.A., S. Belair, M. Faucher, M. Vallee, M.L. Carrera, and A. Glazer. 2016. The pan-Canadian high resolution (2.5 km) deterministic prediction system. *Weather and Forecasting* 31: 1791–1816.
Morrison, H., and A. Gettelman. 2008. A new two-moment bulk stratiform cloud microphysics scheme in the NCAR community atmosphere model (CAM3), Part I: Description and numerical tests. *Journal of Climate* 21 (15): 3642–3659.
Morrison, Hugh, and Jason A. Milbrandt. 2015. Parameterization of cloud microphysics based on the prediction of bulk ice particle properties. Part I: Scheme description and idealized tests. *Journal of the Atmospheric Sciences* 72 (1): 287–311. https://doi.org/10.1175/JAS-D-14-0065.1.
Morrison, H., J.A. Curry, and V.I. Khvorostyanov. 2005. A new double-moment microphysics parameterization for application in cloud and climate models. Part I: Description. *JAS* 62: 1665–1677.
Morrison, Hugh, Renata B. McCoy, Stephen A. Klein, Shaocheng Xie, Yali Luo, Alexander Avramov, Mingxuan Chen, et al. 2009. Intercomparison of model simulations of mixed-phase clouds observed during the ARM mixed-phase arctic cloud experiment. II: Multilayer cloud. *Quarterly Journal of the Royal Meteorological Society* 135 (641): 1003–1019. https://doi.org/10.1002/qj.415.
Neale, Richard B., C.C. Chen, A. Gettelman, P.H. Lauritzen, S. Park, D.L. Williamson, A.J. Conley, et al. 2010. *Description of the NCAR community atmosphere model (CAM5.0).* Boulder, CO, USA: National Center for Atmospheric Research.
Ose, T. 1993. An examination of the effects of explicit cloud water in the UCLA GCM. *Journal of the Meteorological Society of Japan* 71: 93–109.
Rasch, P.J., and J.E. Kristjansson. 1998. A comparison of CCM3 model climate using diagnosed and predicted condensate parameterizations. *JOC* 11: 1587–1614.
Rotstayn, L.D., B.F. Ryan, and J.J. Katzfey. 2000. A scheme for calculation of the liquid fraction in mixed-phase stratiform clouds in large-scale models. *MWR* 128 (4): 1070–1088.
Rutledge, S.A., and P.V. Hobbs. 1984. The mesoscale and microscale structure and organization of clouds and precipitation in midlatitude cyclones. XII: A diagnostic modeling study of precipitation development in narrow cold-frontal rainbands. *Journal of the Atmospheric Sciences* 41: 2949–2972.
Seifert, A., and K.D. Beheng. 2001. A double-moment parameterization for simulating autoconversion, accretion, and self-collection. *Atmospheric Research* 59–60: 265–281.
Shima, S., K. Kusano, A. Kawano, T. Sugiyama, and S. Kawahara. 2009. The super-droplet method for the numerical simulation of clouds and precipitation: A particle-based and probabilistic microphysics model coupled with a non-hydrostatic model. *Quarterly Journal of the Royal Meteorological Society* 135 (642): 1307–1320. https://doi.org/10.1002/qj.441.
Shipway, B.J., and A.A. Hill. 2012. Diagnosis of systematic differences between multiple parametrizations of warm rain microphysics using a kinematic framework. *Quarterly Journal of the Royal Meteorological Society* 138 (669): 2196–2211. https://doi.org/10.1002/qj.1913.
Slingo, A. (ed.). 1985. Handbook of the meteorological office 11-layer atmospheric general circulation model. Rep. DCTN, 29, Meteorol. Pff., Bracknell, U.K.

Song, X., and G.J. Zhang. 2011. Microphysics parameterization for convective clouds in a global climate model: Description and single column model tests. *JGR* 116 (D02201). https://doi.org/10.1029/2010JD014833.

Song, X., G.J. Zhang, and J.L.F. Li. 2012. Evaluation of Microphysics Parameterization for Convective Clouds in the NCAR Community Atmosphere Mode CAM5. J. Climate 25, no. 24 (2012): 8568–8590. https://doi.org/10.1175/JCLI-D-11-00563.

Stier, P., and others. 2005. The aerosol-climate model ECHAM5-HAM. *Atmospheric Chemistry and Physics* 5: 1125–56.

Thayer-Calder, K., A. Gettelman, C. Craig, S. Goldhaber, P.A. Bogenschutz, C.-C. Chen, H. Morrison, et al. 2015. A unified parameterization of clouds and turbulence using CLUBB and subcolumns in the community atmosphere model. *Geoscientific Model Development* 8 (12): 3801–3821. https://doi.org/10.5194/gmd-8-3801-2015.

Thompson, Gregory, and Trude Eidhammer. 2014. A study of aerosol impacts on clouds and precipitation development in a large winter cyclone. *Journal of the Atmospheric Sciences* 71 (10): 3636–3658. https://doi.org/10.1175/JAS-D-13-0305.1.

Thompson, Gregory, Paul R. Field, Roy M. Rasmussen, and William D. Hall. 2008. Explicit forecasts of winter precipitation using an improved bulk microphysics scheme. Part II: Implementation of a new snow parameterization. *Monthly Weather Review* 136 (12): 5095–5115. https://doi.org/10.1175/2008MWR2387.1.

Wetherald, R.T., and S. Manabe. 1988. Cloud feedback processes in a general circulation model. *Journal of the Atmospheric Sciences* 45: 1397–1415.

Wood, R. 2005. Drizzle in stratiform boundary layer clouds. Part II: Microphysical aspects. *JAS* 62 (9): 3034–3050.

Wilson, D.R., and S.P. Ballard. 1999. A Microphysically Based Precipitation Scheme for the UK Meteorological Office Unified Model. *Qjrms* 125: 1607–1636.

Zhang, Junhua, Ulrike Lohmann, and Philip Stier. 2005. A microphysical parameterization for convective clouds in the ECHAM5 climate model: Single-column model results evaluated at the Oklahoma atmospheric radiation measurement program site. *Journal of Geophysical Research: Atmospheres* 110 (D15): D15S07. https://doi.org/10.1029/2004JD005128.

Zhao, Q., T.L. Black, and M.E. Baldwin. 1997. Implementation of the cloud prediction scheme in the Eta model at NCEP. *Weather and Forecasting* 12: 697–712.

Microphysical Representations and Their Consistency with In Situ and Remote-Sensing Observations

Ziad S. Haddad

Abstract The different microphysical schemes developed to date represent hydrometeors using parameters whose statistics—mean values, variances, correlations, and joint probability distribution—have not received much scrutiny, let alone validation against actual observations. This leads to some very problematic inconsistencies, which are described in this chapter, along with guidance on approaches to resolve them objectively.

Keywords Hydrometeors · Cloud microphysics · Precipitation

Microphysical schemes are approximate numerical representations of the cloud processes that take place at scales on the order of a typical time step in a cloud-resolving model. These processes center on the conversion rates of water between its different phases as well as different species of condensed water, and the resulting changes to the local dynamics and thermodynamics. For condensed water, these rates and changes depend not only on the mixing ratios but also on the sizes and habits of the hydrometeors themselves. The schemes with the most realism use a bin-based representation, identifying a discrete spectrum of hydrometeor sizes (the bins) and then, accounting for the evolution of hydrometeors between the bins as well as across species. These schemes are computationally very expensive, prohibitively so for three-dimensional simulations over a significant domain. The more efficient options use bulk representations, which identify a handful of species of condensed water (e.g., small cloud water, larger cloud water, cloud ice, rain, snow, snow aggregates, graupel, hail), then assume that the characteristics of the hydrometeors within each species are entirely parametrized by a scalar size parameter, typically the hydrometeor diameter D, whose probability density function p_{D} has an a priori fixed analytic form. For nonspherical hydrometeors, the size parameter can represent either the maximum diameter or the melted equivalent diameter.

As far as the model is concerned, the bulk representation works is as follows: at a time step when the microphysical scheme is called, the scheme ingests from

Z. S. Haddad (✉)
Jet Propulsion Laboratory, California Institute of Technology, La Cañada Flintridge, USA
e-mail: zsh@jpl.nasa.gov

D. A. Randall et al. (eds.), *Current Trends in the Representation of Physical Processes in Weather and Climate Models*, Springer Atmospheric Sciences,
https://doi.org/10.1007/978-981-13-3396-5_5

the model the amount of condensed water for a given species in a resolved volume element, and distributes this amount into hydrometeors of different sizes so that the proportion of water with diameters between D_- and D_+ is given by

$$\int_{D_-}^{D_+} p_D(D)dD$$

(and in particular, when $D_- = 0$ and $D_+ = \infty$, the proportion must be 1, i.e., the integral must equal 1). In turn, the diameter is the main variable used by the scheme in the calculation of the conversion rates between different species as well as the sedimentation rate.

1 Single Moment Assumption

Single moment schemes are those where p_D is entirely determined by a single parameter, for a given species. In that case, one might as well make that single parameter the mean of the distribution, whence the designation "single moment". The most commonly used single moment form that is consistent with the observations of Marshall and Palmer (1948) is the exponential density function $p_D(D) = \Lambda exp(-\Lambda D)$, whose mean D_1 can be readily computed and seen to equal $1/\Lambda$. Whether one considers the single parameter to be Λ or D_1 is not consequential—what is, is that this type of scheme assumes a single parameter. At a given time step, once the model specifies a water content of Q g/m^3 for a given species in a given volume, this amount then has to be distributed among the possible diameters so that the two constraints are satisfied: the proportion must be given by p_D and the total mass per unit volume must add up to Q. These two constraints answer the most basic question, namely what is the total number N_T of hydrometeors per unit volume: indeed, the number of hydrometeors whose size falls within dD of a given value D must be

$$N_T p_D(D)dD$$

and their mass is $\frac{4}{3}\pi\,(D/2)^3\rho$. For liquid water, where the density ρ is independent of size, the total mass is then the integral of $4/3\ \pi\ (D/2)^3\ \rho\ N_T p_D(D)dD$, which comes to $\pi N_T \rho/\Lambda^3$. Thus,

$$Q = \pi N_T \rho/\Lambda^3$$

or equivalently

$$N_T = Q\Lambda^3/(\pi\rho) \tag{1}$$

Most working definitions use the number distribution $N(D)dD$ rather than the density function $p_D(D)dD$ to represent the hydrometeor sizes—the former is the number of hydrometeors per unit volume that are within dD of D, and the latter is the proportion. By their very definitions, they are related by $N(D) = N_T\, p_D(D)$. This product formula, combined with the derivation (in the previous paragraph) of N_T in terms of Q and Λ, shows that under the exponential distribution assumption, we have

$$N(D) = Q\Lambda^4(\pi\rho)\exp(-\Lambda D)$$

or equivalently

$$N(D) = \frac{Q}{D_1^4\pi\rho}\exp(-D/D_1)$$

with no information as to whether or how Q and Λ (or Q and D_1) may be correlated. One way to implement this representation in practice is to assume a priori that the single (first) moment D_1 is known and constant. The value of Q at any time step then completely specifies the entire number distribution. This is not the way that the representation is implemented in most schemes. Instead, most schemes generalize the conclusion of Marshall and Palmer's about the factor in front of the exponential, and assume that it is the coefficient $Q/(D_1^4\pi\rho)$ that is constant. Marshall and Palmer had based that assumption on their analysis of a very particular small set of data collected in Canada during the summer of 1946 (Marshall and Palmer 1948). This coefficient is almost universally referred to as the vertical axis intercept "N_0" of the distribution (as it is indeed the value $N(0)$ of N when $D = 0$).

The implications of these seemingly straightforward assumptions, that N_0 is constant or (instead) that D_1 is constant, are important and vastly different. The former is equivalent to postulating that

$$D_1 = \text{constant}\, Q^{0.25}$$

that is, that the mean diameter is an increasing function of water content—more specifically, a 1/4 power law. The latter is equivalent to assuming that the mean diameter is constant. Under either assumption, Eq. (1) shows that the total number of hydrometeors grows with Q, linearly in the case of constant D_1 (because Λ is then constant), or as a 1/4 power law in the case of constant N_0. Neither assumption is consistent with the possibility that the total number of hydrometeors does not grow with Q at all, but instead remains constant—as it would need to be if the only process governing the conversion between water species were accretion without any breakup or collision/coalescence, etc.

These vastly different implications, and the fact that they are inconsistent with basic process properties and observations, highlight the need to characterize the joint behavior of the parameters used in the microphysical representation, to ensure that the simplifying assumptions that will reduce the dimensionality of the representation are consistent with the characteristics of observed microphysical distributions. This need will be discussed in the more general context of double-moment representations, thought to allow more realism in the microphysical representation.

2 Double-Moment Assumption

Double-moment schemes are those that allow two parameters for the description of p_{D} for a given hydrometeor species. The most straightforward two-moment extension of the single moment exponential form is the Γ form:

$$p_D(D) = cD^{\mu} \exp(-\Lambda D)$$

the constant c necessarily being the value that makes p_{D} integrate to 1, i.e., $c = \Lambda^{(\mu+1)}/\Gamma(\mu + 1)$. As in the single moment case, if N_T is the total number of hydrometeors per unit volume, the number of hydrometeors whose size falls within dD of a given value D must be $N_T\ p_{\mathrm{D}}(D)\ dD$, whose total mass is the integral of $\frac{4}{3}\pi(D/2)^3 N_T p_D(D)dD$, which in this double-moment case is

$$Q = \pi N_T \rho(\mu + 3)(\mu + 2)(\mu + 1)/\left(6\Lambda^3\right)$$

(and one can verify that if μ is 0, this would reduce exactly to the single moment equation 1). The first moment, the mean (diameter) D_1, is $\int Dp_D(D)dD/\int p_D(D)dD$, which equals $(\mu+1)/\Lambda$ in this case

$$D_1 = (\mu + 1)/\Lambda \tag{2}$$

Instead of the literal first moment itself, one can consider replacing it by the mass-weighted mean D_m, which equals

$$\int DD^3 p_D(D)dD/\int D^3 p(D)dD, \text{ i.e. } (\mu + 4)/\Lambda.$$

As to the second moment, one can similarly consider the diameter standard deviation σ, which equals

$$D_1/(\mu + 1)^{1/2}$$
$$\sigma = D_1/(\mu + 1)^{1/2} \tag{3}$$

or the mass-weighted standard deviation σ_m of the diameter, which equals $D_m/(\mu + 4)^{1/2}$.

Thus, in this case, the number distribution $N(D)dD$ can be described with different triples of parameters. To describe these options without losing generality, one might as well replace the assumption that the density is constant, with the assumption that the mass is a power law in the diameter, mass $(D) = \alpha D^{\beta}$ with β possibly different from 3. Under this assumption, the simplest representation, esthetically, is

$$N(D) = N_0 D^{\mu} exp(-\Lambda D) \quad (4)$$

where the three parameters are μ, Λ, and

$$N_0 = N_T \Lambda^{\mu+1} / \Gamma(\mu + 1), \quad (5)$$

i.e.,

$$N_0 = Q \Lambda^{\mu+\beta+1} / \Gamma(\mu + \beta + 1) \quad (6)$$

A second, different option for the triple of parameters that one can use, can be obtained from (4) by writing μ and Λ as

$$\mu = \left(\frac{D_1}{\sigma}\right)^2 - 1 - \beta$$
$$\Lambda = D_1/\sigma^2,$$

so that one can then use the zeroth, first, and second moments, (N_T, D_1, σ) instead of (N_0, μ, Λ) as the parameters of the representation (and the relation between the two sets of parameters is given by the bijective set of relations 5, 2, 3). Similarly, one can replace the first and second moments with their mass-weighted counterparts to obtain a third alternate triple of parameters. Perhaps more practically, from the perspective of the microphysical scheme, one would want to use the water content Q itself as one of the three parameters, since the model has a dynamical evolution equation for Q, and one would then need to identify the two additional parameters to be used in the representation.

This frames the question of which parameter set to use in the context of the application. Because the model does calculate and track the evolution of Q in space and time, it makes sense to use Q as one of the parameters. And because it would be very difficult to postulate a dynamical equation for the evolution of the unphysical parameters N_0, or μ, using either would require one to make some assumption such as spatial and temporal constancy, and this, in turn, would be very dangerous if the supposedly constant parameter turns out, in reality, to be highly correlated with the highly variable Q, or if its constancy assumption turns out to have some unverified implications.

What implicit correlations are there in the simplest assumptions made by the schemes developed to date? To answer this question, consider Eq. (6). To better understand the right-hand side, note that under the general power law mass-diameter relation mass $(D) = \alpha D^{\beta}$, the mass-weighted mean diameter D_m is given by

$$D_m = \int D D^{\beta} p_D(D) dD / \int D^{\beta} p_D(D) dD = (\mu + \beta + 1)/\Lambda \quad (7)$$

so that (6) can be rewritten as

$$N = Q \cdot \frac{1}{\alpha} \cdot (\mu + \beta + 1)^{\mu+\beta+1} / \left[D_m^{\mu+\beta+1} \cdot \Gamma(\mu + \beta + 1)\right] \qquad (6')$$

It is now straightforward to see that if a scheme assumes that N_0 and μ are constant, the scheme is implicitly accepting a very specific relation between D_m and Q, namely

$$D_m = \text{constant } Q^{1/(\beta+\mu+1)}$$

If the scheme were to assume, instead, that N_T is constant (instead of N_0), a similar derivation shows that the scheme is then implicitly accepting a different relation between D_m and Q, namely

$$D_m = \text{constant } Q^{1/\beta}$$

The conclusion is that the assumption that N_0 (or N_T) is constant is equivalent to assuming a specific value for the exponent in the corresponding power law relation between D_m and Q. The exponent corresponding to the assumption that N_0 is constant is different from the exponent corresponding to the assumption that N_T is constant. As we shall see in the next subsection, neither is consistent with in situ observations. The "N_T constant" assumption has a physical interpretation, namely that the condensation is by accretion only (the total number of hydrometeors remaining fixed), whereas "N_0 constant" assumption does not have a straightforward physical justification.

As to the search for any empirical evidence justifying the assumption that N_0 is constant, one should cite the extensive statistics compiled by Ulbrich and Atlas (1998). The results are illustrated in Fig. 1, showing the observed strong correlation in the scatter diagrams between μ and N_0. Note that when $\mu = 0$, the corresponding values of $\log_{10}(N_0)$ span a range whose width is greater than 3, clearly showing that N_0 varies by over 3 orders of magnitude. Manifestly, empirical evidence strongly contradicts the hypothesis that N_0 is constant.

3 In Situ Observed Correlations

Two types of in situ observations can be considered to evaluate the microphysical scheme assumptions. One type is distrometer observations, and another is rain gage observations that are nearly simultaneous and nearly coincident with high-spatial-resolution ground radar backscatter measurements.

Radar reflectivity from precipitation has long been used as a measurement directly sensitive to precipitation amount and hydrometeor size. This relation is easiest to study when one can approximate the radar backscattering cross section $\sigma^0(D)$ of a drop of diameter D by a power law

$$\sigma^0(D) \sim \text{a } D^b \qquad (8)$$

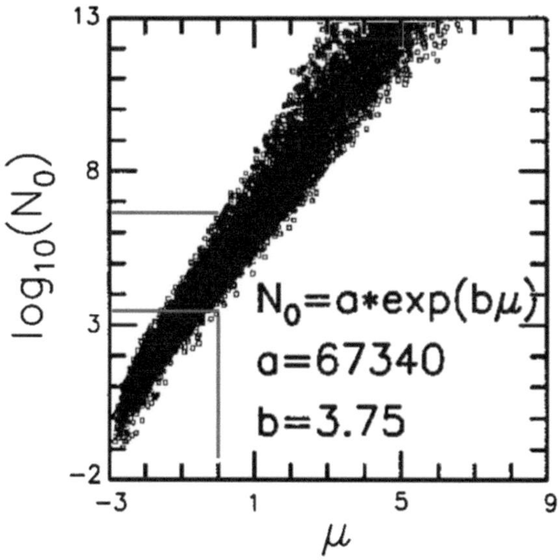

Fig. 1 Excerpt from the results of Ulbrich and Atlas (1998), showing the correlation between the parameters μ (horizontal axis) and N_0 (vertical axis in logarithmic scale) for rain observed during the TOGA-COARE campaign with the airborne "Knollenberg" 2D precipitation probe mounted on the NCAR Electra aircraft. Note that, the three orders of magnitude spanned by N_0 when $\mu = 0$

a relation that applies for longer wavelengths lambda where the Rayleigh approximation is valid. Non-Rayleigh effects become significant only when D/Λ exceeds about 0.05, when resonances can arise within the hydrometeor. Since surface tension constraints confine realistic drop diameters to remain below about 8 mm, one can assume that the Rayleigh approximation applies to all drops if $8/\Lambda < 0.05$, i.e., if $\Lambda > 16$ cm. For such longer wavelengths, (8) is a reasonable approximation, and therefore, the radar reflectivity factor $Z = \int \sigma^0(D)N(D)dD$ can be rewritten using (8) as

$$Z = \mathrm{a}\Gamma(\mu + \beta + 1)/\Lambda^{\mu+\beta+1} N_0$$

which, using (7), is equivalent to

$$Z = \mathrm{a}/\alpha\Gamma(\mu + b + 1)/\,\Gamma(\mu + \beta + 1)[1/(\mu + \beta + 1)^{b-\beta} D_m^{b-\beta}]Q \tag{9}$$

This expression for the dependence of the radar reflectivity factor on the condensed water content and the mass-weighted mean diameters is very useful, because it allows one to evaluate directly the consequences of any constancy assumption about the parameters of the microphysical representation. Indeed, the previous section showed that the assumption that N_0 and μ are constant implied that D_m = constant $Q^{1/(\beta+\mu+1)}$. Putting this together with (9) implies that

$$Z = \text{constant}\ Q^{(b+\mu-1)/(\beta+\mu+1)}$$

(so for the typical values $\mu=2$, b ~ 6, and β ~ 3, this would say that Z ~ $Q^{1.17}$), whereas the assumption that N_T and μ are constant, together with (9), implies that

$$Z = \text{constant}\ Q^{b/\beta}$$

(so for the typical values b ~ 6 and β ~ 3, this would say that Z ~ Q^2). More generally, any assumption on the microphysical parameters that leads to a correlation power law between D_m and Q of the form $D_m = \delta\ Q^{\gamma}$ implies that

$$Z = \text{constant}\ Q^{1+\gamma(b-\beta)}$$

These three possible relations between Z and Q can be compared to the numerous so-called Z–R relations (R being the rain rate) that have been derived for different regimes from nearly coincident and nearly simultaneous radar and rain gage obser-

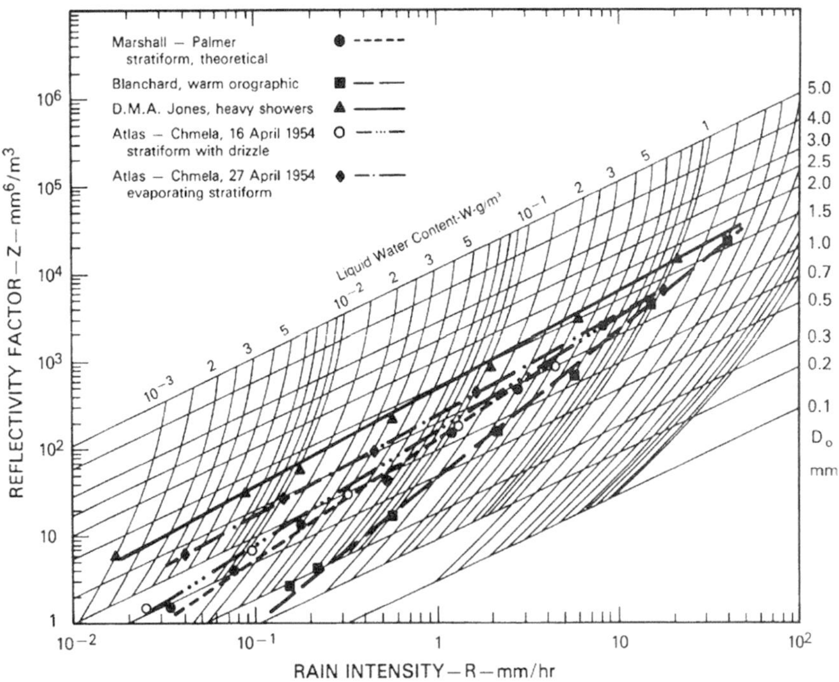

Fig. 2 Excerpt from the results of Atlas (1990), showing the different coefficients a (vertical axis intercepts) and exponents b (slopes) of the nominal $Z = aR^b$ fits obtained in different analyses

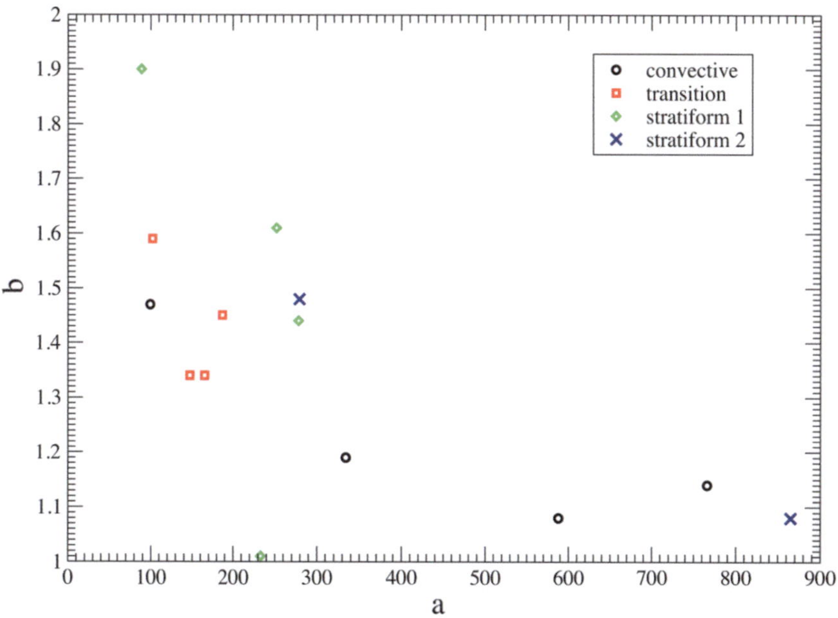

Fig. 3 Illustration of the results of Atlas et al. (1999), showing the different coefficients a and exponents b of the nominal $Z = aR^b$ fits obtained from analyses of distrometer data obtained at Kapingamarangi atoll (in the Federated States of Micronesia) after rain type classification

vations. Figure 2 illustrates the range of the resulting coefficient–exponent pairs, with a significant spread about the nominal $Z = 300\, R^{1.4}$ and $Z = 200\, R^{1.6}$ relations used by the US National Weather Service. Distrometer observations can also be used to calculate simultaneously the parameters of the size distribution and the rain rate, as well as the expected values of radar reflectivity, so that power law fits can then be made between the calculated Z and R derived from the distrometer observations. Perhaps, most interesting in this approach is the analysis of Atlas (1990), showing a wide range of values in the coefficient and exponent of the power law even after classification of the distrometer samples according to rain type (see Fig. 3).

As to the distrometer observations, some of the analyses of Ulbrich and Atlas have already been cited above, and clearly show strong empirical correlations between the parameters of the Gamma distribution. A recent comprehensive study by Williams et al. (2014) strongly supports a power law relation between mass-weighted first and second moments D_m and σ_m for liquid rain,

$$\sigma_m = 0.3 D_m^{1.36}$$

along with power law relations between D_m and the rain rate R, namely $D_m = 0.956 R^{0.32}$(see Fig. 4). The main use of such relations is not only to allow one to replace correlated parameters, in this case (D_m, σ_m), with uncorrelated ones, in

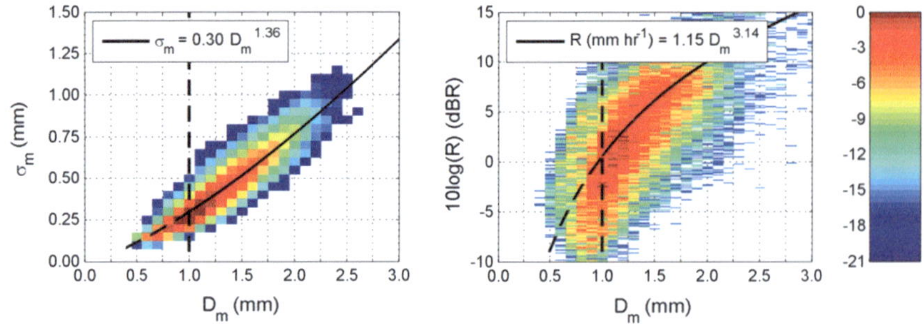

Fig. 4 Excerpt from the results of Williams et al. (2014), showing the correlations between σ_m and D_m, and between D_m and the rain rate R itself, from analysis of a large dataset of distrometer observations made near Huntsville, Alabama

this case $\left(D_m, \sigma_m/\left[0.3D_m^{1.36}\right]\right)$. Rather, it is to establish how the assumption that one parameter, say N_0, is constant, which it very well may be over a confined location and short time interval, necessarily leads to a predictable confined range of empirically realistic values for the other parameters to which it is correlated. For example, earlier analyses of different tropical datasets, obtained from ground distrometers as well as flight-level airborne probes (Haddad et al. 1996 and 1997), had established power law relations between σ_m, D_m, and Q, which were subsequently used to constrain single- (respectively dual-) frequency radar retrievals of one (respectively two) underlying unknown(s), namely the rain rate (respectively, the rain rate and the mass-weighted mean drop size), without violating the empirically observed correlations in the joint behavior of the microphysical parameters.

More recent campaigns at different latitudes and seasons have illustrated the range of variability of these power laws, especially as one considers solid hydrometeors in different regimes. Figure 5 compares the joint distributions of Q, D_m, and σ_m, obtained during the GPM Cold-Season Precipitation Experiment (GCPEX) with those obtained during the tropical campaigns Kwajex and TOGA-COARE. Figure 6 illustrates the joint behavior of D_m and σ_m in GCPEX for different microphysical species.

4 Consistency with Satellite Observations

In addition to ground and airborne measurements, satellite observations can be used to evaluate the empirical consistency of microphysical assumptions. Indeed, top-of-the-atmosphere microwave radiances measured from space over clouds are sensitive to the upwelling microwave radiation and its modulation by the hydrometeors within the radiometer field of view. The modulations are due to two fundamental mechanisms, absorption/emission, and scattering. The first is the process by which upwelling radiation is absorbed and re-emitted by the hydrometeors, while the second represents the scattering of the radiation out of the radiometer beam, with a resulting systematic

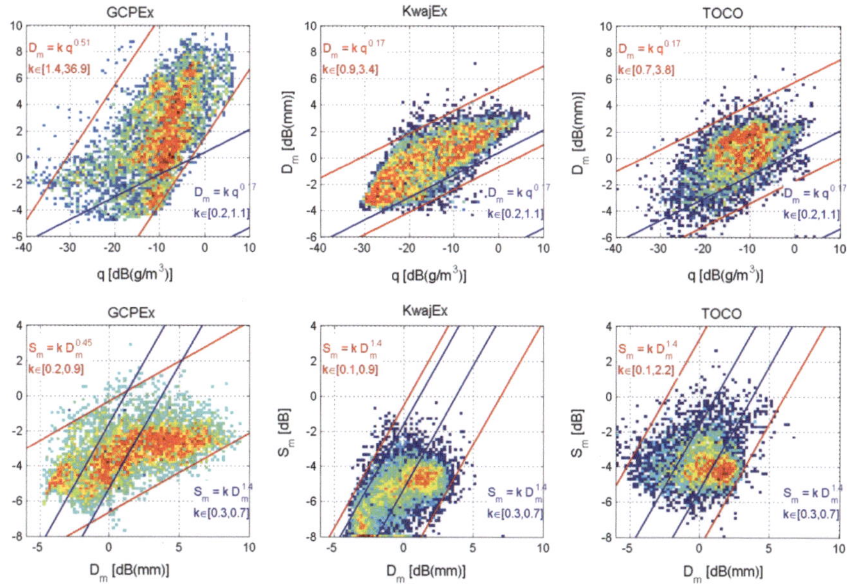

Fig. 5 Sample joint distributions of Q and D_m (top row) and σ_m and D_m (bottom row) for three campaigns, from left to right GCPEx, KwajEx, and TOGA-COARE

reduction in the measured brightness temperature. The higher the microwave frequency, the greater the attenuation due to the absorption by hydrometeors at lower levels—implying that higher frequencies will be less sensitive to the complex effects of hydrometeors from deep within the cloud, and will instead be more sensitive to the systematic cooling by hydrometeors at the upper levels. Because we now have several years' worth of higher frequency microwave observations over clouds in the tropics, it is quite reasonable to compare these observations with those that one can synthesize from model simulations with different microphysical schemes, to compare the joint behavior of the synthetic radiances (at different microwave frequencies, reflecting the corresponding differences in the hydrometeor scattering signatures) with those actually observed.

To that end, a database of coincident observations made by the Tropical Rainfall Measuring Mission (TRMM) Precipitation Radar (TRMM-PR) and the cross-track-scanning Microwave Humidity Sounder radiometer with channels at 89, 157, 183.3 ± 1, 183.3 ± 3, and 190.3 GHz, was compiled between 2009 and 2010 in a mostly tropical band between 25°S and 25°N. For each radiometer beam, the radar beam whose surface footprint center was closest to the radiometer beam center was identified, then the 11×11 neighboring radar beams, and finally the radar reflectivities in only those beams whose 6 dB footprints fell entirely within the ellipse approximating the radiometer footprint was averaged (without accounting for the relatively small vertical change due to the increasing slant in the radiometer beams as the radiometer scan moves toward the outer edges of the swath). All cases where

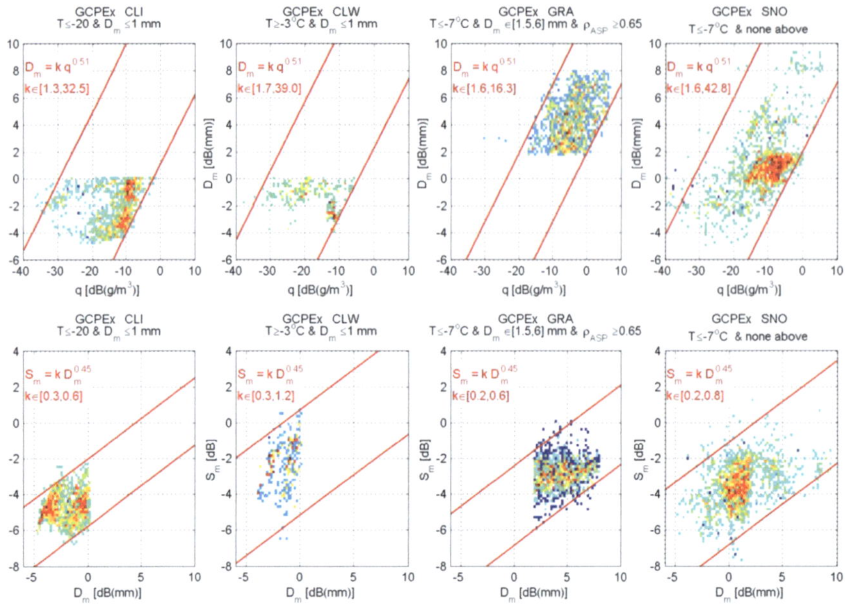

Fig. 6 GCPEx joint distributions of the moments of different hydrometeor species, from left to right cloud ice, cloud liquid, graupel, and snow. The top row shows the joint distribution of (Q, D_m), the bottom row that of (D_m, σ_m)

the two instruments observed within 3 min of one another were kept. This produced 79,090 MHS beams which the radar classified as "rainy". A Bayesian analysis was then performed to (1) derive the formulas, in five-dimensional space, for the principal components (PCTb1, ..., PCTb5) of the five simultaneous rainy radiances (in the five channels), based on their empirical covariance, and (2) estimate, from the top two radiometer principal components, the top two vertical principal components (PCQ1, PCQ2) of the vertical distribution of water content Q, as determined by the radar (basically, accepting the radar as "truth"). Figure 7 illustrates the result, showing the joint distribution of (PCTb1, PCTb2) as well as the estimated value of PCQ1 for each discretized (PCTb1, PCTb2) pair. The obvious question is: to what extent can a model microphysical scheme reproduce this joint behavior?

The answer, of course, depends on the storms being simulated. It is prohibitively resource-intensive to attempt to simulate 2 years' worth of tropical storms with microphysics-resolving accuracy for any single scheme, let alone several. Nevertheless, one can start by simulating a "typical" tropical storm, with different microphysical representations, then forward calculate the brightness temperatures for each representation, apply the condensed joint distribution analysis as in the previous paragraph, and compare the results. This is exactly what was done in Haddad et al. (2017). WRF was used to simulate the tropical depression that gave rise to Hurricane Isabel in 2003. The model was configured with five nested grids, with respective horizontal

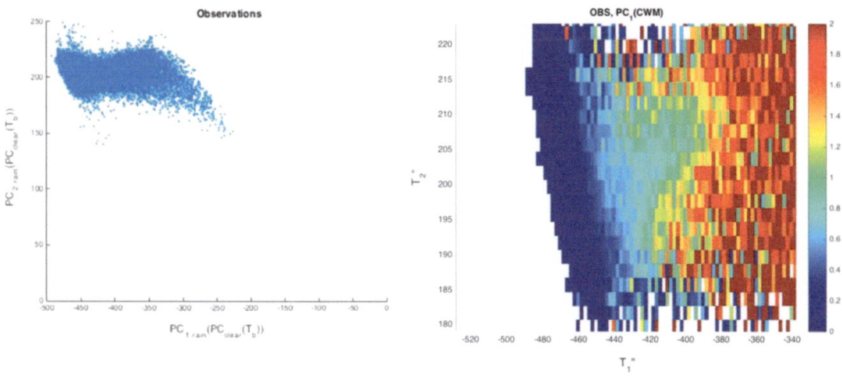

Fig. 7 Left panel: scatterplot of the pair of derived sounder observation made up of the top two rainy brightness temperature principal components PCTb1 and PCTb2, applied to the MHS observations over the tropics from 2009 to 2010. Right panel: Zoom in on the area of the joint observations in the left panel, with the colors representing the conditional mean of the first vertical principal component of the condensed water mass, conditioned on the pair (PCTb1, PCTb2)

resolutions of 12 km (for the outer grid), 4 km, 1.333 km (the "d03" grid), 444 m, and 148 m. The d03 grid roughly spanned a domain from longitudes 36°W to 26°W and from latitudes 6°N to 16°N. Initializing WRF with the GFS FNL analysis on September 5, 2003 at 12:00 UTC, WRF output was collected every 10 s from 06:00 to 06:05 on 6 September, then every 30 s from 06:05 to 07:00 UTC. This output was ingested by the SOI forward radiative transfer model (Heidinger et al. 2006) to calculate MHS brightness temperatures. The results are illustrated in Fig. 8. Rather than dwell on the discrepancies in the results, the remainder of this subsection describes how the one representation—call it MC—that produces the results (on a single storm) that are most consistent with the (two years' worth of) observations, was constructed, mainly to build into it the microphysical correlations derive from in situ observation as described in the previous sections.

Because the high-frequency microwave radiances are most sensitive to the solid condensation aloft, the representation is specifically concerned with accurately reflecting the correlations in snow and aggregates. The starting point is the synthetic hydrometeors created by Kuo et al. (2016). Using Gravner and Griffeath's Snowfake (2009), about 10,000 realistic solid hydrometeors were constructed, including both pristine and aggregate types, with maximum diameter ranging from about 100 μm to about 15 mm. Then, using the discrete dipole approximation code DDSCAT, the microwave signatures of each individual hydrometeor (extinction efficiency, absorption efficiency, scattering efficiency, scattering asymmetry parameter, and radar backscattering cross section) were calculated at the different relevant frequencies, and incorporated into a database consisting of one entry for each synthesized hydrometeor, specifying the hydrometeor's defining physical characteristics and the associated radiometric signatures. The representation problem thus becomes one of

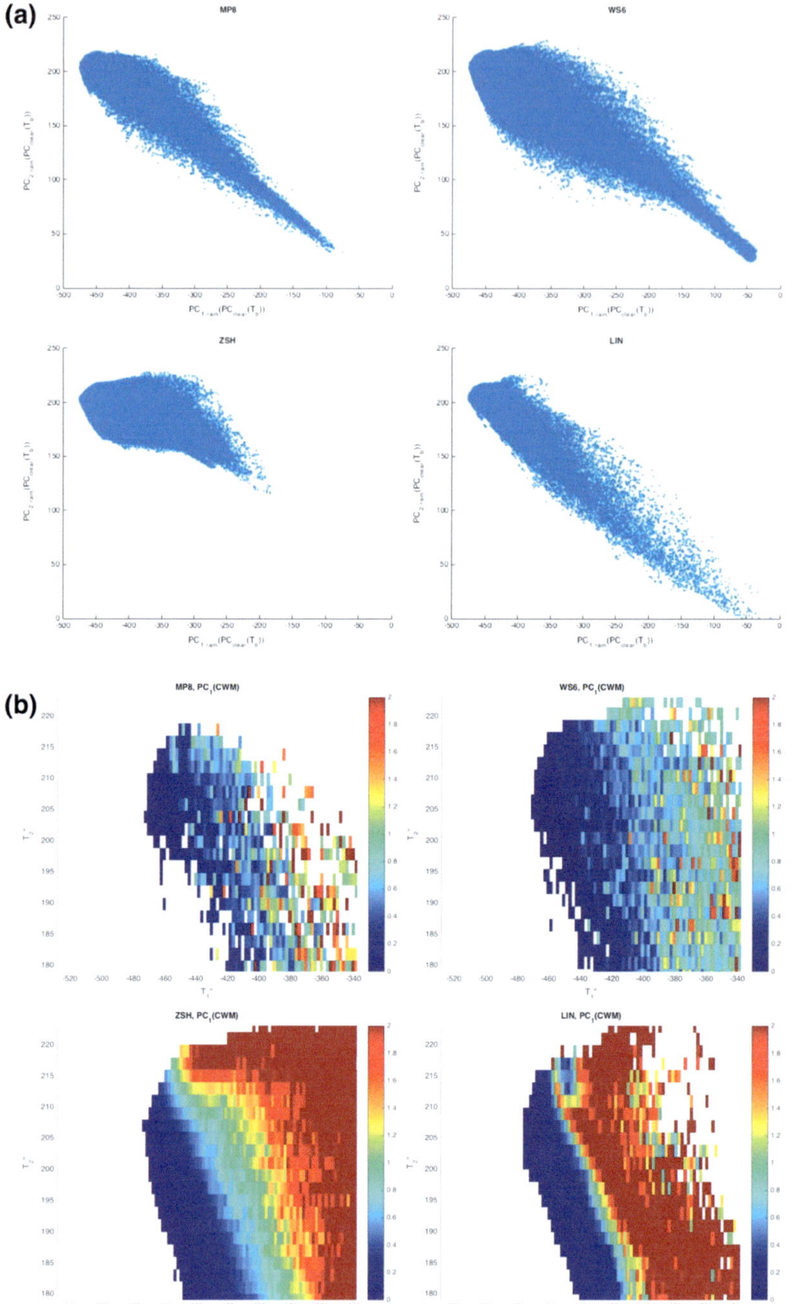
(a)
MP8
WS6
ZSH
LIN
(b)
MP8, PC1(CWM)
WS6, PC1(CWM)
ZSH, PC1(CWM)
LIN, PC1(CWM)

◀**Fig. 8** **a** Scatterplots as the left panel of Fig. 7 but for the WRF simulation of pre-Hurricane Isabel, with brightness temperatures calculated using four different microphysical representations. Note how the scatterplot in the lower left panel is most consistent ("MC") with the left panel of Fig. 7. **b** Zoom in on the area of the joint synthetic observations of figure (**a**), showing the conditional mean of the condensed water mass first principal component (defined by the empirical observations as for Fig. 7) but calculated here for the simulated data with the four different microphysical representations. Note that, MC (the lower left panel) is most consistent with the right panel of Fig. 7

deciding how to assemble these hydrometeors into realistic "candidate" distributions of hydrometeors, and how to parametrize these distributions.

The requirements are that the distribution is efficiently parametrized (to allow one to estimate the parameters from remote-sensing observations) and that the parameters reflect the correlations that have been empirically determined to exist from in situ observations. The approach to do this is inspired by the solution of the far simpler case of spherical hydrometeors, where the single parameter that differentiates the hydrometeors is their diameter. In order for this approach to be applicable to the more complex case of distributions of solid hydrometeors, one needs to address the fact that the latter are not spherical and indeed come in different habits whose characterization requires more than a single "diameter" variable. A. Heymsfield's extensive analyses and literature reviews (Heymsfield et al. 2010; Schmitt and Heymsfield 2010) point to a range of power law relations between the maximum dimension D_{M} of individual observed hydrometeors and each hydrometeor's mass m, and these power laws do appear to characterize the particle habit—the power being the fractal dimension of the hydrometeors in question.

For microphysical representations, one needs to represent not just the mean relation, but rather the spread about this mean too. Therefore, analytically, the representation MC assumes that

$$m = aD_M^b \tag{10}$$

where b is the fractal dimension characterizing the type of solid hydrometeor (Schmitt and Heymsfield 2010), and where the ranges of values that a and b can take are

$$0.004 < a < 0.009 \text{ and } 1.8 < b < 2.4 \tag{11}$$

Rather than force possibly unrealistic constraints a priori, the first step in the construction of MC starts with the database of individual synthetic hydrometeors (pristine and aggregates), loops over combinations of values of (a, b) as above, and chooses, for each empirical power law all the individual hydrometeors that obey that fractal power law—specifically every hydrometeor for which $\left|aD_M^b/m - 1\right| \leq 0.1$ is retained. Thus, the first step avoids any a priori assumptions about the correlation between the coefficient a and the fractal dimension b (realistic constraints will be applied in the final step 3 of the construction). To obtain distributions that are representative of the mean relations in Heymsfield's analyses the values, the values a = either 0.003, 0.004, 0.005, 0.006, 0.007, 0.008, 0.009, or 0.01 are used with b

= either 1.8, 2, 2.2, or 2.4 independently, knowing that not all of the $8 \times 4 = 32$ (*a*, *b*) pairs will survive the consistency test in step 3. Still, at the end of step 1, each hydrometeor is classified as having one of the 32 habits parametrized by (10) × (11). The second step consists in associating, with each collection of hydrometeors, a count for each hydrometeor representing the relative frequency of that hydrometeor versus the others in the collection. This is akin to the Γ distribution for spherical raindrops, and the second step is indeed implemented by considering different Γ distributions with different shape parameters μ and Λ chosen from a priori independent uniform distribution, in each of the 32 collections obtained in step 1. By the end of this second step, each one of the (*a*, *b*) pairs from step 1 will have produced M × L candidate distributions, M and L being the number of values of μ and Λ that were used—the third parameter, the overall multiplier N_0, only comes into play in the third (and final) step.

Indeed, step 3 consists in sifting through all the candidate distributions created in step 2 and retaining only those distributions that meet the observed relations between Q, D_m and σ_{m}—more specifically, the D_m–Q and D_m–σ_{m} relations derived in Turk et al. (2011)

$$D_m = \kappa Q^{0.17} \text{ and } \sigma_m = \lambda D_m^{1.3} \text{ mmm}$$

where the coefficients κ and λ turned out to lie in the ranges

$$0.2 < \kappa < 2 \text{ and } 0.15 < \lambda < 0.6$$

Pending analysis of data from different regimes, step 3, therefore, consists in calculating, for every candidate distribution produced by step 2, the moments (Q, D_m, σ_m), and removing those candidate distributions whose ratios $D/Q^{0.17}$ and $\sigma_m / D_m^{1.3}$ are not within the observed ranges. This effectively removes distributions whose (*a*, *b*) pair was not realistic to begin with, or whose Γ-distribution parameters are not consistent with the bulk correlations derived from the relevant in situ observations. The main advantage of this approach is that it leverages what reliable in situ observations one has (of the correlation between the moments of the observed hydrometeor distributions) to produce a realistic effective parametrization. Figure 9 confirms the potential of this approach and illustrates its issues. It shows the maximum D_{M} of the members of our input single hydrometeor database, plotted against their mass (or, actually the melted equivalent sphere's diameter). The curves show a sampling of the (*a*, *b*) power laws proposed above, the pink curves corresponding to $a = 0.003$ (with $b = 1.8$, 2.0, 2.2, and 2.4 from left to right), the green to $a = 0.005$, the red to $a = 0.007$, and the single blue curve to $a = 0.009$ with $b = 2.4$. Note that the simulated hydrometeors are generally consistent with the proposed relation, though they do appear to be biased toward the denser side, implying that inclusion of aggregates that are synthesized with larger center-to-center aggregation distance is still desirable. Conversely, the simulation-derived relations do appear generally consistent with the simulated 3D hydrometeors, though the smaller fractal coefficients may well be the

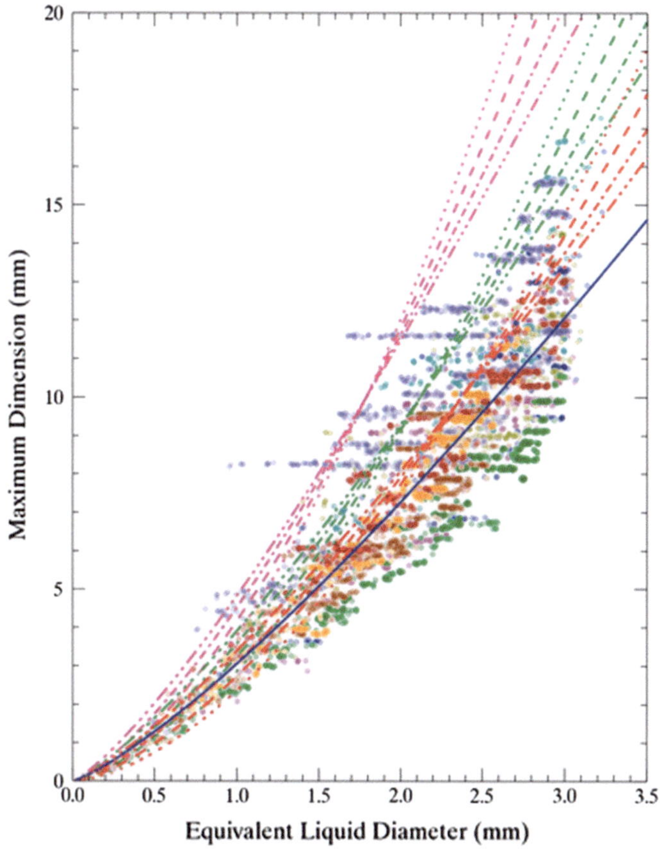

Fig. 9 Illustration of the relations between hydrometeor mass (replaced here by the melted equivalent diameter, on the horizontal axis) and hydrometeor maximum apparent dimension (on the vertical axis). The parametrized power law relations (the curves) are superposed on the scatter diagram calculated for the synthetic hydrometeors in the Kuo database

result of a too generous 2D to 3D fractal dimension scaling factor (the factor "S" in Schmitt and Heymsfield 2010).

Most importantly, this approach illustrates how replacing ad hoc assumptions (such as "N_0 constant") by simplifying assumptions based on in situ observations of the joint behavior of the microphysical moments, can indeed lead to a drastic improvement in the realism of one's microphysical representation, as evidenced by comparisons with hydrometeor-sensitive satellite microwave observations.

Acknowledgements This work was performed at the Jet Propulsion Laboratory, California Institute of Technology, under contract with the National Aeronautics and Space Administration.

References

Atlas, D. 1990. In *Radar in Meteorology—Battan memorial and 40th Anniversary Radar Meteorology Conference*, ed. D. Atlas, 86–97.

Atlas, D., C.W. Ulbrich, F.D. Marks, E. Amitai, and C.R. Williams. 1999. Systematic variation of drop size and radar-rainfall relations. *Journal Geophysical Research* 104: 6155–6169.

Gravner, J., and D. Griffeath. 2009. Modeling snow-crystal growth: A three-dimensional mesoscopic approach. *Physical Review E* 79: 011601. https://doi.org/10.1103/physreve.79.011601.

Haddad, Z.S., S.L. Durden, and E. Im. 1996. Parametrizing the raindrop size distribution. *Journal of Applied Meteorology and Climatology* 35: 3–13.

Haddad, Z.S., D.A. Short, S.L. Durden, et al. 1997. A new parametrization of the rain drop size distribution. *IEEE Trans. Geosci. Rem. Sens.* 35: 532–539.

Haddad, Z.S., R.C. Sawaya, S. Kacimi, O.O. Sy, F.J. Turk, and J. Steward. 2017. Interpreting millimeter-wave radiances over tropical convective clouds. *Journal of Geophysical Research A* 122: 1650–1664. https://doi.org/10.1002/2016jd025923.

Heidinger, A.K., C. O'Dell, R. Bennartz, and T. Greenwald. 2006. The Successive-Order-of-Interaction radiative transfer model. Part I: Model development. *Journal of Applied Meteorology and Climatology* 45: 1388–1402.

Heymsfield, A.J., C. Schmitt, A. Bansemer, and C.H. Twohy. 2010. Improved representation of ice particle masses based on observations in natural clouds. *Journal of the Atmospheric Sciences* 67: 3303–3318.

Kuo, K.-S., W.S. Olson, B.T. Johnson, M. Grecu, L. Tian, T.L. Clune, B.H. van Aartsen, A.J. Heymsfield, L. Liao, and R. Meneghini. 2016. The microwave radiative properties of falling snow derived from nonspherical ice particle models. Part I: An extensive database of simulated pristine crystals and aggregate particles, and their scattering properties. *Journal of Applied Meteorology and Climatology* 55: 691–708. https://doi.org/10.1175/jamc-d-15-0130.1.

Marshall, J.S., and W.M.K. Palmer. 1948. The distribution of raindrops with size. *Journal of Meteorology* 5: 165–166.

Schmitt, S.G., and A.J. Heymsfield. 2010. The dimensional characteristics of ice crystal aggregates from fractal geometry. *Journal of the Atmospheric Sciences* 67: 1605–1616.

Turk, F.J., K.-W. Park, Z.S. Haddad, P. Rodriguez, and D. Hudak. 2011. Constraining CloudSat-based snowfall profiles using AMSU/MHS and C-band ground radar. *Journal Geophysical Research* 116: D23205. https://doi.org/10.1029/2011jd016126.

Ulbrich, C.W., and D. Atlas. 1998. Rainfall microphysics and radar properties: Analysis methods for drop size spectra. *Journal of Applied Meteorology and Climatology* 37: 912–923.

Williams, C.W., V.N. Bringi, L.D. Carey, V. Chandrasekar, P.N. Gitlin, Z.S. Haddad, R. Meneghini, S.J. Munchak, S.W. Nesbitt, W.A. Petersen, S. Tanelli, A. Tokay, A. Wilson, and D.B. Wolff. 2014. Describing the shape of raindrop size distributions using uncorrelated raindrop mass spectrum parameters. *Journal of Applied Meteorology and Climatology* 53: 1282–1296.

Convective Available Potential Energy and Precipitation in a Cloud-Resolving Model Simulation of Indian Summer Monsoon

Deepeshkumar Jain, Arindam Chakraborty and Ravi S. Nanjundiah

Abstract Relationship between convective available potential energy (CAPE) and precipitation is explored in a season-long cloud-resolving model (CRM) simulation of Indian summer monsoon. The location of maximum precipitation and CAPE does not always coincide in a CRM simulation. The diurnal land surface heating is shown to have an effect on CAPE and precipitation over ocean. Convective inhibition energy is shown to have a significant effect on the location of precipitation. It is shown that mass flux parameterizations which depend on CAPE consumption do not get the location or magnitude of precipitation right at CRM resolution. It is emphasized that once the model resolution starts approaching cloud scale, the basic assumption of convective quasi-equilibrium is not sufficient and representation of organized mesoscale convective systems becomes imperative. Present-day cumulus parameterizations do not include any representation of organized mesoscale convective systems. We show that CAPE consumed by these systems not only triggers vertical motion but also contributes to horizontal motion of the system.

Keywords CRM · CAPE · CINE · LCL · LFC · MCS

1 Introduction

South Asian summer monsoon (monsoon hereafter) precipitation contributes to more than 80% of the total annual precipitation over the Indian region. A large percentage of Indian population depends on monsoon precipitation for their agricultural

D. Jain (✉) · A. Chakraborty · R. S. Nanjundiah
Indian Institute of Science, Bangalore, India
e-mail: deepeshkumar@tropmet.res.in

D. Jain · R. S. Nanjundiah
Institute of Tropical Meteorology, Pune, India

D. A. Randall et al. (eds.), *Current Trends in the Representation of Physical Processes in Weather and Climate Models*, Springer Atmospheric Sciences,
https://doi.org/10.1007/978-981-13-3396-5_6

water supply. The importance of correct monsoon precipitation prediction cannot be overemphasized. The monsoon is a remarkably regular annual phenomenon characterized by seasonal reversal in the direction of the wind. However, it does show a considerable variability in time and space (Gadgil 2003).

Present-day general circulation models (GCMs) usually have negative bias over land and positive bias over ocean with respect to observed precipitation over Indian subcontinent (Sperber et al. 2013). In a GCM, convective parameterizations determine the amount of precipitation produced by the model. Convective parameterizations also determine the amount and intensity of sub-grid-scale convective clouds in a model. Understanding the parameterizations which produce clouds and precipitation is central to improving the prediction. Clouds play significant role in the redistribution of sensible and latent heat and momentum. They have significant impact on the radiative balance of the earth. They also modify planetary boundary layer by influencing boundary layer convergence and saturated downdrafts which enter boundary layer and results in cooling and drying (Arakawa 2004). Convection parameterization is one of the most challenging problems in a GCM. Many assumptions are used in convection parameterization such as lateral entrainment/detrainment, convective trigger, microphysics, rainwater conversion, closure, cloud area compared to grid size, etc. (Arakawa 2004). As the model resolution increases, most of these assumptions need to change.

One of the least understood and probably the one having the most impact on simulations is the closure assumption. The closure assumption determines how much convection occurs in a given large-scale condition. Low-level moisture convergence (Tiedtke 1989) and convective available potential energy (Arakawa and Schubert 1974) are two of the most widely used quantities as basis for closure assumption. Majority of the present-day cumulus parameterizations in GCMs use mass flux scheme as a cloud model and convective quasi-equilibrium (CQE) (Yano and Plant 2012) as a closure assumption. CQE relates the rate of change of CAPE to the cloud base mass flux. An assumed cloud entrainment and detrainment rates then determine the vertical distribution of cloud mass flux. The model uses this vertical mass flux to calculate updraft, downdraft, and environmental properties. Precipitation is parameterized as a fraction of condensate mass flux at a particular level. In CQE by Arakawa and Schubert (1974) the CAPE produced by large-scale conditions is consumed instantaneously by cloud-scale processes. While in the relaxed versions of Arakawa and Schubert (1974) such as Moorthi and Suarez (1992), the cloud-scale processes consume large-scale CAPE in a certain longer time called a relaxation time.

This relaxation time is shown to have a significant impact on the seasonal to interannual predictability of Indian monsoon precipitation. Jain et al. (2012) found that a longer relaxation time produced better season precipitation over India. They also emphasized the need for a cloud type dependent relaxation time in improving predictability further. Jain et al. (2013) found that the CAPE consumption rate had significant impact on the interannual variability of monsoon. They also found that the monsoon teleconnection with El Nino Southern Oscillations

(Webster and Yang 1992) and Indian Ocean Dipole (Saji et al. 1999; Ashok et al. 2001) was sensitive to the relaxation time.

It is important to understand the relationship between CAPE consumption and precipitation over India. Subrahmanyam et al. (2015) conducted an observational study of phase relationship between CAPE and precipitation over Indian region. They found that the CAPE preceded precipitation by 2–6 h. Another way to understand the relationship between these two quantities is to explicitly simulate cumulus convection. To that end, we carried out a season-long (June to September of 2008) weather research and forecast (WRF) simulation of monsoon at a cloud system resolving horizontal resolution of 3 km. The CAPE and precipitation spatial and temporal variability are then studied for this simulation. To contrast these results with the ones produced by conventional mass flux based cumulus parameterization we carried out additional simulations in which we used Kain–Fritsch cumulus parameterization (Kain 2004) and without a microphysics scheme. We first discuss the CAPE and convection inhibition energy (CINE) calculations in our simulations. Following which, we briefly discuss convective quasi-equilibrium.

1.1 CAPE and CINE Calculations

CAPE is calculated by vertically integrating the local buoyancy of a parcel from the level of free convection (LFC) to the equilibrium level (EL) shown in Fig. 1 as

$$CAPE = \int_{Z_{lfc}}^{Z_{EL}} g\frac{T_{v,parcel} - T_{v,env}}{T_{v,env}}dz \qquad (1)$$

CINE is calculated as

$$CINE = -\int_{Z_{surface}}^{Z_{LFC}} g\frac{T_{v,parcel} - T_{v,env}}{T_{v,env}}dz \qquad (2)$$

where $Z_{surface}$ is usually zero. However, this can be any starting parcels height. In our calculations, we tend to find out maximum CAPE and CINE (in J/kg) only for the parcel with max equivalent potential temperature in the column. An air parcel in our calculation refers to a 500 m deep parcel and the actual temperature and moisture are averaged over that depth.

1.2 Convective Quasi-Equilibrium

Convective quasi-equilibrium proposed by Arakawa (2004) assumes that the large-scale production of CAPE is consumed instantaneously by cloud-scale processes.

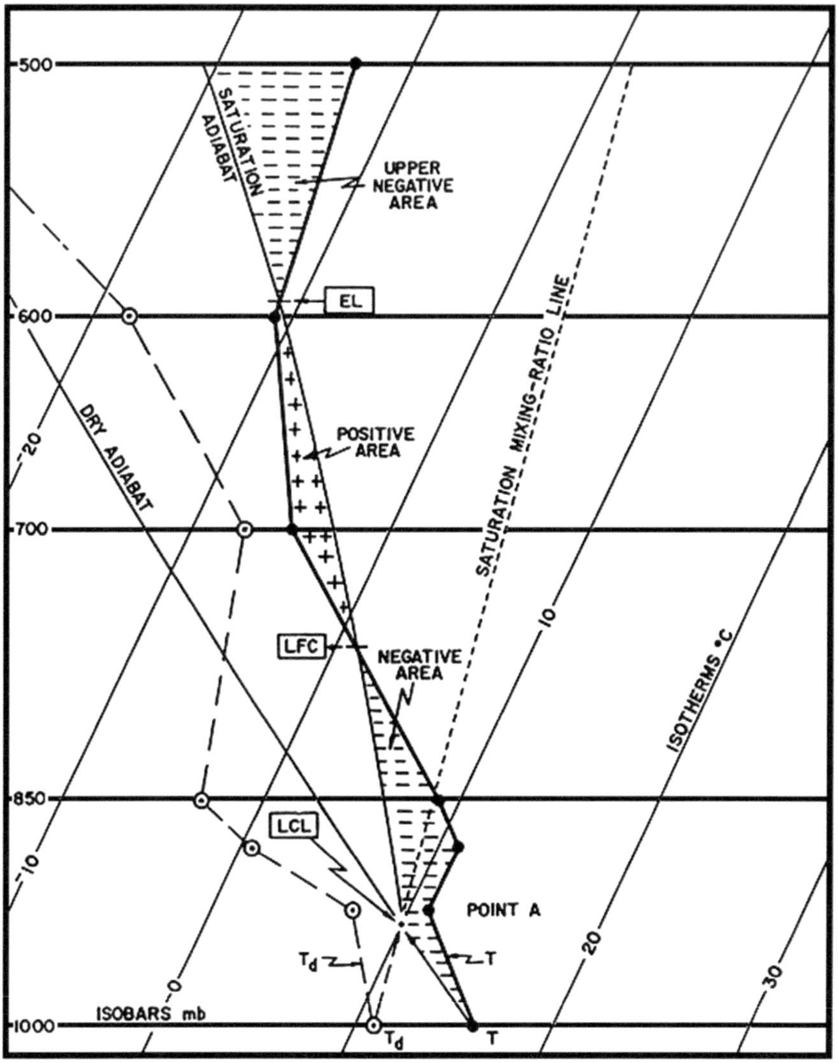

Fig. 1 Skew-T diagram with important features like lifting condensation level (LCL), level of free convection (LFC), equilibrium level (EL). CAPE corresponds to the positive area and CINE corresponds to lower negative area

It assumes that the typical cloud size (cloud area) is much smaller compared to the typical horizontal model grid size. Accordingly, the following equation holds true.

$$\left[\frac{dA}{dt}\right]_{ls} \ll \left[\frac{dA}{dt}\right]_{c} \tag{3}$$

where A is cloud work function (or CAPE normalized with cloud base mass flux), $\left[\frac{dA}{dt}\right]_{ls}$ is the rate of large-scale production of CAPE, and $\left[\frac{dA}{dt}\right]_{c}$ is its cloud-scale rate of consumption. The above equation states that the rate of CAPE consumption by cloud scale is much faster compared to its production due to large scale. In other simplifications of this assumptions, the cloud-scale processes are assumed to dissipate CAPE in a specified relaxation time. CQE is used to find the cloud base mass flux of a spectrum of clouds.

CQE is a spatial and temporal scale-dependent assumption and to test its applicability, in additional simulations for the month of June, we changed the horizontal resolution in the model. CAPE is also sensitive to diurnal land surface heating and we carried out a simulation for June in which the land surface temperature prediction was switched off while the sea surface temperature (SST) was allowed to evolve. Section 2 discusses model details and experimental setup for the model. In Sect. 3 we discuss results from the model simulations. Section 4 concludes the results.

2 Model Simulation Details

We use WRF version 3.4 for the present study. WRF is a non-hydrostatic mesoscale model developed by Mesoscale and Microscale Meteorology Division of National Centre for Atmospheric Research (Skamarock and Klemp 2008). The model has fully compressible and non-hydrostatic equations. It uses mass-based terrain-following coordinate system. The vertical grid spacing varies with height. The horizontal grids follow Arakawa C-grid staggering. The time stepping follows Runge–Kutta third-order time step. The lateral boundary conditions for the model are specified with relaxation zone of four grid points. Upper boundary has an absorbing layer with Rayleigh relaxation damping.

The primary control simulation was carried out at cloud system resolving horizontal resolution of 3 km and explicit microphysics for the duration of June to September of 2008 over the Indian region. The selection of spatial domain for the simulation was dictated by our requirement to simulate convection over Central Indian (CI) landmass and over Bay of Bengal (BoB). Figure 2 shows the Indian subcontinental orography and the domain for the model simulation. The most important geographic features are highlighted in Fig. 2. While we are also interested in convection over oceans, we have selected the domain such that the core of the domain is the Central Indian landmass. Some part of the Himalayan foothills are included in the domain at around 27N, the Western Ghats running parallel all across the western coast, the Eastern Ghats on the eastern coast around 18N, and the Satpura ranges over CI land can be seen in the domain. The control simulation has 1000×1000 grid points in the horizontal and 100 levels in the vertical with eta-coordinate system. The initial condition is provided at the start of the simulation while the boundary conditions are provided every 6 h using NCEP FNL analysis dataset (NCEP 2000).

We use WRF single moment class-3 (WSM3) microphysics scheme by Hong et al. (2004) for bulk microphysical processes in the model. WSM3 is a 3-class

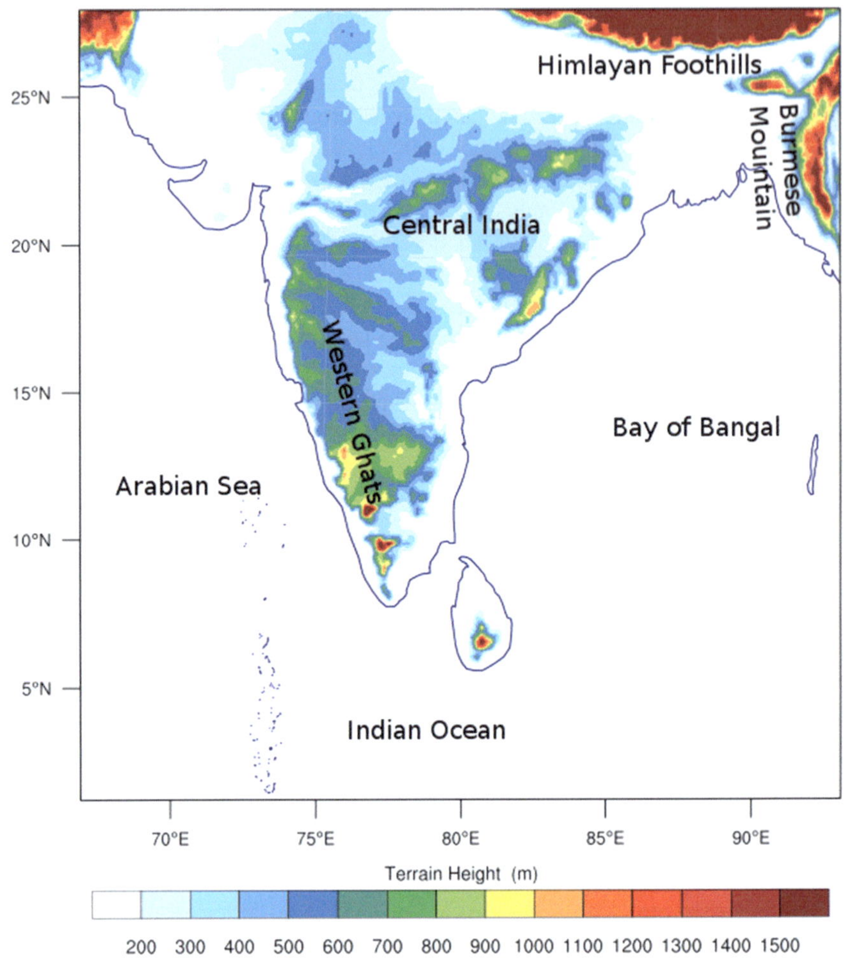

Fig. 2 Domain of model simulation showing prominent orographic regions such as Western Ghats at 10–20N, 75E, Sri Lankan orography at 7N, 80E, and Himalayan foothills north of 25N. The inclined line on the top right corner refers to the section along which the dynamic and thermodynamic conditions are analyzed

microphysics scheme which treats water vapor, cloud water, and rainwater mixing ratio above 0 °C and water vapor, ice water, and snow water mixing ratio below 0 °C. The model uses Yonsei University (YSU) scheme Hong et al. (2006) to represent planetary boundary layer processes. The model time step is 5 s and the model output is saved every 3 h.

As discussed earlier, we perform additional simulations at different horizontal resolutions and with explicit microphysics and Kain–Fritsch cumulus parameterization (Kain and Fritsch 1993; Kain 2004). The details of all the simulations performed

Table 1 Different model resolutions and convective representations used in the simulations in WRF

Simulation details					
Case name	Horizontal resolution (km)	Number of vertical levels	Microphysics	Cumulus parameterization	Duration of simulation
3Mi	3	100	WSM3	None	June to September
3Cu	3	100	None	Kain–Fritsch	June
12Mi	12	50	WSM3	None	June
12Cu	12	50	None	Kain–Fritsch	June
30Mi	30	50	WSM3	None	June
30Cu	30	50	None	Kain–Fritsch	June
3const	3	100	WSM3	None	June
12CuMicro	12	50	WSM3	Kain–Fritsch	June (not shown)
30CuMicro	30	50	WSM3	Kain–Fritsch	June (not shown)

for the present study can be found in Table 1. We first evaluate the models capability in simulating mean IMP features usually observed. To compare the temporal and spatial features of the precipitation, we use the 3 hourly tropical rainfall measuring mission (TRMM) 3B42 version 7 data (Huffman et al. 2007).

3 Results

3.1 Mean IMP Simulation

Figure 3 shows TRMM satellite-estimated and model-simulated monthly mean precipitation from June to September of 2008. The mean surface winds during monsoon are southwesterly. On the west coast, these winds carry moisture from Arabian sea to Indian landmass. These winds are orthogonal to the coast and are blocked by Western Ghats (WG) shown in Fig. 2. Due to orographic lift, most of the precipitation occurs on the windward side of WG and air loses moisture. Compared to TRMM estimates, the model simulates this precipitation satisfactorily.

The lost of moisture is evident on the leeward side of WG from the simulated precipitation and TRMM precipitation estimates. The model tends to over precipitate over Equatorial Indian Ocean (EIO, near 5N) compared to TRMM. Most of the general circulation models suffer from this problem of overestimating precipitation over ocean and underestimating precipitation over land. Late in the monsoon season

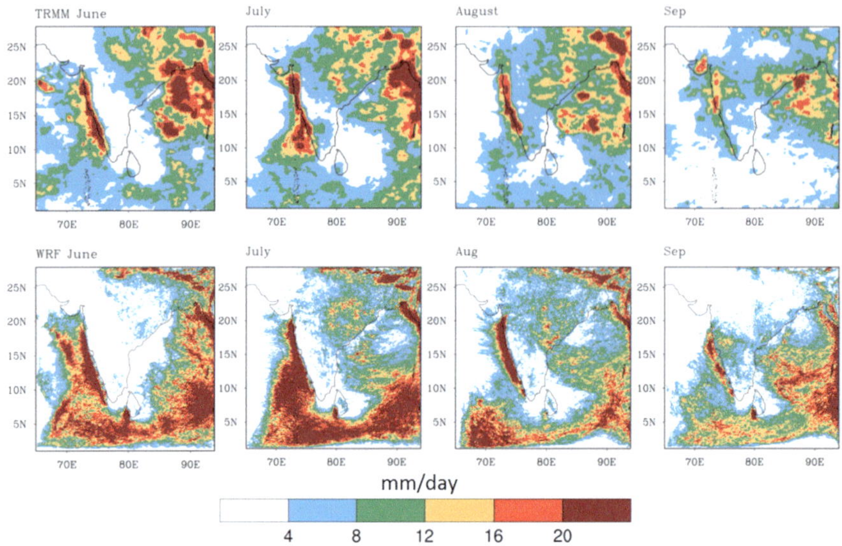

Fig. 3 June to September monthly precipitation (mm/day) from satellite observations (TRMM) and from model (WRF) simulations showing high precipitation over BoB, west of WG, and near the Himalayan foothills

(late September to October) is characterized by northeast monsoon over peninsular India. This is also called retreating southwest monsoon season. In TRMM and model, this is characterized by a lot of precipitation over south India including the leeward side of WG.

There is a region northeast of Sri Lanka which is characterized by comparatively lower SST [which is higher than the critical value for convection triggering found by Rajendran et al. (2012)]. This region which is called BoB cold pool (5–10N, 80–85E) receives very less precipitation. This is simulated very well by the model. The moisture-laden surface winds travel northeast over BoB and are blocked by Himalaya. This causes orographic lift and combined with the diurnal land surface heating, this moist air triggers a lot of moist convection and is responsible for frequent precipitation episodes near the Himalaya foothills. This can be seen as high precipitation in both TRMM and model in Fig. 3.

Indian monsoon is also characterized by wet and dry spells associated with the intraseasonal oscillations (Goswami et al. 2013; Lau et al. 1988; Goswami and Ajaya Mohan 2001). To evaluate model capability in simulating these oscillations, we take two regions, one over CI and one over BoB (Fig. 4, left panel). These wet–dry spells are very prominent in TRMM over BoB (right panel). The model seems to get the phase of these spells right, while the amplitude is mostly underestimated. Over CI, these spells are not very clearly separated as are over BoB. The model again captures the phase correctly and underestimates the amplitude.

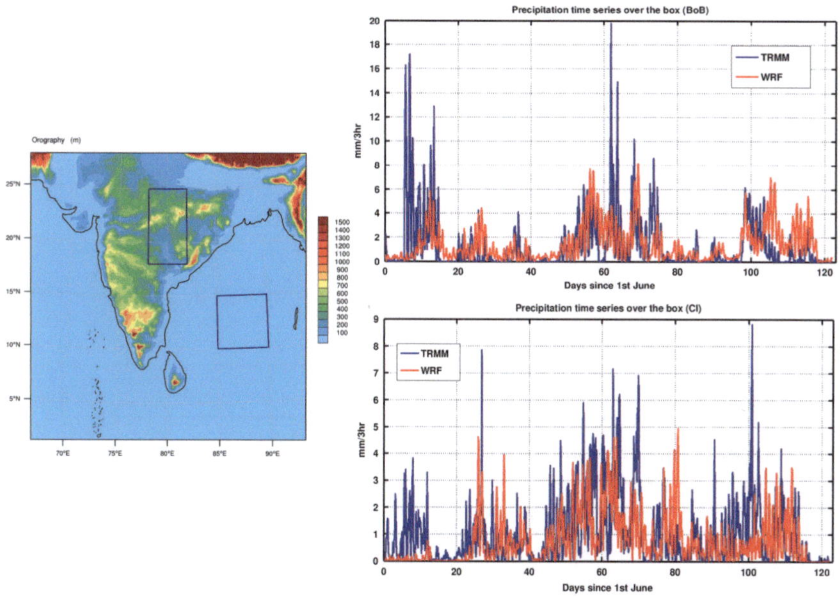

Fig. 4 Left panel shows the regions over which the time series of precipitation is calculated. The box over landmass refers to Central India while the box over ocean is BoB. Right panels show time series of mean precipitation from TRMM and model

3.2 CAPE and Precipitation

We also evaluated model capability in simulating diurnal cycle of precipitation over CI and BoB. We show mean June–September diurnal cycle from TRMM and model in Fig. 5. Over CI the model simulates precipitation maxima 3 h earlier than TRMM while the minima are correctly simulated. Over BoB the maxima and minima in the model occur 3–6 h before that in TRMM.

CQE proposed by Arakawa and Schubert (1974) assumes that precipitation (cloud-scale processes) consumes CAPE instantaneously in response to its large-scale production. While others like Moorthi and Suarez (1992) and Kain and Fritsch (1993) assume some relaxation time between large-scale CAPE production and its cloud-scale consumption. To see the mean diurnal cycle of precipitation and CAPE in the model, we show their diurnal cycle over CI and BoB in Fig. 6. It can be seen that the maxima in CAPE occur 6 h prior to precipitation over land and 3 h prior to it over ocean (BoB). Precipitation causes CAPE to reduce and the minima in CAPE occurs 3–6 h before precipitation minima over both CI and BoB. It is important to note that precipitation is not the only factor which changes CAPE. In the model, the land surface temperature is updated 4 times and the SST is updated once a day. This diurnal land heating (DLH) modifies CAPE, CINE, LCL, and LFC (CCLL hereafter)

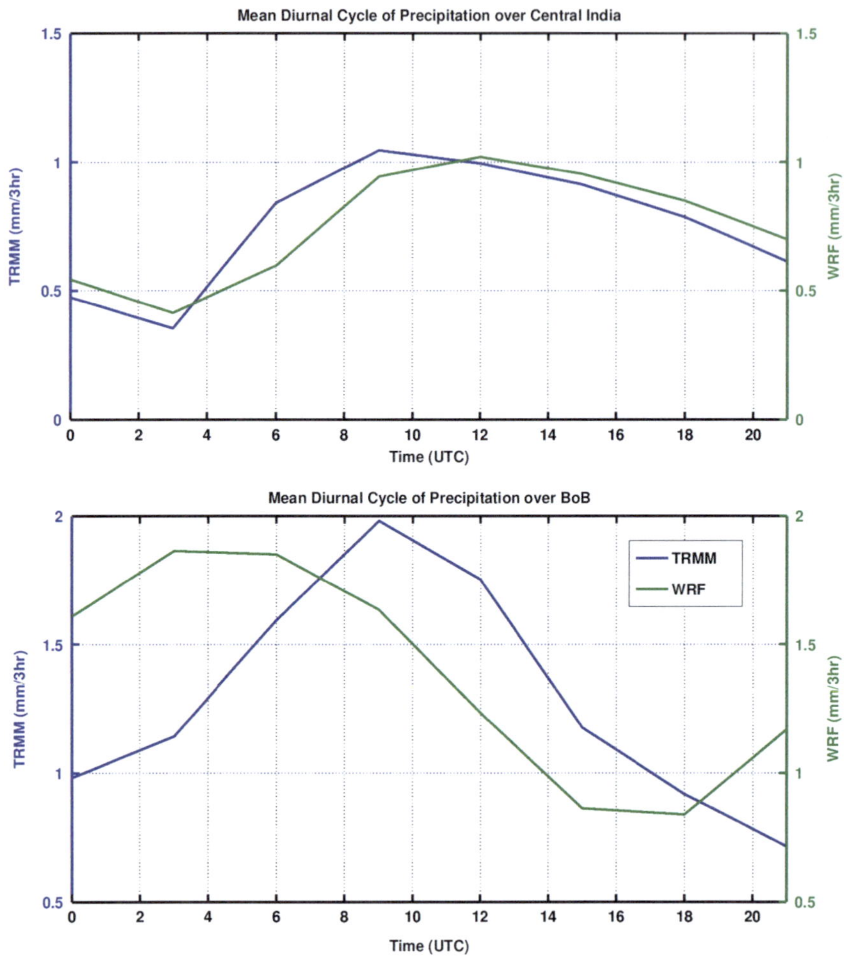

Fig. 5 Mean JJAS diurnal cycle of precipitation from TRMM and model over CI (top) and BoB (bottom)

significantly. Before we investigate the role of DLH in CAPE control, it is helpful to study the monthly mean spatial structure of CCLL.

Figure 7 shows June to September monthly mean spatial CCLL. The BoB has maximum CAPE in June and it decreases as the season progresses. It is important to note that the precipitable water is substantial over most of the ocean from June to September. Hence, intuitively, BoB should have a lot of precipitation. However, this is where CINE comes into play. It can be seen that the low precipitation region over BoB is characterized by high CINE, whereas, over EIO, the CINE values are nearly zero. This results in lot of precipitation in the model over EIO. As the season progresses, the CAPE over land increases, while CINE decreases. Thus, the precipitation over

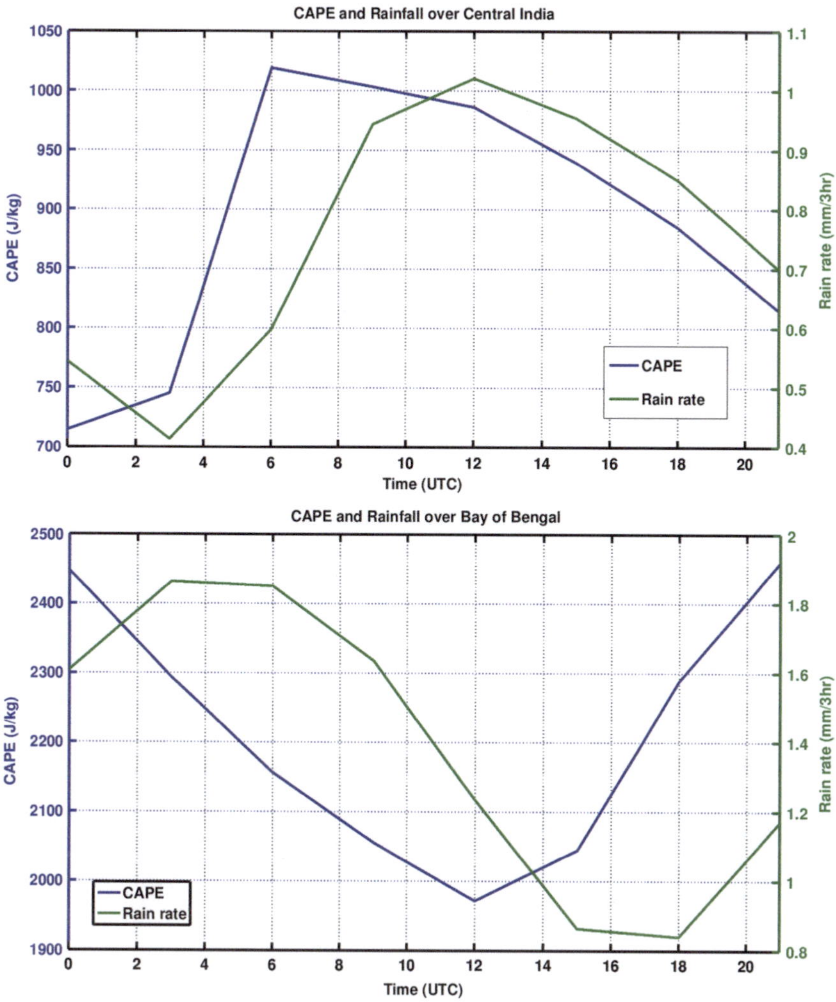

Fig. 6 Mean JJAS diurnal cycle of precipitation and CAPE from model over CI (top) and BoB (bottom)

BoB and CI increases. This motivates us to have a convective closure based on the ratio of CINE and CAPE with boundary layer turbulent kinetic energy as one of the control parameters. In Fig. 7, the LFC and LCL values are just the reflection of CINE values. One of the interesting features is the higher value of LFC on the leeward side of WG throughout the season.

Figure 8 shows monthly anomaly from JJAS mean of CCLL. During June and July, CAPE anomalies are negative over most landmass. It is positive during August and September indicating comparatively favorable condition for moist convection as far as CAPE is considered. Positive CINE anomaly is present for June over most of

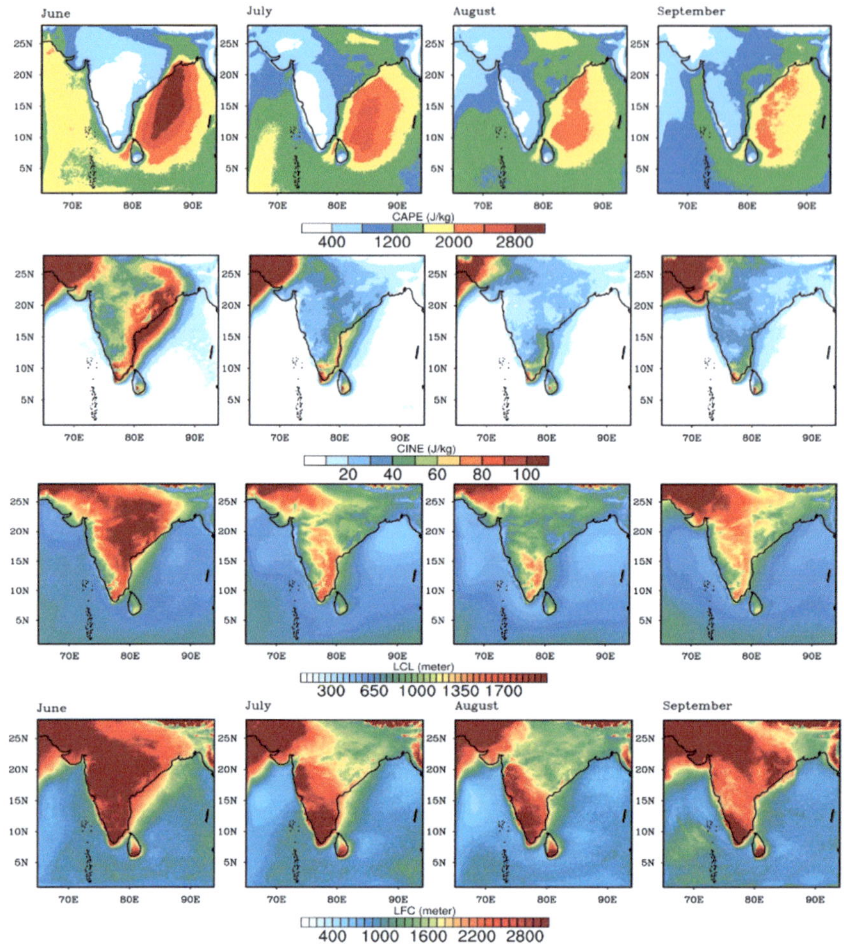

Fig. 7 June to September monthly mean of model-simulated CAPE, CINE, LCL, and LFC

the domain and is very prominent over the east coast region. The positive anomaly indicates unfavorable condition for convection to occur. The larger CINE also manifests as positive anomaly value of LFC. As LFC and LCL increases, a surface parcel needs more lifting to trigger a feasible cloud. In the subsequent months (July to September), we see negative anomalies for CINE, LCL, and LFC, indicating that it is easier for the surface parcels to trigger moist convection. However, anomalies do not indicate why we should have so much precipitation over EIO. Significantly lower values of CINE and reasonable higher values of CAPE might be responsible for that.

Having had a feel of the spatial and temporal variability of precipitation and CCLL, we now look at the correlation between CAPE and precipitation in the model. The

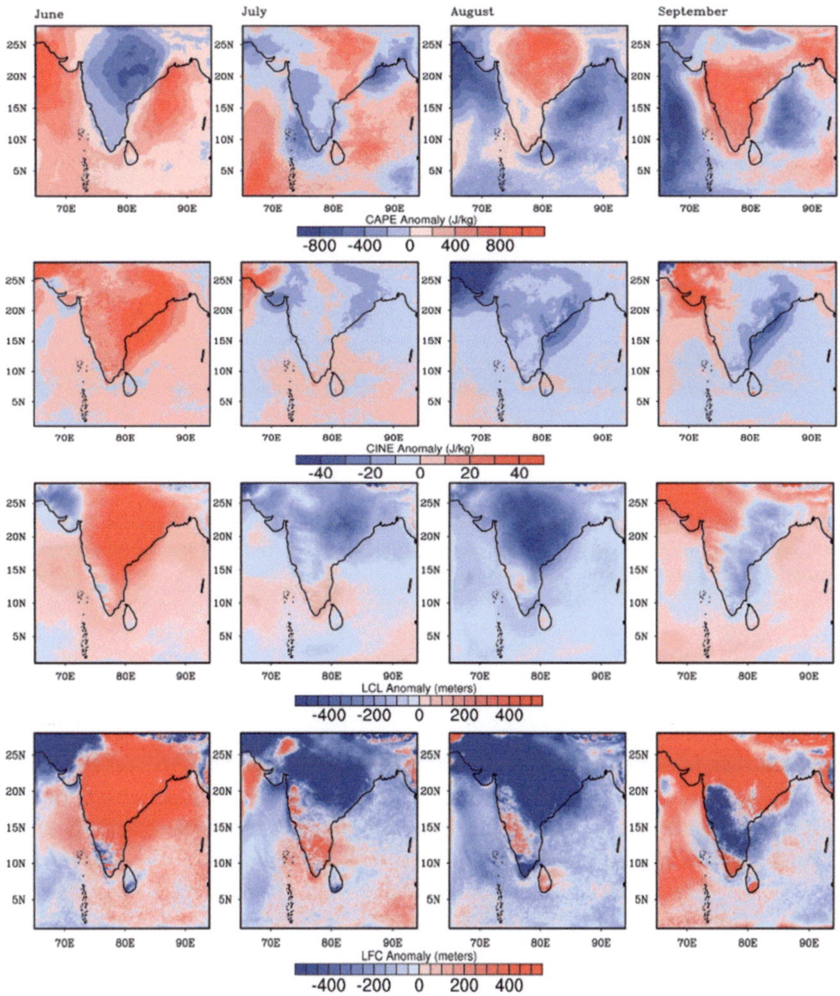

Fig. 8 June–September monthly anomaly from JJAS mean of model-simulated CAPE, CINE, LCL, and LFC

model produces accumulated precipitation and since we are archiving data every 3 h we get 3 hourly accumulated precipitation. The CCLL values are instantaneous though. In Fig. 9, we show temporal correlation coefficients of two types. The first one (left panel) is between CAPE value at the present instant and precipitation in the next 3 h. The second one is between CAPE value at the present instant and precipitation in the previous 3 h. Intuitively, we expect the CAPE values to show maxima before precipitation and we expect the left panel to show positive correlation everywhere. We also expect the CAPE to go down after precipitation and this should show up as negative correlation in the right panel. We actually find that the correlation coefficient

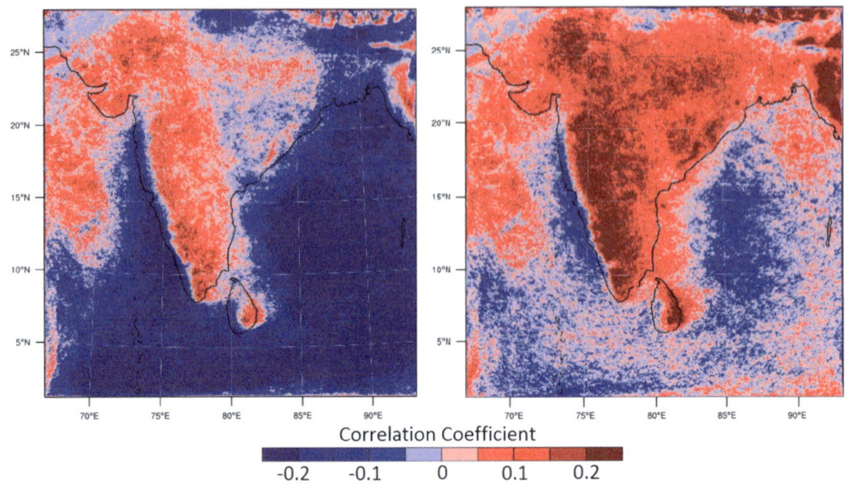

Fig. 9 Correlation coefficient of CAPE and 3 hourly model-simulated precipitation. Left panel shows correlation between CAPE now and precipitation in next 3 h. The right panel shows correlation between CAPE now and precipitation 3 h prior

has opposite sign over land and ocean in both the cases. In the first case, the CAPE at the present instant is negatively correlated to precipitation in the next 3 h over ocean and positively correlated over land. In the second case as well CAPE at the present instant is mostly negatively correlated to precipitation in the next 3 h over ocean and positively correlated over land. This brings us back to the original conception of CQE in which the rate of CAPE consumption is related to convective activity rather than the actual value itself.

CAPE builds up before precipitation and the resulting precipitation consumes that CAPE. After precipitation, the final CAPE should be lesser than the initial CAPE before precipitation. We take different time gaps (3–24 h in steps of 3 h) and calculate the change in CAPE and accumulated precipitation during that duration. Figure 10 shows correlation coefficient between the two for various time gaps. It can be seen that almost entire region shows negative correlation and verifies the hypothesis that rate of change of CAPE is negatively correlated to convective activity. The correlation changes as the time gap are changed (lower panel of Fig. 10 which shows mean values over the entire domain). The best correlation is for the least time gap and correlation decreases almost linearly as the duration increases. The correlation coefficient is not negative everywhere though. The western region of Arabian Sea and the southernmost tip of India show positive correlation.

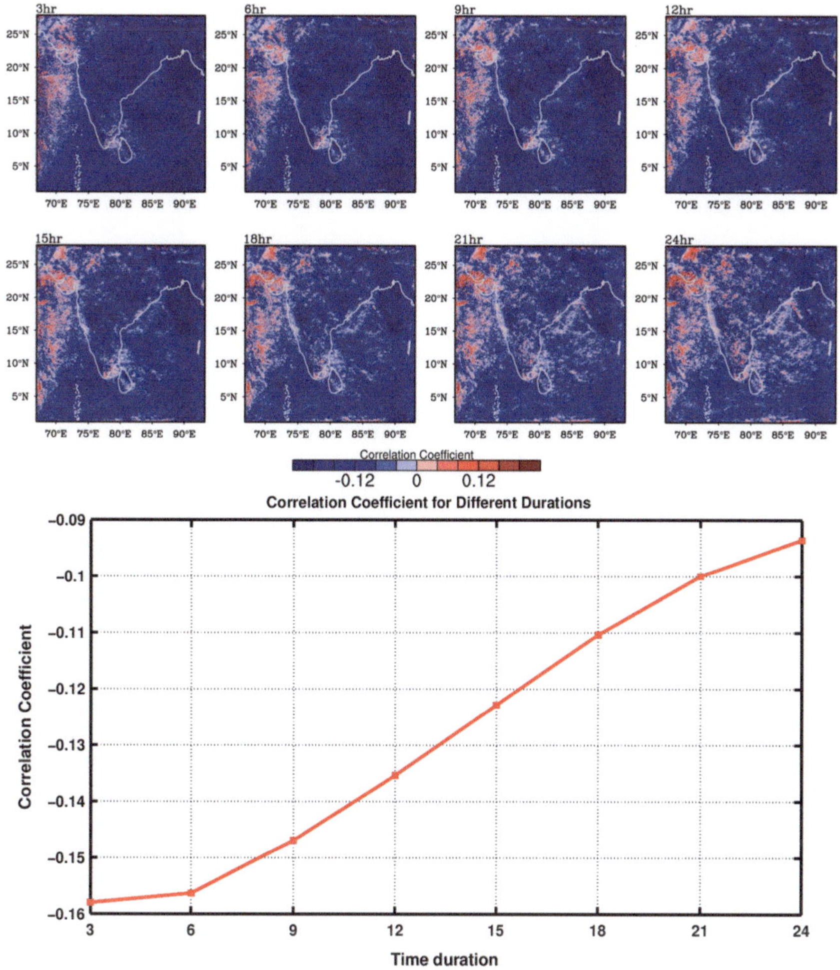

Fig. 10 Correlation coefficient between rate of change of CAPE and precipitation with different time intervals

3.3 *Effects of Diurnal Land Heating*

Boundary layer plays a major role in controlling CAPE (Donner and Phillips 2003) and the diurnal surface heating plays a major role in boundary layer stability, and hence CAPE. To study the effects of diurnal land heating only, we carried out a simulation for the month of June 2008 where we switched of land surface temperature prediction. The SST was allowed to evolve though. The model physics was kept the same as that of the control simulation. The simulated precipitation and its difference from control are shown in Fig. 11. We can see that diurnal land heating (DLH)

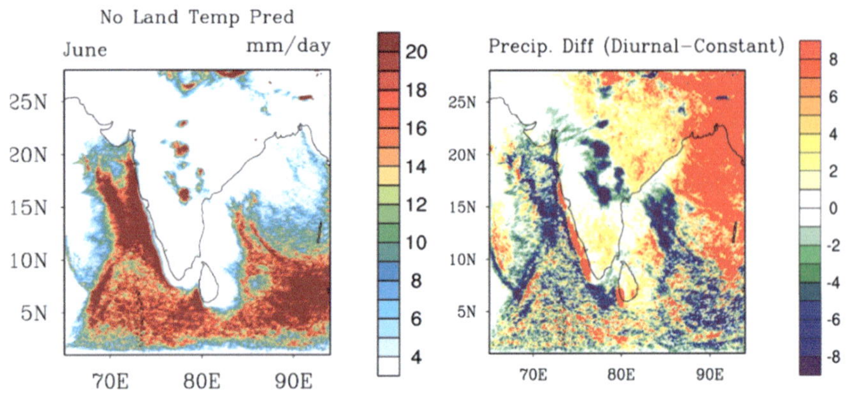

Fig. 11 Left panel shows simulated precipitation when land surface temperature prediction was switched off in the model. Right panel shows the difference between simulated precipitation with constant land temperature from control run

affects the precipitation over both land and ocean. The biggest difference is seen over Himalayan foothills and the north BoB. In these regions, it seems that the DLH accounted for at least half the precipitation (difference is 8 mm while control precipitation is around 16 mm). The DLH reduces precipitation over southwest BoB and Arabian Sea while it increases precipitation over WG. This motivates us to study the contributions of DLH and orographic lift on the overall convection over WG and Himalayan foothills. It also motivates us to investigate the relationship between convective systems which are triggered over land region north of BoB (Himalayan foothills) and those that occur over north BoB sea surface. These studies will be carried out in the future. In the present study, we would continue to investigate the CAPE production in response to DLH.

Figure 12 compares CCLL values simulated by constant land temperature (CLT) with DLH. A lot of CAPE over BoB is produced due to DLH. This should explain the decrease in precipitation in CLT compared to that in DLH over north BoB. However, that is not a complete picture since it does not explain the increase in precipitation in the southwest of BoB seen in Fig. 11. We are inclined to believe that the precipitating systems over north BoB are triggered over land. The CINE values, on the other hand, are nearly zero over most of the ocean and especially over eastern coast in CLT simulation. The CAPE values are reasonably higher in these regions in CLT simulation. This makes the southwest BoB more favorable for convection in CLT compared to DLH. LCL and LFC values are lower over the east coast in CLT case compared to DLH case. DLH has a significant effect on the simulated CINE, LCL, and LFC over Indian landmass. All the three quantities show an increase (LCL and LFC are implicitly connected to CINE) over the landmass. Switching off DLH shows that there is a lot of effect of orography on LFC. This can be seen in Fig. 12 as localized increase in LFC in response to local orography.

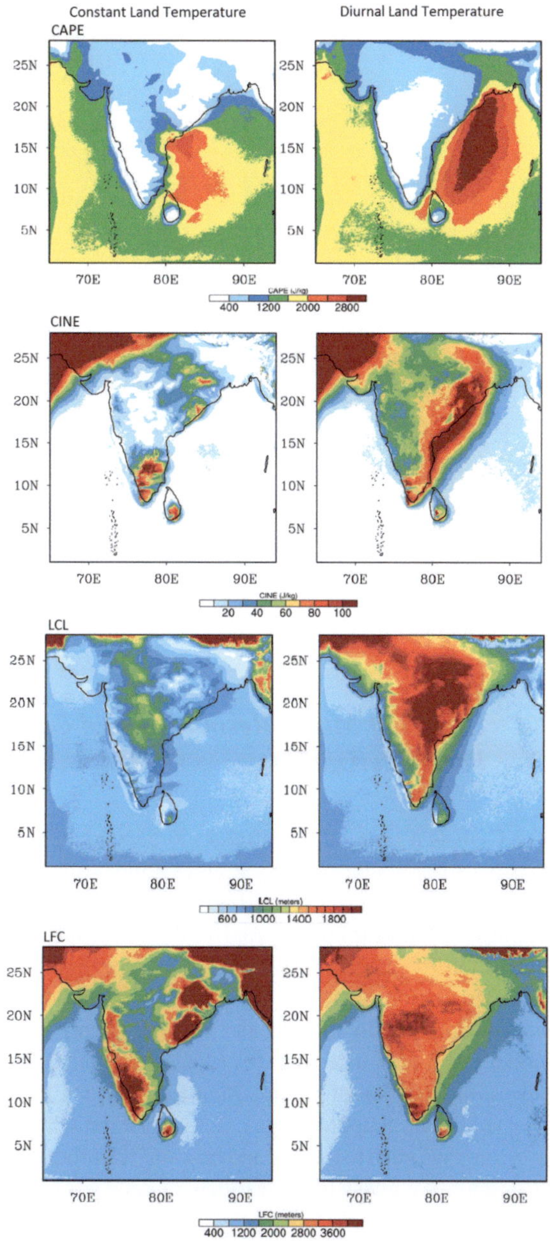

Fig. 12 CCLL from constant land temperature and from diurnal land surface temperature simulation

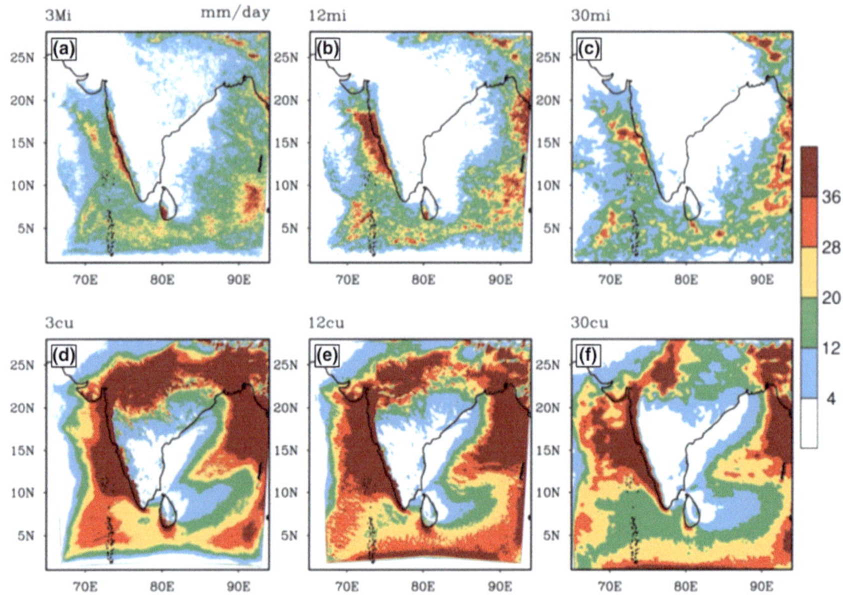

Fig. 13 June precipitation simulated by various model configurations mentioned in Table 1

3.4 Effects of Model Resolution and Cumulus Parameterization

The CQE is a time- and spatial-scale-dependent assumption. Hence, we study the effects of horizontal resolution and cumulus parameterization in the model simulations. We carried out five additional simulations (Table 1) at different horizontal resolutions with explicit microphysics and with Kain–Fritsch (KF) cumulus parameterization (Kain and Fritsch 1993). Figure 13 shows simulated precipitation from these simulations. The simulations with KF produces a lot of precipitation everywhere compared to TRMM and explicit microphysics case (Fig. 3) irrespective of model resolution. The cu cases (3cu, 12cu, and 30cu) do not get either the magnitude or the location of precipitation right. Over CI, the precipitation in cu cases gets affected by the Satpura ranges. To understand why the model gets the precipitation so wrong, we need to understand how the KF produces precipitation. In the following discussion, we give a brief overview of convective trigger and closure assumption in KF.

The goal of any cumulus parameterization is to predict the effects of convective clouds (in a statistically averaged sense) inside the model grid box in response to large-scale conditions. KF assumes that an average cloud horizontal size is much smaller compared to the horizontal grid size of the model. The various cloud processes shown in Fig. 14 are modeled in KF. The first shortcoming of KF is that as the model resolution increases, these processes start occupying more than one grid box. In

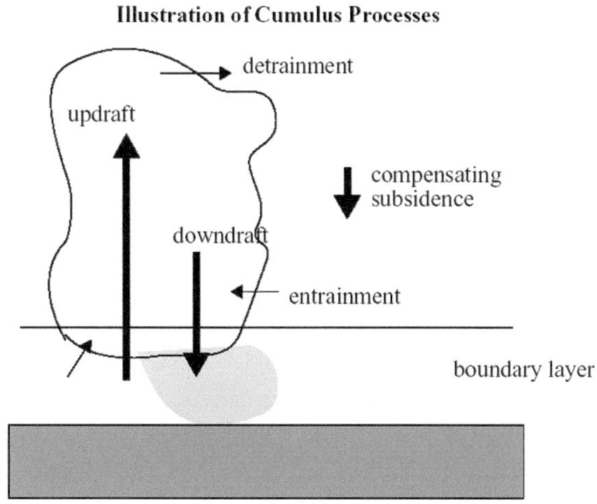

Fig. 14 Cloud processes parameterized by conventional mass flux based cumulus parameterization like Kain–Fritsch

order to correctly represent processes then, the parameterization assumptions need to change (Arakawa and Wu 2013). To figure out why the model produces so much precipitation, we have to understand how KF triggers and intensifies convection.

The trigger for a cumulus convection determines if the environment is favorable enough to trigger a new convection. In KF, a potential updraft source layer (USL) is identified by vertically mixing (starting from surface) the adjacent layers till the overall thickness reaches 60 hPa. This is the first potential USL. The thermodynamic characteristics (like the temperature at LCL (T_{lcl})) of this undiluted USL parcel is calculated at its LCL. Then, a perturbation temperature (T_p) is calculated based on grid-scale resolved vertical velocity that is added to T_{lcl}. The sum ($T_{lcl} + T_p$) is then compared to environmental temperature (T_{env}). If $T_{lcl} + T_p > T_{env}$, then the convection is triggered. If not, then the starting layer is moved up by one model level and the calculations are repeated till either a USL is found or a prescribed height is reached after which no convection is triggered.

Once the convection is triggered, the intensity and likelihood of convection depend on the value of CAPE. In KF, the calculation of vertical cloud mass flux based on CAPE and an assumed entrainment/detrainment profile is called closure assumption. In KF, the transports of mass in a model column is carried out using updrafts, downdrafts, and environmental large-scale mass fluxes till 90% of the CAPE is consumed. The above discussion implies two things, the first is that convection is more likely to occur in the regions of higher CAPE. The second is probably more important when the model resolution increases. It is that KF assumes all the cloud processes are happening inside the single grid column.

Figure 15 shows the simulated CAPE from different model configurations shown in Table 1. We can see that irrespective of the way in which convection is represented, the CAPE values are always high over oceans. The 3cu case shows large CAPE everywhere in the region. Though the CAPE values decrease with coarser resolution, they are still quite high. All the cumulus cases show large CAPE north of

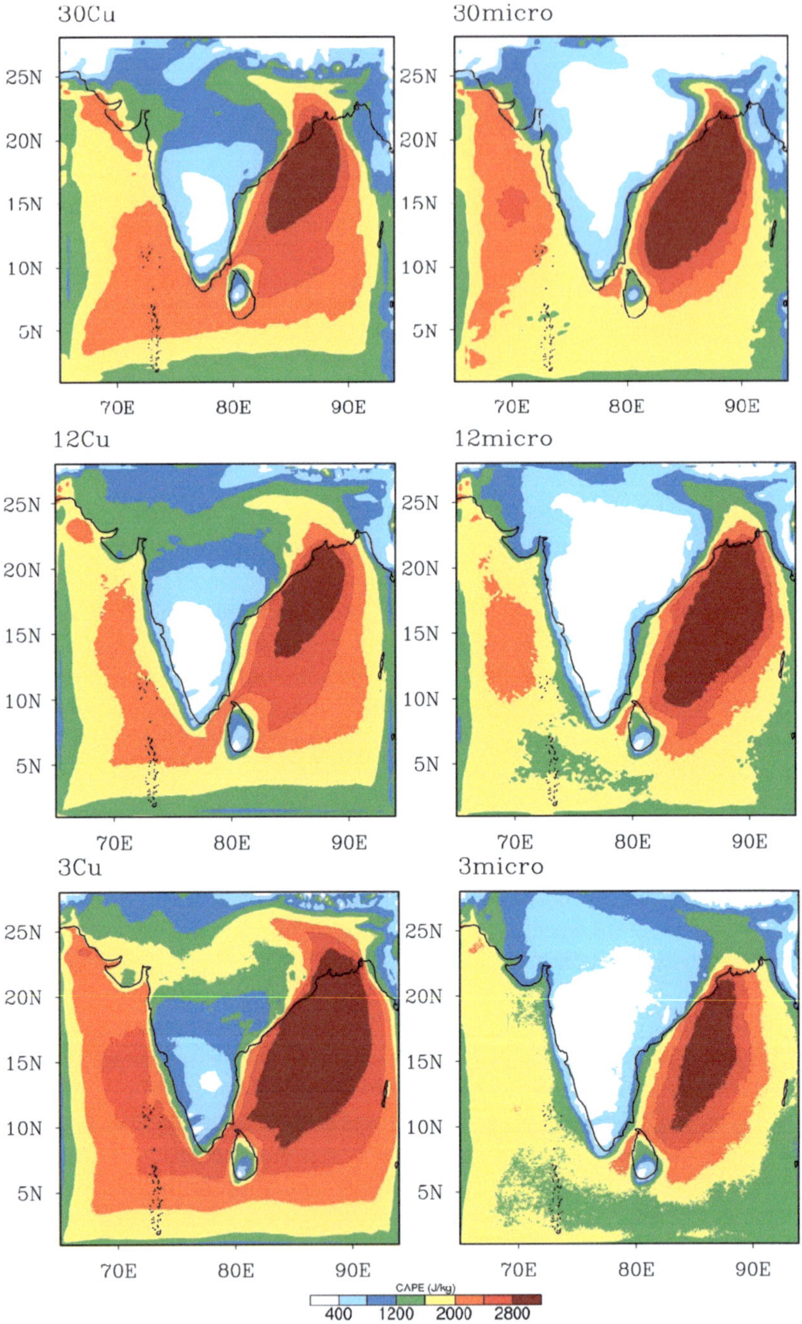

Fig. 15 Simulated CAPE for different model configurations (Table 1)

Satpura ranges (the CI orography shown in Fig. 2). If we look at the precipitation and CAPE produced by cu cases, we can see that the region of highest CAPE is also the region of highest precipitation. While this is not true in micro-cases. The region of highest CAPE is not always the region of highest precipitation in 3, 12, and 30 μ. The problem with the convective trigger used in KF is that it tries really hard to find a USL iteratively. Since CAPE values are higher everywhere, KF ends up triggering a lot of convection and precipitation everywhere. This is not a problem for explicit microphysics cases (3, 12, and 30 μ) since the microphysical constituents are explicitly calculated and the resultant heating/cooling due to phase change is what determines buoyancy and convection. The micro-cases do not implicitly assume either updraft, downdraft, or environmental mass fluxes. These quantities are explicitly simulated.

One thing to note from Figs. 13 and 15 is that although the CAPE is high everywhere over BoB, the precipitation is not. The east coast region shows very less precipitation compared to the rest of the domain. This is because of a large amount of CINE over this region (Fig. 16). In cu cases, the CINE values seem to be model resolution dependent. The spatial extent as well as magnitude of CINE increases as the resolution of the model decreases from 3cu to 30cu. The CINE in the micro-cases does not seem to be very sensitive to the horizontal resolution.

As discussed earlier, the KF assumed cloud processes to be much smaller compared to the model grid size. This presents a problem in simulating organized convection such as mesoscale convective systems (MCSs) and mesoscale convective complexes which extend from 10 to 100 km. The present-day cumulus parameterizations do not include any representation for MCSs (Randall et al. 2003). Attempts are being made to have MCSs in cloud parameterizations (Yano and Moncrieff 2016). One of the examples of propagating MCS and the associated CAPE is shown in Fig. 17. It can be seen that as the MCS propagates, it consumes CAPE. We found that a large fraction of the precipitation produced by 3 μ case was due to MCSs. Hence, any parameterization which wants to predict correct precipitation over Indian region has to include the effects of organized MCSs.

It is worthwhile to discuss the CAPE consuming characteristics of an MCS. Houze (2014) simplified momentum equation on the two-dimensional streamlines of an MCS for the quantities which are conserved over the streamlines. We rewrite their Eq. 9.14 here.

$$\frac{1}{2}(u^2 + w^2) + \frac{p^*}{\rho_0} - \int_{Z_{in}}^{Z_{out}} g\frac{\theta^*}{\hat{\theta}}dz = \text{ Constant on a Streamline} \tag{4}$$

where u = horizontal velocity, w = vertical velocity, p* = pressure perturbation (deviation from hydrostatic pressure), θ^* = perturbation potential temperature, $\hat{\theta}$ = environmental potential temperature, and Z_{in} and Z_{out} refer to height of streamline. The above equation states that the CAPE (or energy from buoyancy) does not necessarily manifest itself as vertical motion. It may get converted to any combination of enthalpy ($\frac{p^*}{\rho_0}$), horizontal motion, and vertical motion.

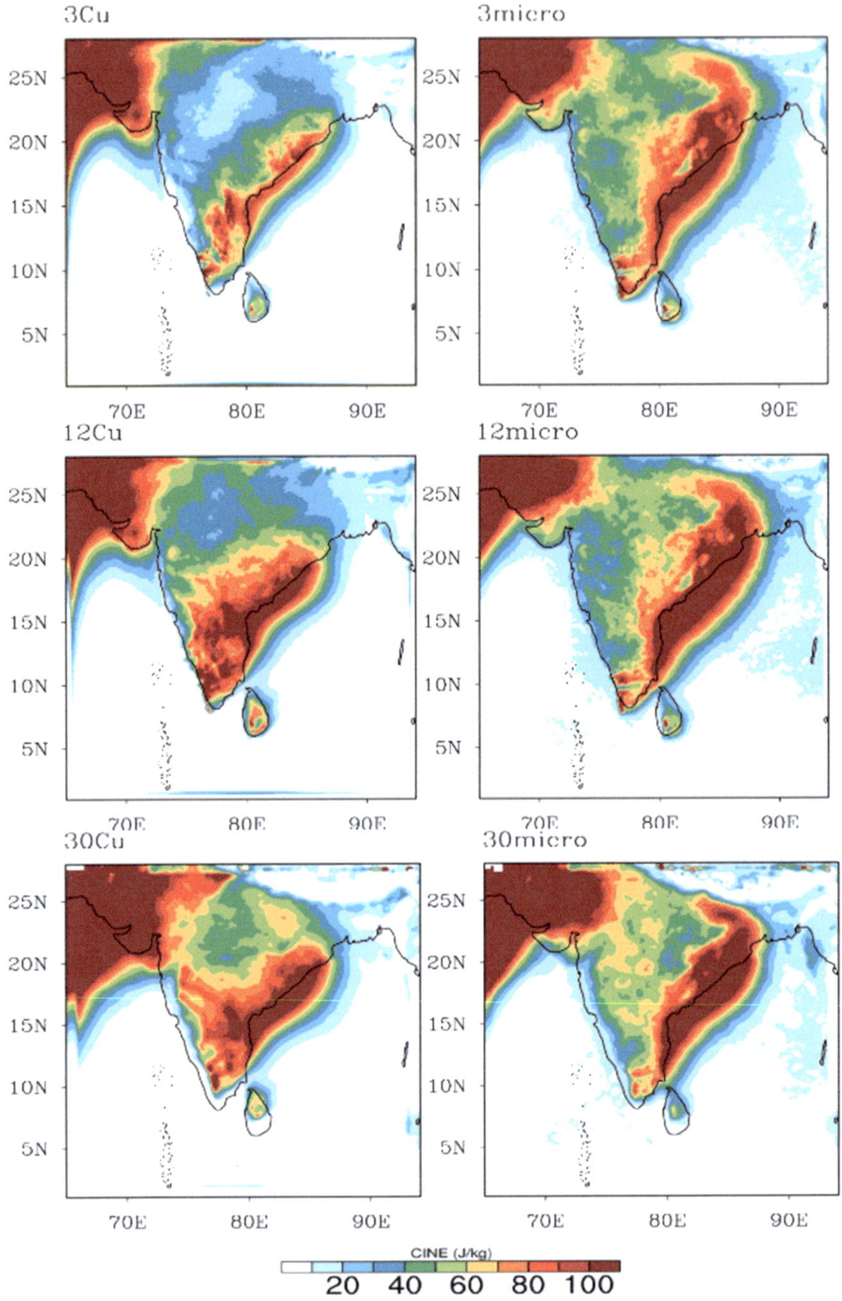

Fig. 16 Simulated CINE for different model configurations (Table 1)

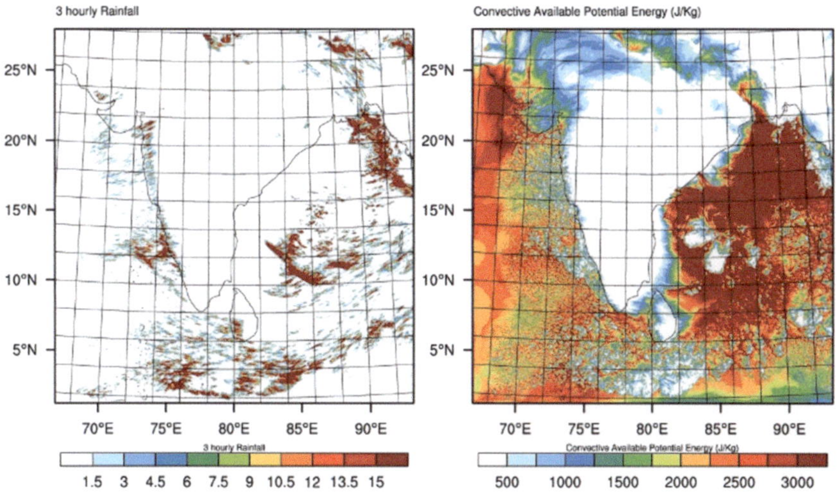

Fig. 17 MCS simulate by 3 μ case on June 6, 2008. Left panel shows precipitation (mm/3 h), the right panel shows CAPE

4 Conclusions

In the present study, we analyzed CAPE and precipitation produced by a season-long cloud-resolving simulation of Indian monsoon. The model reproduces most of the observed spatial mean precipitation structures such as orographic precipitation and central Indian landmass precipitation correctly. It overestimates precipitation over EIO. The model diurnal cycle of precipitation shows peak 3–6 h earlier compared to TRMM. The model gets the phase of dry and wet spells associated with intraseasonal oscillations right over CI and BoB while it underestimates the amplitude.

We found that the regions of high CAPE and precipitation do not always coincide and CINE plays an important role in controlling the location of precipitation. We also found that CAPE and precipitation over BoB were very sensitive to diurnal land heating over India. But the locations of precipitation and CAPE affected by DLH do not coincide. Diurnal land heating affects precipitation over north BoB, while it affects CAPE over southwest BoB. Correlation coefficient between actual values of instantaneous CAPE and precipitation shows sign reversal from land to ocean. However, the rate of change of CAPE is negatively correlated to precipitation verifying CQE.

Mass flux scheme (KF) based on CAPE closure performs poorly over India. The precipitation in KF is too dependent on CAPE and gets the location and magnitude of precipitation wrong irrespective of the horizontal resolution. KF (or any of the present-day cumulus parameterizations) does not include representation for organized convection such as MCSs and MCS. These systems typically span more than one grid box. They consume CAPE to trigger new convection and move forward.

The resolution between explicitly resolving convection (as in CRMs) and heavily parameterized cases (as in cumulus parameterization for GCMs) is the gray area in parameterization and is a deadlock in convection representation in numerical weather prediction models (Arakawa and Wu 2013; Wu and Arakawa 2014). The mesoscale non-hydrostatic models suffer from the fact that there is no proper convection representation available in the gray area of parameterization (5–30 km horizontal resolution). Using only microphysics at coarser resolution is problematic. Similarly, using cumulus parameterizations at these resolutions is also not correct because of all the complexities of convection happening at this resolution and the inability of parameterizations to adapt to varying horizontal resolutions. We can say that future convection schemes should be able to transform themselves smoothly as the resolution of the model is changed. The model should be able to simulate mesoscale propagations even at coarser resolution.

References

Arakawa, A. 2004. The cumulus parameterization problem: Past, present, and future. *Journal of Climate* 17 (13): 2493–2525.

Arakawa, A., and W.H. Schubert. 1974. Interaction of a cumulus cloud ensemble with the large-scale environment, part I. *Journal of the Atmospheric Sciences* 31 (3): 674–701.

Arakawa, A., and C.M. Wu. 2013. A unified representation of deep moist convection in numerical modeling of the atmosphere, part I. *Journal of the Atmospheric Sciences* 70 (7): 1977–1992.

Ashok, K., Z. Guan, and T. Yamagata. 2001. Impact of the Indian ocean dipole on the relationship between the Indian monsoon rainfall and enso. *Geophysical Research Letters* 28 (23): 4499–4502.

Donner, L.J., and V.T. Phillips. 2003. Boundary layer control on convective available potential energy: Implications for cumulus parameterization. *Journal of Geophysical Research: Atmospheres* 108 (D22).

Gadgil, S. 2003. The Indian monsoon and its variability. *Annual Review of Earth and Planetary Sciences* 31 (1): 429–467.

Goswami, B.N., and R. Ajaya Mohan. 2001. Intraseasonal oscillations and interannual variability of the Indian summer monsoon. *Journal of Climate* 14 (6): 1180–1198.

Goswami, B.B., P. Mukhopadhyay, M. Khairoutdinov, and B. Goswami. 2013. Simulation of indian summer monsoon intraseasonal oscillations in a superparameterized coupled climate model: Need to improve the embedded cloud resolving model. *Climate Dynamics* 41 (5–6): 1497–1507.

Hong, S.Y., J. Dudhia, and S.H. Chen. 2004. A revised approach to ice microphysical processes for the bulk parameterization of clouds and precipitation. *Monthly Weather Review* 132 (1): 103–120.

Hong, S.Y., Y. Noh, and J. Dudhia. 2006. A new vertical diffusion package with an explicit treatment of entrainment processes. *Monthly Weather Review* 134 (9): 2318–2341.

Houze Jr. R.A. 2014. *Cloud dynamics*, vol. 104. New York: Academic Press.

Huffman, G.J., D.T. Bolvin, E.J. Nelkin, D.B. Wolff, R.F. Adler, G. Gu, Y. Hong, K.P. Bowman, and E.F. Stocker. 2007. The trmm multisatellite precipitation analysis (tmpa): Quasi-global, multiyear, combined-sensor precipitation estimates at fine scales. *Journal of Hydrometeorology* 8 (1): 38–55.

Jain, D.K., A. Chakraborty, and R.S. Nanjundiah. 2012. On the role of cloud adjustment time scale in simulating precipitation with relaxed arakawa-schubert convection scheme. *Meteorology and Atmospheric Physics* 115 (1–2): 1–13.

Jain, D.K., A. Chakraborty, and R.S. Nanjundiah. 2013. Role of the cloud adjustment time scale in simulation of the interannual variability of indian summer monsoon. *Meteorology and Atmospheric Physics* 122 (3–4): 159–173.

Kain, J.S., and J.M. Fritsch. 1993. Convective parameterization for mesoscale models: The kain-fritsch scheme, pp. 165–170.
Kain, J.S. 2004. The kain-fritsch convective parameterization: An update. *Journal of Applied Meteorology* 43 (1): 170–181.
Lau, K., G. Yang, and S. Shen. 1988. Seasonal and intraseasonal climatology of summer monsoon rainfall over East Asia. *Monthly Weather Review* 116 (1): 18–37.
Moorthi, S., and M.J. Suarez. 1992. Relaxed Arakawa-Schubert. A parameterization of moist convection for general circulation models. *Monthly Weather Review* 120 (6): 978–1002.
National Centers for Environmental Prediction NUDoC National Weather Service. 2000. NCEP FNL operational model global tropospheric analyses, continuing from July 1999. https://doi.org/10.5065/D6M043C6.
Rajendran, K., R.S. Nanjundiah, S. Gadgil, and J. Srinivasan. 2012. How good are the simulations of tropical SST–rainfall relationship by IPCC AR4 atmospheric and coupled models? *Journal of Earth System Science*, pp. 1–16.
Randall, D., M. Khairoutdinov, A. Arakawa, and W. Grabowski. 2003. Breaking the cloud parameterization deadlock. *Bulletin of the American Meteorological Society* 84 (11): 1547–1564.
Saji, N., B. Goswami, P. Vinayachandran, and T. Yamagata. 1999. A dipole mode in the tropical Indian ocean. *Nature* 401 (6751): 360–363.
Skamarock, W.C., and J.B. Klemp. 2008. A time-split nonhydrostatic atmospheric model for weather research and forecasting applications. *Journal of Computational Physics* 227 (7): 3465–3485.
Sperber, K.R., H. Annamalai, I.S. Kang, A. Kitoh, A. Moise, A. Turner, B. Wang, and T. Zhou. 2013. The Asian summer monsoon: An intercomparison of CMIP5 vs. CMIP3 simulations of the late 20th century. *Climate Dynamics* 41 (9–10): 2711–2744.
Subrahmanyam, K.V., K.K. Kumar, and A. Narendra Babu. 2015. Phase relation between cape and precipitation at diurnal scales over the Indian summer monsoon region. *Atmospheric Science Letters* 16 (3): 346–354.
Tiedtke, M. 1989. A comprehensive mass flux scheme for cumulus parameterization in large-scale models. *Monthly Weather Review* 117 (8): 1779–1800.
Webster, P.J., and S. Yang. 1992. Monsoon and enso: Selectively interactive systems. *Quarterly Journal of the Royal Meteorological Society* 118 (507): 877–926.
Wu, C.M., and A. Arakawa. 2014. A unified representation of deep moist convection in numerical modeling of the atmosphere, part II. *Journal of the Atmospheric Sciences* 71 (6): 2089–2103.
Yano, J.I., and R. Plant. 2012. Convective quasi-equilibrium. *Reviews of Geophysics* 50 (4).
Yano, J.I., and M.W. Moncrieff. 2016. Numerical archetypal parameterization for mesoscale convective systems. *Journal of the Atmospheric Sciences* 73 (7): 2585–2602.

A Gray Zone GCM with Full Representation of Cloud Microphysics

In-Sik Kang and Min-Seop Ahn

Abstract This chapter describes the development procedure of general circulation model (GCM) with full representation of cloud microphysics at a medium-range horizontal resolution of 50 km and discusses the simulation results of their precipitation climatology and Madden and Julian Oscillation (MJO). One issue of developing such a GCM is to modify the cloud microphysics suitable to the horizontal resolution. In the present study, the modification is made based on sensitivity experiments for the parameters of the important processes sensitive to the model resolution, particularly the condensation process and the terminal velocity. It is demonstrated that shallow convection and scale-dependent deep convection are still needed in the present model of 50 km resolution with cloud microphysics. The present GCM is shown to simulate the precipitation statistic such as the light and heavy precipitation frequencies and the MJO reasonably well, although the MJO intensity is rather strong. Both cloud microphysics and scale-dependent deep convection play important roles in simulating a realistic MJO in the present GCM. Also noted is that the precipitation climatologies of the present atmospheric GCM (AGCM) and the coupled ocean–atmosphere GCM (CGCM) are quite different from each other, indicating that the air–sea interaction plays an important role in determining the climatology, and this result suggests us to tune the model physics and their parameters with CGCM rather than with AGCM.

Keywords Gray zone GCM · Cloud microphysics · Scale-dependent convection MJO

I.-S. Kang (✉)
Indian Ocean Operational Oceanographic Research Center, SOED,
Second Institute of Oceanography, Hangzhou, China
e-mail: insik.kang1@gmail.com

I.-S. Kang
Center of Excellence of Climate Change Research, King Abdulaziz University,
Jeddah, Saudi Arabia

M.-S. Ahn
Department of Atmospheric Sciences, University of Washington, Seattle, WA, USA

M.-S. Ahn
Department of Oceanography, Chonnam National University, Gwangju, Korea

D. A. Randall et al. (eds.), *Current Trends in the Representation of Physical Processes in Weather and Climate Models*, Springer Atmospheric Sciences,
https://doi.org/10.1007/978-981-13-3396-5_7

1 Introduction

Although moist physical parameterizations have been improved substantially in recent years, most of recent general circulation models (GCMs) still have problems in simulating the precipitation statistics (Kang et al. 2015) and the MJO as evaluated by Hung et al. (2013) and Ahn et al. (2017) with the CMIP5 models. In order to overcome the limitation of parameterized convection, several recent studies have attempted to include cloud microphysical processes, so-called "explicit convection", in regional and global models (Miura et al. 2007; Benedict and Randall 2009; Kang et al. 2015, 2016). Moncrieff and Klinker (1997) showed that the explicit convection produced a more realistic simulation of superclusters than the parameterized convection did. More recently, Holloway et al. (2013, 2015) performed MJO simulations with parameterized and explicit convection with varying horizontal resolutions and found better performance with explicit moist physics.

The global cloud resolving model (CRM), which expresses the cloud microphysical processes explicitly by GCM state variables, has been developed by a Japanese group since early 2000 (Tomita et al. 2005; Satoh et al. 2014, and many others). With several modifications and improvement of their earlier version of the global CRM, they recently reported that the model is able to reproduce the eastward propagation of the observed MJO and typhoon genesis reasonably well (Miura et al. 2007; Oouchi et al. 2009; Miyakawa et al. 2014; Kodama et al. 2015). One benefit of using CRM is to permit turbulent motions associated with clouds, and the other benefit of using CRMs is to permit a direct coupling between the fluid dynamics and the cloud microphysics. However, this approach requires very heavy computing resources due to an ultra-high horizontal resolution of an order of km, and therefore their climatological behaviors are not well described and many aspects of the global CRM are not known yet. The so-called "superparameterization" is a promising alternative strategy for representing the effects of moist convection explicitly through a cloud resolving model (CRM) embedded within each grid of a GCM with a horizontal resolution of about 100 km (Iorio et al. 2004; DeMott et al. 2007). The superparameterized GCM has been shown to simulate the MJO reasonably well with computational efficiency (Benedict and Randall 2009; Zhu et al. 2009), although the MJO intensity is somewhat exaggerated. It is pointed out that the superparameterization has some deficiencies in that it does not consider the interaction between clouds in neighboring GCM grids and in that the cloud properties of each grid point are quickly adjusted to the boundary conditions prescribed by GCM state variables.

Recently, Kang et al. (2015, 2016) presented a GCM with full representation of cloud microphysics at 50 km horizontal resolution. The cloud microphysics in their GCM was modified to be suitable to 50 km horizontal resolution. With their model, Kang et al. (2015) demonstrated that the cloud microphysics is essential to simulate the heavy and extreme precipitation statistics, which can not be simulated with a GCM with convective parameterization, and better simulates the MJO compared to the conventional GCM (Kang et al. 2016). In particular, Kang et al. (2016) demonstrated that the cloud microphysics itself is not enough to simulate the vertical profile of moisture field, particularly in the lower troposphere, in a 50 km resolution GCM, and a shallow convection is required to simulate the mean low-level moisture and its anomalies with MJO time scales. Recently, they found that the GCM used by Kang et al. (2016) still produces dry bias in middle and upper troposphere. For further improvement of their GCM, they have tried to add a scale-dependent deep convection to enhance the vertical transport of moisture and temperature, which helps improve the dry bias in middle and upper troposphere and MJO simulation.

In the present chapter, we describe a medium-range resolution GCM with cloud microphysics and a scale-dependent cumulus parameterization, based on the recent studies by Kang et al. (2015, 2016) and Ahn and Kang (2018). In particular, the development strategies of this GCM, and the simulation qualities of their precipitation climatology and MJOs are described and discussed. The target horizontal resolution is 50 km in the present study. The GCM and the cloud resolving model used are described in the next section, and Sect. 3 describes the cloud microphysics modified for a 50 km horizontal resolution and the performance of the GCM with the cloud microphysics. Section 4 describes a simple approach for developing a scale-dependent cumulus convection. Section 5 describes the AGCM and CGCM with the modified cloud microphysics and the scale-dependent cumulus convection with a horizontal resolution of 50 km and discusses their simulation results of precipitation climatology and MJO statistics. A summary and concluding remarks are given in Sect. 6.

2 The GCM and Cumulus Resolving Model

The Atmospheric GCM (AGCM) utilized is a Seoul National University (SNU) model. The dynamical core adapted in the model is a finite volume method with a hybrid sigma-pressure vertical coordinate developed by Lin (2004). The model is represented by a 50 km horizontal resolution and 20 vertical levels. The convective parameterizations include a deep convection represented by a simplified version of relaxed Arakawa-Schubert cumulus convection scheme (Moorthi and Suarez 1992) and a diffusion-type shallow convection scheme developed by Tiedtke (1984). Precipitation processes are formulated in terms of a convective precipitation expressed in the deep convection scheme and a large-scale condensation scheme based on Le Trent and Li (1991). Radiation processes are represented by the two-stream k-distribution scheme developed by Nakajima et al. (1995). A detailed description of the physical parameterizations of the AGCM can be found in Lee et al. (2001) and Kim and Kang (2012). The SNU CGCM is a coupled version of the SNU AGCM and the MOM2.2 Ocean GCM developed at Geophysical Fluid Dynamics Laboratory. The CGCM includes the mixed layer model developed by Noh and Kim (1999). The ocean zonal resolution of the CGCM is 1.0° and the meridional grid spacing is 1/3° between 8S and 8N, gradually increases to 3° between 30S and 30N, and 3° poleward. A detail description of the CGCM can be found in Ham et al. (2010).

The cloud resolving model (CRM) of the present study is obtained from the Goddard Cumulus Ensemble (GCE) Model developed at the Goddard Space Flight Center of National Aeronautic Space Administration (Tao et al. 2003). The GCE model includes a dynamical core, microphysics, radiation, surface flux, and sub-grid-scale turbulence scheme. It uses the compressible equations (Klemp and Wilhelmson 1978) with periodic lateral boundary conditions. The cloud microphysics includes the Kessler-type two-category liquid water scheme and the three-category ice-phase scheme developed by Lin et al. (1983). In this study, a two-dimensional and a three-dimensional versions of the model are used for the tropical domain of 300 km with the horizontal grid size of 1 km. The vertical resolution is about 80 m near the surface and gradually increases with height up to about 700 m near the 10 km level. A two-dimensional CRM is used for the sensitivity experiments for various horizontal resolutions in Sect. 3, and a three-dimensional CRM is used for developing a scale-dependent convection scheme for the GCM in Sect. 4.

3 A Modified Cloud Microphysics at 50 km Horizontal Resolution

The resolution dependency of cloud microphysics has been examined by a number of authors (Weisman et al. 1997; Grabowski et al. 1998; Bryan et al. 2003; Jung and Arakawa 2004; Pauluis and Garner 2006; Arakawa et al. 2011; Bryan and Morrison 2012; Kang et al. 2015). In particular, Kang et al. (2015) have demonstrated a significance resolution dependency of the cloud microphysics, and they have developed a modified version of cloud microphysics suitable for the 50 km resolution. The major parts of modification are for the condensation process and the terminal velocity. The original CRM condensation formula is replaced by the large-scale condensation formula of Le Trent and Li (1991), which is commonly used in a GCM, except that the relative humidity criterion for condensation is changed to 90%. Also, the terminal velocity formula is modified by changing the coefficient determining its speed to a half of its original value. The details are described in Kang et al. (2015).

Figure 1 shows the vertical profiles of cloud hydrometers and various microphysical processes simulated by the CRMs with the original cloud microphysics of 1 and 50 km resolutions and with the modified microphysics at 50 km resolution. The modified cloud microphysics at 50 km resolution (red lines) produces the vertical profiles close to those of 1 km resolution (black lines). Figure 1a–d shows the vertical profiles of cloud water, graupel/hail, rainwater, and cloud ice from the model of 1 km resolution (black line) and the models of 50 km resolution with (red line) and without modification (blue dotted line). Figure 1e–h shows the vertical profiles of various microphysical processes: the accretion of cloud water by rainwater (Fig. 1e), accretion of cloud water by graupel (Fig. 1f), melting of graupel (Fig. 1g), and condensation (Fig. 1h) from the models with 1 and 50 km resolutions. It is noted that the water species (cloud water and rainwater) are less sensitive to the horizontal res-

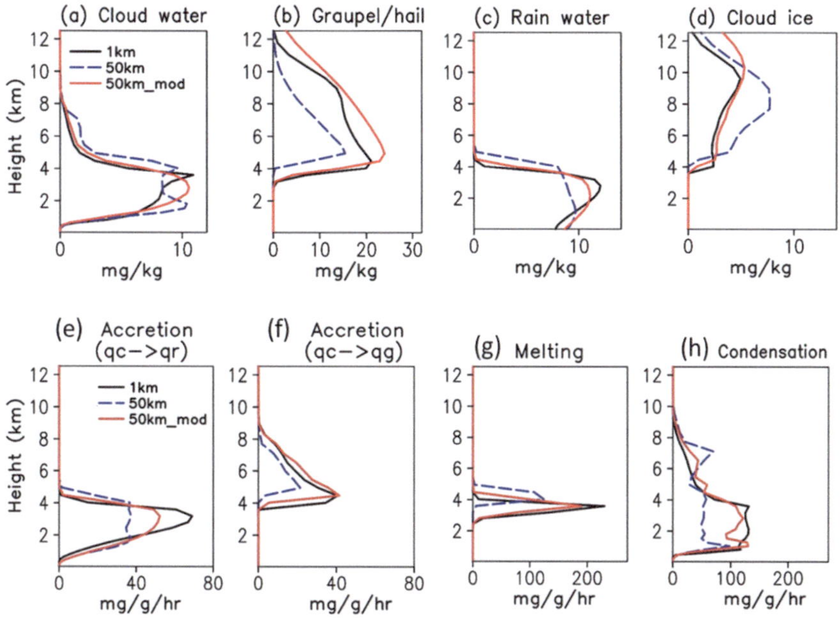

Fig. 1 Vertical profiles of domain averaged **a** cloud liquid water, **b** graupel/hail, **c** rainwater, and **d** cloud ice water, **e** accretion of cloud water by rainwater, **f** accretion of cloud water by graupel, **g** melting of graupel, and **h** condensation from the CRM with 1 km horizontal resolution (black solid line), 50 km horizontal resolution (blue dashed line), and the CRM with 50 km horizontal resolution but modified with GCM condensation formula (Le Trent and Li 1991) and terminal velocity reduced by 50% (red solid line)

olution compared to those of ice species (graupel and cloud ice). As the horizontal resolution becomes coarse, most cloud microphysics processes weaken and become underestimated, especially the accretion and condensation processes (Fig. 1e–h). But, as mentioned above, the modification improves the simulation of the cloud microphysical processes at 50 km resolution.

The modified cloud microphysics was implemented in a SNU AGCM at 50 km resolution, where the convective parameterizations (both deep convection and large-scale condensation schemes) were replaced by the modified cloud microphysics described above. In this GCM, the cloud hydrometers are treated as prognostic variables, and the cloud microphysics are computed explicitly by GCM state variables. As expected and shown by Kang et al. (2015), the model with the modified cloud microphysics at 50 km resolution produced relatively large moisture biases at the low troposphere, due to insufficient vertical transport of moisture from the surface to the free atmosphere. It is due to relatively weak vertical velocity in the coarse horizontal resolution of the model, and thus they added a diffusion-type shallow convection to remove the excessive low-level moisture bias. An experiment with the AGCM with the cloud microphysics and shallow convection was made for 4

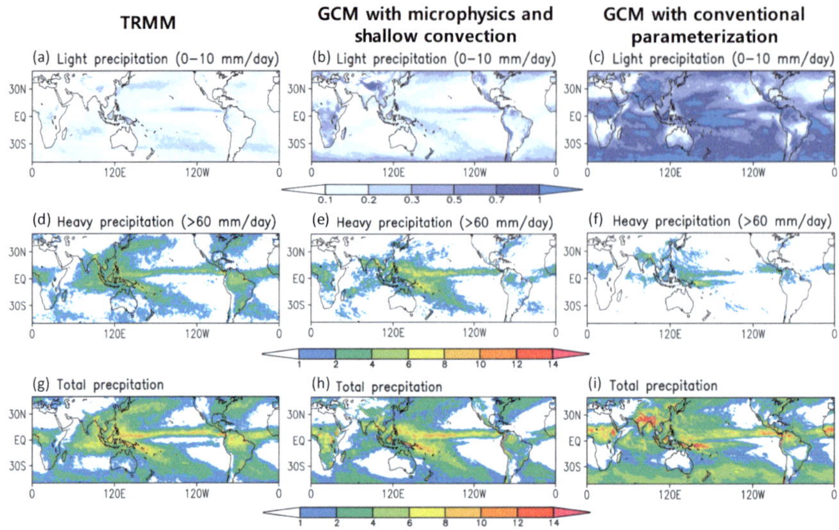

Fig. 2 Global distributions of light (upper), heavy (middle) and total mean precipitation (lower) from TRMM (**a**, **d** and **g**), AGCM with the modified cloud microphysics and shallow convection (**b**, **e** and **h**), and AGCM with a BULK convection scheme (**c**, **f** and **i**). 3-hourly precipitation data for 4 years are used. The light and heavy precipitation shown above are, respectively, the average of precipitation less than 10 mm/day and the average of heavy precipitation exceeding 60 mm/day

years with climatologically varying SST condition over the globe. Figure 2 shows the global distributions of light (Fig. 2b), heavy (Fig. 2e), and total (Fig. 2h) precipitation averaged for a simulation period. Here, three hourly data is used, and the light and heavy precipitation are the precipitation less than 10 mm/day and that more than 60 mm/day, respectively. Corresponding figures obtained with the TRMM precipitation data are shown in Fig. 2a, d, g, and the corresponding figures obtained with a conventional GCM with the parameterized convection (original SNU AGCM) are shown in Fig. 2c, f, i. As seen in figure, the convectional AGCM produces too much light rain and much less heavy precipitation compared the TRMM counterparts, which is a common feature of other convectional GCMs (Chen and Knutson 2008; Wehner et al. 2010; Li et al. 2011). However, the GCM with cloud microphysics simulates the light precipitation (Fig. 2b) and the heavy precipitation amount (Fig. 2e) close to those of TRMM observation. It is well known that the conventional GCMs produce the convection too frequently to release the convective instability, resulting in too much light rain. On the other hand, as shown by Kang et al. (2015), a major portion of heavy precipitation is from not only the cloud liquid water but also various ice species, particularly graupel. The graupel is made mainly from the accretion process of cloud water, and some of the graupel is made by freezing process of rainwater and by accretion of snow. In most of conventional GCMs without those cloud microphysics, however, the rainwater is made directly by the cloud water with an auto-conversion time scale and partly by the cloud ice without any delayed time scale, meaning that those GCMs do not include the major rain processes for heavy precipitation, such as growth of graupel and its melting process. In order words, the GCM needs to include the microphysical processes associated with ice phases of cloud water, such as snow and graupel, for simulating heavy precipitation statistics realistically.

4 A GCM with Resolution Dependent Cumulus Convection

In a subsequent study of Kang et al. (2015, 2016) examined the MJO structure of the model developed by Kang et al. (2015), which was much improved compared to that of a conventional GCM with convective parameterizations. But, they found that the moisture anomalies associated with the MJO have large dry biases in the upper troposphere. This problem motivated us to think of including a certain portion of the deep convection generated by the convection scheme in the original AGCM. The convection scheme, a simplified Arakawa-Schubert (SAS) scheme, in the present GCM was developed for a GCM with a horizontal resolution of an order of several hundred kilometers. However, the sub-grid-scale convective mixing, which is expressed by a convective parameterization, should be dependent on the horizontal grid size; we need to consider modifying the convection scheme suitable to the 50 km resolution model. The resolution dependency of sub-grid-scale vertical mixing is examined using a three-dimensional CRM of 1 km horizontal resolution in a radiative–convective equilibrium condition (Ahn and Kang 2018).

The moist static energy $\mathrm{h = CpT + gz + Lq}$ of an air partial is a conserved property with respect to the vertical motion, total vertical transport of moisture static energy over a GCM grid $\overline{w\frac{\partial \mathrm{h}}{\partial \mathrm{z}}}$ is the sum of the grid-scale vertical transport $\overline{w}\frac{\partial \overline{h}}{\partial z}$ and the sub-grid-scale vertical transport $\overline{w'\frac{\partial h'}{\partial z}}$. All the variables used here are expressed with conventional notations, and the bar indicates the area average over the GCM grid. The area average is done only over the updraft domains $\overline{w} > 0$ and $\frac{\partial \overline{h}}{\partial z} < 0$. In a GCM, the sub-grid vertical transport is usually obtained by a cumulus parameterization. Here both the sub-grid-scale vertical transport and the total vertical transport at 850 hPa for various grid sizes, e.g., 50 km $\times$ 50 km, are calculated using the CRM simulation data of 1 km horizontal resolution, and the ratios of the sub-grid-scale vertical transport to the total vertical transport with respect to the grid size are shown in Fig. 3. As expected, as the grid size increases, the fraction of grid-scale vertical transport decreases (Fig. 3a), the sub-grid-scale transport increases (Fig. 3b), and thus the ratio increases (Fig. 3c). The result indicates that the convective mixing plays a more important role to the total vertical mixing as the grid size increases. The value of the ratio is close to 1 for 280 km and about 0.62 for 50 km resolution, indicating that about 60% of the convective condensation and heating produced by the SAS scheme developed for a grid size of about 300 km could be applied to the model of 50 km resolution. However, as shown by the blue bars in Fig. 3c, without any change (no resolution dependency) of the present convection scheme, the ratio of

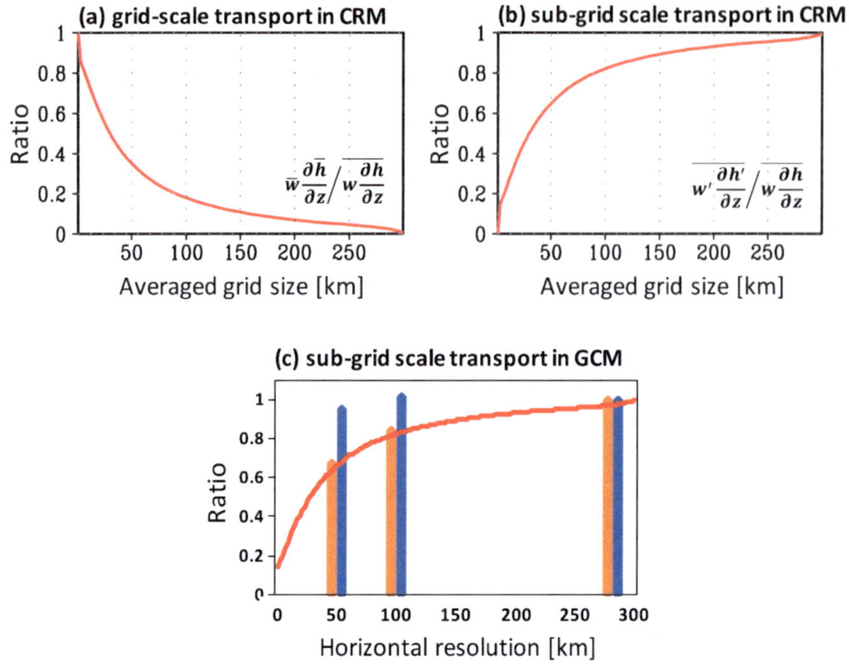

Fig. 3 Resolution dependency of the vertical transport of moist static energy at 850 hPa. **a** For the ratio of the grid-scale vertical transport to the total vertical transport for various grid sizes, **b** the ratio of the sub-grid-scale vertical transport to the total vertical transport for various grid sizes, and **c** the ratio of the sub-grid-scale vertical transport to the total vertical transport for the domains of 50, 100, and 280 km. The data used is from the CRM simulation with 1 km horizontal resolution. The blue bar in (**c**) is for the GCM simulations without any change of convection scheme, and the orange bar for the GCM simulations with a scale-dependent convection scheme. Red curve in (**c**) is same as that of (**b**). The GCM ratio is the normalized value with respect to the 280 km GCM simulation result

the sub-grid-scale convective transport to the total vertical transport does not change for the GCMs with different horizontal resolutions.

In the present study, the resolution dependency of deep convection from the SAS scheme is controlled by changing the cloud base mass flux. In the SAS scheme, the convection is determined not only by the cloud base mass flux but also more importantly by the buoyance inside the cloud. Therefore, the convection is not simply proportional to the cloud base mass flux. In the present study, the reduction factor (multiplication coefficient) of the cloud base mass flux for a GCM with a different resolution (e.g., 50 km) is determined by many sensitivity experiments with different values of the reduction coefficient. For example, the reduction coefficient for the GCM with 50 km resolution is determined when the convective ratio of the GCM is close to 60% of the ratio of the GCM of 280 km resolution. See the details in Ahn and Kang (2018). In Fig. 3c, the red bars show the convection ratios of the

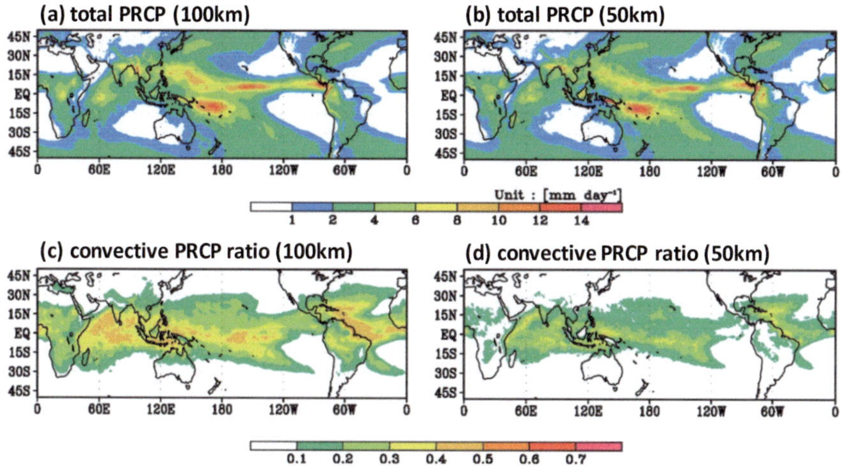

Fig. 4 Annual mean precipitation of the GCM with a scale-dependent convection scheme for **a** 100 km and **b** 50 km horizontal resolution. **c** and **d** are the ratio of the convective precipitation to the total precipitation for the GCMs of (**a**) and (**b**), respectively

GCMs with horizontal resolutions of 50 km and 100 km, which are obtained by the reduction coefficients of the cloud base mass flux of 0.092 and 0.16, respectively. Here, we refer the convection scheme with scale-dependent cloud mass flux to as "scale-dependent convection scheme." The resolution dependency of the convective rain amount is examined with the original GCM (without the cloud microphysics) but with the scale-dependent convection scheme. As shown in Fig. 4a, b, the precipitation climatology of the GCM is not sensitive to the resolution. However, the convective rain ratio to the total precipitation is sensitive to the model resolution. As shown in Fig. 4c, d, the present GCM with a scale-dependent convection scheme produces the convective rain ratio about 20–50% in the Indian and western Pacific oceans between 15°S and 15°N for 100 km resolution and 10–40% for 50 km resolution. It is also noted that the GCM with the scale-dependent convection scheme simulates the MJO and heavy rain statistics much better than those of the original GCM, as documented by Ahn and Kang (2018).

5 GCMs with Cloud Microphysics and Scale-Dependent Cumulus Convection

The scale-dependent convection scheme is added in the GCM with the modified cloud microphysics and the shallow convection (MS-GCM), and the GCM with all of the components including the scale-dependent convection will be referred to as the "MC-GCM". In the MC-GCM, the condensation produced by the convection

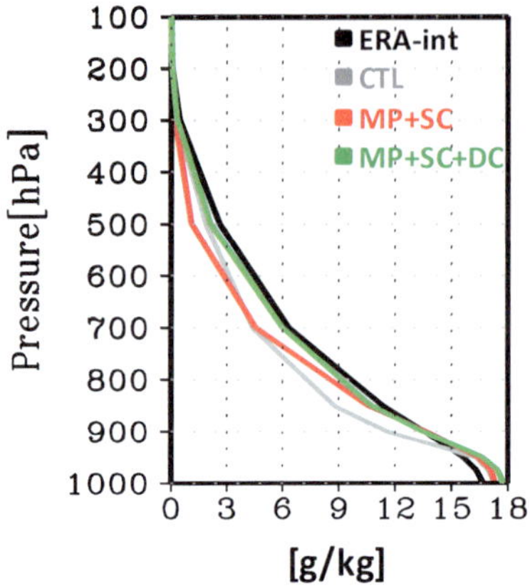

Fig. 5 Vertical profiles of specific humidity averaged over 60°E–180°E and 15°S–15°N for ERA-interim (black curve), the conventional GCM (gray curve), the GCM with modified cloud microphysics and shallow convection (red curve), and the GCM with modified cloud microphysics, shallow convection and scale-dependent convection (green curve). 5-year simulation data is used for the models and 10-year data is used for the ERA-interim

scheme is passed to the cloud microphysics scheme, so that the moisture, hydrometers, and precipitation are all produced through the cloud microphysical processes, and the convective heating is added in the temperature equation. It is also noted that the MC-GCM of 50 km resolution does not use a sub-time interval for the cloud microphysics calculation but the time interval of model integration is reduced to 600 s for all GCM and microphysics variables except the terms with the terminal velocity computed every 20 s. The MC-AGCM with climatologically varying SST prescribed was integrated for 5 years with 50 km horizontal resolution. The vertical profile of tropical mean moisture simulated with the MC-GCM is shown in Fig. 5 (green line) along with those of observation (black line), the original GCM without cloud microphysics (CNT, gray line), and the MS-GCM (red line). The dry bias of moisture in the middle and upper troposphere appearing in the MS-GCM, mentioned in the previous section, is improved in the MC-GCM, whose vertical profile (green) is close to the observed one (black). The climatological mean precipitation simulated by the MC-GCM is shown in Fig. 6. The global pattern of the figure is, in general, similar to that of the TRMM observation shown in Fig. 2g, but there are differences in regional details: the MC-GCM produces more precipitation in the western Pacific particularly in the Southern Hemisphere and India and Indo-China region but less in the eastern Indian Ocean. It is noted that those precipitation biases are also appearing in the original GCM with convective parameterization shown in Fig. 2i. The common biases of the models with different precipitation processes may indicate that the biases could be related to problems in the physical parameterizations other than the precipitation processes and the air–sea interaction missing in both models, which will be discussed later in this section.

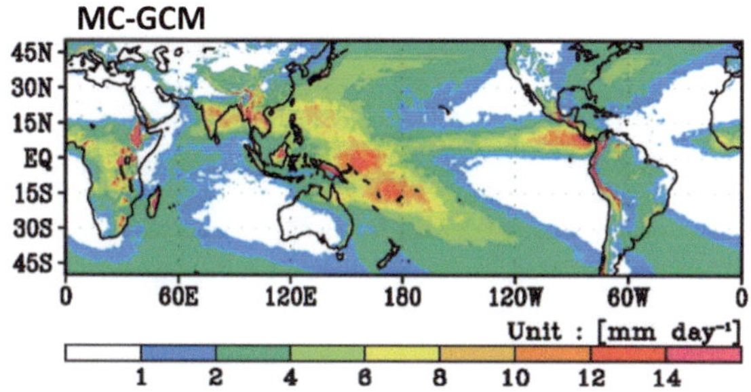

Fig. 6 Spatial distribution of annual mean precipitation for the AGCM with modified cloud microphysics, shallow convection, and scale-dependent convection. 5-year simulation data is used

The impact of cloud microphysics, shallow and deep convections on the MJO is examined by comparing the simulation results of the models with those components added one by one in the GCM. Figure 7 shows the Hovmöller diagrams of the precipitation along the equation (10°S–10°N average) for the three months of 1 January–31 March. Figure 7a shows the TRMM precipitation and Fig. 7b is for the original GCM with convective parameterizations (CNT). The zonal propagation characteristics of precipitation simulated by the original GCM is different from that of TRMM and its MJO is very weak. Figure 7c is the precipitation simulated by the GCM with the modified cloud microphysics only (MP-GCM) without the shallow and deep convection schemes. The MP-GCM simulates the eastward propagating signal clearly but with the propagation speed relatively fast and the eastward propagations happen too much regularly. The comparison of the results shown in Fig. 7b, c indicates that the cloud microphysics contributes to generating the eastward propagating precipitation signal. But the precipitation pattern is relatively small scale and is scattered (not shown), and therefore a mechanism of aggregating the individual clouds into large-scale cloud cluster may be needed to simulate the MJO realistically. Adding the shallow convection in the MP-GCM (MS-GCM, Fig. 7d) does not much change the propagation characteristics shown in Fig. 7c. But, the MC-GCM with additional deep convection produces the large-scale precipitation cluster slowly moving eastward (Fig. 7e). The result indicates that the deep convective mixing would be needed to generate the large-scale cloud cluster and contributes to slow down the eastward propagation. But, the speed is too much slow compared to the observed. This slow-moving problem appears to be related to the increase of the residence time of hydrometers within the air by adding the deep convection. In the present, modified cloud microphysics of 50 km resolution, the terminal velocity is reduced to the half of the value from the original formula to adjust the falling speed of hydrometers appropriate to the 50 km resolution. But, adding the deep convection, the water vapor and hydrometers are transported to the upper troposphere without a time delay, resulting in additional suspension of the hydrometers in the air. We attempted to increase the terminal velocity in the MC-GCM. The precipitation shown in Fig. 7f is for the MC-GCM but with the terminal velocity increased. Now the original formula of terminal velocity (without multiplying 0.5) is used in the MC-GCM. The propagation characteristics of precipitation, shown in Fig. 7f, appears to be close to

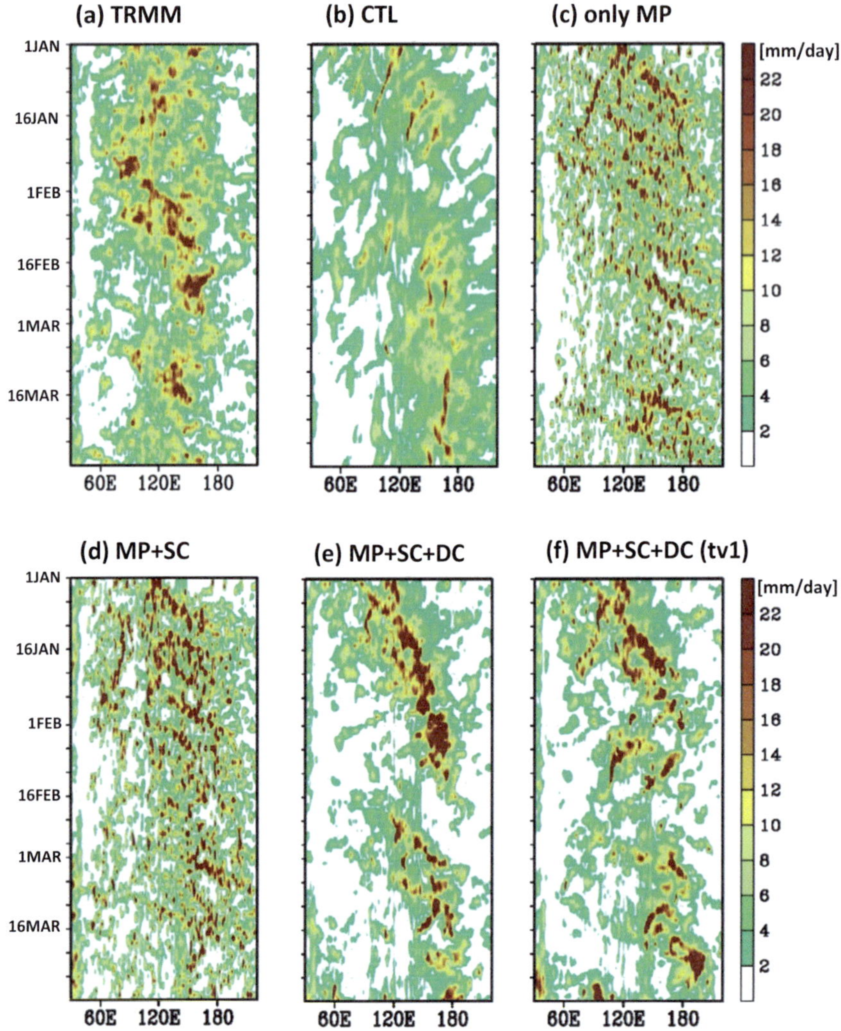

Fig. 7 Hovmöller diagrams of 10°S–10°N averaged precipitation for **a** TRMM, **b** the conventional GCM, **c** the GCM with modified cloud microphysics, **d** the GCM with modified cloud microphysics and shallow convection, **e** the GCM with modified cloud microphysics, shallow convection and scale-dependent deep convection, and **f** as in (**e**) except the terminal velocity computed with the CRM original formula, which is twice bigger than that used in (**e**)

the TRMM observation, indicating that the terminal velocity and residence timescale of hydrometers could play an important role in simulating the observed MJO.

As discussed above, some of biases of the precipitation simulated by the MC-AGCM may be related to the lack of air–sea coupling. To examine this possibility, the MC-AGCM is coupled with the ocean components of the SNU CGCM, which are

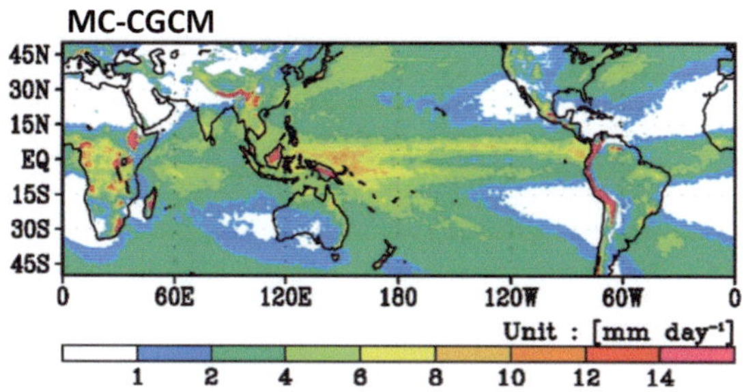

Fig. 8 Spatial distribution of annual mean precipitation for the CGCM with modified cloud microphysics, shallow convections and scale-dependent deep convection. Shown is the mean of 5-year simulation

the MOM2 GFDL global ocean model and the ocean mixed layer model developed by Noh and Kim (1999). Interestingly, the MC-CGCM (Fig. 8) produces a quite different distribution of precipitation from that of MC-AGCM (Fig. 6) over the tropical oceans, particularly the western Pacific, indicating that the air–sea interaction influences the precipitation climatology over the tropical oceans significantly. Overall, the coupled model appears to better simulate the precipitation intensity over the tropical oceans, particularly in the western Pacific, compared to the AGCM does, although the MC-GCM still has large differences from the observation. The double ITCP, which is a common problem in most CGCMs, still appears in the MC-CGCM. It is noted that the heavy precipitation biases over mountain regions, appearing in the present GCM, are related to a topography-related problem of the finite volume dynamical core adapted in the present model. It is noted that the coupled model with a spectral dynamical core does not have such a problem (not shown). The topography-related problem is not overcome yet. Here we have shown a result of the MC-CGCM, an ocean coupled version of MC-AGCM without any tuning, in order to examine the effect of ocean–atmosphere coupling on the simulation of precipitation. The MJO of the MC-CGCM is examined based on the simulation data for 5 years boreal winter (November–April). Figure 9a and 9b show the wavenumber–frequency power spectra of precipitation for the TRMM observation and the MC-CGCM, respectively. The propagation speed in the MC-GCM is somewhat faster and the intensity is stronger than those of the TRMM observation. However, the MC-CGCM simulates the MJO reasonably well compared to other CMIP5 models (Ahn et al. 2017). Whereas a large power spectrum of the convectional CGCM with parameterized convection schemes is shifted to the low frequency longer than 100 days (Fig. 9d), which is not appearing in the observation, the MC-GCM have two spectral peaks, a relatively small peak at about 100 days and a major peak at about 30–40 days. This spectrum-frequency characteristic of the coupled model is not much different from that of the AGCM (Fig. 9c), although the eastward propagating low-frequency spectrum and westward propagating signal are somewhat reduced and closer to the observed counterparts in the coupled model. Overall, the results discussed above indicate that both cloud microphysics and ocean–atmosphere coupling contribute to improving the MJO, but a combination of the cloud microphysics and the scale-dependent deep convection appears to play a major role for a realistic simulation of the MJO.

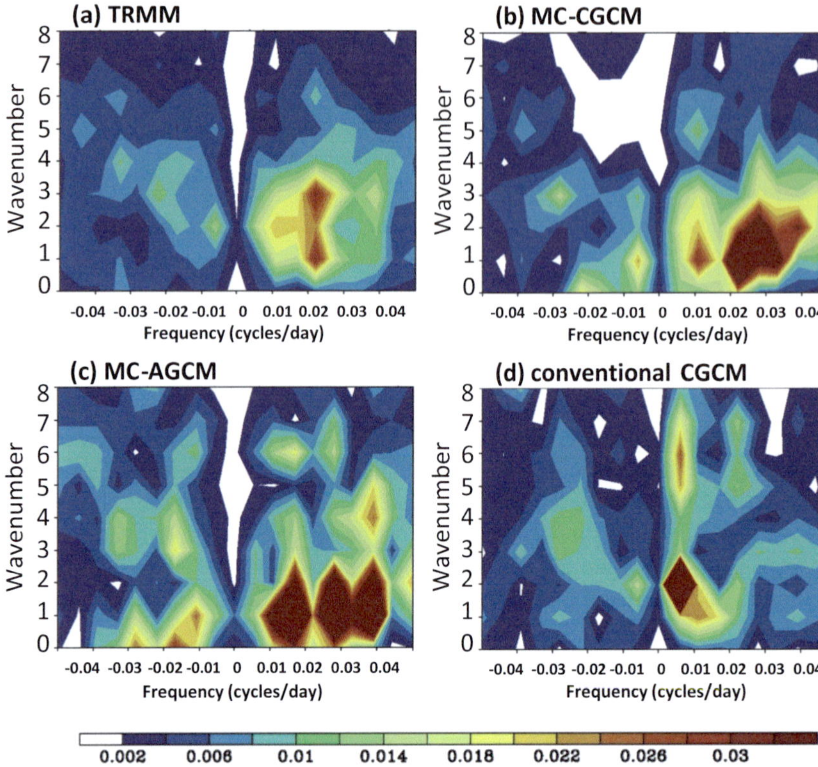

Fig. 9 November–April wavenumber–frequency power spectra of 10°S–10°N averaged precipitation for **a** TRMM, **b** MC-CGCM, **c** MC-AGCM, and **d** the conventional CGCM. The MC indicates the model with modified cloud microphysics, shallow convection, and scale-dependent deep convection. 5-year simulation data is used for the models and 13-year data is used for the TRMM

6 Summary and Conclusion Remarks

This chapter described the development procedure of general circulation models with full representation of cloud microphysics at a medium-range horizontal resolution of 50 km and examined their simulated precipitation climatology and the Madden and Julian Oscillation. One issue of developing an order of 10 km resolution GCM is to modify the cloud microphysics suitable to the horizontal resolution. For the present model, the modification is made based on sensitivity experiments for the parameters of the important processes sensitive to the model resolution, particularly the condensation process and the terminal velocity. It is demonstrated that the shallow and deep convections are still needed in the present model of 50 km resolution with cloud microphysics. The present AGCM with the cloud microphysics and paramterized convections is shown to simulate the precipitation statistic, such as the light and heavy precipitation frequencies, reasonably well (Kang et al. 2015), and the MJO is simulated with a rather stronger intensity than the observed. The MJO characteristics of the present CGCM is similar to that of the AGCM. However, the precipitation climatologies of the AGCM and CGCM are quite different from each other, indicating that the air–sea interaction plays an important role in determining the climatology, and this result suggests us to tune the model physics and their parameters with CGCM rather than with AGCM.

The conventional AGCM and CGCM show a lack of eastward propagating power within the MJO wavenumber and period as shown in Fig. 7b and Fig. 9d, respectively. The MC-AGCM (Fig. 9c) and MC-CGCM (Fig. 9b), on the other hand, produce strong eastward propagation in the wavenumbers and frequencies close to the observed counterparts, although the MJO-like signals of both modes are somewhat stronger than the observed. It is shown that the implementation of the cloud microphysics alone helps simulate eastward wave propagation along the equator, but the propagation speed is relatively fast and precipitation is more or less scattered and not well organized in the tropics. In the present study, the scale-dependent convection is added to enhance the deep convective mixing in the GCMs with cloud microphysics and shallow convection, which results in the MJO-like eastward propagation of precipitation somewhat stronger than the observed. It is noted that the vertical profile of moisture field is better simulated in the MC-GCMs in the upper troposphere, compared to those of the models without the deep convection. The better moisture field induces enhancement of convection-moisture coupling, resulting in the stronger MJO. Ahn et al. (2017) demonstrated that a sufficient degree of convective-moisture coupling is an important factor of MJO simulation in the CMIP5 models. It is also noted that a number of authors after Lee et al. (2001) demonstrated that the long-wave radiation associated with convection plays an important role on the MJO. From those studies, we may anticipate that the cloud-radiation feedback, which is better represented in the present GCMs with full representation of cloud microphysics, also helps simulate the MJO with a strong intensity. This subject would be a good topic for future research.

References

Ahn, M.-S., and I.S. Kang. 2018. A practical approach to scale-adaptive deep convection in a GCM by controlling the cumulus base mass flux. *npj Climate and Atmospheric Science* (in revision).

Ahn, M.-S., D. Kim, K.R. Sperber, I.S. Kang, E. Maloney, D. Waliser, H. Hendon on behalf of WGNE MJO Task Force. 2017. MJO simulation in CMIP5 climate models: MJO skill metrics and process oriented diagnosis. *Climate Dynamics* 49: 4023–4045.

Arakawa, A., J.-H. Jung, and C.-M. Wu. 2011. Toward unification of the multiscale modeling of the atmosphere. *Atmospheric Chemistry and Physics* 11 (8): 3731–3742.

Benedict, J.J., and D.A. Randall. 2009. Structure of the Madden-Julian oscillation in the superparameterized CAM. *Journal of the Atmospheric Sciences* 66 (11): 3277–3296.

Bryan, G.H., and H. Morrison. 2012. Sensitivity of a simulated squall line to horizontal resolution and parameterization of microphysics. *Monthly Weather Review* 140 (1): 202–225.

Bryan, G.H., J.C. Wyngaard, and J.M. Fritsch. 2003. Resolution requirements for the simulation of deep moist convection. *Monthly Weather Review* 131 (10): 2394–2416.

Chen, C.-T., and T. Knutson. 2008. On the verification and comparison of extreme rainfall indices from climate models. *Journal of Climate* 21: 1605–1621.
DeMott, C.A., D.A. Randall, and M. Khairoutdinov. 2007. Convective precipitation variability as a tool for general circulation model analysis. *Journal of Climate* 20: 91–112.
Grabowski, W.W., X. Wu, M.W. Moncrieff, and W.D. Hall. 1998. Cloud-resolving modeling of cloud systems during Phase III of GATE. Part II: Effects of resolution and the third spatial dimension. *Journal of the Atmospheric Sciences* 55 (21): 3264–3282.
Ham, Yoo-Geun, Jong-Seong Kug, In-Sik Kang, Fei-Fei Jin, and Axel Timmermann. 2010. Impact of diurnal atmosphere-ocean coupling on tropical climate simulations using a coupled GCM. *Climate Dynamics* 34: 905–917.
Holloway, C.E., S.J. Woolnough, and G.M.S. Lister. 2013. The effects of explicit versus parameterized convection on the MJO in a large-domain high-resolution tropical case study. Part I: Characterization of large-scale organization and propagation*. *Journal of the Atmospheric Sciences* 70: 1342–1369.
Holloway, C.E., S.J. Woolnough, and G.M.S. Lister. 2015. The effects of explicit versus parameterized convection on the MJO in a large-domain high resolution tropical case study. Part II: Processes leading to differences in MJO development. *Journal of the Atmospheric Sciences* 72: 2719–2743.
Hung, M.-P., J.-L. Lin, W. Wang, D. Kim, T. Shinoda, and S.J. Weaver. 2013. MJO and convectively coupled equatorial waves simulated by CMIP5 climate models. *Journal of Climate* 26: 6185–6214.
Iorio, J., P. Duffy, B. Govindasamy, S. Thompson, M. Khairoutdinov, and D. Randall. 2004. Effects of model resolution and subgrid-scale physics on the simulation of precipitation in the continental United States. *Climate Dynamics* 23 (3–4): 243–258.
Jung, J.-H., and A. Arakawa. 2004. The resolution dependence of model physics: Illustrations from nonhydrostatic model experiments. *Journal of the Atmospheric Sciences* 61 (1): 88–102.
Kang, I.-S., Y.-M. Yang, and W.-K. Tao. 2015. GCMs with implicit and explicit representation of cloud microphysics for simulation of extreme precipitation frequency. *Climate Dynamics* 45: 325–335.
Kang, I.-S., M.-S. Ahn, and Y.-M. Yang. 2016. A GCM with cloud microphysics and its MJO simulation. *Geoscience Letters* 3: 16.
Kim, D., and I.-S. Kang. 2012. A bulk mass flux convection scheme for climate model: Description and moisture sensitivity. *Climate Dynamics* 38: 411–429.
Klemp, J.B., and R. Wilhelmson. 1978. The simulation of three-dimensional convective storm dynamics. *Journal of the Atmospheric Sciences* 35: 1070–1096.
Kodama, C., Y. Yamada, A.T. Noda, K. Kikuchi, Y. Kajikawa, T. Nasuno, T. Tomita, T. Yamaura, H.G. Takahashi, M. Hara, Y. Kawatani, M. Satoh, and M. Sugi. 2015. A 20-year climatology of a NICAM AMIP-type simulation. *Journal of the Meteorological Society of Japan* 93: 393–424.
Le Trent, H., and Z.-X. Li. 1991. Sensitivity of an atmospheric general circulation model to prescribed SST changes: Feedback effects associated with the simulation of cloud optical properties. *Climate Dynamics* 5 (3): 175–187.
Lee, M.-I., I.-S. Kang, J.-K. Kim, and B.E. Mapes. 2001. Influence of cloud-radiation interaction on simulating tropical intraseasonal oscillation with an atmospheric general circulation model. *Journal Geophysical Research* 106 (14): 219–233.
Li, F., W.D. Collins, M.F. Wehner, D.L. Williamson, J.G. Olson, and C. Algieri. 2011. Impact of horizontal resolution on simulation of precipitation extremes in an aqua-planet version of community atmospheric model (CAM3). *Tellus A* 63: 884–892.
Lin, S.-J. 2004. A "vertically Lagrangian" finite-volume dynamical core for global models. *Monthly Weather Review* 132 (10): 2293–2307.
Lin, Y.-L., R.D. Farley, and H.D. Orville. 1983. Bulk parameterization of the snow field in a cloud model. *Journal of Applied Meteorology and Climatology* 22 (6): 1065–1092.
Miura, H., M. Satoh, T. Nasuno, A.T. Noda, and K. Oouchi. 2007. A Madden-Julian oscillation event realistically simulated by a global cloud-resolving model. *Science* 318 (5857): 1763–1765.

Miyakawa, T., M. Satoh, H. Miura, H. Tomita, H. Yashiro, A.T. Noda, Y. Yamada, C. Kodama, M. Kimoto, and K. Yoneyama. 2014. Madden-Julian oscillation prediction skill of a new-generation global model. *Nature Communications* 5: 3769.

Moorthi, S., and M.J. Suarez. 1992. Relaxed Arakawa–Schubert: A parameterization of moist convection for general circulation models. *Monthly Weather Review* 120: 978–1002.

Moncrieff, M.W., and E. Klinker. 1997. Organized convective systems in the tropical western Pacific as a process in general circulation models. *Quarterly Journal of the Royal Meteorological Society* 123: 805–828.

Nakajima, T., M. Tsukamoto, Y. Tsushima, and A. Numaguti. 1995. Modelling of the radiative processes in an AGCM. *Climate System Dynamics and Modelling* 3: 104–123.

Noh, Y., and H.J. Kim. 1999. Simulations of temperature and turbulence structure of the oceanic boundary layer with the improved nearsurface process. *Journal of Geophysical Research* 104: 15621–15634.

Oouchi, K., A.T. Noda, M. Satoh, H. Miura, H. Tomita, T. Nasuno, and S. Iga. 2009. A simulated preconditioning of typhoon genesis controlled by a boreal summer Madden-Julian Oscillation event in a global cloud-system-resolving model. *SOLA* 5: 65–68.

Pauluis, O., and S. Garner. 2006. Sensitivity of radiative-convective equilibrium simulations to horizontal resolution. *Journal of the Atmospheric Sciences* 63 (7): 1910–1923.

Satoh, M., and Coauthors. 2014. The non-hydrostatic icosahedral atmospheric model: Description and development. *Progress in Earth and Planetary Science* 1: 18.

Tao, W.-K., J. Simpson, D. Baker, S. Braun, M.-D. Chou, B. Ferrier, D. Johnson, A. Khain, S. Lang, and B. Lynn. 2003. Microphysics, radiation and surface processes in the Goddard Cumulus Ensemble (GCE) model. *Meteorology and Atmospheric Physics* 82 (1): 97–137.

Tiedtke, M. 1984. *Sensitivity of the time-mean large-scale flow to cumulus convection in the ECMWF model*, 297–316.

Tomita, H., H. Miura, S. Iga, T. Nasuno, and M. Satoh. 2005. A global cloud-resolving simulation: Preliminary results from an aqua planet experiment. *Geophysical Reseach Letters* 32: 1–4. https://doi.org/10.1029/2005GL022459.

Wehner, M.F., R.L. Smith, G. Bala, and P. Duffy. 2010. The effect of horizontal resolution on simulation of very extreme US precipitation events in a global atmosphere model. *Climate Dynamics* 34: 241–247.

Weisman, M.L., W.C. Skamarock, and J.B. Klemp. 1997. The resolution dependence of explicitly modeled convective systems. *Monthly Weather Review* 125 (4): 527–548.

Zhu, H., H. Hendon, and C. Jakob. 2009. Convection in a parameterized and superparameterized model and its role in the representation of the MJO. *Journal of the Atmospheric Sciences* 66 (9): 2796–2811.

The Stochastic Multi-cloud Model (SMCM) Convective Parameterization in the CFSv2: Scopes and Opportunities

B. B. Goswami, B. Khouider, R. Phani, Parthasarathi Mukhopadhyay and A. J. Majda

Abstract A stochastic multi-cloud model (SMCM) convective parameterization is incorporated in the National Centers for Environmental Predictions' Climate Forecast System version 2 (CFSV2). The resulting model is referred to here as CFSsmcm. Two 15-year-long climate simulations of the CFSsmcm, differing only by one SMCM parameter, namely, the mid-tropospheric dryness parameter, MTD0 are analyzed and interpreted here. This particular parameter is chosen because not only it plays a crucial role in the SMCM formulation, but also is observed to be critical for triggering tropical convection. In one case, we have used a single homogeneous MTD0 value for the entire globe and in the other run two different MTD0 values are used for land and ocean. The global precipitation climatology significantly improves in the inhomogeneous MTD0 case without significantly affecting the excellent performance of the CFSsmcm in terms of the intraseasonal and synoptic variability as documented in previous publications.

Keywords Stochastic · Parameterization · Convection · Cloud

B. B. Goswami (✉) · B. Khouider
Department of Mathematics and Statistics, University of Victoria, Victoria, BC, Canada
e-mail: bgoswami@uvic.ca

R. Phani · P. Mukhopadhyay
Indian Institute of Tropical Meteorology, Pune, India

A. J. Majda
Department of Mathematics and Center for Atmosphere and Ocean Sciences,
Courant Institute for Mathematical Sciences, New York University, New York City, USA

A. J. Majda
Center for Prototype Climate Models, New York University-Abu Dhabi,
Abu Dhabi, United Arab Emirates

D. A. Randall et al. (eds.), *Current Trends in the Representation of Physical Processes in Weather and Climate Models*, Springer Atmospheric Sciences,
https://doi.org/10.1007/978-981-13-3396-5_8

1 Introduction

The importance of the role played by a convective parameterization (CP) scheme can never be overemphasized in a global climate model (GCM). Most of the biases in a simulated climate originate from the inaccuracy in representing the subgrid-scale convective elements (Randall 2013; Arakawa 2004). Quest for an efficient CP scheme has been on for a few decades now (Kuo 1965; Arakawa and Schubert 1974; Betts and Miller 1986; Kain and Fritsch 1990; Gregory and Rowntree 1990; Zhang and McFarlane 1995). The assumptions these CP schemes are based on, stem from our understanding of atmospheric convection. However, there is one feature common to all these different schemes, they are all deterministic in nature. Or in other words, these schemes do not account for the subgrid-scale variability among the different convective elements. The basis for a deterministic convective parameterization is the underlying assumption that a typical GCM grid size is large enough to encompass a large ensemble of the clouds, which are in quasi-equilibrium with the large scales and that the large-scale mean ensemble is uniquely determined (Arakawa and Schubert 1974). However, with the increasing resolution of the present-day GCMs, the validity of this assumption needs to be reevaluated (Palmer 1996). Consequently, there is an undeniable possibility that neglecting the variability of the subgrid-scale convective elements may lead to biases in the mean climate (Palmer 2001). Efforts to adequately represent these convective systems in GCMs have led the scientific community to think beyond conventional CP schemes. Superparameterized GCMs (SP-GCM) (Grabowski 2001; Khairoutdinov and Randall 2001) and global cloud-resolving models (GCRM) (Satoh et al. 2005) (also see Randall (2013) for a review) are such promising approaches. However, SP-GCMs and GCRMs are computationally expensive and definitely unlikely candidates for operational centers, especially for ensemble predictions. Nevertheless, the success of these approaches highlighted the importance of accurate representation of the subgrid-scale (SGS) variability collectively while realizing the individual behavior of the convective elements, in the GCMs and their impact on the large resolved scales. In the spirit of superparameterization, a computationally significant less expensive approach was introduced in Khouider et al. (2010), the authors termed it as the stochastic multi-cloud model convective parameterization. This was the descendant of the same multi-cloud model introduced in Khouider and Majda (2006) but with the added feature of stochasticity.

Driven by the general consensus that a faithful representation in some way of the subgrid-scale convective variability is probably the only way forward, stochastic approaches to the convective parameterization problem are getting more attention in the recent times than ever before (Buizza et al. 1999; Lin and Neelin 2000, 2002, 2003; Palmer 2001; Majda and Khouider 2002; Khouider et al. 2003; Plant and Craig 2008; Teixeira and Reynolds 2008; Deng et al. 2015, 2016; Ajayamohan et al. 2016; Davini et al. 2016). In order to introduce stochasticity to an existing deterministic convective parameterization, different methods have been adopted. The perturbed parameterization tendencies approach introduced by Buizza et al. (1999) consists of multiplying the CP outputs by correlated or non-correlated random numbers at each

GCM column (Davini et al. 2016 and references therein). Teixeira and Reynolds (2008) followed a similar technique as Buizza et al. (1999) but they multiplied only the convective tendencies. Lin and Neelin (2000) added stochasticity to a deterministic scheme by adding a zero-mean red noise to its closure equation, namely, the convectively available potential energy (CAPE) closure equation. In the study by Lin and Neelin (2002), a distribution of precipitation is assumed a priori to control the statistics of the overall convective heating. Lin and Neelin (2003) tested a stochastic deep convective parameterization in a general circulation model for the first time. Plant and Craig (2008) used equilibrium statistical mechanics to derive a Poisson distribution for convective plumes based on radiative–convective equilibrium cloud-resolving simulations. Majda and Khouider (2002) and Khouider et al. (2003) used a Markov process on a lattice for convective inhibition. The stochastic lattice approach has been extended in Khouider et al. (2010) to derive the stochastic multi-cloud model (SMCM) designed to mimic the interactions at subgrid scales of multiple cloud types in the tropics. The SMCM has been extensively used and evaluated in simple models for organized convection and convectively coupled equatorial waves (CCEW) (Frenkel et al. 2012, 2013; Peters et al. 2013; De La Chevrotière et al. 2015; De La Chevrotière and Khouider 2017). Moreover, the SMCM has been successfully adopted as a cumulus parameterization in an aquaplanet GCM to simulate the Madden–Julian oscillation (MJO), CCEWs, and Indian summer monsoon intraseasonal oscillations (MISOs) (Deng et al. 2015, 2016; Ajayamohan et al. 2016). In this chapter, we present the highlights of the simulated climate when the SMCM is incorporated into the National Centers for Environmental Prediction (NCEP) Climate Forecast System version 2 (CFSv2) model (referred to as CFSsmcm hereafter) in lieu of the pre-existing simplified Arakawa-Schubert (SAS) cumulus scheme.

A first insight into the CFSsmcm-simulated climate is provided in Goswami et al. (2017a). They demonstrated that while retaining an equally good mean state (if not better) as the parent model (CFSv2), CFSsmcm significantly improved the synoptic and intraseasonal variability; provided a better account of convectively coupled equatorial waves and the Madden–Julian oscillation (MJO) exhibited better northward and eastward propagation of intraseasonal oscillation of convection including the MJO propagation beyond the maritime continent barrier. The distribution of precipitation events was also found to be better simulated in CFSsmcm which was severely biased toward too much drizzling precipitation in the parent model. An overview of the SMCM formulation, and the development and tuning of the CFSsmcm in detail can be found in Goswami et al. (2017b), where the model's sensitivity to the key parameters of the SMCM formulation is reported through a comparative analysis of a few 5-year-long climate simulations in order to distinguish the best possible set of SMCM parameters for the CFSsmcm model. The model was found to be the most sensitive to the mid-tropospheric dryness parameter (MTD) and to the stratiform cloud decay timescale (τ_{30}). MTD was more effective in controlling the global mean precipitation and its distribution while τ_{30} had more effect on the organization of convection as noticed in the simulation of the Madden–Julian oscillation (MJO). This is consistent with the fact that, in the SMCM formulation, mid-tropospheric humidity controls the deepening of convection and stratiform clouds control the backward tilt of tropospheric heating and the strength of unsaturated downdrafts which cool and dry the boundary layer and trigger the propagation of organized convection (Ajayamohan et al. 2016; Deng et al. 2016). Noteworthy, the CFSsmcm model was found to be robust in the sense that the simulated mean climate appeared resilient to small changes in the parameter values. A detailed analysis of the tropical intraseasonal variability (TISV) and convectively coupled equatorial waves (CCEW), in comparison with the parent GCM and with observations, was presented in Goswami et al. (2017c). Significant improvements were noted in the simulation of the Madden–Julian oscillation (MJO) and most of the CCEWs as well as the Indian summer monsoon (ISM) intraseasonal oscillation (MISO). The authors also demonstrated these improvements to be a result of improved mechanisms and physical structure of these oscillations. They also found that improved representation of interaction of the multiple clouds in the SMCM formulation holds the basis of this improved climate simulation by the CFSsmcm model. The SMCM has been used to modify the triggering of deep convection in the German GCM ECHAM4 and noticeable improvements are seen, especially in terms of the ability of the model to represent tropical rainfall variability (Peters et al. 2017). A variant of the SMCM has also been adopted and used to stochastisize an existing CP in Dorrestijn et al. (2016).

Upon the implementation of CFSsmcm, an extensive parameter testing has been conducted by making several short 5-year runs. A few of these simulations codified with whole numbers are reported in Goswami et al. (2017b) providing a first-hand analysis of the model's parameter sensitivity and behavior. In this chapter, we take Run 139 from Table 1 of Goswami et al. (2017b) and run it to simulate a 15-year-long climate and then compare the results with that of Run 129. It should be noted that Run 129 is the only CFSsmcm run which has been extensively analyzed and reported in detail in Goswami et al. (2017a, c). The Run 129 was selected from a number of 5-year-long simulations based on a few basic metrics reported in Goswami et al. (2017b). Some changes, more often good, in the simulated mean state and variability were noted when we ran Run 129 for 15 years. In Goswami et al. (2017b), the closest competitor to Run 129 was Run 139. The only difference between the two runs resides in the way the mid-tropospheric dryness parameter, MTD0, is prescribed. The physical significance of MTD0 is that it decides how moist the middle atmosphere needs to be to initiate convection. In the SMCM formulation, a small MTD0 means that the middle troposphere needs to be very moist to allow deep convection. From Goswami et al. (2017b), we recall that Run 129 uses a single uniform value of MTD0 $= 5$ for the entire globe, while in Run 139 we have set that MTD0 $= 5$ over the oceans and MTD0 $= 25$ over the continents. In other words, in Run 139, the atmosphere over the oceans wait longer to initiate convection than it does over continents. However, one MTD0 value for the entire globe implies no such distinction in Run 129. Goswami et al. (2017b) briefly discussed the benefits of using distinct MTD0 values, one over land and one over the oceans, over using a single MTD0 value for the entire globe. A few crucial improvements were achieved with the variable MTD0 runs including the precipitation climatology. In particular, the dry bias in the simulated Indian summer monsoon rainfall was significantly reduced. As a consequence, the poleward migrations of convection bands over the Indian monsoon region also had improved while the Takayabu–Wheeler–Kiladis (TWK) spectra (Takayabu 1994; Wheeler and Kiladis 1999) remained almost unchanged. They explained this improvement by the fact that the variable MTD0 affects primarily the mean, while the intraseasonal and synoptic variability are mostly affected by convection over the oceans.

The motivation behind this exercise is not to find out the better one between Run 129 and 139. Rather, we want to highlight the possibilities offered by the CFSsmcm model as a virtual laboratory to study the interaction between convection and cloud and the climate system. More than summarizing the results in a review mode, we wish to explore the scopes and opportunities of SMCM. Comparing the Runs 129 and 139, which differ only by one parameter value, may seem like just a simple tuning exercise, but because of the role played by that particular parameter, MTD0 (the scaling value for the middle tropospheric dryness), in the SMCM formulation, we expect to get valuable guidance toward improving the SMCM formulation further.

The rest of this chapter is organized as follows: the SMCM framework, including the developmental and implementation aspects, is explained in Sect. 2. Section 3 presents and compares the numerical results obtained with the two MTD0 configurations. Finally, a concluding discussion is provided in Sect. 4.

2 The SMCM Formulation

2.1 Parameterization of the Total Heating

The stochastic multi-cloud model uses three prescribed profiles for convective heating, ϕ_c, ϕ_d, and ϕ_s associated with cumulus congestus cloud decks (which warm and moisten the lower troposphere and cool the upper troposphere through radiation and detrainment), deep cumulonimbus clouds (which heat up the whole atmospheric column), and stratiform cloud types lagging deep convection (which heat the upper troposphere and cool the lower troposphere due to the evaporation of stratiform rain), respectively.

While in the original multi-cloud model (Khouider and Majda 2006, 2008), simple sine functions were used to set up the basis functions and Khouider et al. (2011) used the vertical mode eigenstructure of Kasahara and Puri (1981), the CFSsmcm implementation combines observational studies with theory on tropical heating profiles to construct ϕ_c, ϕ_d, ϕ_s. The shape of the deep heating basis function is designed based on the average heating profile in Fig. 3 of Stachnik et al. (2013). The stratiform basis function is designed following the stratiform heating profile reported in Fig. 1 of Schumacher et al. (2007). The congestus heating profile is designed following Khouider and Majda (2006), but slightly modified to represent the lower level peak (around 700 hPa) noted in the convective heating profiles plotted from the CFSR data (not shown here). When constructing ϕ_c, we have also consulted the work of Schumacher et al. (2007) (The "Shallow convective" and the "Strongly detraining Cu congestus" profiles in their Fig. 1). Incidentally (as it can be seen in Fig. 1), the

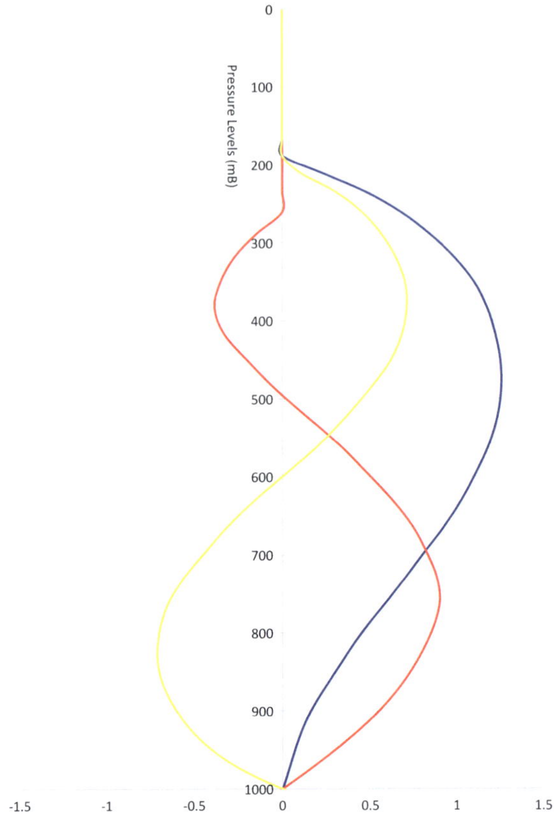

Fig. 1 Basis functions. Cumulus congestus profile in red, deep cumulus profile in blue, and stratiform profile in yellow

heating basis functions were clipped to zero at or slightly below 200 hPa. This is somewhat arbitrary as there are instances where the tropopause level is higher and it is not clear how much the results would change if this level was a bit higher or lower. This will be the subject of future studies.

The total convective heating is

$$Q_{tot}(z) = H_d\phi_d(z) + H_c\phi_c(z) + H_s\phi_s(z). \tag{1}$$

Here, ϕ_c, ϕ_d, and ϕ_s are the three basis functions shown in Fig. 1; and H_c, H_d, and H_s are the associated heating rates, which are parameterized using the corresponding stochastic area fractions, σ_d, σ_c, and σ_s, respectively, and the large-scale dynamical variables as given below:

$$H_d = \frac{\sigma_d}{\bar{\sigma_d}} Q_d, \tag{2}$$

$$H_c = \frac{\sigma_c}{\bar{\sigma_c}} \alpha_c Q_c, \tag{3}$$

$$\frac{\partial H_s}{\partial t} = \frac{1}{\tau_s}\left[\frac{\sigma_s}{\bar{\sigma_s}} \alpha_s H_d - H_s\right]. \tag{4}$$

Here, $\bar{\sigma_c}$, $\bar{\sigma_d}$, and $\bar{\sigma_s}$ are the background values of σ_c, σ_d, and σ_s, respectively, α_c and α_s are the congestus and stratiform adjustment coefficients, respectively, and τ_s is the stratiform convection adjustment timescale.

According to Khouider et al. (2010), the cloud area fractions σ_c, σ_d, and σ_s describe a Markov jump stochastic process in the form of a multidimensional birth–death process whose transition probabilities depend explicitly on the mid-tropospheric dryness (MTD), convective available potential energy (CAPE), and convective inhibition (CIN) and vertical velocity (W). The formulation of the transition rates from one cloud type to the other are the same as prescribed in Deng et al. (2015), except for the formation of congestus and deep convection from clear sky condition. This change occurs due to the inclusion of CIN and W in the transition rules. The inclusion of CIN and W in the transition rates has driven the desire to make the deep convection parameterization aware of the shallow convection scheme in the sense that in the event of strong subsidence and/or strong CIN, deep convection is inhibited leaving "space" for shallow convection which is naturally promoted in such circumstances. The modified transition rates (formation rates of congestus and deep clouds are highlighted in bold) are given in Table 1. The values of the transition timescales, on the last column of Table 1, are from De La Chevrotière et al. (2015), who used a systematic Bayesian inference technique to learn these parameters from large eddy simulation data (Khairoutdinov et al. 2009).

In Eqs. (2)–(4), the potentials for deep (Q_d) and congestus (Q_c) convection are computed, using the following equations:

$$Q_d = \left[\bar{Q}_d + \frac{1}{\tau_q}\frac{L_v}{C_p} q_m^{'} + \frac{1}{\tau_c}\left(\theta_{eb}^{'} - \gamma_c \theta_m^{'}\right)\right]^{+}, \tag{5}$$

$$Q_c = \left[\bar{Q}_c + \frac{1}{\tau_c}\left(\theta_{eb}^{'} - \gamma_c \theta_m^{'}\right)\right]^{+}. \tag{6}$$

In Eqs. (5) and (6), $\bar{Q}_d$, $\bar{Q}_c$, $\bar{Q}_s$ are prescribed background potentials for deep, congestus, and stratiform convection, respectively. They are inferred from CFS reanalysis data (Saha et al. 2010) by projecting the climatological convective heating, Q_1 onto the three basis functions in (1). The parameters τ_q and τ_c are the convective adjustment timescales of moisture and temperature, respectively. The quantities $q_m^{'}$, $\theta_{eb}^{'}$, $\theta_m^{'}$ are given by

Table 1 SMCM transition rules. The transition rates are given in terms of the large-scale predictors CAPE, $C = CAPE/CAPE0$, low-level CAPE, $C_L = LCAPE/LCAPE0$, dryness, $D = \mathcal{H}/MTD0$, where $\mathcal{H}$ is the relative humidity, large-scale subsidence, $W_N = -\min(0, W/W0)$, and $C_N = -CIN/CIN0$

Description	Transition rate, where $\Gamma(x) = \begin{cases} (1-e^{-x}), & \text{if } x > 0 \\ 0, & \text{otherwise} \end{cases}$	Timescale (h)
Formation of congestus	$R_{01} = \frac{1}{\tau_{01}}(\Gamma(C_L)\Gamma(D)^{\frac{(1-\Gamma(W_N))+(1-\Gamma(C_N))}{2}}$	$\tau_{01} = 32$
Decay of congestus	$R_{10} = \frac{1}{\tau_{10}}\Gamma(D)$	$\tau_{10} = 2$
Conversion of congestus to deep	$R_{12} = \frac{1}{\tau_{12}}\Gamma(C)(1-\Gamma(D))$	$\tau_{12} = 0.25$
Formation of deep	$R_{02} = \frac{1}{\tau_{02}}(\Gamma(C)(1-\Gamma(D))^{\frac{(1-\Gamma(W_N))+(1-\Gamma(C_N))}{2}}$	$\tau_{02} = 12$
Conversion of deep to stratiform	$R_{23} = \frac{1}{\tau_{23}}$	$\tau_{23} = 0.25$
Decay of deep	$R_{20} = \frac{1}{\tau_{20}}(1-\Gamma(C))$	$\tau_{20} = 9.5$
Decay of stratiform	$R_{30} = \frac{1}{\tau_{30}}$	$\tau_{30} = 1$

Here, $LCAPE$ is the part of the CAPE integral between LFC and the freezing level. We note that CIN is by definition a negative definite quantity, so that when CIN is large, $\Gamma(C_N) \longrightarrow 1$

$$q_m^{'} = q_m - \bar{q_m}, \tag{7}$$

$$\theta_{eb}^{'} = \theta_{eb} - \bar{\theta_{eb}}, \tag{8}$$

$$\theta_m^{'} = \theta_m - \bar{\theta_m}. \tag{9}$$

They are the deviations of the model's middle troposphere moisture, equivalent potential temperature, in the planetary boundary layer (PBL) and middle troposphere potential temperature, respectively, from their background states denoted by over bars. These background values are set according to the climatology of 20-year CFSR data averaged in space over distinct regional boxes in Fig. 2.

Earlier theoretical studies with the SMCM (Khouider and Majda 2006; Khouider et al. 2010; Deng et al. 2015; and the relevant references therein) rely on the radiative–convective equilibrium (RCE) solution (space-time homogeneous solution) of the governing equations to construct the background to set up the parameterization in Eqs. (2)–(10). However, such solution is not practical in the context of a comprehensive climate model because of existence of various inhomogeneities like land–ocean, tropics–mid-latitude, etc. To overcome this conundrum, we have used climate data to compute surrogates for the RCE solution as time and spatial means for a set of boxes centered over different areas of relatively homogeneous climatologies. These different areas (boxes) of relatively homogeneous climatologies are shown in Fig. 2 overlaid over the shows of the long-term mean of specific humidity at the sur-

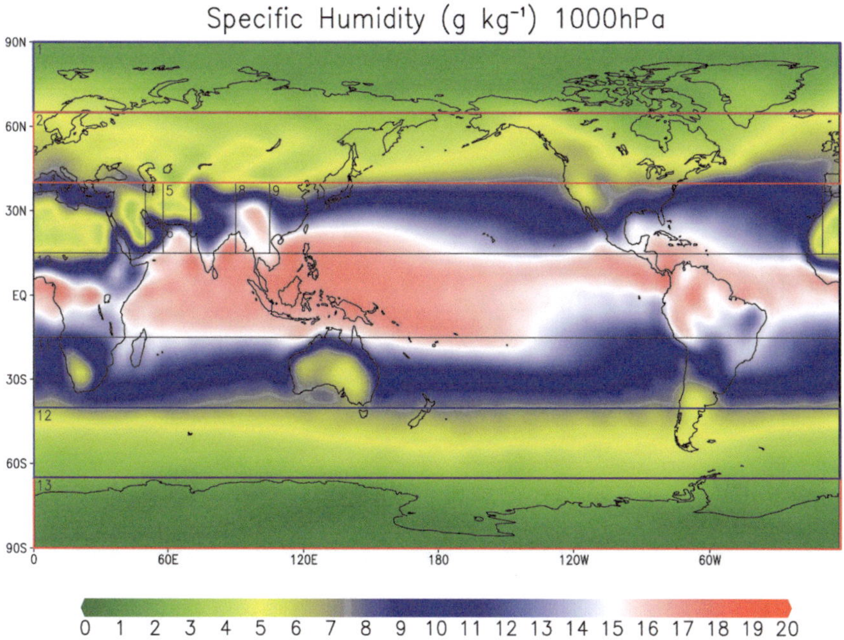

Fig. 2 Long-term mean of specific humidity at the surface (computed from climate forecast system reanalyses product)

face, which provide a rationale behind choosing these boxes. We have plotted other thermo-dynamical fields as well (not shown here), before deciding on the partition boxes. Noteworthy, we have smoothed each background field before prior to the implementation into CFSsmcm. As an example of a background field, the middle-level specific humidity is shown in Fig. 3.

Furthermore, the CFSsmcm includes an unsaturated downdraft mass flux which serves to cool and dry the lower troposphere due to the evaporation of stratiform rain in the lower troposphere. It is given by

$$D_c = \mu \left[\frac{H_s - H_c}{\bar{Q}_c} \right]^+, \tag{10}$$

where $\mu = 1.25$ cm s^{-1} is the downdraft reference scale. Here and elsewhere in the paper X^+ denotes the positive part of the variable X, i.e., $\max(X, 0)$.

In the equations in (5) and (6), the subscript b indicates variables averaged over the boundary layer height defined as $X_b = \frac{1}{h}\int_0^h X(z)\,dz$. The PBL height, h, is inputted from CFSv2. The height of the stable PBL, h, is estimated iteratively from ground up using bulk Richardson number (R_b) until a critical value $R_{bc} = 0.25$ is reached (Troen and Mahrt 1986). Incidentally, h is the height of the mixed layer which is consistent with the design of the multi-cloud model (Khouider and Majda 2006, 2008; Waite

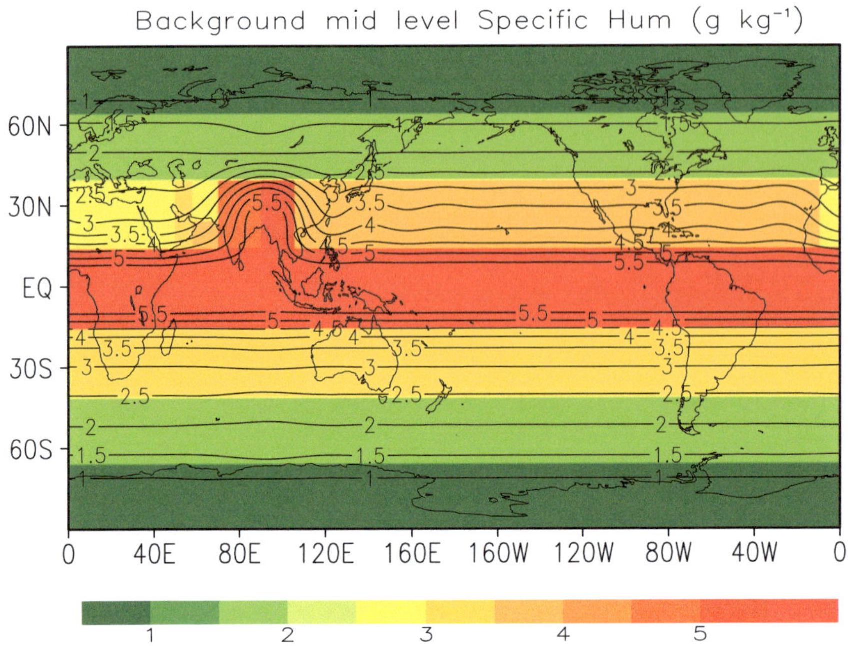

Fig. 3 Background mid-level (700 hPa) specific humidity (g/kg). In shading is the non-smoothed box-wise values. Smoothed field for the same is shown by the overlaid contours

and Khouider 2009). The subscript m indicates values of the variables taken at the middle troposphere. The middle-level specific humidity is chosen at 700 hPa. This is based on the long-term mean (obtained from CFSR 20-year reanalysis data) of moist static energy (MSE) profile. We have chosen the level where a minimum in the MSE profiles is noted (for those boxes which lie within 40S–40N in Fig. 2). Based on the climatological profiles of equivalent potential temperature (θ_e) and convective heating (not shown here), we have defined the middle- and low-troposphere value of θ_e at 500 hPa and 700 hPa, respectively.

2.2 *Prescribed Vertical Profiles of Moistening and Drying*

The moisture sink is set to

$$P(z) = -\frac{c_p}{L_v}\langle Q_{tot}\rangle Q_2(z), \tag{11}$$

where $\langle Q_{tot}\rangle$ is the vertical average of the total heating $Q_{tot}(z)$, c_p is the specific heat at constant pressure, L_v is the latent heat of vaporization, and $H = 16$ km is the

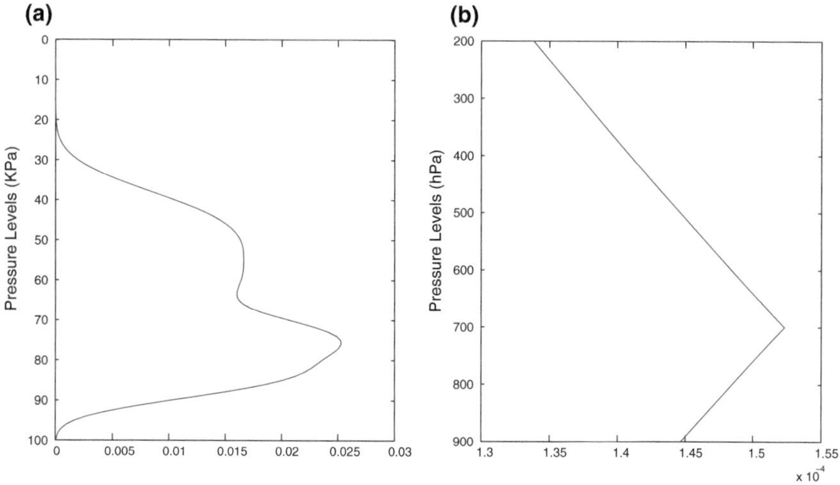

Fig. 4 **a** The shape of the $Q_2(z)$ structure function in the moisture sink Eq. 11 (drying). **b** The shape of the $\delta_m(z)$ structure function in Eq. 13 (moistening)

rough estimate of the tropospheric height. Moreover, $Q_2(z)$ is a prescribed moisture sink function whose exact shape is given in Fig. 4 and its vertical average is unity.

The surface precipitation is given by

$$P = \frac{c_p}{L_v} \int_0^H Q_{tot}(z)dz. \tag{12}$$

The introduction of the structure function $Q_2(z)$ is a new feature of this current version of SMCM. The shape of $Q_2(z)$ (Fig. 4a) is inspired by the Yanai moisture sink profile (Yanai et al. 1973). In earlier versions of the SMCM, the moisture sink is set according to the fact that only the column-integrated water vapor is integrated, i.e, the free tropospheric moisture is represented by one single vertical grid point. In this context, under the constraint of the conservation of vertically integrated moist static energy, the precipitation rate reduces to the vertical integral of the convective heating potential temperature tendency, renormalized by the ratio of the latent heat of vaporization and the specific heat at constant pressure. The moisture sink closure provided in Eq. 11 was derived under the same constraint of moist static energy conservation.

The evaporation rate is given by

$$E(z) = \left(\delta_m(z)\frac{D_c}{H}\right)\Delta_m\theta_e, \tag{13}$$

where $\triangle_m X = X_b - X_m$, with the suffixes b and m indicating, respectively, the PBL and middle troposphere values of the variable X. The structure function $\delta_m(z)$ (Fig. 4b) is defined by

$$\delta_m(z) = \begin{cases} 2\exp\left(-\alpha_m \frac{|P(z)-P_{MID}|}{P_{BOT}-P_{TOP}}\right), & \text{if } z \geq h \\ 0, & \text{if } z < h \end{cases} \tag{14}$$

where α_m is a constant, so that, $\frac{1}{H}\int_0^H \delta_m(z) = 1$.

The expression $P(z) = -\frac{c_p}{L_v}\langle Q_{tot}\rangle Q_2(z)$ ensures that the vertically averaged convective heating balances the total amount of precipitation reaching the ground while that of the evaporation rate, $E(z)$, is designed to balance the drying and cooling of the PBL by downdrafts so that the vertically averaged moist static energy is conserved as anticipated.

The SMCM feeds back onto the dynamical core variables through the temperature and moisture convective tendencies given by

$$\left[\frac{\partial}{\partial t}\theta(z)\right]_{SMCM} = Q_{tot}(z) - D_{b\theta}, \tag{15}$$

$$\left[\frac{\partial}{\partial t}q(z)\right]_{SMCM} = -P(z) + E(Z) - D_{bq}. \tag{16}$$

Here, $D_{b\theta}$ and D_{bq} represent the effect of unsaturated downdraft which results in cooling and drying below the PBL (h, which is a variable imported from the CFSv2 boundary layer scheme into the SMCM module), and they are given by

$$D_{bq}(z) = \begin{cases} \frac{D_c}{h}\triangle_m q & \text{if } z < h, \text{ results drying,} \\ 0 & \text{if } z > h, \end{cases} \tag{17}$$

$$D_{b\theta}(z) = \begin{cases} \frac{D_c}{h}\triangle_m \theta & \text{if } z < h, \text{ results cooling,} \\ 0 & \text{if } z > h. \end{cases} \tag{18}$$

While the GCM dynamical core has time step of 10 min, the SMCM convective tendencies in Eqs. (13) and (14) are updated every 10 s, i.e., 60 times per GCM time step in order to ensure stability, due to the fast convective timescale.

A comprehensive list of parameters that appear in Eqs. (1)–(18) and their values are provided in Table 2. The first column of Table 2 refers to the number of equation where the parameters appear for first.

Table 2 List of parameters

Reference equation no.	Parameter	Value	Remarks
3	α_c	0.1	Congestus adjustment coefficient
4	τ_s	96 h	Stratiform convection adjustment timescale
4	α_s	0.2	Stratiform adjustment coefficient
5	τ_q	144 h	Moisture adjustment timescale
5	L_v	2.5×10^6 J kg^{-1}	Latent heat of condensation
5	C_p	1004.6 J kg^{-1} K^{-1}	Specific heat of air at constant pressure
5	τ_c	240 h	Congestus convection adjustment timescale
5	γ_c	0.1	Adjustment coefficient for relative contribution of congestus to deep heating
10	μ	1.25 cm s^{-1}	Downdraft reference scale
12	H	16 km	Height of the tropical troposphere
14	α_m	0.22	–

3 Results

In this section, we compare two longtime simulations corresponding to Runs 129 and 139 from Goswami et al. (2017b), by comparing their mean state and their intraseasonal variability against observational benchmarks consisting of TRMM rainfall (Huffman et al. 2010) and NCEP reanalysis data (Kalnay et al. 1996) for temperature, moisture, and wind fields. The CFSsmcm simulations are based on a T126 horizontal resolution combined with 64 vertical levels and a 10 min time step. The SMCM birth–death process is simulated via Gillespie's exact Monte Carlo algorithm, which is run in parallel at every GCM time step (see Khouider et al. (2010), for details). As mentioned earlier, the key difference between the two SMCM simulations presented here resides in the way the mid-troposphere dryness parameter, MTD0, appearing in the caption of Table 1, is set. Run 129 uses a single value of MTD0 = 5 globally, while Run 139 uses two values simultaneously by setting MTD0 = 25 over land and MTD0 = 5 over the ocean.

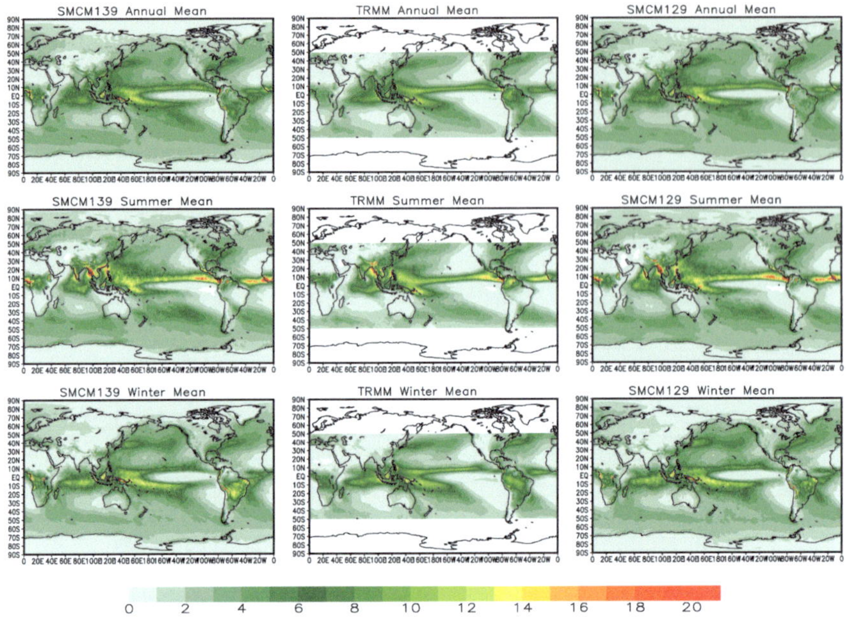

Fig. 5 Annual and seasonal (summer: JJAS and winter: O-M) mean rainfall fields (mm day^{-1}) for Run 139 (left-hand side panels), TRMM (middle panels), and Run 129 (right-hand side panels)

3.1 *Mean State*

The simulated precipitation fields are shown in Fig. 5, where the annual mean and the summer and winter seasonal means are displayed separately. It is evident from the annual mean, and individually for the two seasons as well, the precipitation field has significantly improved in Run 139. The wet bias over the oceans and the dry bias over the continents have significantly reduced. Overall, the geographical distribution of the precipitation looks much better in Run 139. Improvement in the simulation of the Indian summer monsoon mean state is one of the major gains of Run 139 over Run 129. As a more convincing evidence, we have plotted the climatological annual cycle of rainfall over the Indian summer monsoon domain in Fig. 6. We consider three different boxes over the ISM domain and plotted the climatological annual cycle of rainfall over each box. Clearly, Run 139 simulation matches the observation fairly well and much better than Run 129 and the CFSv2 control run. For both the boxes "Central India" and "Extended IMR", Run 139 looks consistent with observation. Importantly, the improved annual cycle is actually due to good distribution of rainfall and not due to any compensation of rainfall covering the dry bias over India by wet bias over the mountainous terrains. However, Run 139 still looks dry over the box "Monsoon trough" accompanied by an early withdrawal.

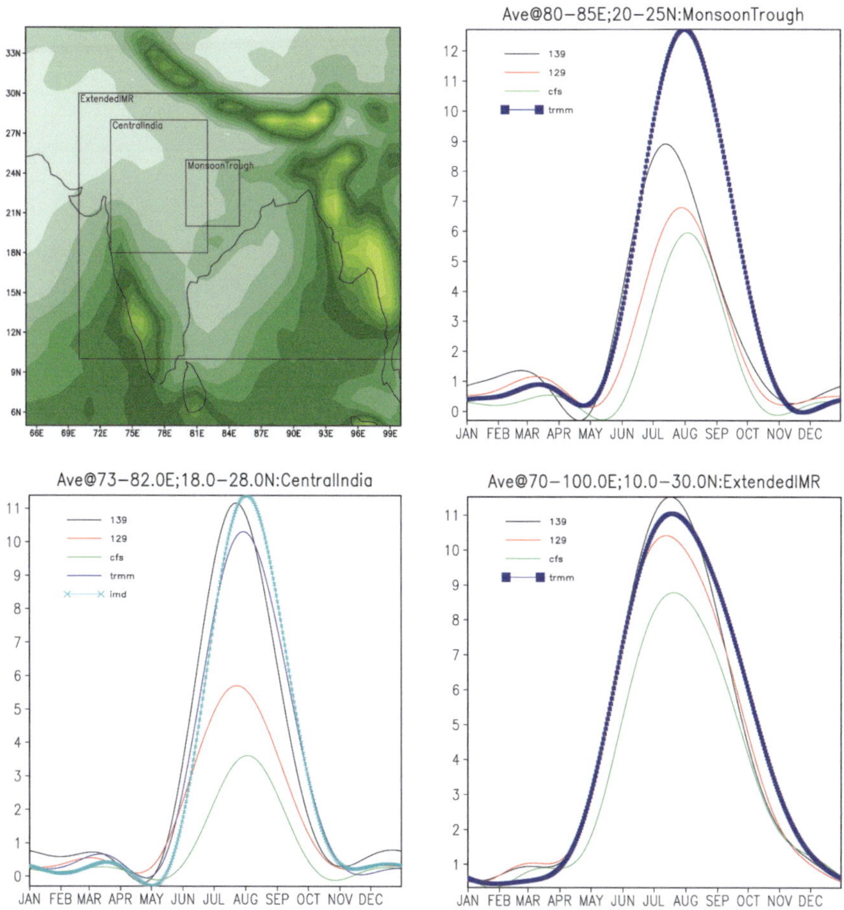

Fig. 6 Annual cycle of climatological daily mean rainfall (mm day^{-1}) over the Indian monsoon region (different boxes)

3.2 Intraseasonal Variability

Capturing the intraseasonal variability has been a key achievement of the SMCM effort in its idealized simulations (Deng et al. 2015, 2016; Ajayamohan et al. 2016) as well as when implemented in CFS (Goswami et al. 2017a, c). Another feature of the CFSsmcm has been its resilience in terms of minor changes to its parameter values (Goswami et al. 2017b). Since changing the value of the middle tropospheric dryness parameter is a considerable change from SMCM's perspective, we examined the response of the CFSsmcm Run 139 in capturing the intraseasonal variability. In Fig. 7, we have plotted the standard deviation of 10–90-day bandpass-Lanczos-filtered rainfall anomalies for the two seasons. This gives an overview of the

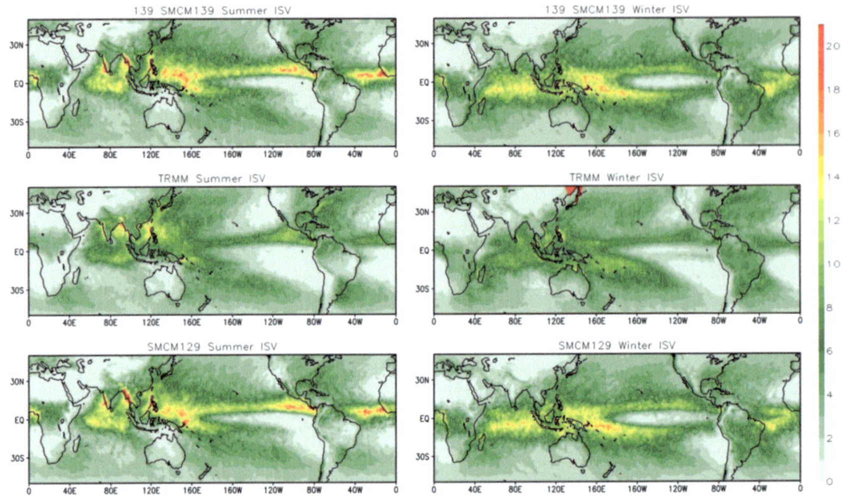

Fig. 7 Intraseasonal variability (standard deviation of the 10–90-day bandpass-filtered rainfall anomalies) for Run 139 (top panels), TRMM (middle panels), and Run 129 (bottom panels) for the summer and winter seasons

intraseasonal variability in the simulated precipitation fields. Comparing the Runs 129 and 139, the intraseasonal variability does not change significantly. As we have already mentioned, resilience to changes in parameter values has been a hallmark of the CFSsmcm throughout its development (Goswami et al. 2017b). Nevertheless, there are slight increases in variability observed over the Western Pacific and the Indian landmass. This increase in variability is consistent with the increase in the mean seasonal rainfall.

3.3 *Tropical Wave Spectrum*

When implementing the SMCM in CFSv2, the simulation of the tropical intraseasonal variability (TISV) improved significantly compared to the default CFSv2 simulation, evident from the Takayabu–Wheeler–Kiladis (TWK) diagram (Goswami et al. 2017c). Therefore, we plotted the same for Run 139 to see if the improvements are retained or changed. In Fig. 8, the TWK diagram for the outgoing longwave radiation (OLR) is shown for the whole length of the 10-year climate for the Runs 129 and 139 and observation [OLR from the National Oceanic and Atmospheric Administration; Liebmann and Smith (1996)]. As evident from the faded color shading, Run 139 is relatively less skillful compared to Run 129. However, Run 139 still outperforms the control CFSv2 run [see Fig. 1b, c of Goswami et al. (2017c)]. Except the equatorial Rossby waves, there is a loss of power in all other modes of the tropical wave

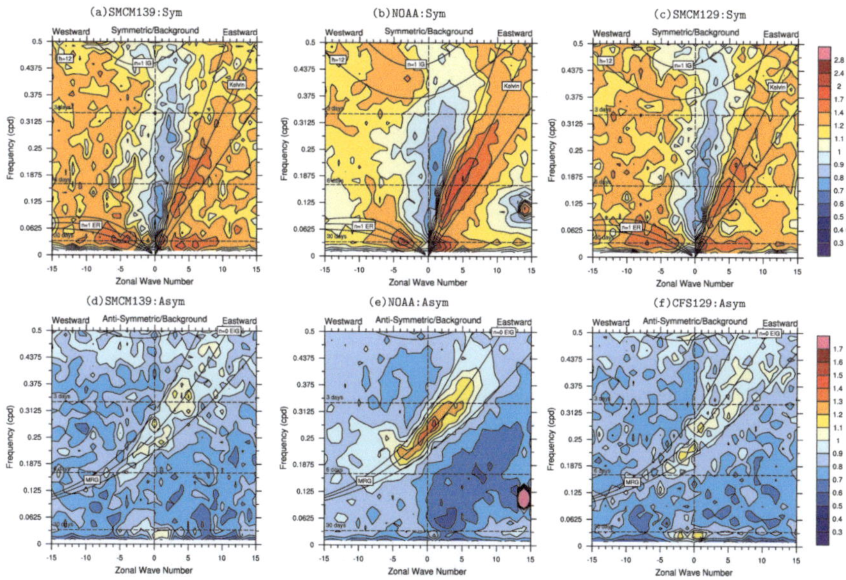

Fig. 8 Wheeler–Kiladis spectra of OLR from **a** Run 139, **b** NOAA OLR, and **c** Run 129 for the symmetric component. The corresponding anti-symmetric spectra are shown in panels **d**, **e**, and **f**, respectively

spectrum. Especially, the MJO mode appears somewhat weak with unrealistic power in higher wavenumber regime.

3.4 MJO Variability and Propagation

The MJO is the major mode of variability in the tropics on the intraseasonal timescales. Also, it is notoriously difficult to simulate realistically by coarse resolution climate models. Hence, it can be treated as a metric for the fidelity of a climate model in simulating the tropical variability at such scales. Goswami et al. (2017c) showed that CFSsmcm simulates the MJO significantly better compared to the default CFSv2 model. Now, as we have already seen in Sect. 3.2 that the TWK plot has slightly deteriorated in Run 139, it is of obvious curiosity to explore the MJO features in this run. Following the exact same methodology adopted to plot Figs. 2, 3, and 4 of Goswami et al. (2017c), we have plotted here the fields in Figs. 9, 10, and 11, respectively. We note that in Fig. 9, the MJO variance has somewhat deteriorated in Run 139. Particularly, the meridional span of the region of strong variance has narrowed, the variance over the Western Pacific has unrealistically strengthened and that over the California coast has weakened. The variance along the oceanic intertropical convergence zone toward the south of the Western Pacific, which was already poorly

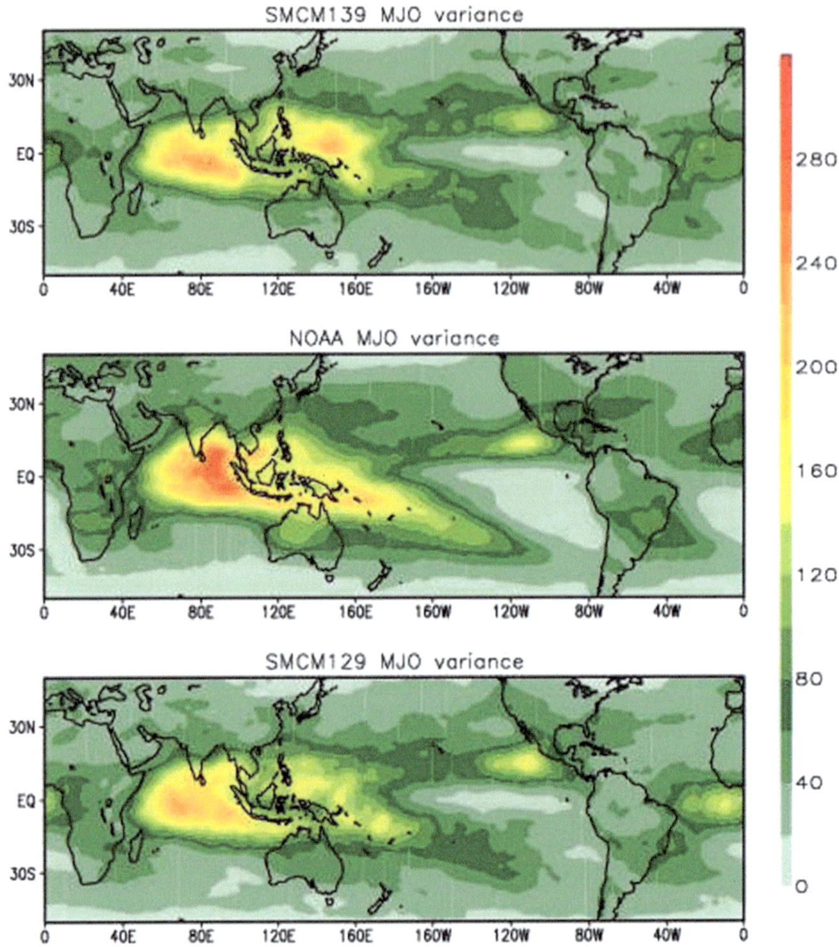

Fig. 9 Daily variance of the MJO filtered (wavenumber 1–9 and 36–90 days) OLR ($(W\ m^{-2})^2$) anomalies: Run 139 (top), OBS (NOAA OLR) (middle), and Run 129 (bottom)

simulated in Run 129 has further worsened. Consistent with the observed limitation in the simulation of the MJO variance in Fig. 9, the propagation features are also simulated with limited fidelity as seen from Fig. 10. Although both CFSsmcm runs are better compared to the CFSv2-simulated propagation features, it is debatable to claim for any improvement or its lack thereof in Run 139 compared to Run 129. Instead, it would be proper to say that both the CFSsmcm runs have their own strengths and weaknesses in simulating the MJO propagation features, especially passed the maritime continent. A more detailed picture of the MJO propagation is shown in Fig. 11, where the lag–lead composite of OLR anomalies, with respect to the MJO peak defined over a box region in the Bay of Bengal (82.5°E–90°E and Eq-8.5°N)

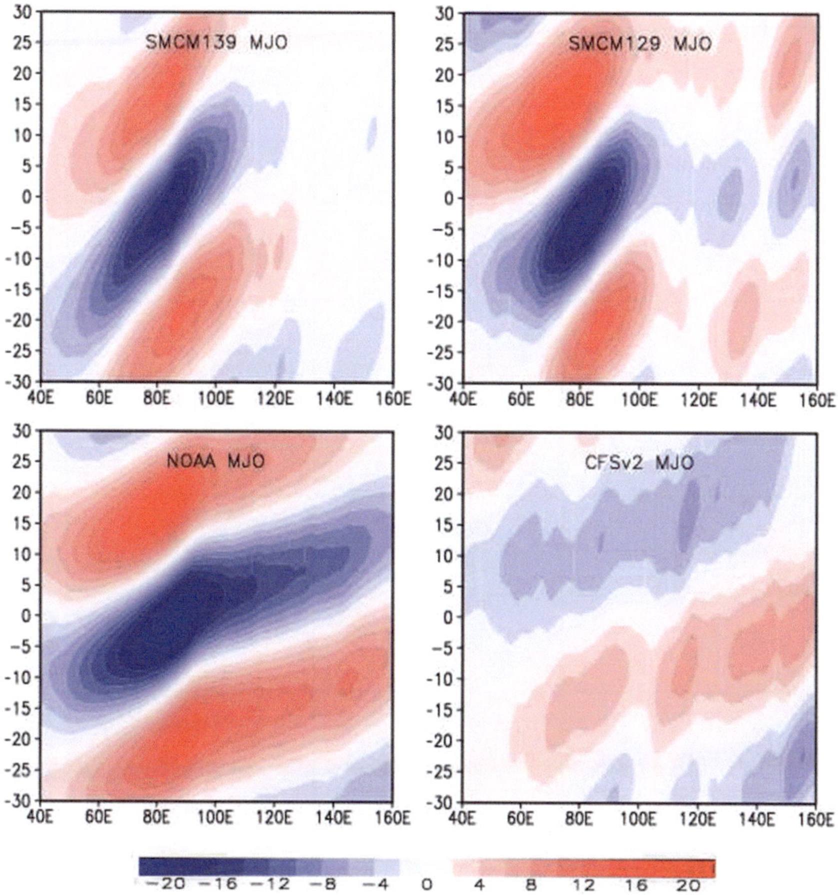

Fig. 10 Hovmöller (averaged from 5°S–5°N) plots showing MJO propagation for the MJO filtered OLR (W m^{-2}) anomalies [composite based on MJO peak over the box bounded by 82.5°E–90°E and Eq-8.5°N]

are plotted. Consistent with the results shown in Fig. 10, the MJO structure is not as prominent as in the observations with a hint of a smaller spatial structure in the model simulations. Nevertheless, the simulated MJO structure in both the CFSsmcm Runs, 129 and 139, looks significantly better than in CFSv2 MJO, shown on the bottom right corner of Fig. 11.

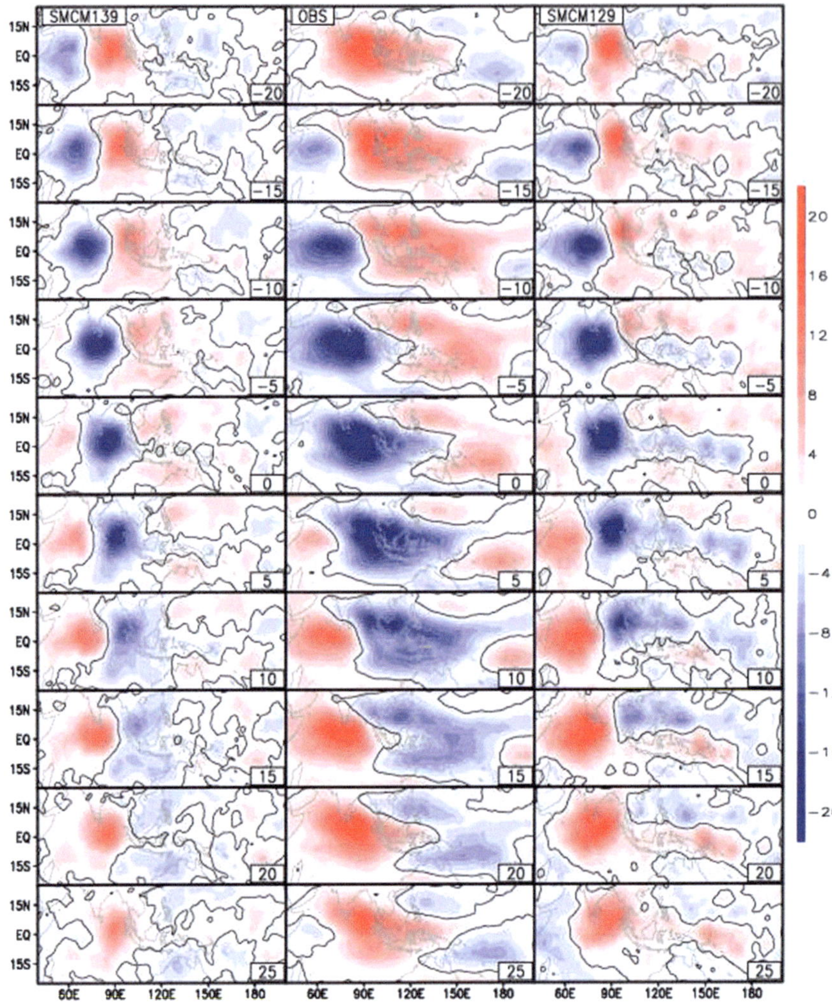

Fig. 11 MJO phase propagation. Composite of different phases of the MJO filtered OLR (W m^{-2}) anomalies constructed based on an MJO index averaged over 82.5°E–90°E and Eq-8.5°N. Run 139 is shown in the left-hand side column, OBS in the middle, and Run 129 in the right-hand side column. Phase-lag stamps are seen in the right-hand bottom corner

3.5 Indian Summer Monsoon Intraseasonal Oscillation (MISO)

Analogous to the TWK spectra along the east–west direction in the tropics, the north–south version of the same diagram plotted for the boreal summer data over the Indian monsoon domain provides a first hand overview of the major modes of oscillation of

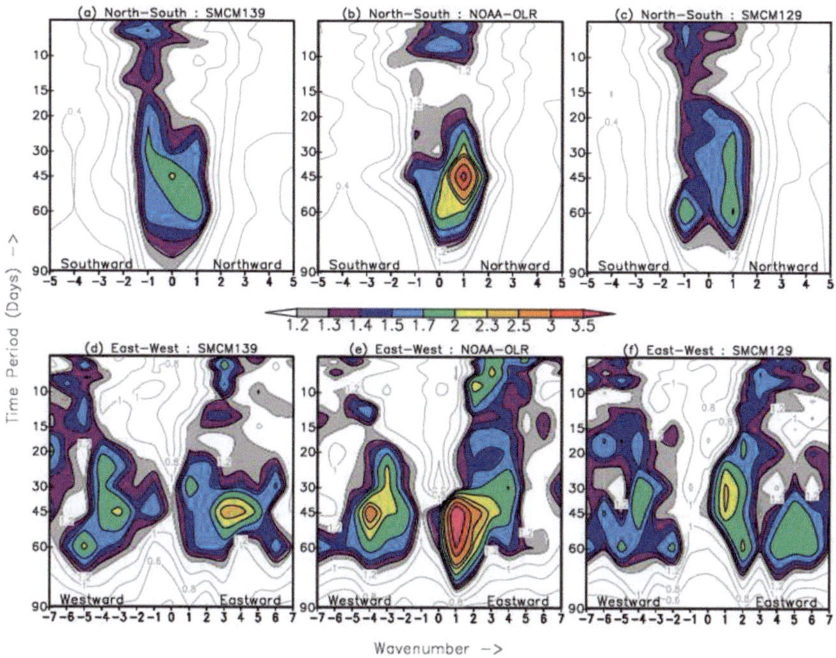

Fig. 12 Wavenumber–frequency spectra of OLR (divided by the background red spectrum) computed for the boreal summer season (JJAS). The top three panels show the north–south spectra (wavenumber 1 corresponds to the largest wave that exactly fits into 50° latitudes, from 20°S to 30°N; computed over 60°E to 100°E). The bottom three panels show the east–west spectra (wavenumber 1 corresponds to the length of the equator)

the Indian summer monsoon (ISM). For the north–south TWK spectra (Fig. 12a, c and e), wavenumber 1 corresponds to 50° of latitude (from 20°S to 30°N). As we had seen for the TWK spectra in Fig. 8, the north–south wavenumber–frequency spectra also has deteriorated in Run 139. Interestingly, the MISO power in the north–south spectra in Run 139 deteriorates, whereas the seasonal mean precipitation improves. We need to recall here that the SMCM parameter responsible for the stratiform convection decay time was found to be crucial for organization of convection in the CFSsmcm (Goswami et al. 2017b) and MTD0 is influential for controlling the mean precipitation only. So, this is consistent with the formulation of the SMCM. Finding a balanced pair of values for MTD0 for continents/oceans and an adequate stratiform convection decay timescale to complement this pair of MTD0 values calls for further tuning of CFSsmcm parameters.

4 Discussion

We have run two different versions of the CFSsmcm model: one with one middle tropospheric dryness parameter (MTD0) value for the entire globe (Run 129) and the other with two separate values of MTD0 for continents and oceans (Run 139). For the sake of ease of discussion, let us call these two Runs 129 and 139 as MTD_G and MTD_L/O, respectively. We performed some standard analyses to examine the difference in mean climate and its variability, based on 10-year-long climate simulations. The motive behind doing this exercise is to highlight the sensitivity and resilience of the CFSsmcm, to changes in parameter values. Thereby exposing the scopes of improving the CFSsmcm model to the climate modeling community.

The CFSsmcm mean rainfall has already been demonstrated to be sensitive to the MTD0 parameter in Goswami et al. (2017b). As a consequence, the mean rainfall of the MTD_G and MTD_L/O runs are significantly different, especially over the rain abundant regions in the tropics like the Indian summer monsoon, West Pacific, Amazonia, etc. In the MTD_L/O run, the MTD0 values are chosen in such a way that the atmosphere over the continents triggers precipitation relatively quickly compared to that over the oceans. This adjustment has resulted in reducing the dry bias over the continents. As per our analyses, the simulation of the Indian summer monsoon (ISM) mean rainfall has improved the most in the MTD_L/O run. However, intraseasonal variability has not shown much improvement. In fact, at times, it has worsened. The tropical wave spectrum (as seen from the TWK diagram in Fig. 8) looks a bit deteriorated in the MTD_L/O run. The same can be said for the MJO variability and propagation (Figs. 10, 11, and 12). As can be seen from Fig. 12, power in the desired modes of variability of the ISM climate has also slightly deteriorated. The parameters responsible for organization of convection, especially the stratiform convection decay time parameter (Goswami et al. 2017b) needs to be adjusted to suit the MTD_L/O run in order to simulate better intraseasonal variability. However, the results obtained from this single attempt with a varying MTD0 look very encouraging. A thorough tuning of the CFSsmcm model for the MTD_L/O run bears promise to lead us to an even better version of the CFSsmcm model with better seasonal mean rainfall and perhaps better intraseasonal- and synoptic-scale variability.

References

Ajayamohan, R.S., B. Khouider, A.J. Majda, and Q. Deng. 2016. Role of stratiform heating on the organization of convection over the monsoon trough. *Climate Dynamics*, 1–20. https://doi.org/10.1007/s00382-016-3033-7.

Arakawa, A. 2004. The cumulus parameterization problem: Past, present, and future. *Journal of Climate* 17 (13): 2493–2525. https://doi.org/10.1175/1520-0442(2004)017<2493:RATCPP>2.0.CO;2.

Arakawa, A., and W.H. Schubert. 1974. Interaction of a cumulus cloud ensemble with the large-scale environment. *Part I. Journal of Atmosperic Science* 31 (3): 674–701. https://doi.org/10.1175/1520-0469(1974)031<0674:IOACCE>2.0.CO;2.

Betts, A.K., and M.J. Miller. 1986. A new convective adjustment scheme. Part II: Single column tests using GATE wave, BOMEX, ATEX and arctic air-mass data sets. *Quarterly Journal of the Royal Meterological Society* 112 (473): 693–709. https://doi.org/10.1002/qj.49711247308.

Buizza, R., M. Milleer, and T.N. Palmer. 1999. Stochastic representation of model uncertainties in the ECMWF ensemble prediction system. *Quarterly Journal of the Royal Meteorological Society* 125 (560): 2887–2908. https://doi.org/10.1002/qj.49712556006.

Davini, P., J. von Hardenberg, S. Corti, H.M. Christensen, S. Juricke, A. Subramanian, P.A.G. Watson, A. Weisheimer, and T.N. Palmer. 2016. Climate SPHINX: Evaluating the impact of resolution and stochastic physics parameterisations in climate simulations. *Geoscientific Model Development Discussion* 2016: 1–29. https://doi.org/10.5194/gmd-2016-115.

De La Chevrotière, M., and B. Khouider. 2017. A zonally symmetric model for the monsoon-Hadley circulation with stochastic convective forcing. *Theoretical and Computational Fluid Dynamics* 31 (1): 89–110. https://doi.org/10.1007/s00162-016-0407-8.

De La Chevrotière, M., B. Khouider Michèle, and A. Majda. 2015. Stochasticity of convection in Giga-LES data. *Climate Dynamics* 47 (5): 1845–1861. https://doi.org/10.1007/s00382-015-2936-z.

Deng, Q., B. Khouider, and A.J. Majda. 2015. The MJO in a coarse-resolution GCM with a stochastic multicloud parameterization. *Journal of Atmospheric Science* 72 (1): 55–74. https://doi.org/10.1175/JAS-D-14-0120.1.

Deng, Q., B. Khouider, A.J. Majda, and R.S. Ajayamohan. 2016. Effect of stratiform heating on the planetary-scale organization of tropical convection. *Journal of Atmospheric Science* 73 (1): 371–392. https://doi.org/10.1175/JAS-D-15-0178.1.

Dorrestijn, J., D.T. Crommelin, A.P. Siebesma, H.J. Jonker, and F. Selten. 2016. Stochastic convection parameterization with Markov Chains in an intermediate-complexity GCM. *Journal of Atmospheric Science* 73 (3): 1367–1382.

Frenkel, Y., A.J. Majda, and B. Khouider. 2012. Using the stochastic multicloud model to improve tropical convective parameterization: A paradigm example. *Journal of Atmospheric Science* 69 (3): 1080–1105. https://doi.org/10.1175/JAS-D-11-0148.1.

Frenkel, Y., A.J. Majda, and B. Khouider. 2013. Stochastic and deterministic multicloud parameterizations for tropical convection. *Climate Dynamics* 41 (5–6): 1527–1551. https://doi.org/10.1007/s00382-013-1678-z.

Goswami, B.B., B. Khouider, R.P.M. Krishna, P. Mukhopadhyay, and A.J. Majda. 2017a. Improving synoptic and intra-seasonal variability in CFSv2 via stochastic representation of organized convection. *Geophysics Research Letters* 44 (2): 1104–1113. https://doi.org/10.1002/2016GL071542.

Goswami, B.B., B. Khouider, R.P.M. Krishna, P. Mukhopadhyay, and A.J. Majda. 2017b. Implementation and calibration of a stochastic convective parameterization in the NCEP climate forecast system. *Journal of Advances in Modeling Earth Systems* (p. submitted).

Goswami, B.B., B. Khouider, R.P.M. Krishna, P. Mukhopadhyay, and A.J. Majda. 2017c. Improved tropical modes of variability in the climate forecast system model (version 2) via a stochastic multicloud model. *Journal of Atmospheric Science* (submitted).

Grabowski, W.W. 2001. Coupling cloud processes with the large-scale dynamics using the cloud-resolving convection parameterization (CRCP). *Journal of Atmospheric Science*. 58 (9): 978–997. https://doi.org/10.1175/1520-0469(2001)058<0978:CCPWTL>2.0.CO;2.

Gregory, D., and P.R. Rowntree. 1990. A mass flux convection scheme with representation of cloud ensemble characteristics and stability-dependent closure. *Monthly Weather Review* 118 (7): 1483–1506. https://doi.org/10.1175/1520-0493(1990)118<1483:AMFCSW>2.0.CO;2.

Huffman, G.J., R.F. Adler, D.T. Bolvin, and E.J. Nelkin. 2010. The TRMM multi-satellite precipitation analysis (TMPA). In *Satellite rainfall applications for surface hydrology*, 3–22. Berlin: Springer. https://doi.org/10.1007/978-90-481-2915-7_1.

Kain, J.S., and J.M. Fritsch. 1990. A one-dimensional entraining/detraining plume model and its application in convective parameterization. *Journal of Atmospheric Science* 47 (23): 2784–2802. https://doi.org/10.1175/1520-0469(1990)047<2784:AODEPM>2.0.CO;2.

Kalnay, E., M. Kanamitsu, R. Kistler, W. Collins, D. Deaven, L. Gandin, M. Iredell, S. Saha, G. White, J. Woollen, Y. Zhu, A. Leetmaa, R. Reynolds, M. Chelliah, W. Ebisuzaki, W. Higgins, J. Janowiak, K.C. Mo, C. Ropelewski, J. Wang, R. Jenne, and D. Joseph. 1996. The NCEP/NCAR 40-Year reanalysis project. *Bulletin of the American Meteorological Society* 77 (3): 437–471. https://doi.org/10.1175/1520-0477(1996)077<0437:TNYRP>2.0.CO;2.

Kasahara, A., and K. Puri. 1981. Spectral representation of three-dimensional global data by expansion in normal mode functions. *Monthly Weather Review* 109 (1): 37–51. https://doi.org/10.1175/1520-0493(1981)109<0037:SROTDG>2.0.CO;2.

Khairoutdinov, M.F., and D.A. Randall. 2001. A cloud resolving model as a cloud parameterization in the NCAR community climate system model: Preliminary results. *Geophysics Research Letters* 28 (18): 3617–3620. https://doi.org/10.1029/2001GL013552.

Khairoutdinov, M.F., S.K. Krueger, C.-H. Moeng, P.A. Bogenschutz, and D.A. Randall. 2009. Large-eddy simulation of maritime deep tropical convection. *Journal of Advances in Modeling Earth Systems* 1 (4).

Khouider, B., and A.J. Majda. 2006. A simple multicloud parameterization for convectively coupled tropical waves. Part I: Linear analysis. *Journal of Atmospheric Science* 63 (4): 1308–1323. https://doi.org/10.1175/JAS3677.1.

Khouider, B., and A.J. Majda. 2008. Multicloud models for organized tropical convection: Enhanced congestus heating. *Journal of Atmospheric Science* 65 (3): 895–914. https://doi.org/10.1175/2007JAS2408.1.

Khouider, B., A.J. Majda, and M.A. Katsoulakis. 2003. Coarse-grained stochastic models for tropical convection and climate. *Proceedings of the National Academy of Sciences United States of America* 100 (21): 11,941–11,946. https://doi.org/10.1073/pnas.1634951100.

Khouider, B., J. Biello, and A.J. Majda. 2010. A stochastic multicloud model for tropical convection. *Communications in Mathematical Sciences* 8 (1): 187–216.

Khouider, B., A. St-Cyr, A.J. Majda, and J. Tribbia. 2011. The MJO and convectively coupled waves in a coarse-resolution GCM with a simple multicloud parameterization. *Journal of Atmospheric Science* 68 (2): 240–264. https://doi.org/10.1175/2010JAS3443.1.

Kuo, H.L. 1965. On formation and intensification of tropical cyclones through latent heat release by cumulus convection. *Journal of Atmospheric Science* 22 (1): 40–63. https://doi.org/10.1175/1520-0469(1965)022<0040:OFAIOT>2.0.CO;2.

Liebmann, B., and C. Smith. 1996. Description of a complete (interpolated) outgoing longwave radiation dataset. *Bulletin of the American Meteorological Society* 77: 1275–1277.

Lin, J.W.-B., and J.D. Neelin. 2000. Influence of a stochastic moist convective parameterization on tropical climate variability. *Geophysics Research Letters* 27 (22): 3691–3694. https://doi.org/10.1029/2000GL011964.

Lin, J.W.-B., and J.D. Neelin. 2002. Considerations for stochastic convective parameterization. *Journal of Atmospheric Science* 59 (5): 959–975. https://doi.org/10.1175/1520-0469(2002)059<0959:CFSCP>2.0.CO;2.

Lin, J. W.-B., and J.D. Neelin. 2003. Toward stochastic deep convective parameterization in general circulation models. *Geophysics Research Letters* 30 (4): https://doi.org/10.1029/2002GL016203.

Majda, A.J., and B. Khouider. 2002. Stochastic and mesoscopic models for tropical convection. *Proceedings of the National Academy of Sciences of the United States of America* 99 (3): 1123–1128. https://doi.org/10.1073/pnas.032663199.

Palmer, T. 1996. On parameterizing scales that are only somewhat smaller than the smallest resolved scales, with application to convection and orography. In *Workshop on New Insights and Approaches to Convective Parametrization*, 4–7 Nov 1996, 328–337. ECMWF, Shinfield Park, Reading: ECMWF.

Palmer, T.N. 2001. A nonlinear dynamical perspective on model error: A proposal for non-local stochastic-dynamic parametrization in weather and climate prediction models. *Quarterly Journal of the Royal Meteorological Society* 127 (572): 279–304. https://doi.org/10.1002/qj.49712757202.

Peters, K., C. Jakob, L. Davies, B. Khouider, and A.J. Majda. 2013. Stochastic behavior of tropical convection in observations and a multicloud model. *Journal of Atmospheric Science* 70 (11): 3556–3575. https://doi.org/10.1175/JAS-D-13-031.1.

Peters, K., T. Crueger, C. Jakob, and B. Möbis. 2017. Improved MJO-simulation in ECHAM6.3 by coupling a stochastic multicloud model to the convection scheme. *Journal of Advances in Modeling Earth Systems*. https://doi.org/10.1002/2016MS000809.

Plant, R.S., and G.C. Craig. 2008. A stochastic parameterization for deep convection based on equilibrium statistics. *Journal of Atmospheric Science* 65: 87–105. https://doi.org/10.1175/2007JAS2263.1.

Randall, D.A. 2013. Beyond deadlock. *Geophysics Research Letter* 40 (22): 5970–5976. https://doi.org/10.1002/2013GL057998.

Saha, S., S. Moorthi, H.-L. Pan, X. Wu, J. Wang, S. Nadiga, P. Tripp, R. Kistler, J. Woollen, D. Behringer, H. Liu, D. Stokes, R. Grumbine, G. Gayno, J. Wang, Y.-T. Hou, H.-Y. Chuang, H.-M.H. Juang, J. Sela, M. Iredell, R. Treadon, D. Kleist, P. Van Delst, D. Keyser, J. Derber, M. Ek, J. Meng, H. Wei, R. Yang, S. Lord, H. Van Den Dool, A. Kumar, W. Wang, C. Long, M. Chelliah, Y. Xue, B. Huang, J.-K. Schemm, W. Ebisuzaki, R. Lin, P. Xie, M. Chen, S. Zhou, W. Higgins, C.-Z. Zou, Q. Liu, Y. Chen, Y. Han, L. Cucurull, R.W. Reynolds, G. Rutledge, and M. Goldberg. 2010. The NCEP climate forecast system reanalysis. *Bulletin of the American Meteorological Society* 91 (8): 1015–1057. https://doi.org/10.1175/2010BAMS3001.1.

Satoh, M., H. Tomita, H. Miura, S. Iga, and T. Nasuno. 2005. Development of a global cloud resolving model a multi-scale structure of tropical convections. *Journal of the Earth Simulator* 3 (September): 11–19.

Schumacher, C., M.H. Zhang, and P.E. Ciesielski. 2007. Heating structures of the TRMM field campaigns. *Journal of Atmospheric Science* 64 (7): 2593–2610. https://doi.org/\penalty-\@M10.1175/JAS3938.1.

Stachnik, J.P., C. Schumacher, and P.E. Ciesielski. 2013. Total heating characteristics of the ISCCP tropical and subtropical cloud regimes. *Journal of Climate* 26 (18): 7097–7116. https://doi.org/10.1175/JCLI-D-12-00673.1.

Takayabu, Y.N. 1994. Large-scale cloud disturbances associated with equatorial waves. Part II: Westward-propagating inertio-gravity waves. *Journal of the Meteorological Society of Japan* 72 (3): 451–465.

Teixeira, J., and C.A. Reynolds. 2008. Stochastic nature of physical parameterizations in ensemble prediction: A stochastic convection approach. *Monthly Weather Review* 136 (2): 483–496. https://doi.org/10.1175/2007MWR1870.1.

Troen, I.B., and L. Mahrt. 1986. A simple model of the atmospheric boundary layer; Sensitivity to surface evaporation. *Boundary Layer Meteorology* 37 (1): 129–148. https://doi.org/10.1007/BF00122760.

Waite, M.L., and B. Khouider. 2009. Boundary layer dynamics in a simple model for convectively coupled gravity waves. *Journal of Atmospheric Science* 66 (9): 2780–2795. https://doi.org/10.1175/2009JAS2871.1.

Wheeler, M., and G.N. Kiladis. 1999. Convectively coupled equatorial waves: Analysis of clouds and temperature in the wavenumber-frequency domain. *Journal of Atmospheric Science* 56 (3): 374–399. https://doi.org/10.1175/1520-0469(1999)056<0374:CCEWAO>2.0.CO;2.

Yanai, M., S. Esbensen, and J.-H. Chu. 1973. Determination of bulk properties of tropical cloud clusters from large-scale heat and moisture budgets. *Journal of Atmospheric Science* 30 (4): 611–627.

Zhang, G., and N.A. McFarlane. 1995. Sensitivity of climate simulations to the parameterization of cumulus convection in the canadian climate centre general circulation model. *Atmosphere-Ocean* 33 (3): 407–446. https://doi.org/10.1080/07055900.1995.9649539.

The IITM Earth System Model (ESM): Development and Future Roadmap

R. Krishnan, P. Swapna, Ramesh Vellore, Sandeep Narayanasetti, A. G. Prajeesh, Ayantika Dey Choudhury, Manmeet Singh, T. P. Sabin and J. Sanjay

Abstract This article provides a brief account of the development of the IITM Earth System Model (IITM-ESM) at the Centre for Climate Change Research, Indian Institute of Tropical Meteorology, in order to address key questions pertaining to the science of Climate Change. The IITM-ESM has been developed by transforming a state-of-the-art seasonal prediction model into a radiatively balanced climate modeling framework suitable for investigating long-term climate variability and change. The IITM-ESM is the first climate model from India to contribute to the Coupled Modeling Intercomparison Programme—Phase 6 (CMIP6) for the Intergovernmental Panel for Climate Change (IPCC) sixth assessment report (AR6). The IITM-ESM has shown promising capabilities required for making reliable assessments of the impacts of climate change on the (a) Global and regional monsoon hydroclimate, (b) Regional weather and climate extremes, (c) Global and Indian Ocean sea level, (d) Marine primary productivity and mechanisms controlling the ocean carbon cycle, and (e) Global and Himalayan cryosphere, to name a few important ones. Future plans for the development of high-resolution climate change projections and the next-generation community version of the IITM-ESM are also briefly discussed.

Keywords IITM-ESM · Global climate change
Indian and regional monsoon systems

1 Introduction

Earth's climate is dominated by a variety of phenomena and variability across multiple scales (both space and time) that arise as a natural consequence of the evolving nature of the Earth system processes and their adjustments to intersys-

R. Krishnan (✉) · P. Swapna · R. Vellore · S. Narayanasetti · A. G. Prajeesh · A. D. Choudhury
M. Singh · T. P. Sabin · J. Sanjay
Centre for Climate Change Research, Indian Institute of Tropical Meteorology,
Pune 411008, India
e-mail: krish@tropmet.res.in

D. A. Randall et al. (eds.), *Current Trends in the Representation of Physical Processes in Weather and Climate Models*, Springer Atmospheric Sciences,
https://doi.org/10.1007/978-981-13-3396-5_9

tem feedbacks. Earth System Models (ESMs) are important tools that allow us to understand the interactions among the Earth System components (viz., atmosphere, ocean, land, cryosphere, and biosphere) and quantify the physical, chemical and biological processes governing the rates of change of elements within and across the different components of the Earth system. ESMs are essentially coupled global numerical models that describe the physical and natural laws governing the Earth System processes which are represented in the form of mathematical equations. Numerical simulations to understand the evolution of ESM coupled processes across different spatiotemporal scales can be constructed using either finite difference methods or Spectral/Galerkin methods, in which the numerical equations are integrated in time. Since numerical integrations of the state-of-the-art ESMs typically involve several hundreds of millions of computations at each time-step, ESMs are computationally intensive and can be time consuming. Major advances in High Power Computing (HPC) and computational techniques have enabled substantial progress in ESM development during the recent decades.

The concept of Earth System Modeling is a powerful approach to advance our understanding of the climate system, its nonlinear behavior and multi-scale variability, as well as to make predictions of future states of the climate system under a wide variety of conditions. In particular, ESMs are extensively used in recent times to understand the impacts of human induced perturbations (e.g., enhanced greenhouse gas and aerosol emissions, land-use and land-cover changes …) on the climate system. Climate Change poses a major threat to the world today in view of its far-reaching implications for environment, agriculture, water availability, natural resources, ecosystems, biodiversity, economy and social well-being. There is strong scientific evidence pointing to the role of human activities in altering the Earth's climate through a rapid rise in the concentration of atmospheric greenhouse gases (GHG) since the nineteenth century (IPCC 2013). Observations of the climate system based on direct measurements and remote sensing from satellites and other platforms indicate that the warming of the climate system has been unequivocal since the 1950s and many of the observed changes are unprecedented over decades to millennia (Stocker et al. 2013). The series of assessment reports of the United Nations Intergovernmental Panel on Climate Change (IPCC), with the Fifth Assessment Report (AR5) being the latest, provides reliable scientific evidence which shows that "*Human influence on the climate system is clear, and recent anthropogenic emissions of greenhouse gases (GHG) are the highest in history. The atmosphere and ocean have warmed, the amounts of snow and ice have diminished, and sea level has risen (Climate Change 2013, IPCC)". Recent climate changes have had widespread impacts on human and natural systems.*

Long-term climate model simulations, that took part in the IPCC AR5 report, provide very high confidence in interpreting the observed global-mean surface temperature trends during the twentieth and early twenty-first centuries; and the human influence on the climate system (Stocker et al. 2013). The Coupled Modeling Intercomparison Project (CMIP3; Meehl et al. 2007, CMIP5; Taylor et al. 2012) coordinated by the World Climate Research Programme (WCRP) form the basis of the climate projections in the IPCC Assessment reports. The influence of human-induced

GHG forcing on the rising trend of global mean temperatures has been convincingly demonstrated by the CMIP models (Flato et al. 2013). On the other hand, there are major challenges in quantifying the impacts of climate change at regional levels. For example, the twentieth-century simulations and future projections of the South Asian monsoon rainfall based on the IPCC models exhibit a wide range of variations and uncertainties (Kitoh 2017; Krishnan et al. 2016; Sharmila et al. 2015; Sperber et al. 2013; Turner and Annamalai 2012), which pose huge challenges to policy makers and for development of adaptation strategies. In the backdrop of this underlying recognition, it is essential to build human capacity to address all challenging issues relating to the science of global warming and climate change.

The philosophy behind the development of an ESM at the Indian Institute of Tropical Meteorology (IITM) is to create capabilities in global modeling, with special emphasis on the South Asian monsoon, to address the science of climate change, including detection, attribution and future projections of global and regional climate. With this view, the first version (IITM-ESM version 1) was developed at the Centre for Climate Change Research, IITM, Pune (CCCR-IITM) by transforming a state-of-the-art seasonal prediction model—the Climate Forecast System version 2 (CFSv2, Saha et al. 2010) into a model suitable for investigating long-term climate variability and change (Swapna et al. 2015). An updated version of the IITM Earth System Model (IITM-ESM version 2) has been developed recently, through incorporation of various refinements leading to a radiatively balanced global climate modeling framework appropriate for addressing the science of climate change (Swapna et al. 2018). The IITM-ESMv2 will be first climate model from South Asia to participate in the CMIP model intercomparison project and contribute to the IPCC AR6 assessment report. This article provides an overview of the current capabilities of the IITM-ESMv2 to address key scientific questions relating to long-term climate variability and change, together with a roadmap for model development and scientific investigations in future.

2 Elements of the IITM-ESMv2

Basic platform for the IITM-ESM: The atmosphere component of the IITM-ESMv2 is based on the spectral version of the National Center for Environmental Prediction (NCEP) Global Forecasting System (GFS) model (Moorthi 2001) and is configured at a triangular truncation of 62 waves (T62) in the horizontal (~2° grid) and finite differencing in the vertical with 64 sigma-pressure hybrid layers with top model layer extending up to 0.2 hPa. A time-step of 10 min is used for the model integrations, while the atmospheric radiation calculations are performed every 1 h. The parameterization of cumulus convection (deep and shallow) in the model follows the revised version of the Simplified Arakawa-Schubert (SAS) scheme (Han and Pan 2011). More details about the representation of atmospheric physical processes can be found in Saha et al. (2010). The land surface processes are represented in the

IITM-ESMv2 based on the Noah land surface model (LSM) with four layers (Ek et al. 2003).

The ocean component in IITM-ESMv2 is the GFDL Modular Ocean Model Version 4p1 (MOM4p1, Griffies 2009). The MOM4p1 is a hydrostatic model and uses the Boussinesq approximation and a rescaled geopotential vertical coordinate with 50 vertical levels. The zonal resolution is 1° and the meridional resolution is 0.33° between 10°S and 10°N and gradually becoming 1° poleward of 30°S and 30°N. Key physical parameterizations include the K-Profile Parameterization (KPP) surface boundary layer scheme similar to that of Large et al. (1994) for computing vertical diffusivity, vertical viscosity, and nonlocal transport as a function of the flow and surface forcing. Interactive modules for ocean biogeochemistry and marine ecosystem processes are also included in MOM4p1. More details about MOM4p1 are described in Griffies (2009). The sea ice component of IITM-ESMv2 is the Sea Ice Simulator (SIS, Winton 2000), which is a dynamical model with three vertical layers, one snow and two ice, and five ice thickness categories.

Transforming CFSv2 into IITM-ESMv2

The IITM-ESM, developed at the CCCR-IITM, emerged by transforming the CFSv2 operational seasonal prediction model to a long-term climate model suitable for addressing climate change related scientific issues. Incorporating certain critical ingredients in a climate model is essential in order to satisfy the quest of having an ESM for long-term climate investigations. Keeping this in view, the transformation strategy of the IITM-ESM has evolved through a series of model developments which are briefly summarized below. More details about the capabilities achieved from the development of the IITM-ESM can be found in Swapna et al. (2015)

- **IITM-ESMv1**: The first version of the ESM was developed by incorporating the GFDL MOM4P1 ocean component with biogeochemistry in CFSv2, followed by long simulations (>100 years) and detailed model evaluation (Swapna et al. 2015). Major improvements were notable in the IITM-ESMv1 simulations vis-à-vis CFSv2, which include (a) Significant reduction of cold bias of global mean SST by ~0.8 °C (b) Robust reproduction of major climate drivers like El Nino/Southern Oscillation (ENSO) and Pacific Decadal Oscillation (PDO) (c) Teleconnections between ENSO and the Indian summer monsoon precipitation.
- **IITM-ESMv2**: For reliable assessments of climate change simulations, it is imperative that the global radiation energy balance at the top-of-the-atmosphere (TOA) and at the earth's surface are realistically represented in climate models. This was a major challenging issue with the IITM-ESMv1 which exhibited a large radiation imbalance ~6 W m^{-2} at the TOA. This problem was addressed in the IITM-ESMv2 by incorporating various improvements, which led to significant reduction in the radiation imbalance (Fig. 1). The radiation imbalance at the TOA in the IITM-ESMv2 is about 0.8 W m^{-2}, which is well within the acceptable range of imbalance simulated across multiple CMIP models (Forster et al. 2013). A significant part of this improvement in IITM-ESMv2 came from incorporating the effects of heating from viscous dissipation of turbulent kinetic energy (TKE) (Fiedler 2000;

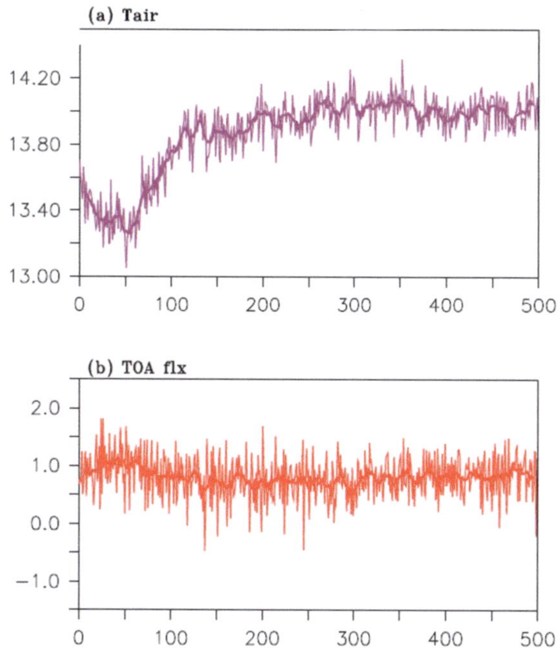

Fig. 1 Time-series from the Pre-Industrial (PI) control simulation of IITM-ESMv2 **a** global mean surface air-temperature (°C) **b** net radiation flux at the top-of-atmosphere (TOA). The PI control experiment is a multi-century simulation uses GHG, aerosols, land-use and land-cover and other forcing corresponding to 1850. Note that the time-series tend towards quasi-equilibrium with mean values of global mean surface air-temperature ~14 °C and TOA net radiation flux ~0.8 W m^{-2}

Han et al. 2016). Additionally several other refinements have been incorporated in the IITM-ESMv2 and the improvements include (a) Realistic sea ice distribution in the Arctic (b) Reproduction of the Atlantic Meridional Overturning Circulation (AMOC) (c) Improved mean monsoon precipitation over India in IITM-ESMv2 vis-à-vis IITM-ESMv1 (d) Robust teleconnections of the Indian Ocean Dipole (IOD) and the Indian monsoon rainfall (e) Incorporation of time-varying properties of aerosols (natural and anthropogenic) and land-use land-cover changes for the CMIP6 experiments.

Synthesis of improvements in the IITM-ESMv2

While the transition from ESMv1 to ESMv2 is an outcome of several modifications in the climate model, it is essential to comprehend the processes that contributed to two major improvements viz, (a) Impressive reduction of global average radiative imbalance at the TOA from 6 to 0.8 W m^{-2} and (b) Realistic simulations of the polar sea ice distribution and the Atlantic Meridional Overturning Circulation (AMOC)—an important component of the global climate system. Although a significant fraction of the radiation imbalance was remedied by including the effects of

TKE dissipative heating in the IITM-ESMv2, achievement of a radiatively balanced framework largely came from the recognition that the processes associated with the improvements (a) and (b) are closely linked to each other. Earlier studies have noted that the strength of the AMOC, which is characterized by northward flow in the upper ocean and returning southward flow in the deep ocean, is generally weak in almost all of the CMIP models; which in turn leads to cold sea surface temperature (SST) biases in the Northern Hemisphere (Wang et al. 2014; Zhang and Zhao 2015) [e.g.,. Basically, reduced northward oceanic heat transport associated with a weak AMOC induces cooling in the extra-tropical North Atlantic Ocean which also tends to have remote impacts on SST biases in the North Pacific through atmosphere-ocean interactions (Zhang and Zhao 2015)]. SST biases also tend to alter the radiation balance at the TOA through changes in the outgoing long-wave radiation. This implies that cold biases in global SSTs can result in a positive net radiation imbalance at the TOA. In the IITM-ESMv1, the simulated AMOC was extremely weak due to excessive melting of the Arctic sea ice and the associated freshening of the North Atlantic. In particular, the sea ice concentration in the summer hemisphere (i.e., June–August in the Arctic and Dec–Feb in the Antarctic) was severely underestimated in ESMv1. Therefore, it was envisaged that the realism of the Arctic sea ice distribution and the AMOC simulation would be the key remedy for improving the radiation balance in the IITM-ESMv2 (Figs. 2 and 3). This overarching hypothesis was the fundamental basis that aided in realistically reproducing the Arctic sea ice distribution (Figs. 2 and 3), the strength of the AMOC and the net radiation balance at the TOA in the IITM-ESMv2. The improved sea ice simulation in IITM-ESMv2 was achieved through incorporation of Sea Ice Simulator (SIS) for flux computations over ice-covered regions and implementation of fractional grids for the exchange of fluxes between atmosphere and other component models (Swapna et al. 2018).

The realism of simulating the seasonal monsoon precipitation over India is of paramount importance, being a monsoon centric region. The summer monsoon (June–September) precipitation, which accounts for more than 75% of the annual rainfall, has profound influence on the agriculture and economy of the country (Parthasarathy et al. 1994). The large-scale pattern of the boreal summer monsoon precipitation over South and Southeast Asia and adjoining oceanic regions and extending into the Maritime continent and tropical west Pacific are reasonably captured well both in ESMv1 and ESMv2 (Swapna et al. 2018). While the precipitation amounts over the Indian region are underestimated in both the versions, the dry bias is significantly smaller in ESMv2 as compared to ESMv1 (Swapna et al. 2018). The use of the revised SAS parameterization for cumulus convection (Han and Pan 2011) in ESMv2 improves the simulation of monsoon precipitation over the Indian subcontinent (Fig. 4). Recent studies also corroborate similar improvements in representing the summer monsoon precipitation and its sub-seasonal variability over the Indian region through the use of the revised SAS cumulus convection scheme in the CFSv2 coupled model (Abhik et al. 2017; Ganai et al. 2015).

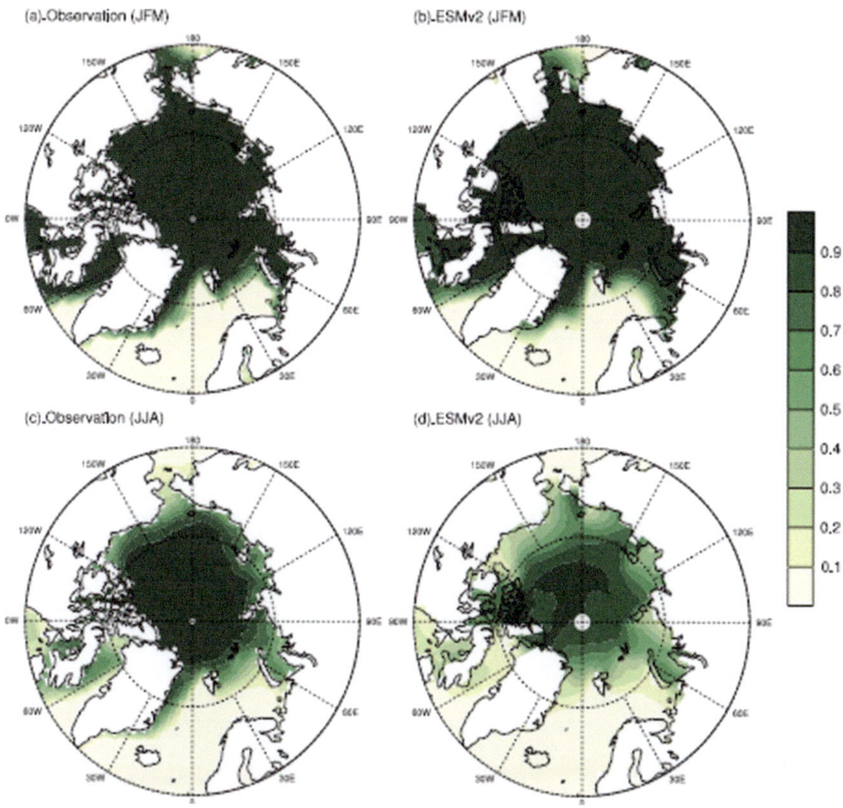

Fig. 2 Spatial map of mean sea ice concentration (%) over the Arctic from observations (Hadley Centre) and IITM-ESMv2 simulations **a**, **b** Winter and early Spring (Jan–Feb–Mar) **c**, **d** Northern Summer (Jun–Jul–Aug)

3 Key Scientific Questions and Future Directions

The current capabilities of the IITM-ESMv2 enable us to investigate some of the key scientific questions relating to the science of climate change, as follows:

- *The radiatively balanced modeling framework is a valuable tool for making reliable assessments of the climate sensitivity, as well as the global and regional hydrological responses to the radiative effects of anthropogenic forcing (e.g. GHG, aerosols, land-use changes, ...).*
- ***Reliable assessments of the impacts of climate change:*** *Monsoon hydrological cycle, Regional weather and climate extremes, sea-level changes in the global and Indian Ocean, Marine primary productivity and mechanisms controlling the ocean carbon cycle, and the global and Himalayan cryosphere.*
- *Separating the roles of natural and anthropogenic forcing on the global climate system, atmosphere-ocean coupled phenomena and the monsoons, and providing*

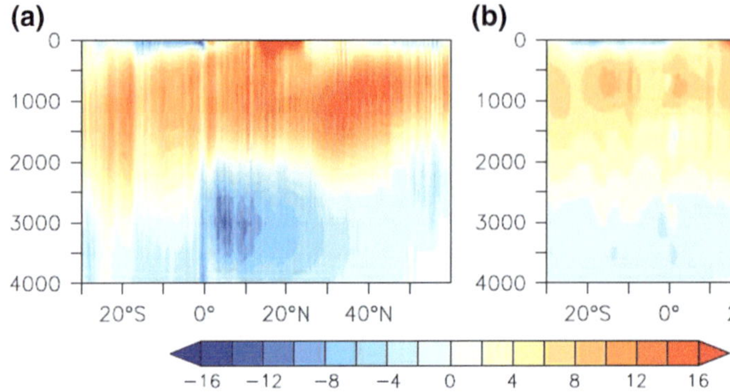

Fig. 3 Latitude-depth section showing Atlantic Meridional Overturning Circulation (AMOC expressed in Sverdrups (1 SV = 10^6 m^3 s^{-1}) from **a** GODAS **b** ESMv2

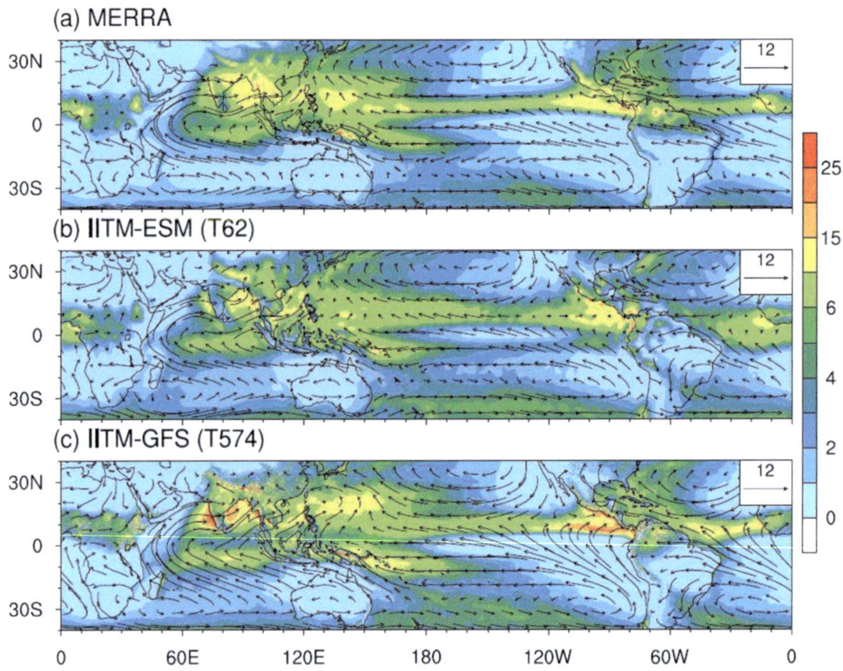

Fig. 4 Spatial maps of climatological mean precipitation (mm day^{-1}) and 850 hPa winds during the boreal summer monsoon (June–September) season **a** TRMM precipitation and MERRA reanalysis winds **b** IITM-ESMv2 (PI control simulation) and **c** high-resolution (T574: 27 km grid) atmospheric-only version of IITM-ESMv2. The simulated means are based on the last 50 years of the PI Control experiment and 10 years from high-resolution atmospheric-only version of IITM-ESMv2

reliable assessments of the South Asian monsoon precipitation response to climate change.

- ***Aerosol forcing and Monsoons:*** *IITM-ESMv2 incorporates the radiative effects of aerosols, both natural (e.g., dust, sea-salt, volcanic emissions…) and anthropogenic (sulfate, nitrate, organic carbon, black carbon, …) on the climate system. Atmospheric aerosols affect climate through scattering and absorption of the incoming solar radiation (direct effect) and through modification of cloud properties (indirect effect). Rapid industrialization during the last 5–6 decades has increased atmospheric aerosol loading* (IPCC 2013). *Recent studies have pointed to the role of anthropogenic aerosol forcing on radiation, monsoon rainfall, and regional climate* (Bollasina et al. 2011; Krishnan et al. 2016; Ramanathan et al. 2005). *Unlike GHGs, the space-time variability of aerosols is large. Furthermore, monsoon precipitation variability over South Asia is strongly influenced by monsoon internal dynamics. Furthermore, recent studies have drawn attention to the role of stratospheric aerosol forcing from large explosive volcanic eruptions on the ENSO and monsoon rainfall variability* (Khodri et al. 2017; Liu et al. 2016; Ning et al. 2017). *Therefore, it is necessary to have reliable attributions of aerosol forcing on regional monsoon precipitation changes. The IITM-ESMv2 is a valuable tool to address the aerosol and monsoon connection.*
- ***Land-use and land-cover changes:*** *The IITM-ESMv2 has capabilities to address the effects of land-use and land-cover changes (LULC) on the climate system. This is of particular interest to the Asian region which has undergone major changes in forest cover, agricultural land and vegetation types since pre-industrial times. The IITM-ESMv2 provides a great opportunity to investigate the role of LULC on the regional monsoon precipitation pattern.*
- ***Response of AMOC to global warming:*** *Realistic representation of AMOC is an essential ingredient in climate models in order to study long-term climate variability on decadal, centennial and longer time-scales. The significant improvements in simulating the AMOC and polar sea ice distribution in the IITM-ESMv2 present a great opportunity to quantify the impacts of global warming on the polar ice caps and the strength of the AMOC. These impacts can in turn affect the global climate, including the Asian monsoon, through alterations in the large-scale atmospheric and oceanic circulation patterns. The AMOC sensitivity experiments will require model simulations extending over at least a few thousand years, which is realizable in the next 4–5 years using the latest multi-petaflop high-performance computing (HPC) facility at IITM, Pune.*
- ***Paleo-climate modeling:*** *In the coming 5–10 years, we plan to conduct paleo-climatic modeling studies using the IITM-ESM to understand the past variations of the Asian and Indian monsoons during the last 20,000 years (from the Last Glacial Maximum through the Holocene). The modeling experiments will focus on the influence of the Sun–Earth orbital variations on the Earth's climate and monsoons and these simulations should provide key insights into the understanding of the future evolution of the Asian monsoon climate in a warming world.*

  High-resolution climate change projections: *In the next 3–5 years, it is planned to generate global high-resolution (grid size ~ 27 km) climate change simulations and future projections using the atmospheric-only component of the IITM-ESMv2* (Figs. 4 and 5). *High-resolution climate change projections are crucial*

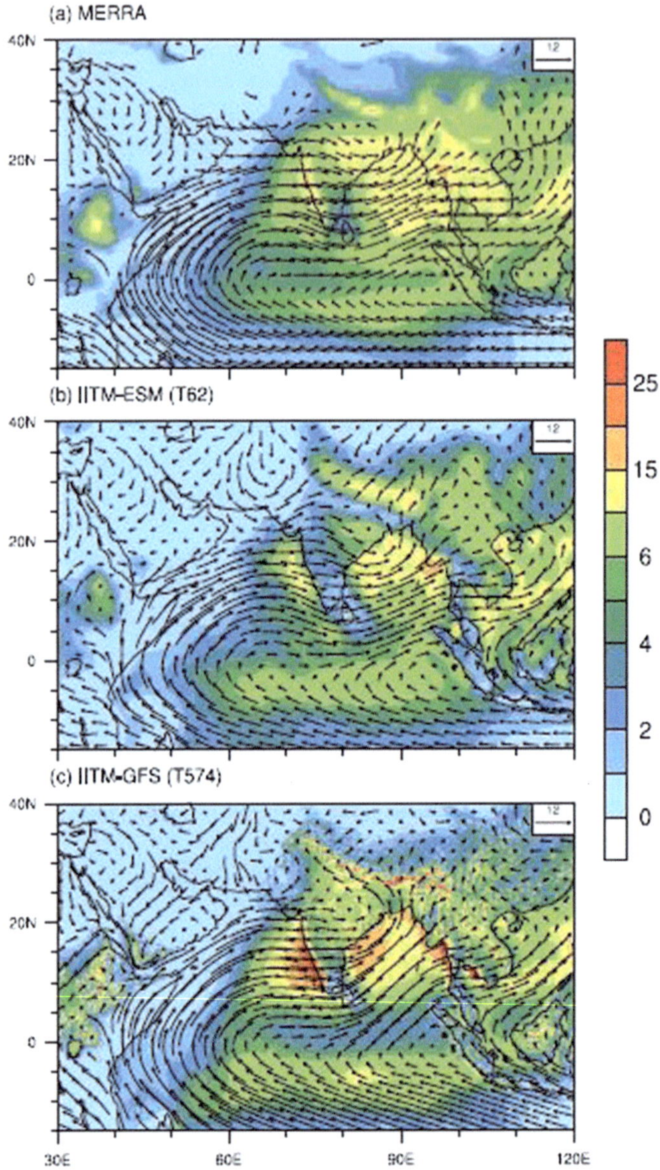

Fig. 5 Spatial maps of climatological mean precipitation (mm day^{-1}) and 850 hPa winds during the boreal summer monsoon (June–September) season over the South Asian/Indian Ocean region **a** TRMM precipitation and MERRA reanalysis winds **b** IITM-ESMv2 (PI control simulation) and **c** High-resolution (T574: 27 km grid) atmospheric-only version of IITM-ESMv2. The high-resolution version shows improvements in mean monsoon rainfall over the Western Ghats, Bay of Bengal and northcentral India as compared to the low-resolution IITM-ESMv2

for assessment of changes in weather and climate extremes, changes in mean monsoon summer precipitation and extreme precipitation occurrences, which are generally not captured by low-resolution models (Kitoh 2017; Krishnan et al. 2016). *The construction of surface boundary conditions (SST and sea ice) for the high-resolution atmospheric-only (IITM-ESMv2-HR-Atmos) model would be based on bias-corrected SST and sea ice fields obtained from the coarse resolution IITM-ESMv2 simulations. Similar strategies for producing high-resolution simulations and climate change projections of the Asian monsoon have been adopted using the Meteorological Research Institute (MRI, Japan) and the Laboratoire de Météorologie Dynamique Zoom (LMDz, Paris and IITM, Pune) climate models (*Kitoh 2017; Kitoh and Kusunoki 2008; Krishnan et al. 2013, 2016; Rajendran and Kitoh 2008). *High-resolution climate change projections also serve as important inputs to drive other modeling applications, e.g., hydrology, glaciers agriculture, forest fire, health, etc.*

- ***Tropical and monsoon precipitation:*** *Although the IITM-ESMv2 shows some improvements in capturing the mean summer monsoon precipitation over the Indian region as compared to ESMv1, there is a need to further improve the realism of simulating the Intertropical Convergence Zone (ITCZ) and monsoon precipitation distribution over the Indian and Asian monsoon land region and the adjoining Indo-Pacific oceanic areas; as well as the humidity fields and large-scale tropical divergent circulations* (Swapna et al. 2018). *Focused activities will be taken up to address these scientific issues during the next few years.*
- ***Next generation IITM-ESM:*** *Finally, we envisage the development of the next generation IITM-ESM, in the next 7–15 years, having a new dynamical core, interactive aerosols, atmospheric chemistry, and carbon cycle. The next generation IITM-ESM is intended to serve as a community model for further model developments, climate change studies and assessments and monsoon prediction across different time-scales.*

References

Abhik, S., R.P.M. Krishna, M. Mahakur, M. Ganai, P. Mukhopadhyay, and J. Dudhia. 2017. Revised cloud processes to improve the mean and intraseasonal variability of Indian summer monsoon in climate forecast system: Part 1. *Journal of Advances in Modeling Earth Systems* 9: 1–28. https://doi.org/10.1002/2016MS000819.

Bollasina, M.A., Y. Ming, and V. Ramaswamy. 2011. Anthropogenic aerosols and the weakening of the South Asian Summer Monsoon. *Science* 334 (6055): 502–505. https://doi.org/10.1126/science.1204994.

Ek, M.B., K.E. Mitchell, Y. Lin, E. Rogers, P. Grunmann, V. Koren, and J.D. Tarpley. 2003. Implementation of Noah land surface model advances in the National Centers for Environmental Prediction operational mesoscale Eta model. *Journal of Geophysical Research: Atmospheres* 108 (D22): n/a-n/a. https://doi.org/10.1029/2002JD003296.

Fiedler, B.H. 2000. Dissipative heating in climate models. *Quarterly Journal of the Royal Meteorological Society* 126 (564): 925–939. https://doi.org/10.1256/smsqj.56407.

Flato, G., J. Marotzke, B. Abiodun, P. Braconnot, S. C. Chou, W. Collins, and M. Rummukainen. 2013. Evaluation of climate models. Climate change 2013: The physical science basis. Contribution of Working Group I to the Fifth Assessment Report of the Intergovernmental Panel on Climate Change, 741–866. https://doi.org/10.1017/CBO9781107415324.

Forster, P.M., T. Andrews, P. Good, J.M. Gregory, L.S. Jackson, and M. Zelinka. 2013. Evaluating adjusted forcing and model spread for historical and future scenarios in the CMIP5 generation of climate models. *Journal of Geophysical Research: Atmospheres* 118 (3): 1139–1150. https://doi.org/10.1002/jgrd.50174.

Ganai, M., R. Phani Murali Krishna, P. Mukhopadhyay, and M. Mahakur. 2015. The impact of revised simplified Arakawa-Schubert scheme on the simulation of mean and diurnal variability associated with active and break phases of Indian summer monsoon using CFSv2. *Journal of Geophysical Research Atmospheres* 9301–9323. https://doi.org/10.1002/2014JD021636.

Griffies, S.M., M. Schmidt, and M. Herzfeld. 2009. Elements of mom4p1. GFDL Ocean Group Technical Report 6, 444 pp. http://data1.gfdl.noaa.gov/~arl/pubrel/r/mom4p1/src/mom4p1/doc/guide4p1.pdf.

Han, J., and H.-L. Pan. 2011. Revision of convection and vertical diffusion schemes in the NCEP global forecast system. *Weather and Forecasting* 26 (4): 520–533. https://doi.org/10.1175/WAF-D-10-05038.1.

Han, J., M.L. Witek, J. Teixeira, R. Sun, H.-L. Pan, J.K. Fletcher, and C.S. Bretherton. 2016. Implementation in the NCEP GFS of a hybrid eddy-diffusivity mass-flux (EDMF) boundary layer parameterization with dissipative heating and modified stable boundary layer mixing. *Weather and Forecasting* 31 (1): 341–352. https://doi.org/10.1175/WAF-D-15-0053.1.

IPCC, 2013. Climate Change 2013. The Physical Science Basis. In *Contribution of Working Group I to the Fifth Assessment Report of the Intergovernmental Panel on Climate Change*, ed. Stocker, T. F., Qin, D., Plattner, G.-K., Tignor, M., Allen, S. K., Boschung, J., Nauels, A., Xia, Y., Bex, V., Midgley, P.M., 1535 pp. , United Kingdom and New York, NY, USA: Cambridge University Press, Cambridge. https://doi.org/10.1017/CBO9781107415324.

Khodri, M., T. Izumo, J. Vialard, S. Janicot, C. Cassou, M. Lengaigne, and M.J. McPhaden. 2017. Tropical explosive volcanic eruptions can trigger El Ninõ by cooling tropical Africa. *Nature Communications* 8(1): 1–12. https://doi.org/10.1038/s41467-017-00755-6.

Kitoh, A. 2017. The Asian monsoon and its future change in climate models: A review. 95. https://doi.org/10.2151/jmsj.2017-002.

Kitoh, A., and S. Kusunoki. 2008. East Asian summer monsoon simulation by a 20-km mesh AGCM. *Climate Dynamics* 31 (4): 389–401. https://doi.org/10.1007/s00382-007-0285-2.

Krishnan, R., T.P. Sabin, D.C. Ayantika, A. Kitoh, M. Sugi, H. Murakami, and K. Rajendran. 2013. Will the South Asian monsoon overturning circulation stabilize any further? *Climate Dynamics* 40(1–2): 187–211. https://doi.org/10.1007/s00382-012-1317-0.

Krishnan, R., T.P. Sabin, R. Vellore, M. Mujumdar, J. Sanjay, B.N. Goswami, and P. Terray. 2016. Deciphering the desiccation trend of the South Asian monsoon hydroclimate in a warming world. *Climate Dynamics* 47(3–4): 1007–1027. https://doi.org/10.1007/s00382-015-2886-5.

Large, W.G., J.C. McWilliams, and S.C. Doney. 1994. Oceanic vertical mixing: A review and a model with a nonlocal boundary layer parameterization. *Reviews of Geophysics*. https://doi.org/10.1029/94RG01872.

Liu, F., J. Chai, B. Wang, J. Liu, X. Zhang, and Z. Wang. 2016. Global monsoon precipitation responses to large volcanic eruptions. *Scientific Reports* 6 (April): 1–11. https://doi.org/10.1038/srep24331.

Meehl, G.A., T.F. Stocker, W.D. Collins, P. Friedlingstein, A.T. Gaye, J.M. Gregory, and I.G. Watterson. 2007. Global climate projections. Climate change 2007: The physical science basis. Contribution of Working Group I to the Fourth Assessment Report of the Intergovernmental Panel on Climate Change.

Moorthi, S., H.L. Pan, and P. Caplan. 2001. Changes to the 2001 NCEP operational MRF/AVN global analysis/forecast system, NWS Tech. *Procedures Bulletin* 484: 14. NCEP, Silver Spring, Md.

Ning, L., J. Liu, and W. Sun. 2017. Influences of volcano eruptions on Asian Summer Monsoon over the last 110 years. *Scientific Reports* 7: 3–8. https://doi.org/10.1038/srep42626.
Parthasarathy, B., A.A. Munot and D.R. Kothawale. 1994. All-India monthly and seasonal rainfall series: 1871–1993. *Theoretical Applied Climatology* 49:217–224.
Rajendran, K., and A. Kitoh. 2008. Indian summer monsoon in future climate projection by a super high-resolution global model.
Ramanathan, V., C. Chung, D. Kim, T. Bettge, L. Buja, J.T. Kiehl, and M. Wild. 2005. Atmospheric brown clouds: Impacts on South Asian climate and hydrological cycle. *Proceedings of the National Academy of Sciences* 102(15): 5326–5333. https://doi.org/10.1073/pnas.0500656102.
Saha, S., S. Moorthi, H.L. Pan, X. Wu, J. Wang, S. Nadiga, M. Goldberg. 2010. The NCEP climate forecast system reanalysis. *Bulletin of the American Meteorological Society* 91(8): 1015–1057. https://doi.org/10.1175/2010BAMS3001.1.
Sharmila, S., S. Joseph, A.K. Sahai, S. Abhilash, and R. Chattopadhyay. 2015. Future projection of Indian summer monsoon variability under climate change scenario: An assessment from CMIP5 climate models. *Global and Planetary Change* 124: 62–78. https://doi.org/10.1016/j.gloplacha.2014.11.004.
Sperber, K.R., H. Annamalai, I.S. Kang, A. Kitoh, A. Moise, A. Turner, and T. Zhou. 2013. The Asian summer monsoon: An intercomparison of CMIP5 vs. CMIP3 simulations of the late 20th century. *Climate Dynamics* 41. https://doi.org/10.1007/s00382-012-1607-6.
Stocker, T.F., D. Qin, G.-K. Plattner, M. Tignor, S.K. Allen, J. Boschung, and P.M. Midgley. 2013. Summary for policymakers. In Climate change 2013: The physical science basis. Contribution of Working Group I to the Fifth Assessment Report of the Intergovernmental Panel on Climate Change, 33. https://doi.org/10.1017/CBO9781107415324.
Swapna, P., M.K. Roxy, K. Aparna, K. Kulkarni, A.G. Prajeesh, K. Ashok, and B.N. Goswami. 2015. The IITM earth system model: Transformation of a seasonal prediction model to a long-term climate model. *Bulletin of the American Meteorological Society* 96 (8): 1351–1368. https://doi.org/10.1175/BAMS-D-13-00276.1.
Swapna, P., R. Krishnan, N. Sandeep, Prajeesh, Ayantika, Manmeet, and V. Ramesh. 2018. Long-term climate simulations using the IITM-ESMv2 with focus on the South Asian monsoon. *Journal of Advances in Modeling Earth Systems*. https://doi.org/10.1029/2017ms001262.
Taylor, K.E., R.J. Stouffer, and G.A. Meehl. 2012. An overview of CMIP5 and the experiment design. *Bulletin of the American Meteorological Society* 93 (4): 485–498. https://doi.org/10.1175/BAMS-D-11-00094.1.
Turner, A.G., and H. Annamalai. 2012. Climate change and the South Asian summer monsoon. *Nature Climate Change* 2 (8): 587–595. https://doi.org/10.1038/nclimate1495.
Wang, C., L. Zhang, S.K. Lee, L. Wu, and C.R. Mechoso. 2014. A global perspective on CMIP5 climate model biases. *Nature Climate Change* 4 (3): 201–205. https://doi.org/10.1038/nclimate2118.
Winton, M. 2000. A reformulated three-layer sea ice model. *Journal of Atmospheric and Oceanic Technology* 17 (4): 525–531. https://doi.org/10.1175/1520-0426(2000)017%3c0525:ARTLSI%3e2.0.CO;2.
Zhang, L., and C. Zhao. 2015. Processes and mechanisms for the model SST biases in the North Atlantic and North Pacific: A link with the Atlantic meridional overturning circulation. *Journal of Advances in Modeling Earth Systems* 7 (2): 739–758. https://doi.org/10.1002/2014MS000415.

Cumulus Friction in the Asian Monsoon of a Global Model with 7 km Mesh

Suvarchal K. Cheedela and Brian E. Mapes

Abstract Vertical transport of horizontal momentum by convective eddies (CMT) in the 7–400 km size range is examined comprehensively in data from the GEOS-5 Nature Run (G5NR), a 2-year global simulation with a 7 km horizontal mesh. This diagnosis is possible because NASA offers a coarse-grained dataset of the quadratic flux terms wu and wv in addition to the model velocity variables u, v, w. We assess the time tendency of large-scale vertically integrated shear kinetic energy (SKE) due to CMT. Negative values of a few tenths of $1\,Wm^{-2}$ prevail on average over warm tropical oceans, indicating that explicit convection on these scales exerts a viscous or frictional or downgradient transport effect on wind shear. However, positive as well as negative values do occur locally, based on spatial correlations u′w′ and v′w′ in the arrangement ("organization") of convective motions. In the Asian monsoon, where convection and shear are both strong, the viscosity can be characterized by a regression coefficient with values of about 5% cm^{-1}, meaning that convection which yields 1 cm of precipitation decrements SKE by about 5%. Adjustment of balanced monsoon flow to such a viscous effect implies adiabatic ascent to the north of existing convection, a mechanism that may be relevant to northward-propagating large-scale variability.

Keywords Tropical convection · Momentum flux · G5NR · Mesoscale Convection

S. K. Cheedela · B. E. Mapes (✉)
Department of Atmospheric Sciences, Rosenstiel School of Marine and Atmospheric Science (RSMAS), University of Miami, 4600 Rickenbacker Causeway, Miami, FL 33149, US
e-mail: mapes@miami.edu

D. A. Randall et al. (eds.), *Current Trends in the Representation of Physical Processes in Weather and Climate Models*, Springer Atmospheric Sciences,
https://doi.org/10.1007/978-981-13-3396-5_10

1 Introduction

While most scalars are mixed downgradient by turbulent convection, tilted updrafts in squall lines "can feed energy into the mean flow" (LeMone 1983), due to "upgradient" convective momentum transport (CMT). Because CMT depends on the structure of convective motions down to the finest scales, its assessment is among the most compelling scientific uses for costly convection-permitting simulations—if the challenge of data volume can be met sensibly.

Here we summarize CMT's impact on the 4-degree scale mean flow in a unique global convection-permitting simulation. Rather than speaking of upgradient versus downgradient flux (a vertically local question), here we summarize the broader question of whether CMT "can feed energy into the mean flow": Specifically, we compute $<S\dot{K}E>$ as derived in Sect. 2.

In strictly two-dimensional convection-resolving models, spontaneous domain-mean shear can arise from squalls whose tilt is itself shear-modulated (Held et al. 1993; Nishimoto et al. 2016). In the early growth stages of such a large-scale shearflow, upgradient CMT must have been active. However, when simulated in realistic shears, 3D convection and even 2D convection tend to exhibit an overall viscous effect, decreasing the kinetic energy (Mapes and Wu 2001). However, those results may be distorted by the cyclic boundary conditions and domain size limitations. Tung and Yanai (2002) also concluded from a large-scale budget estimate using observations that "The product of eddy momentum flux with the large-scale vertical wind shear shows that the momentum transport is, on the average, downgradient; that is, kinetic energy is converted from the large-scale motion to convection and turbulence."

This momentum effect of convection (CMT) is in addition to convection's impacts on the mass field through latent heat release, radiative heating, and eddy fluxes of thermodynamic quantities. The quantitative impact of CMT, in the context of all other terms, is beyond the scope of this study, but an estimation tool in the form of a regression coefficient will facilitate such assessment theoretically.

2 Derivation and Data

2.1 *Eddy Momentum Flux, Its Convergence, and Impact on Coarse-Grained Flow*

The horizontal momentum budget, averaged over some horizontal filter scale (denoted by an overbar $\overline{\psi}$) has a vertical eddy flux convergence term, among its other tendency terms such as large-scale advection (ADV), the pressure gradient force (PGF), and horizontal eddy flux convergence

$$\frac{\partial}{\partial t}(\overline{\vec{V}}) = -\frac{1}{\rho_0}\frac{\partial}{\partial z}(\overline{\rho_0 \vec{V}'w'}) + ADV + PGF - \nabla_h \cdot (\vec{V}'\vec{V}') + \cdots$$

Here we take $\rho_0(z)$ to be a constant reference profile, and use hydrostatic pressure (simply denoted p) as a vertical coordinate, so $dp = -\rho_0 g dz$ and $\omega \equiv dp/dt = -\rho_0 g w$. The first term, Eddy Momentum Tendency (EMT), can then be written in terms relevant to our pressure-coordinate model data

$$EMT = -\frac{1}{\rho_0}\frac{\partial}{\partial z}(\overline{\rho_0 \vec{V}'w'}) = g\frac{\partial}{\partial p}(\overline{\rho_0 \vec{V}'w'})$$

The eddy flux and its convergence are most meaningful when considered in light of the mean flow. In particular, we want to measure how deep convective systems on the 7 km mesh act as eddies, fluxing momentum among the layers of a vertically sheared, filtered (4-degree averaged) version of mean flow they are embedded in. The choice of 4-degree scale is arbitrary, but places much of mesoscale convective structure in the "eddy" category while distorting only modestly the depiction of the large-scale monsoon flow. This EMT, so defined, simply transports momentum between levels in the atmosphere: it does not include surface friction, so it cannot impact the mass-weighted mean momentum of a 4-degree air column (the mass-weighted vertical average of EMT is zero).

Since we are interested in the sheared part of $\vec{V}$ which EMT can change, we subtract the vertically averaged $<\vec{V}>$ from $\vec{V}$, to isolate the baroclinic or sheared flow:

$$\vec{V}_s(p) = \vec{V}(p) - <\vec{V}>$$

The inner product of EMT and $\vec{V}_s$, integrated over the mass of the troposphere, is a term in the budget of *shear kinetic energy* (SKE) (that is, the variance budget of $V_s = |\vec{V}_s|$):

$$\frac{\partial}{\partial t}_{EMT} < \overline{V_s^2}/2 > = < EMT \cdot \vec{V}_s > \equiv < S\dot{K}E >$$

Actually, $\vec{V}$ need not be separated this way, as the product term involving vertically constant $<\vec{V}>$ and EMT vanishes in the vertical integral. We define $< \dot{K}E >$ as:

$$\frac{\partial}{\partial t}_{EMT} < \overline{V^2}/2 > = \frac{1}{p_s}\int_0^{p_s} g\frac{\partial}{\partial p}(\overline{\rho_0 \vec{V}'w'}) \cdot \overline{\vec{V}} dp \equiv < \dot{K}E > \equiv < S\dot{K}E > \quad (1)$$

From the computed $< \dot{K}E >$ and $< SKE >$, we fit a simple model of the form

$$< \dot{K}E > = \alpha P < SKE > + \epsilon \quad (2)$$

where P is 4-degree surface rain rate, by minimizing the mean square of the residual ϵ. Coefficient α will be called cumulus friction.

2.2 Data

The data for this project were obtained through the Web site www.gmao.gsfc.nasa.gov/global_mesoscale/G5NR/. Details of the model configuration are in Putman et al. (2014), and an extensive scientific assessment of the simulation is available in (Gelaro and Coauthors 2015).

To summarize the key specifications: The Ganymede version of the GEOS-5 model (Putman and Suarez 2011), on a cubed-sphere global grid with about 7 km grid spacing, was integrated for 2 years, using sea surface temperature boundary conditions from May 15, 2005 to June 16, 2007. This simulation was called a "Nature Run", intended as a resource for observing system simulation experiments. Besides its visually compelling imagery archive (https://gmao.gsfc.nasa.gov/global_mesoscale/7km-G5NR/images/), detailed data are offered through an ftp download server, and also an OpenDAP portal for returning hypercubes requested by client software.

While convection is parameterized in GEOS-5 using the Relaxed Arakawa-Schubert (RAS), its action is restrained by high entrainment rates, so that most deep convection is explicit, albeit under-resolved, on the 7 km mesh.

We downloaded the 0.5-degree gridded hourly data for state variables u, v, w, ρ, and the product terms uw and vw. These fields were further coarse-binned using Climate Data Operators cdo (obtained from http://www.code.zmaw.de), using a 90×45 conservative regridding to yield global fields on a 4-degree mesh. This regridding procedure defines the overbar in the equations above, as well as the primes (deviations therefrom). Total precipitation and other fields were also obtained and similarly processed.

3 Results

The 2-year mean map of $< \dot{KE} >$ is shown in Fig. 1. Readers should ignore values over steep terrain, as these are not the topic here, and may contain artifacts of the procedures used to convert model data to a pressure coordinates and then coarse-grain the pressure-level fields. Over the oceans, $< \dot{KE} >$ is predominantly negative and in patterns that correspond to precipitation rate (contours). Values are a few tenths of 1 W m^{-2}, comparable to values from Steinheimer et al. (2008). The global mean of $< \dot{KE} >$ is about -0.06 Wm^{-2}, a value that is not too sensitive to screening the extreme values around topography. A faint positive shear production (red shading) is seen in the stratocumulus cloud regions of the eastern subtropical basins.

Two-year time series of hourly $< \dot{KE} >$ at three contrasting 4-degree grid cells are shown in Fig. 2. The monsoon point (light blue) has the expected seasonality (strong in boreal summer), and is predominantly negative, albeit with some positive excursions.

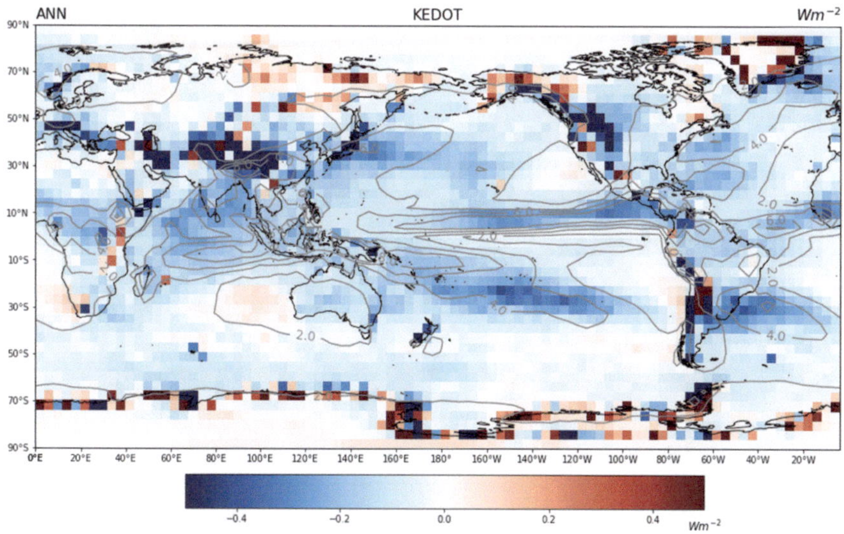

Fig. 1 Global map of 2-year time mean of KEdot (color shading) and precipitation (contours). The global mean is -0.06 W m^{-2}

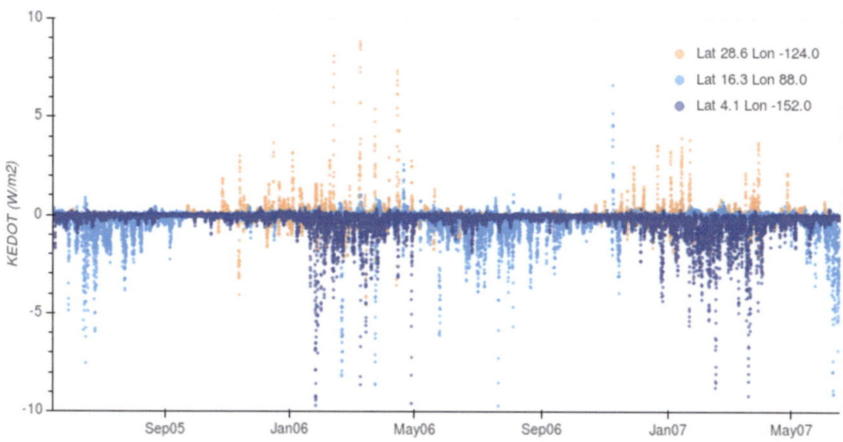

Fig. 2 Hourly time series of $< \dot{KE} >$ at selected grid cells (see legend)

Regressing such local time series onto precipitation times SKE yields the cumulus friction coefficient α from Eq. 2. The South Asian monsoon is especially interesting for such a fit, since precipitating convection is embedded in strong shear. An example of this regression fit, using pooled data from all the 4-degree cells in the South Asian monsoon area, is shown in Fig. 3. A linear fit is a reasonable model, and the regression line overplotted in the figure has $\alpha = -0.05$ cm^{-1}. This value is an order of magnitude smaller than the value assessed from higher resolution cyclic limited-area convection models by Mapes and Wu (2001). The smaller values of cumulus friction might be

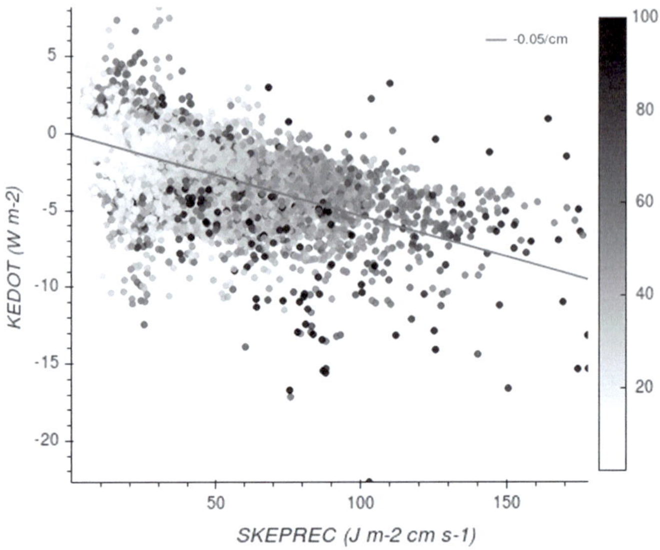

Fig. 3 Scatter plot of $< \dot{KE} >$ vs. SKE times precipitation, for all hours at all 4-degree grid cells where precipitation is greater then 2 mm day^{-1} in the Indian monsoon area (60–90E, 0–25N). Shading indicates precipitation (color bar, in mm day^{-1} units)

because of the definition of $<SKE>$ used in Eq. 2 SKE was integrated though the mass of the column, whereas Mapes and Wu (2001) used a time varying vertical integral of SKE, extending only through the layer where cloud mass flux exceeds 3 hPa h^{-1}. Resolution differences and other possible reasons for discrepancy are discussed further in our longer paper, Cheedela et al. (2019).

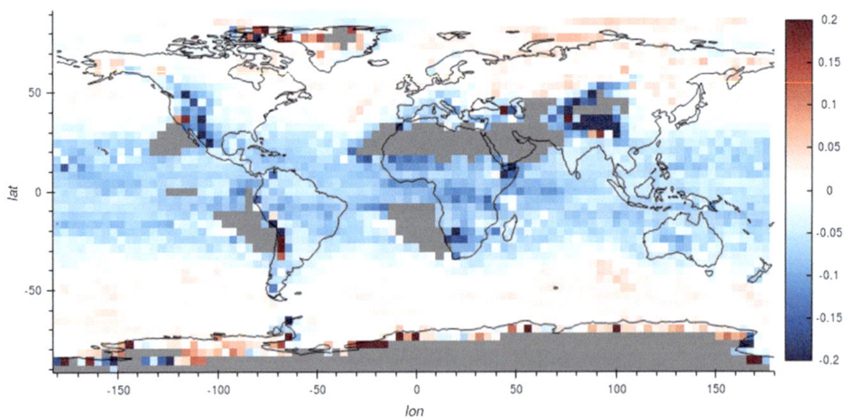

Fig. 4 Global map of cumulus friction $\alpha(\text{cm}^{-1})$ estimated from Eq. 2. Values where the number of precipitating hours is less than 1000 were masked out

This same regression was repeated at each 4-degree cell, yielding the global map of α in Fig. 4. The pattern is predominantly zonal, with negative values in the tropics and positive values in middle and higher latitudes. Values are of comparable magnitude to the example in Fig. 3.

4 Summary and Conclusions

While positive values of $\dot{KE}$ are occasionally observed, the predominant effect of momentum flux by the 7–400 km scale eddies in precipitating areas (our definition of CMT here) is clearly not to "feed energy into the mean flow" (LeMone 1983), but rather to damp the vertical wind shear. The values here are weak, but are relevant to the mesoscale organization arguments of Lemone and Moncrieff (1992). Finer scale eddies in convection, forbidden by the relatively coarse grid here, apparently contribute considerably to CMT, but also in a negative direction, based on results from Mapes and Wu (2001).

It will be interesting to compare other global convection-permitting simulations as resolution gradually increases. Far less than 2 years of data suffice to achieve stable statistics, based on our experimentation with the data here.

It is useful to consider qualitatively the likely impact of this shear-damping effect of convection in the Indian monsoon context. To do this, we will invoke the logic of balanced flow and its secondary circulations, as embodied in textbook quasi-geostrophic theory. The logic is illustrated in Fig. 5.

Panels a, b show thermal wind balance in monsoon flow and the disruption of that balance by CMT. Panel c shows that the induced secondary circulation, arising

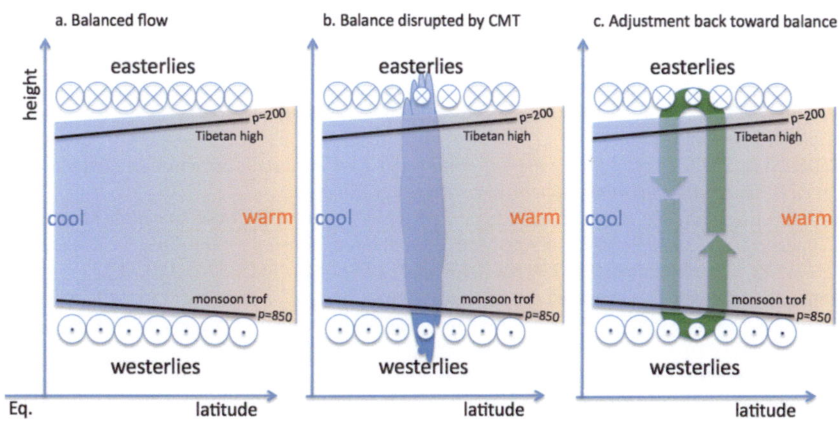

Fig. 5 Illustration of **a** thermal wind balance in monsoon flow, **b** CMT's effect in disrupting balance by damping shear, and **c** the resulting imbalance-induced adiabatic secondary circulation, which acts to adjust the flow back toward balance

in response to the balance-disrupting effect of cumulus friction in a given latitudinal belt, will have its ascending branch on the poleward (north) side of the existing convection. This secondary circulation is adiabatic: it is mechanically driven by the unbalanced Coriolis force. In a dry atmosphere, part of its adjustment effect occurs through decreasing the temperature gradient, via adiabatic cooling occurring on the north flank of the belt of CMT-reduced shear. In a moist atmosphere, such adiabatic cooling can enhance moist convection on the north flank of existing convection, a mechanism that will therefore contribute to poleward propagation of deep convective rain belts. This is an effect of the sub-filter-scale eddies, one whose quantitative strength depends on the mesoscale form of those eddies. Its impact must be assessed in light of other mechanisms of meridional propagation (DeMott et al. 2013). The cumulus friction coefficients shown here could help quantify the effect and contribute to better monsoon depictions in models with parameterized convection.

Acknowledgements The authors gratefully acknowledge the financial support given by the Earth System Science Organization, Ministry of Earth Sciences, Government of India (Grant no. / Project no $MM/SERP/Univ\_Miami\_USA/2013/INT-1/002$) to conduct this research under Monsoon Mission. We also acknowledge NASA grant NNX14AR75G, and Dr. Matthew Niznik for his initial work toward this project's goals.

References

Cheedela, S., M. Niznik, W. Putman, and B. E. Mapes, 2019: Mesoscale eddy momentum flux in a nonhydrostatic global atmosphere model: a global survey, cumulus friction coefficients, and negative viscosity cases. Manuscript in preparation.

DeMott, C.A., C. Stan, and D.A. Randall. 2013. Northward propagation mechanisms of the boreal summer intraseasonal oscillation in the ERA-Interim and SP-CCSM. *Journal of Climate* 26: 1973–1992. https://doi.org/10.1175/JCLI-D-12-00191.1.

Gelaro, R., and Coauthors, 2015: Evaluation of the 7-km GEOS-5 Nature Run. NASA Technical report NASA/TM–2014-104606/Vol. 36, available at https://gmao.gsfc.nasa.gov/pubs/docs/TM2014-104606v36.pdf.

LeMone, M.A. 1983. Momentum transport by a line of cumulonimbus. *Journal of the Atmospheric Sciences* 40: 1815–1834.

Mapes, B.E. 2017. *Toward form-function relationships for mesoscale structure in convection: A review*. Japan: *Journal of the Meteorological Society*. (in press).

Mapes, B.E., and X. Wu. 2001. NOTES AND CORRESPONDENCE Convective Eddy Momentum Tendencies in Long Cloud-Resolving Model Simulations. *Journal of Atmospheric Sciences* 58: 517–526. https://doi.org/10.1175/1520-0469(2001)058<0517:NACCEM>2.0.CO;2.

Moncrieff, M.W. 1992. Organized convective systems: Archetypal dynamical models, mass and momentum flux theory, and parametrization. *Quarterly Journal of the Royal Meteorological Society* 118: 819–850. https://doi.org/10.1002/qj.49711850703.

Nishimoto, E., S. Yoden, and H.H. Bui, 2016: Vertical Momentum Transports Associated with Moist Convection and Gravity Waves in a Minimal Model of QBO-like Oscillation. https://doi.org/10.1175/JAS-D-15-0265.1.

Putman, W., A. M. da Silva, L. E. Ott, and A. Darmenov, 2014: Model Configuration for the 7-km GEOS-5.12 Nature Run, Ganymed Release (Non-hydrostatic 7 km Global Mesoscale Simulation), GMAO Office Note, bf 5. (Version 1.0), 86 pp, available from http://gmao.gsfc.nasa.gov/pubs/office_notes.

Putman, W.M., and M. Suarez. 2011. Cloud-system resolving simulations with the NASA Goddard Earth Observing System global atmospheric model (GEOS-5). *Geophysical Research Letters* 38: L16809. https://doi.org/10.1029/2011GL048438.

Steinheimer, M., M. Hantel, and P. Bechtold. 2008. Convection in Lorenz's global cycle with the ECMWF model. *Tellus* 60A: 1001–1022.

Tung, W., and M. Yanai. 2002. Convective Momentum Transport Observed during the TOGA COARE IOP. PartII: Case Studies. *Journal of Atmospheric Science* 59: 2535–2549. https://doi.org/10.1175/1520-0469(2002)059<2535:CMTODT>2.0,CO;2.

Difficulties in the Subgrid-Scale Redistribution of Moisture of a Global Cloud-Resolving Model

Hiroaki Miura

Abstract More than one decade has passed since the first global cloud-resolving simulation was achieved under an aquaplanet condition in 2005. While such high-resolution global simulations have been beneficial not only to advance our knowledge of organized cloud systems but also to give various hints on improvements of traditional global models that depend on a kind of cumulus parameterization, explicit computations of cloud microphysics cannot necessarily ensure realistic representations of clouds and climate. A direct coupling between fluid dynamics and cloud processes is a strong point of the global cloud-resolving approach, but there still remain various rooms of uncertainties. Here, we briefly summarize successful and unsuccessful results of global or near-global simulations with explicit cloud microphysics and discuss a difficulty in the subgrid-scale redistribution of moisture.

Keywords Global cloud-resolving model · Subgrid-scale processes · Moisture

1 Introduction

About two decades ago, developments of a global "cloud-resolving" model (GCRM) called the Nonhydrostatic Icosahedral Atmosphere Model (NICAM; Satoh et al. 2008, 2014) was started for the use on the Earth Simulator. The Earth Simulator was the world's fastest supercomputer during the years from 2002 to 2004. One of the scientific targets of that supercomputer was to enable global simulations of the atmosphere with a horizontal mesh interval of less than 10 km. In the early 2000s, it was speculated that the standard configuration of the Global Climate Models (GCMs) would be unsuitable for those higher resolution global simulations. One of some reasons, for example, was in the computational aspect that the spectral transformation method using the spherical harmonics was believed to become inefficient on

H. Miura (✉)
Department of Earth and Planetary Science, Graduate School of Science,
The University of Tokyo, 7-3-1 Hongo, Bunkyo-ku, Tokyo 113-0033, Japan
e-mail: h_miura@eps.s.u-tokyo.ac.jp

D. A. Randall et al. (eds.), *Current Trends in the Representation of Physical Processes in Weather and Climate Models*, Springer Atmospheric Sciences,
https://doi.org/10.1007/978-981-13-3396-5_11

massively parallel architectures due to the lack of fast Legendre transformation algorithms. Another reason, in the physical aspect, was that cumulus parameterization schemes became unsuitable because the horizontal cell size is too small to justify the assumption of a significant scale separation between larger scale circulation and smaller scale flows of deep convection inside grid cells.

To overcome those foreseen issues, NICAM adopted the icosahedral hexagonal/pentagonal mesh for the horizontal discretization of the basic equations with the Arakawa A-grid staggering (Tomita et al. 2001) that does not require calculations of Poisson solvers and took the approach using a cloud microphysics scheme (Grabowski 1998 or Tomita 2008) instead of cumulus parameterizations. While researches and technical developments about global quasi-uniform mesh were necessary to develop the new model, the basic equations and physics parameterizations adopted by NICAM were fundamentally the same as the conventional regional "cloud-resolving models" such as MRI/NPD-NHM (Saito et al. 2001) and WRF (Skamarock et al. 2008). Some physical parameterization schemes were transported from MIROC climate model (e.g., Watanabe et al. 2011) to NICAM.

Currently, the spectral transformation models are still widely used. Additionally, needs for a cumulus parameterization are still controversial if a horizontal mesh size with the length scale of about O (km) is used. In terms of these points, we may conclude that the NICAM development was based on overly advanced concerns around the year 2000. While the NICAM's approach seems too radical, its configuration that does not depend on cumulus parameterizations gives us a benefit that results of two qualitatively different designs of models can be compared.

When we talk about NICAM, the configuration with or without a cumulus parameterization or the word "cloud-resolving" itself is sometimes controversial. It may be obvious that NICAM is not a "cloud-resolving" model even if a horizontal grid spacing will be O (m). But, at least in the early 2000, a generic name of the models that adopted configurations like NICAM but used in much smaller domains was the "cloud-resolving" model (e.g., Xu et al. 2002). I speculate that the word "cloud-resolving" has been accepted besides its physical inaccuracy because it can clearly elucidate the qualitative difference between the models with or without cumulus parameterizations.

In my understanding, the largest difference between the GCRMs and GCMs is in the tolerance of the remote redistribution of heat and moisture in a vertical column. On one hand, a cumulus parameterization of GCMs assumes a scene that single or multiple clouds develop or decay within a vertical column. The coexistence of convective updrafts and downdrafts enables redistribution of heat and moisture within a column. Thus, we can say that GCMs are the models that permit a remote influence on the heat and moisture fields. In the subgrid-scale, some properties of clouds are assumed as a physical phenomenon that justifies the remote influence. On the other hand, if a cloud microphysics scheme is used instead of cumulus parameterizations, a remote influence on the heat and moisture fields are almost inhibited, except for the falling processes of rain, snow, and cloud ice. Most of the sensible and latent heat transports are coupled with wind fields computed by a system of fluid equations. Note that the turbulent transports are also an exception of the direct coupling with the explicitly simulated flow fields. Obviously, the latter approach is not necessarily better than the former because the decoupling of the time and spatial scales between the dynamics and physics, which are intentionally or unintentionally set in parameter tunings, exists in both approaches and possibly become serious in the latter if horizontal and vertical resolution is too coarse to justify the use of cloud microphysics schemes.

In this chapter, we will discuss the possible issues of the current generation of GCRMs, which is still suffered from the difficulty due to the vertical redistribution of heat and moisture in a vertical column, along with some results of global O (km) resolution simulations made by NICAM.

2 Some Results of NICAM and Issues of GCRMs

Higher horizontal resolution could be helpful for improving the fidelity of Madden–Julian Oscillation (MJO) simulations. An example was provided by Miura et al. (2007a), showing that a global "cloud-resolving" simulation by NICAM reproduced realistic eastward movement of the convectively active region of the MJO event that initiated over the Indian Ocean in mid-December 2006. In the simulation using a global quasi-uniform 7-km mesh, not only the envelope of convective clouds but also the region of strong low-level westerly developed spontaneously over the Indian Ocean in mid-December 2016 and extended to the western Pacific until mid-January 2017. Liu et al. (2009) further examined the fidelity of this 7-km simulation. The one-week simulation using a global 3.5-km mesh replicated the distribution of clouds and its evolution over the maritime continents with some reality. Inoue et al. (2010) showed that the distribution of ice clouds in the 3.5-km simulation was in good agreement with that of high-level clouds classified by the split window data of geostationary meteorological satellites. This kind of visual reality comparable to an image pictured from satellites is an obvious advantage of higher resolution simulations.

One benefit of using CRMs is permitting turbulent motions associated with clouds. The good visual reality of clouds in global higher resolution simulations with a horizontal grid spacing less than 1 km (Miyamoto et al. 2013) is probably accounted for by the explicit computations of turbulent motions. It is questionable, however, if the accurate representations of smaller scale turbulent flows are physically essential for reproducing gross behaviors of synoptic- or planetary-scale phenomena. Typical zonal wavenumbers of the MJO features are one to five, and thus, a significant scale separation between the MJO scale and the scale of each convective cloud seems a reasonable assumption.

The other benefit of using CRMs is to permit a direct coupling between the fluid dynamics and the cloud microphysics in the model. This provides a pathway to avoid difficulties of the "cumulus parameterization deadlock" (Randall et al. 2003). A breakthrough achieved by the superparameterization approach (Grabowski 2001; Khairoutdinov and Randall 2001) seemed to indicate that rough behaviors of large-scale phenomena could be captured if CRMs computed transports and phase changes of water substances directly without through the cumulus parameterization.

What is expected for CRMs in simulating intraseasonal variability of larger scale phenomena such as the MJO is not that each cloud system develops at the correct location and at the correct timing as it is expected in the case of the studies of squall lines or supercells, but is that faster responses of clouds occur without unnecessarily disrupting slower evolutions of larger scale flow fields. We anticipate that the explicit coupling between the mesoscale dynamics and the cloud microphysics allows more accurate representations, in terms of the spatial and temporal scales, of the transports.

The consideration given above means, on one hand, that not only global CRMs but also GCMs using a cumulus parameterization should have an ability to simulate synoptic- and planetary-scale phenomena about their large-scale features. An essential requirement is that a cumulus parameterization or another framework representing vertical redistribution of heat and moisture by unresolved turbulent motions can represent reasonable spatial and temporal scales of moisture in larger spatial and longer temporal scales. Klingaman et al. (2015) showed an interesting result, based on a GCM intercomparison project, that the highest skill models in simulating two MJO cases during the Year of Tropical Convection (YOTC) better reproduce transition from low-level moistening for light rainfall to midlevel moistening at moderate rainfall and upper level moistening for heavy rainfall.

On the other hand, it also means that global CRMs do not always have the ability to reproduce larger scale phenomena if the representation of moisture is unsatisfactory in either spatial or temporal scale or both. Besides this concern, not only NICAM but also several other CRMs have demonstrated the benefits of using CRMs in global or wider domain in simulating intraseasonal behaviors of organized clouds in the tropics. Miyakawa et al. (2014) showed that NICAM maintained a valid MJO skill score for 27 days without a sophisticated data assimilation system. Holloway et al. (2013) used the limited area mode of the Met Office Unified Model with the finest horizontal mesh size of 4 km to conduct two 10-day simulations of an MJO event during YOTC. Wang et al. (2015) showed an excellent example that the WRF model could reproduce MJO events during the Cooperative Indian Ocean Experiment on Intraseasonal Variability in the Year 2011 (CINDY2011)/Dynamics of the MJO (DYNAMO)/ARM MJO Investigation Experiment (AMIE) observation campaign (hereafter, CYNDY2011/DYNAMO for brevity; Yoneyama et al. 2013; Gottschalck et al. 2013) if the lateral and bottom boundary conditions were given appropriately. Pilon et al. (2016) may be the first example of the MJO simulation by the Model for Prediction Across Scales (MPAS; Skamarock et al. 2012), targeting the period of CINDY2011/DYNAMO. Sasaki et al. (2016) followed the work of Miura et al. (2007a) using a coupled atmosphere–ocean model, the Multiscale Simulator for the Geoenvironment (MSSG). Although these examples appear to confirm the robustness of CRM applications, we can also find several examples of failure in large-scale simulations with global or near-global CRMs.

In the first trial of a realistic global CRM simulation (Miura et al. 2007b), NICAM severely overemphasized the organization of clouds. Tropical cyclones

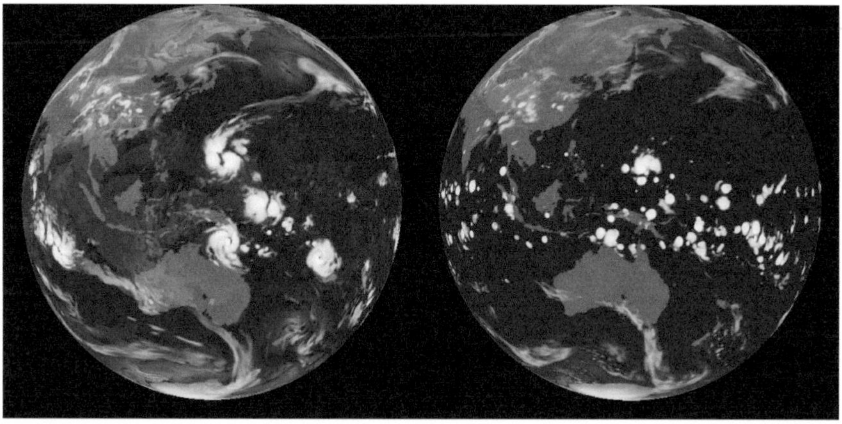

Fig. 1 Outgoing long-wave radiation simulated with a 14-km global mesh. Results of (left) enhanced turbulent mixing case and (right) suppressed turbulent mixing case. Images are adopted from Figure 4 of Miura et al. (2007b).

were exaggerated, while individual convective clouds were muted (Fig. 1). Even the diurnal cycle of clouds over lands forced by strong shortwave radiative heating during daytime was not observed. The cause of this unrealistic result was a significant overestimation of the upward transport of moisture in the subgrid-scale by a turbulence scheme. Holloway et al. (2013) found that their 4-km simulations of an MJO event were sensitive to the choice of a turbulence scheme. The eastward movement of an MJO event was reproduced well when the three-dimensional Smagorinsky mixing was used, while it was degraded when a vertical mixing was computed by another boundary layer scheme. They noted that the standard boundary layer scheme of the Met Office Unified Model was originally designed to work along with a convective parameterization, and the vertical mixing of moisture became insufficient when it was used independently. Ulate et al. (2015) used a channel configuration of WRF to simulate an MJO event during YOTC, but they summarized that "a tropical channel model failed to simulate an observed MJO event." This result is contradictory to the excellent results shown by Wang et al. (2015) with the same model. In their simulation, a cumulus parameterization scheme was used along with a cloud microphysics scheme. Although the reason for the unsatisfactory performance of WRF was not clarified in Ulate et al. (2015), it was likely that this parallel usage of the cumulus parameterization and the cloud microphysics enhanced an efficiency of the vertical redistribution of heat and moisture unrealistically. One of the lessons from these examples is that both the underestimation and overestimation of the vertical subgrid-scale mixing of moisture can lead to unrealistic representations of larger scale phenomena such as the MJO.

This difficulty comes from the fact that the subgrid-scale mixing is not constrained by a balance of the large-scale dynamics, but it can influence the large-scale flows, especially through thermodynamic processes. The vertical distribution of moisture in a CRM simulation is determined by a balance between the explicit and implicit transports represented by the grid-scale flow solver and the turbulence scheme, respectively, and the condensation, redistribution, and evaporation of water substances represented by the cloud microphysics scheme. Therefore, not only the turbulence scheme but also the cloud microphysics scheme can influence the efficiency of the vertical redistribution of moisture in the subgrid-scale, and as a result, can change behaviors of large-scale phenomena.

For the project called "Athena" (Kinter et al. 2013), which addressed a comparison between NICAM and the Integrated Forecast System (IFS) of the European Centre for Medium-Range Weather Forecasts (ECMWF), some physics packages of NICAM had been updated from those of an MJO simulation (Miura et al. 2007a). One significant change was in the cloud microphysics scheme. A simple scheme of Grabowski (1998) was replaced with a one-moment bulk microphysics scheme (NSW6) of Tomita (2008). The parameters of the NSW6 scheme were chosen to increase ice clouds, so that the energy budget at the top of the atmosphere (TOA) was approximately closed. The motivation to use those parameters would be to address the issue of future cloud changes under global warming (Satoh et al. 2012; Tsushima et al. 2014). Although the energy balance at the TOA was improved, this tuning caused severe biases in the fields of temperature, clouds, and the precipitation in the tropics. Corresponding to positive biases of upper tropospheric moisture and clouds, the stability became fairly strong, particularly, in upper troposphere. As a result, convective activity was suppressed over the ocean and was exaggerated over lands because the turbulence was more strongly forced by the surface sensible heat flux over lands. Given these deficiencies, it might be inappropriate to derive a conclusion that there is "the potential for stronger high cloud fraction feedbacks in climate warming scenarios" (Tsushima et al. 2014) based on this version of NICAM. Meanwhile, the stronger convection over lands might contribute to better representations of the diurnal cycle over lands and of the northward progression of the precipitation region associated with the Indian summer monsoon (Kinter et al. 2013). An instruction from this example is that we need to think more than a single index, the energy budget at the TOA in this case, to set microphysics parameters if we are interested in a reality of cloud developments, particularly, over the ocean.

When the first MJO event during CINDY2011/DYNAMO was simulated (Fig. 2a), NICAM with the same configuration as the project Athena almost failed to reproduce the event (Fig. 2c), while the same configuration as Miura et al. (2007a) reproduced it to some extent (Fig. 2b). The first event was initiated over the Indian Ocean about October 26, 2011. In these simulations, the initial conditions of the atmosphere were interpolated from the National Center for Environmental Prediction (NCEP) Final Operational Model Global Tropospheric Analyses data and Sea Surface Temperature (SST) was interpolated from weekly data of the NOAA Optimum Interpolation SST (OISST) (Reynolds et al. 2002). From these tests, it turned out that the set of the NSW6 parameters chosen to increase ice clouds (Satoh et al. 2012; Tsushima et al. 2014) was unsuitable for the MJO simulations. Then, the parameters were retuned (Miura et al. 2012) to recover the MJO event (Fig. 2d). Note that the important parameters were the autoconversion rate of the cloud ice and the fall speeds of rain

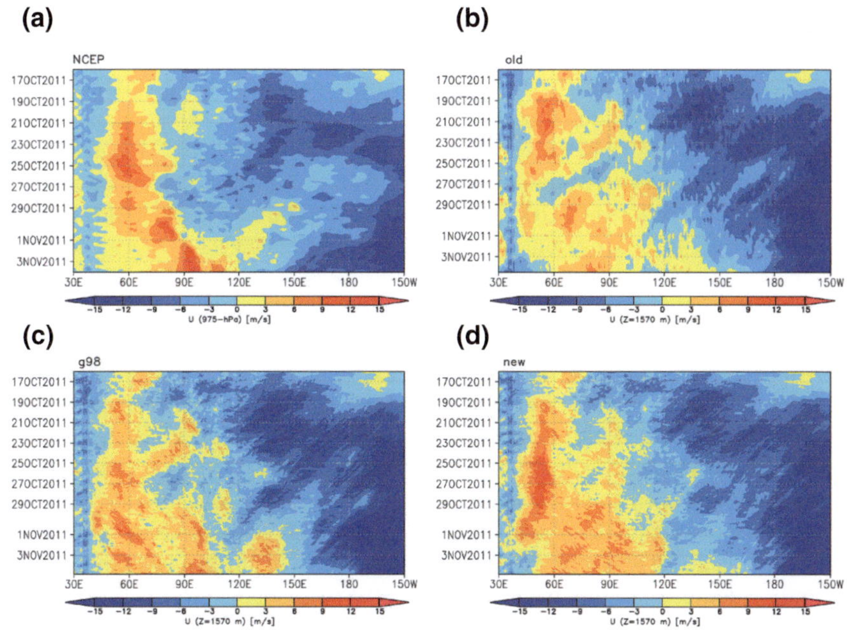

Fig. 2 Longitude-time cross-sections of the zonal wind averaged between 2.5°S and 2.5°N. **a** NCEP Final data, **b** Tomita (2008) scheme with the Athena parameters, **c** Grabowski (1998) scheme and **d** Tomita (2008) scheme with the re-tuned parameters Images are adopted from Figure 4 of Miura et al. (2007b).

and snow. It is obvious that these are parameters that can strongly influence the vertical distribution of water substances and water vapor.

The better performance of the new set of the NSW6 parameters is not limited to the first event of CINDY/DYNAMO, but seems robust for various MJO events. Miyakawa et al. (2014) used NICAM on the K computer with the retuned NSW6 scheme to show that a skill score of the MJO prediction was maintained for 26–28 days, depending on the MJO phase of the time of the model initialization. The YOTC hindcasts performed for an intercomparison project were mentioned as "accurate" by Klingaman et al. (2015). Miura et al. (2015) performed 60-day simulations with different SST settings and showed that the second MJO event during CINDY2011/DYNAMO was at least partially forced by the rapid seasonal warming of the ocean in the southeastern part of the maritime continents.

3 Summary and Discussion

The experience of the loss and the recovery of the MJO in NICAM with the NSW6 scheme inform us that the global or near-global CRMs are still sensitive to the vertical redistribution of moisture that is inevitably represented by the subgrid-scale unresolved processes. It is noted that the MJO simulations of NICAM are also sensitive to the settings of the surface latent heat flux, which is also a part of the subgrid-scale vertical mixing of moisture. These results might disappoint us that the "cloud microphysics deadlock" will emerge, even if we go "beyond the deadlock" (Randall 2013) of the cumulus parameterization. Besides this fear, we can also find a good news in the examples of the loss of MJO signals in CRMs. All failures listed in this chapter are, to a greater or lesser extent, associated with an unintended use or an artificial tuning of physical parameterizations. Hopefully, we may not need to be afraid of the tuning of the subgrid-scale processes unnecessarily if global or near-global CRMs are appropriately constrained by a variety of observations.

Although more than a decade has past since Tomita et al. (2005) performed the first global CRM simulation using a quasi-uniform 3.5-km mesh under an aquaplanet condition, only limited knowledge has been obtained about the characteristics of global CRMs. We need to continue developments of microphysics schemes (Seiki et al. 2015) and their improvements through comparisons with satellite data and in situ observations (Masunaga et al. 2008; Inoue et al. 2010; Dodson et al. 2013; Hashino et al. 2013; Roh et al. 2017). Intercomparisons of global CRMs will also be helpful in promoting further understanding of the behaviors of global CRMs.

References

Dodson, J.B., D.A. Randall, and K. Suzuki. 2013. Comparison of observed and simulated tropical cumuliform clouds by CloudSat and NICAM. *Journal of Geophysical Research Atmospheres* 118: 1852–1867. https://doi.org/10.1002/jgrd.50121.

Gottschalck, J., P.E. Roundy, C.J. Schreck III, A. Vintzileos, and C. Zhang. 2013. Large-scale atmospheric and oceanic conditions during the 2011–12 DYNAMO field campaign. *Monthly Weather Review* 141: 4173–4196. https://doi.org/10.1175/MWR-D-13-00022.1.

Grabowski, W.W. 1998. Toward cloud resolving modeling of large-scale tropical circulations: A simple cloud microphysics parameterization. *Journal of the Atmospheric Sciences* 55: 3283–3298. https://doi.org/10.1175/1520-0469(1998)055%3c3283:TCRMOL%3e2.0.CO;2.

Grabowski, W.W. 2001. Coupling cloud processes with the large-scale dynamics using the cloud-resolving convection parameterization (CRCP). *Journal of the Atmospheric Sciences* 58: 978–997. https://doi.org/10.1175/1520-0469(2001)058%3c0978:CCPWTL%3e2.0.CO;2.

Hashino, T., M. Satoh, Y. Hagihara, T. Kubota, T. Matsui, T. Nasuno, and H. Okamoto. 2013. Evaluating cloud microphysics from NICAM against CloudSat and CALIPSO. *Journal of Geophysical Research Atmospheres* 118: 7273–7292. https://doi.org/10.1002/jgrd.50564.

Holloway, C.E., S.J. Woolnough, and G.M.S. Lister. 2013. The effects of explicit versus parameterized convection on the MJO in a large-domain high-resolution tropical case study. Part I: Characterization of large-scale organization and propagation*. *Journal of the Atmospheric Sciences* 70: 1342–1369. https://doi.org/10.1175/JAS-D-12-0227.1.

Inoue, T., M. Satoh, Y. Hagihara, H. Miura, and J. Schmetz. 2010. Comparison of high-level clouds represented in a global cloud system–resolving model with CALIPSO/CloudSat and geostationary satellite observations. *Journal of Geophysical Research* 115: D00H22. https://doi.org/10.1029/2009jd012371.

Khairoutdinov, M.F., and D.A. Randall. 2001. A cloud resolving model as a cloud parameterization in the NCAR community climate system model: Preliminary results. *Geophysical Reseach Letters* 28: 3617–3620. https://doi.org/10.1029/2001GL013552.

Kinter, J.L., and Coauthors. 2013. Revolutionizing climate modeling with project Athena: A multi-institutional, international collaboration. *Bulletin of the American Meteorological Society* 94: 231–245. https://doi.org/10.1175/BAMS-D-11-00043.1.

Klingaman, N.P., X. Jiang, P.K. Xavier, J. Petch, D. Waliser, and S.J. Woolnough. 2015. Vertical structure and physical processes of the Madden-Julian oscillation: Synthesis and summary. *Journal of Geophysical Research Atmospheres* 120: 4671–4689. https://doi.org/10.1002/2015JD023196.

Liu, P., and Coauthors. 2009. An MJO simulated by the NICAM at 14- and 7-km resolutions. *Monthly Weather Review* 137: 3254–3268. https://doi.org/10.1175/2009MWR2965.1.

Masunaga, H., M. Satoh, and H. Miura. 2008. A joint satellite and global cloud-resolving model analysis of a Madden-Julian Oscillation event: Model diagnosis. *Journal Geophysical Research* 113: D17210. https://doi.org/10.1029/2008JD009986.

Miura, H., M. Satoh, T. Nasuno, A.T. Noda, and K. Oouchi. 2007a. A Madden-Julian oscillation event realistically simulated by a global cloud-resolving model. *Science* 318: 1763–1765. https://doi.org/10.1126/science.1148443.

Miura, H., M. Satoh, H. Tomita, A.T. Noda, T. Nasuno, and S. Iga. 2007b. A short-duration global cloud-resolving simulation with a realistic land and sea distribution. *Geophysical Reseach Letters* 34: L02804. https://doi.org/10.1029/2006GL027448.

Miura, H., T. Miyakawa, T. Nasuno, and M. Satoh. 2012. Simulations of the MJO events during the field campaign of 2011–12 by a global cloud-resolving model NICAM, Abstract A13O-03 presented at 2012 Fall Meeting, AGU, San Francisco, California, 3–7 Dec.

Miura, H., M. Satoh, H. Tomita, A.T. Noda, T. Nasuno, S. Iga, T. Suematsu, and T. Nasuno. 2015. An ensemble hindcast of the Madden-Julian oscillation during the CINDY2011/DYNAMO field campaign and influence of seasonal variation of sea surface temperature. *Journal of the Meteorological Society of Japan Series II* 93A: 115–137. https://doi.org/10.2151/jmsj.2015-055.

Miyakawa, T., and Coauthors. 2014. Madden–Julian oscillation prediction skill of a new-generation global model demonstrated using a supercomputer. *Nature Communications* 5: 3769. https://doi.org/10.1038/ncomms4769.

Miyamoto, Y., Y. Kajikawa, R. Yoshida, T. Yamaura, H. Yashiro, and H. Tomita. 2013. Deep moist atmospheric convection in a subkilometer global simulation. *Geophysical Reseach Letters* 40: 4922–4926. https://doi.org/10.1002/grl.50944.

Pilon, R., C. Zhang, and J. Dudhia. 2016. Roles of deep and shallow convection and microphysics in the MJO simulated by the model for prediction across scales. *Journal of Geophysical Research* 121: 10575–10600. https://doi.org/10.1002/2015JD024697.

Randall, D.A. 2013. Beyond deadlock. *Geophysical Reseach Letters* 40: 5970–5976. https://doi.org/10.1002/2013GL057998.

Randall, D., M. Khairoutdinov, A. Arakawa, and W. Grabowski. 2003. Breaking the cloud parameterization deadlock. *Bulletin of the American Meteorological Society* 84: 1547–1564. https://doi.org/10.1175/BAMS-84-11-1547.

Reynolds, R.W., N.A. Rayner, T.M. Smith, D.C. Stokes, and W. Wang. 2002. An improved in situ and satellite SST analysis for climate. *Journal of Climate* 15: 1609–1625. https://doi.org/10.1175/1520-0442(2002)015%3c1609:AIISAS%3e2.0.CO;2.

Roh, W., M. Satoh, and T. Nasuno. 2017. Improvement of a cloud microphysics scheme for a global nonhydrostatic model using TRMM and a satellite simulator. *Journal of the Atmospheric Sciences* 74: 167–184. https://doi.org/10.1175/JAS-D-16-0027.1.

Saito, K., T. Keenan, G. Holland, and K. Puri, 2001: Numerical simulation of the diurnal evolution of tropical Island convection over the maritime continent. *Monthly Weather Review* 129, 378–400. https://doi.org/10.1175/1520-0493 129<037 8:NSOTDE>2.0.CO;2.

Sasaki, W., R. Onishi, H. Fuchigami, K. Goto, S. Nishikawa, Y. Ishikawa, and K. Takahashi. 2016. MJO simulation in a cloud-system-resolving global ocean-atmosphere coupled model. *Geophysical Reseach Letters* 43: 9352–9360. https://doi.org/10.1002/2016GL070550.

Satoh, M., T. Matsuno, H. Tomita, H. Miura, T. Nasuno, and S. Iga. 2008. Nonhydrostatic icosahedral atmospheric model (NICAM) for global cloud resolving simulations. *Journal of Computational Physics* 227: 3486–3514. https://doi.org/10.1016/j.jcp.2007.02.006.

Satoh, M., S. Iga, H. Tomita, Y. Tsushima, and A.T. Noda. 2012. Response of upper clouds in global warming experiments obtained using a global nonhydrostatic model with explicit cloud processes. *Journal of Climate* 25: 2178–2191. https://doi.org/10.1175/JCLI-D-11-00152.1.

Satoh, M., S. Iga, H. Tomita, Y. Tsushima, A.T. Noda, and Coauthors. 2014. The non-hydrostatic icosahedral atmospheric model: Description and development. *Progress in Earth and Planetary Science* 1: 18. https://doi.org/10.1186/s40645-014-0018-1.

Seiki, T., C. Kodama, A.T. Noda, and M. Satoh. 2015. Improvement in global cloud-system-resolving simulations by using a double-moment bulk cloud microphysics scheme. *Journal of Climate* 28: 2405–2419. https://doi.org/10.1175/JCLI-D-14-00241.1.

Skamarock, W.C., and Coauthors. 2008. A description of the Advanced Research WRF version 3. NCAR Tech. Note NCAR/TN-475 + STR, 125 pp.

Skamarock, W.C., J.B. Klemp, M.G. Duda, L.D. Fowler, S.-H. Park, and T.D. Ringler. 2012. A multiscale nonhydrostatic atmospheric model using centroidal Voronoi tesselations and C-grid staggering. *Monthly Weather Review* 140: 3090–3105. https://doi.org/10.1175/MWR-D-11-00215.1.

Tomita, H. 2008. New microphysical schemes with five and six categories by diagnostic generation of cloud ice. *Journal of the Meteorological Society of Japan* 86A: 121–142. https://doi.org/10.2151/jmsj.86A.121.

Tomita, H., M. Tsugawa, M. Satoh, and K. Goto, 2001: Shallow water model on a modified icosahedral geodesic grid by using spring dynamics. *Journal of Computational Physics* 174: 579–613. https://doi.org/10.1006/JCPH.2001.6897.

Tomita, H., H. Miura, S. Iga, T. Nasuno, and M. Satoh. 2005. A global cloud-resolving simulation: Preliminary results from an aqua planet experiment. *Geophysical Reseach Letters* 32: 1–4. https://doi.org/10.1029/2005GL022459.

Tsushima, Y., S. Iga, H. Tomita, M. Satoh, A.T. Noda, and M.J. Webb. 2014. High cloud increase in a perturbed SST experiment with a global nonhydrostatic model including explicit convective processes. *Journal of Advances in Modeling Earth Systems* 6: 571–585. https://doi.org/10.1002/2013MS000301.

Ulate, M., C. Zhang, and J. Dudhia. 2015. Role of water vapor and convection-circulation decoupling in MJO simulations by a tropical channel model. *Journal of Advances in Modeling Earth Systems* 7: 692–711. https://doi.org/10.1002/2014MS000393.

Wang, S., A.H. Sobel, F. Zhang, Y.Q. Sun, Y. Yue, and L. Zhou. 2015. Regional simulation of the October and November MJO events observed during the CINDY/DYNAMO field campaign at gray zone resolution. *Journal of Climate* 28: 2097–2119. https://doi.org/10.1175/JCLI-D-14-00294.1.

Watanabe, S., T. Hajima, K. Sudo, T. Nagashima, T. Takemura, H. Okajima, T. Nozawa, H. Kawase, M. Abe, T. Yokohata, T. Ise, H. Sato, E. Kato, K. Takata, S. Emori, and M. Kawamiya. 2011. MIROC-ESM 2010: Model description and basic results of CMIP5-20c3m experiments. *Geoscientific Model Development* 4: 845–872. https://doi.org/10.5194/gmd-4-845-2011.

Xu, K., R.T. Cederwall, L.J. Donner, W.W. Grabowski, F. Guichard, D.E. Johnson, M. Khairoutdinov, S.K. Krueger, J. C. Petch, D. A. Randall, C. J. Seman, W. Tao, D. Wang, S. Cheng Xie, J.J. Yio, and M. Zhang. 2002. An intercomparison of cloud-resolving models with the atmospheric radiation measurement summer 1997 intensive observation period data. *Quarterly Journal of the Royal Meteorological Society* 128: 593–624. https://doi.org/10.1256/003590002321042117.

Yoneyama, K., C. Zhang, and C.N. Long. 2013. Tracking pulses of the Madden–Julian oscillation. *Bulletin of the American Meteorological Society* 94: 1871–1891. https://doi.org/10.1175/BAMS-D-12-00157.1.

Challenges of Improving the Stratiform Processes in a Coupled Climate Model with Indian Monsoon Perspective

Parthasarathi Mukhopadhyay, R. Phani Murali Krishna, S. Abhik, Malay Ganai and Kumar Roy

Abstract Stratiform rain and associated cloud processes play an important role in the Indian summer monsoon rainfall propagation and distribution. In spite of improvement in model resolution, the parameterization of stratiform cloud processes remains elusive. An attempt is made here to improve the parameterization of stratiform processes of NCEP (National Center for Environmental Prediction) CFSv2 (climate forecast system version 2.0) coupled model for better simulation of the Indian summer monsoon. Physically more realistic cloud microphysics scheme (WSM6) suitably modified with Indian aircraft observation along with a revised simplified Arakawa Schubert (RSAS) and modified radiation parameterization has been implemented in CFSv2. The simulation of stratiform rainfall and its northward propagation by a modified version of CFSv2 (CFSCR) is compared with the default CFSv2. The improved cloud parameterization enables the model to realistically simulate the stratiform rain and its fraction against the convective rain of the model. The CFSCR is also able to improve the stratiform rain efficiency in the model. This development demonstrates that improved cloud processes can resolve the issue of erroneous convective and stratiform fraction in CFSv2.

Keywords Stratiform processes · Coupled model · Indian monsoon

1 Introduction

Capturing the Indian summer monsoon mean state and its spatiotemporal variability is a challenge (Waliser et al. 2003; Lin et al. 2008; Sperber and Annamalai 2008). While there is significant progress in the past few decades in understanding and

P. Mukhopadhyay (✉) · R. Phani Murali Krishna · M. Ganai · K. Roy
Indian Institute of Tropical Meteorology, Pashan, Pune 411008, India
e-mail: mpartha@tropmet.res.in

S. Abhik
School of Earth Atmosphere & Environment, Monash University,
Wellington Road, Clayton, VIC, Australia

D. A. Randall et al. (eds.), *Current Trends in the Representation of Physical Processes in Weather and Climate Models*, Springer Atmospheric Sciences,
https://doi.org/10.1007/978-981-13-3396-5_12

improving the monsoon prediction, there are still certain areas of challenges which the numerical modelers are thriving to achieve. As described by earlier works, the intraseasonal oscillations are the building blocks of mean monsoon and are responsible for the spatiotemporal variabilities over the Indian subcontinent (Goswami and Ajayamohan 2001). One of the limitations of numerical models is to capture the cloud and convection realistically (Webster et al. 1998; Dai 2006; Yoo et al. 2013). A number of studies (Jiang et al. 2011; Rajeevan et al. 2013; Abhik et al. 2014) emphasized the role of multiscale clouds and convection on different phases of monsoon. Further study by Chattopadhyay et al. (2009) emphasized the role of stratiform rain on the northward propagation of intraseasonal oscillations. These studies indicate the pivotal role of stratiform precipitation and associated cloud processes on the seasonal mean as well as intraseasonal oscillation. Further studies based on observational data by Abhik et al. (2014) and Jiang et al. (2011) threw a new insight on the cloud and its role on the stratiform process and monsoon ISOs. Keeping in tune with the multiscale nature of monsoon convection, many recent developments, e.g., superparameterized CFS by Goswami et al. (2015), the stochastic multi-cloud model in CFSv2 (Goswami et al. 2017), showed much promise in improving the representation of cloud and convection during Indian monsoon vis-à-vis the mean and intra-seasonal oscillations. However, in both these approaches, the improvement of stratiform rain was not addressed specifically. Study by Houze (1997) mentioned that large-scale anvil is the potential source of stratiform rain. Taking the observation-based studies into account and the variety of cloud types, a recent development of Climate Forecast System version 2 (CFSv2) model is reported by Abhik et al. (2017). In this development, the microphysical processes of CFSv2 model are represented with a cloud model having processes required to provide stratiform rainfall. However, the paper did not elaborate much about the stratiform rain and associated process. In this article, a detailed analysis is carried out to document the improvement of stratiform rainfall with a modified version of CFSv2 in comparison to default CFSv2 and also with respect to observation.

2 Data and Methodology

The NCEP CFSv2 (Saha et al. 2014) is a fully coupled ocean–land–atmosphere dynamical modeling system. It uses NCEP Global Forecast System (GFS) atmospheric GCM (Moorthi et al. 2001) and Geophysical Fluid Dynamics Laboratory (GFDL) Modular Ocean Model version 4p0d (Griffies et al. 2004) as an oceanic component. The atmospheric component has a spectral resolution of T126 (~100 km) with 64 sigma-pressure hybrid vertical layers and the oceanic component has a zonal resolution of 0.25°–0.5° with 40 vertical layers. More details about the CFSv2 model and its various physical schemes are documented in Saha et al. (2014). The default CFSv2 has simplified the Arakawa–Schubert (SAS) mass flux-based convective parameterization scheme (Pan and Wu 1995). The revised version of CFSv2 includes RSAS convective parameterization scheme (Han and Pan 2011). The detailed differences

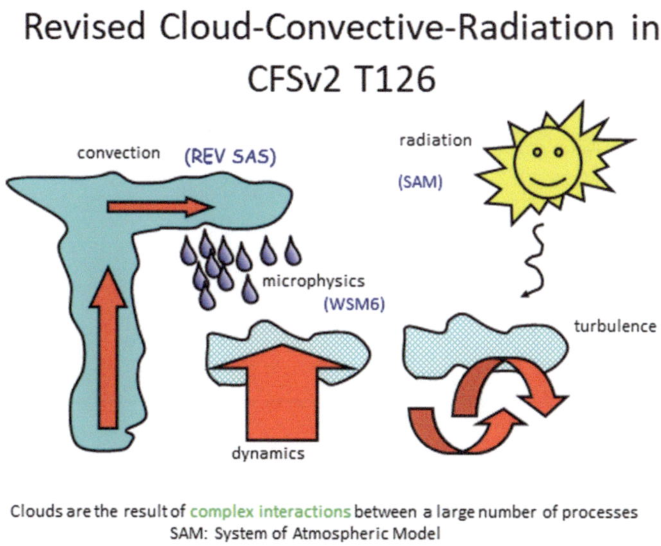

Fig. 1 Schematic of physical parameterizations used in the revised version of CFSv2 (CFSCR) (REV SAS implies Revised Arakawa–Schubert scheme; WSM6 implies WRF single moment 6 class scheme; SAM implies the System of Atmospheric Model.)

between SAS and RSAS are well documented in Han and Pan (2011). The model also uses a simple cloud microphysics scheme with ice physics developed by Zhao and Carr (1997) based on Sundqvist et al. (1989). In addition to RSAS scheme, a six-class WSM6 microphysics scheme (Hong and Lim 2006) is incorporated in place of two-class Zhao and Carr (1997) microphysics scheme in CFSv2. Prognostic water substance variables in WSM6 contain water vapor, cloud water, cloud ice, rainwater, snow, and graupel. More detailed description about WSM6 can be found in Hong and Lim (2006). In order to keep consistency with the modified convective and cloud microphysical processes, the cloud hydrometeors generated by WSM6 are included during the computation of cloud fraction in the RRTM radiation scheme. Further details about the model setup are documented in Abhik et al. (2017) and modified suitably using System of Atmospheric Model (SAM) (Khairoutdinov and Randall 2003) shown in Fig. 1. We have carried out two separate free runs of 15 years of CFSv2 with default SAS and default microphysics scheme (CTRL hereafter) and with RSAS and WSM6 scheme (CFSCR) with the same initial condition. In the present study, we have analyzed the past 12 years of simulation to avoid influences of model spin-up.

To validate the model simulation, the Tropical Rainfall Measuring Mission (TRMM) 3B42 version 7 (V7) (Huffman et al. 2007) daily data at a horizontal resolution of $0.25^\circ \times 0.25^\circ$ for the year 1999–2010 are used. TRMM 3G68 (Kummerow et al. 2001) derived daily convective and large-scale rainfall is analyzed.

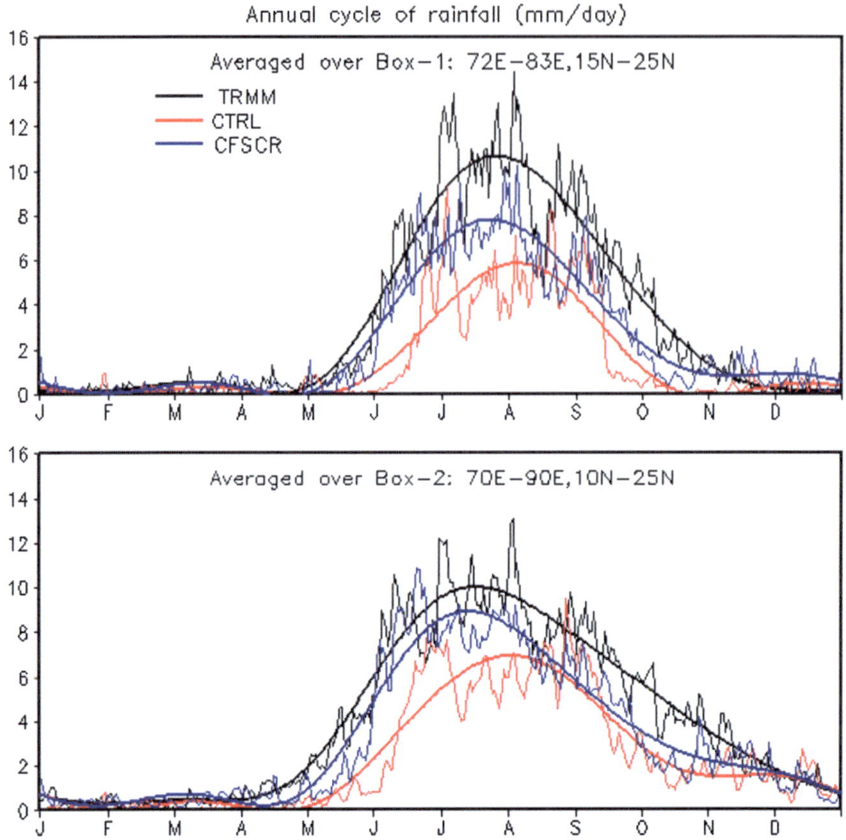

Fig. 2 The area-averaged annual cycle of rainfall (mm day^{-1}), smoothed (first 3 harmonics plus mean) on top of the unsmoothed one, over central Indian region (top figure) and over larger Indian summer monsoon domain (bottom figure) from TRMM (black line), CTRL (red line), and CFSCR (blue line)

3 Results and Discussions

Based on the model simulation, daily climatological rainfall is plotted from observation and also from the control run of CFSv2 (CTRL) and the modified version of CFSv2, i.e., CFSCR. The annual cycle of rainfall is shown in Fig. 2. The smoothed rainfall annual cycle is plotted for two boxes namely 72°E–83°E, 15°N–25°N covering mostly the central Indian monsoon zone and another is over 70°E–90°E and 10°N–25°N covering part of the Bay of Bengal. The annual cycle over both these boxes shows an improvement by CFSCR. Improvements in the length of the monsoon season, onset, and dry bias by the CTRL have been particularly improved.

To depict the gradual northward propagation of the rainfall belt during boreal summer, longitude averaged (70E–90E) rainfall is plotted with latitude versus month.

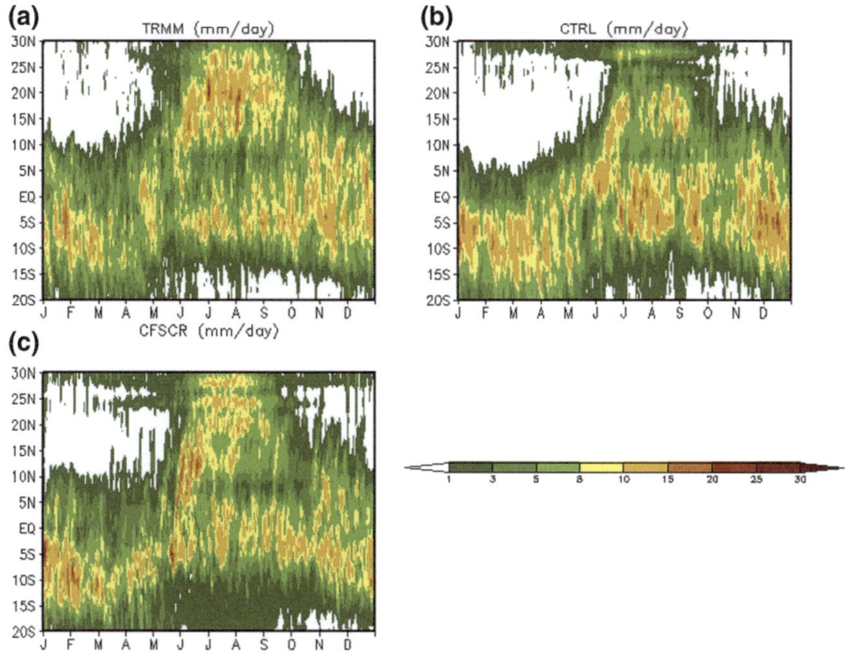

Fig. 3 Time-latitude section of rainfall (mm day^{-1}) from **a** TRMM, **b** CTRL, and **c** CFSCR averaged over 70°E–90°E

Figure 3a shows the northward propagation of the rainfall belt in TRMM data. It is evident from Fig. 3b that CTRL produces a much wet bias over equatorial Indian ocean during Boreal summer (June–July–August–September) and also over the south of the equator during Boreal winter. It may be noted from Fig. 3b that northward propagation is relatively week in CTRL and mostly rainfall belt remain confined over the oceanic region causing wet bias (Goswami et al. 2014). CFSCR (Fig. 3c) overcomes the problem of too much rain over the oceanic region and also improves the wet bias during boreal winter.

To get further insight about the propagation of daily rainfall belt and its improvement in CFSCR, the convective rainfall from TRMM 3G68 data (Fig. 4a) and the corresponding plots from CTRL and CFSCR are shown, respectively, in Fig. 4b and 4c. It is evident from the observation that convective rain propagates northward and there is a region of convective rain around the equator (Fig. 4a). CTRL generates too much convective rain over the equatorial oceanic region. Too much convective rain over the equatorial oceanic region (Fig. 4a) is one of the reasons behind lesser transport of convective rain northward and over the Indian landmass. Although the CFSCR overestimates the convective rain near the equatorial region as compared to observation, it improves the simulation of too much convective rain over the equatorial region and improves the northward propagation of the convective rain as well.

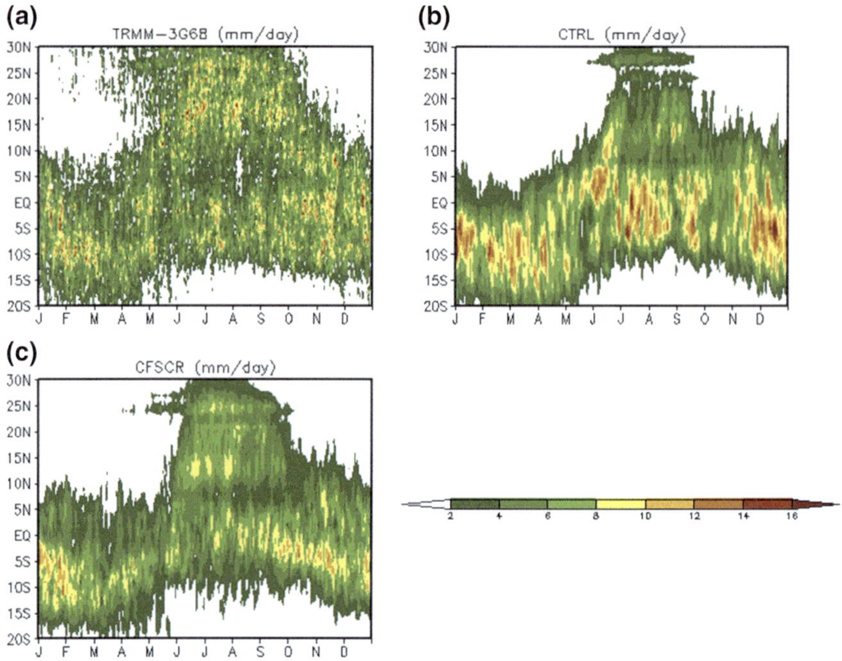

Fig. 4 Time-latitude section of convective rainfall (mm day^{-1}) from **a** TRMM-3G68, **b** CTRL, and **c** CFSCR averaged over 70°E–90°E

Therefore, the improvement seen in the daily rainfall and its propagation is attributed to the improvement of convective rain.

As mentioned by Chattopadhyay et al. (2009), along with convective rain, the stratiform component of the rainfall also shows significant propagation. To explore the simulation of a stratiform component of the rainfall, the longitude-averaged stratiform rainfall is plotted for CTRL (Fig. 5b) and CFSCR (Fig. 5c). While both the models underestimate the stratiform component of rainfall with respect to observation (Fig. 5a). CFSCR has improved the stratiform component and is able to show some northward propagation. It shows that there is further need of improving the cloud and convective process of CFSv2 to improve the stratiform component of the rainfall. Sabeerali et al. (2013) also mentioned the inability of climate models (CMIP5) in realistically capturing the convective and stratiform ratio.

To evaluate the important contribution of convective and stratiform rain not only over the Indian monsoon region but over the global tropics, global mean convective rain, respectively, over land and ocean is shown in Fig. 6a. Similarly, the global mean stratiform rain over land and ocean is shown in Fig. 6b. It is evident from Fig. 6a that over land region, CFSCR produced more convective rain than CTRL and this as such is the reason behind the improvement of dry bias by CFSCR while the convective rainfall over the oceanic region has reduced as compared to CTRL. It is interesting to

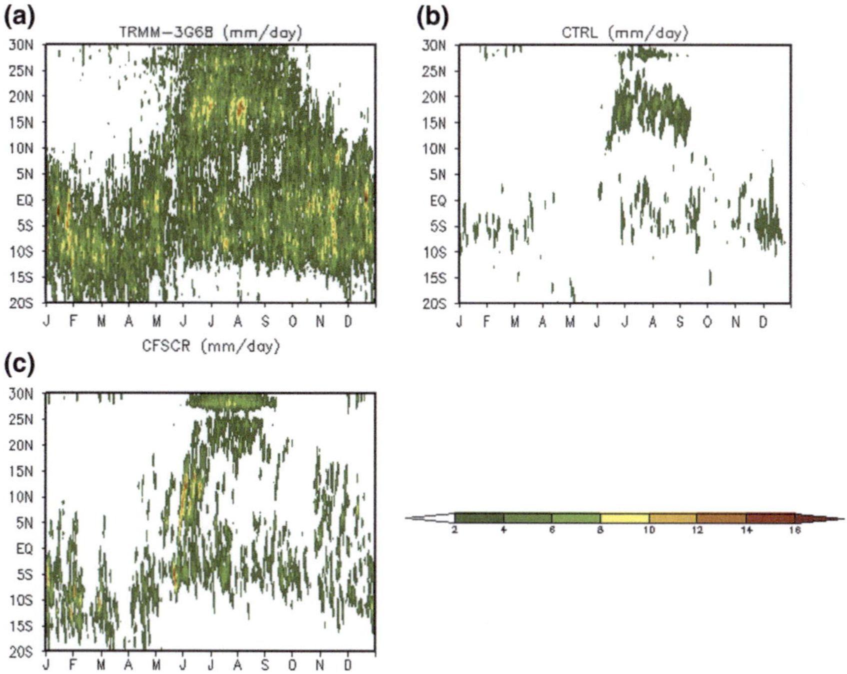

Fig. 5 Time-latitude section of stratiform rainfall (mm day^{-1}) from **a** TRMM-3G68, **b** CTRL, and **c** CFSCR averaged over 70°E–90°E

note that CFSCR is able to improve the stratiform rainfall over the tropical land and oceanic region. In the backdrop of present-day climate model, such improvement is promising. However, as shown in Fig. 5, further improvement is needed.

The possible reason behind the improvement of precipitation in CFSCR appears to be the better efficiency of cloud hydrometeors such as CLW in generating model precipitation (Fig. 7). Following Li et al. (2012), we have calculated precipitation efficiency (=precipitation rate/total grid box cloud water path) for both CTRL and CFSCR for total rain (Fig. 7a), stratiform rain (Fig. 7b), and convective rain (Fig. 7c). CFSCR has shown better precipitation efficiency for total, stratiform, and convective rainfall as compared to CTRL in simulating all the categories of rainfall, in general, and heavier rain in particular. It implies conversion of cloud condensate to precipitation is efficient in CFSCR than that in CTRL for all the categories of rainfall. Thus, proper representation of cloud hydrometeors through WSM6 helps CFSv2 to simulate better distribution of total rain, stratiform rain, and convective rain and improve heavier rain simulation in the model.

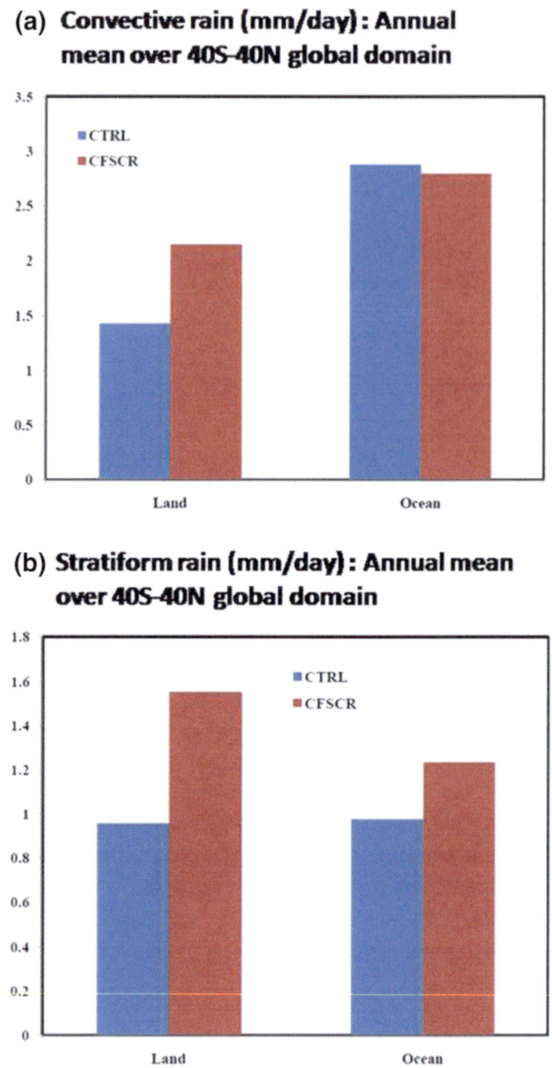

Fig. 6 Globally (40S–40N, 0–360E) mean **a** convective rainfall (mm day^{-1}) and **b** stratiform rainfall (mm day^{-1}) from CTRL (blue bar), and CFSCR (red bar) over land and oceanic regions

4 Conclusions

An attempt is made here to improve the cloud processes of CFSv2 for realistic simulation of convective and stratiform rain distribution. The revised version of the model (CFSCR) shows better fidelity as compared to the default version (CTRL) in the annual cycle of rainfall. It shows improvement in reducing the dry and wet bias of the model. The revised model also shows some improvement in capturing the length of the monsoon season along with monsoon onset. The improvement in the annual cycle of rainfall is attributed to the improvement of convective and to some extent in

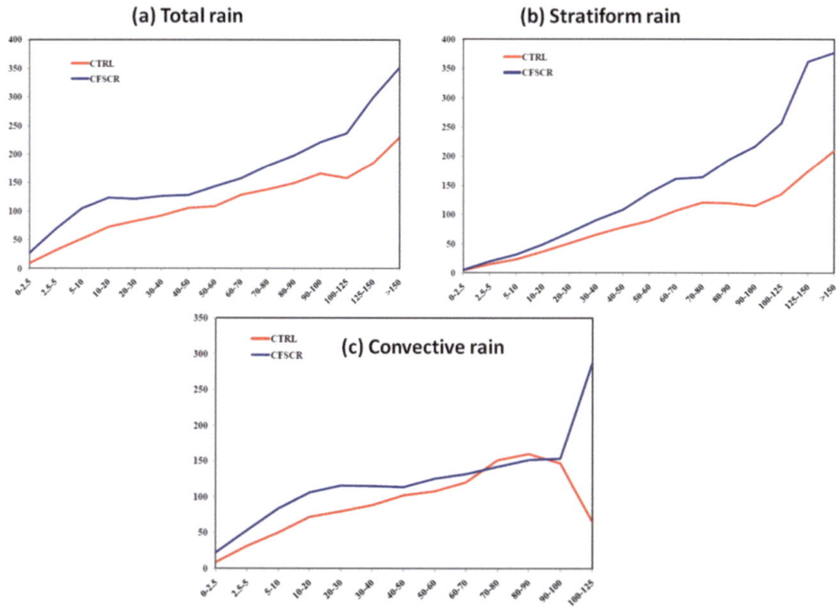

Fig. 7 Precipitation efficiency (day^{-1}) over Central Indian landmass (18N–27N, 74E–85E) for CTRL, and CFSCR for **a** total precipitation, **b** stratiform rainfall, and **c** convective rainfall. Along x-axis indicates various rainfall bins (mm day^{-1}) and y-axis indicates precipitation efficiency (day^{-1})

the stratiform rain. However, there is a need to further improve the stratiform rain and associated cloud processes in the model. The revised version also shows the fidelity in improving the global distribution of convective and stratiform rain, respectively, over land and an oceanic region which remained a challenge for the climate model.

Finally, it is shown that inclusion of more physically based microphysics, i.e., WSM6 has enhanced the efficiency of the climate model CFSv2 to produce improved total, convective, and stratiform rain mainly due to the contributions from CLW. It is also noted that the fidelity of the model in simulating the heavier categories have particularly improved through the modifications of cloud process parameterization.

Acknowledgements The authors are grateful to Director, IITM for the encouragement of the study. The authors are grateful to Ministry of Earth Science, Government of India, for funding and IITM HPC is gratefully acknowledged for allowing the CFSv2 run to be accomplished.

References

Abhik, S., P. Mukhopadhyay, and B.N. Goswami. 2014. Evaluation of mean and intraseasonal variability of Indian summer monsoon simulation in ECHAM5: Identification of possible source of bias. *Climate Dynamics* 43: 389–406. https://doi.org/10.1007/s00382-013-1824-7.

Abhik, S., R.P.M. Krishna, M. Mahakur, M. Ganai, P. Mukhopadhyay, and J. Dudhia. 2017. Revised cloud processes to improve the mean and intraseasonal variability of Indian summer monsoon in climate forecast system: Part 1. *Journal of Advances in Modeling Earth Systems* 9: 1–28. https://doi.org/10.1002/2016MS000819.

Chattopadhyay, R., B.N. Goswami, A.K. Sahai, and K. Fraedrich. 2009. Role of stratiform rainfall in modifying the northward propagation of monsoon intraseasonal oscillation. *Journal of Geophysical Research Atmospheres* 114: 1–15. https://doi.org/10.1029/2009JDO11869.

Dai, A. 2006. Precipitation characteristics in eighteen coupled climate models. *Journal of Climate* 19: 4605–4630. https://doi.org/10.1175/JCLI3884.1.

Goswami, B.N., and R.S. Ajayamohan. 2001. Intraseasonal oscillations and interannual variability of Indian summer monsoon. *Journal of Climate* 14: 1180–1198. https://doi.org/10.1175/1520-0442(2001)014%3c1180:IOAIVO%3e2.0.CO;2.

Goswami, B.B., M. Deshpande, P. Mukhopadhyay, et al. 2014. Simulation of monsoon intraseasonal variability in NCEP CFSv2 and its role on systematic bias. *Climate Dynamics* 43: 2725–2745. https://doi.org/10.1007/s00382-014-2089-5.

Goswami, B.B., R.P.M. Krishna, P. Mukhopadhyay, et al. 2015. Simulation of the Indian summer monsoon in the superparameterized climate forecast system version 2: Preliminary results. *Journal of Climate* 28: 8988–9012. https://doi.org/10.1175/JCLI-D-14-00607.1.

Goswami, B.B., R.P.M. Krishna, P. Mukhopadhyay, et al. 2017. Implementation and calibration of a stochastic multicloud convective parameterization in the NCEP climate forecast system (CFSv2). *Journal of Advances in Modeling Earth Systems*. https://doi.org/10.1002/2017ms001014.

Griffies, S.M., M.J. Harrison, R.C. Pacanowski, and A. Rosati. 2004. A technical guide to MOM4. *GFDL Ocean Group Technical Report* 5: 371.

Han, J., and H.-L. Pan. 2011. Revision of convection and vertical diffusion schemes in the NCEP global forecast system. *Weather Forecast* 26: 520–533. https://doi.org/10.1175/WAF-D-10-05038.1.

Hong, S., and J. Lim. 2006. The WRF single-moment 6-class microphysics scheme (WSM6). *Journal of the Korean Meteorological Society* 42: 129–151.

Houze Jr., R.A. 1997. Stratiform precipitation in the tropics: A meteorological paradox? *Bulletin of the American Meteorological Society* 78: 2179–2196. https://doi.org/10.1175/1520-0477(1997)078%3c2179:SPIROC%3e2.0.CO;2.

Huffman, G.J., D.T. Bolvin, E.J. Nelkin, et al. 2007. The TRMM multisatellite precipitation analysis (TMPA): Quasi-global, multiyear, combined-sensor precipitation estimates at fine scales. *Journal of Hydrometeorology* 8: 38–55. https://doi.org/10.1175/JHM560.1.

Jiang, X., D.E. Waliser, J.L. Li, and C. Woods. 2011. Vertical cloud structures of the boreal summer intraseasonal variability based on CloudSat observations and ERA-interim reanalysis. *Climate Dynamics* 36: 2219–2232. https://doi.org/10.1007/s00382-010-0853-8.

Khairoutdinov, M.F., and D.A. Randall. 2003. Cloud-resolving modeling of the ARM summer 1997 IOP: Model formulation, results, uncertainties and sensitivities. *Journal of Atmospheric Science* 60: 607–625.

Kummerow, C., Y. Hong, W.S. Olson, et al. 2001. The evolution of the Goddard profiling algorithm (GPROF) for rainfall estimation from passive microwave sensors. *Journal of Applied Meteorology* 40: 1801–1820. https://doi.org/10.1175/1520-0450(2001)040%3c1801:TEOTGP%3e2.0.CO;2.

Li, F., D. Rosa, W.D. Collins, and M.F. Wehner. 2012. "Super-parameterization": A better way to simulate regional extreme precipitation? *Journal of Advances in Modeling Earth Systems* 4: M04002. https://doi.org/10.1029/2011MS000106.

Lin, J.-L., K.M. Weickman, G.N. Kiladis, B.E. Mapes, S.D. Schubert, M.J. Suarez, J.T. Bacmeister, and M.-I. Lee. 2008. Subseasonal variability associated with Asian summer monsoon simulated by 14 IPCC AR4 coupled GCMs. *Journal of Climate* 21: 4541–4567. https://doi.org/10.1175/2008JCLI1816.1.

Moorthi, S., H.L. Pan, and P. Caplan. 2001. Changes to the 2001 NCEP operational MRF/AVN global analysis/forecast system. *NWS Technical Procedures Bulletin* 484: 14p.

Pan, H.L., and W.S. Wu. 1995. Implementing a mass flux convective parameterization package for the NMC medium-range forecast model. *NMC Office Note* 409.

Rajeevan, M., P. Rohini, K. Niranjan Kumar, et al. 2013. A study of vertical cloud structure of the Indian summer monsoon using CloudSat data. *Climate Dynamics* 40: 637–650. https://doi.org/10.1007/s00382-012-1374-4.

Sabeerali, C.T., A. Ramu Dandi, A. Dhakate, et al. 2013. Simulation of boreal summer intraseasonal oscillations in the latest CMIP5 coupled GCMs. *Journal of Geophysical Research Atmospheres* 118: 4401–4420. https://doi.org/10.1002/jgrd.50403.

Saha, S., S. Moorthi, X. Wu, et al. 2014. The NCEP climate forecast system version 2. *Journal of Climate* 27: 2185–2208. https://doi.org/10.1175/JCLI-D-12-00823.1.

Sperber, K.R., and H. Annamalai. 2008. Coupled model simulations of boreal summer intraseasonal (30–50 day) variability, Part 1: Systematic errors and caution on use of metrics. *Climate Dynamics* 31: 345–372. https://doi.org/10.1007/s00382-008-0367-9.

Sundqvist, H., E. Berge, and J.E. Kristjansson. 1989. Condensation and cloud parameterization studies with a mesoscale numerical weather prediction model. *Monthly Weather Review* 117: 1641–1657. https://doi.org/10.1175/1520-0493(1989)117%3c1641:CACPSW%3e2.0.CO;2.

Waliser, D.E., K. Jin, I.S. Kang, et al. 2003. AGCM simulations of intraseasonal variability associated with the Asian summer monsoon. *Climate Dynamics* 21: 423–446. https://doi.org/10.1007/s00382-003-0337-1.

Webster, P.J., V.O. Magaña, T.N. Palmer, et al. 1998. Monsoons: Processes, predictability, and the prospects for prediction. *Journal of Geophysical Research: Oceans* 103: 14451–14510. https://doi.org/10.1029/97JC02719.

Yoo, H., Z. Li, Y.T. Hou, et al. 2013. Diagnosis and testing of low-level cloud parameterizations for the NCEP/GFS model using satellite and ground-based measurements. *Climate Dynamics* 41: 1595–1613. https://doi.org/10.1007/s00382-013-1884-8.

Zhao, Q., and F.H. Carr. 1997. A prognostic cloud scheme for operational NWP models. *Monthly Weather Review* 125: 1931–1953. https://doi.org/10.1175/1520-0493(1997)125%3c1931:APCSFO%3e2.0.CO;2.

Current and Future Activities in Unified Modelling and Data Assimilation at NCMRWF

E. N. Rajagopal, A. K. Mitra, Munmun Das Gupta, John P. George, Raghavendra Ashrit, Abhijit Sarkar and A. Jayakumar

Abstract State-of-the-art Numerical Weather Prediction (NWP) models can provide useful weather information in the medium-range timescales (3 to 10 days ahead) which can be applied for decision-making in different sectors like agriculture, power distribution, disaster management and water resource management. Forecasting of monsoon weather system and associated rainfall is one of the most difficult areas in NWP due to complexities in land–ocean–atmosphere interactions and due to interactions between convective systems of cloud scale, mesoscale and synoptic and planetary scales. However, significant improvements can be noticed, in recent years. Some of the chief contributors to the improvement are improved data assimilation methods, enhanced satellite coverage, high-performance computers (HPCs) and high-resolution NWP models. This chapter documents current activities at NCMRWF involving the Unified Model and its Data Assimilation used for summer monsoon forecast in the medium-range timescales. Additionally, the research efforts made for testing and implementation of ensemble models and coupled models are also discussed.

Keywords Medium range · Ensemble model · Data assimilation

1 Introduction

Medium-range weather forecast (prediction of weather 3–10 days ahead) is essential for managing and protecting various weather sensitive activities such as farming operations and helps flood forecasts, water resource management, etc., during summer monsoon season. Numerical weather prediction (NWP) models can provide useful weather information in the medium-range timescales. Forecasting of monsoon weather system and associated rainfall is one of the difficult areas in NWP due to complexities in land–ocean–atmosphere interactions and due to interactions between convective systems of cloud scale, mesoscale and synoptic and planetary

E. N. Rajagopal (✉) · A. K. Mitra · M. D. Gupta · J. P. George · R. Ashrit · A. Sarkar
A. Jayakumar
National Centre for Medium Range Weather Forecasting, Noida, India
e-mail: rajagopal@ncmrwf.gov.in

D. A. Randall et al. (eds.), *Current Trends in the Representation of Physical Processes in Weather and Climate Models*, Springer Atmospheric Sciences,
https://doi.org/10.1007/978-981-13-3396-5_13

scales. However, significant improvements can be noticed, in recent years, in the Indian summer monsoon forecast in the medium-range timescales due to improvements in the NWP systems.

NCMRWF was established in 1988 for developing medium-range weather forecasting techniques using dynamical models. Tremendous progress has been achieved in NWP capability at NCMRWF since its inception due to in-house R&D efforts and various national and international collaborations. List of global models and data assimilation systems used at NCMRWF is given in Table 1.

In 1989, NCMRWF implemented the ECMWF global atmospheric model (Cycle-30 version). Several test runs were carried out and different components of a weather forecasting system were linked up with this model at T79L19 resolution. The experimental medium-range weather forecasts were prepared and evaluated. The same modelling system was further used to carry out simulations of perpetual July climate. As a special support by the international community to help establish an operational medium-range forecast system, scientists from COLA, USA implemented the COLA R40 model at the NCMRWF. Scientists from NCMRWF and COLA jointly implemented all components of NWP system which included an optimum interpolation (OI) based analysis scheme, diagnostic and graphics packages. Later during 1992–93, as a part of a special understanding with the NCEP (then NMC), USA, NCMRWF implemented global data assimilation-forecast system at T80L18 resolution.

NCMRWF currently operates numerical models—global and regional—to make numerical weather forecast up to medium-range timescales. The centre also operates the ensemble prediction system (EPS) for providing probabilistic forecasts up to medium-range timescales. Coupled ocean–atmosphere model is also getting ready for extended range weather prediction. Recent advances in the medium-range forecast capability at NCMRWF using the Unified Model (UM) system (UK Met Office Unified Model) is only described in detail in this Chapter. The Unified Model is based on the emerging paradigm of seamless prediction of weather and climate and NCMRWF is a core UM partner.

Table 1 History of operational global NWP systems at NCMRWF

Year	Model	Resolution	DA
1989	T79L19 (ECMWF)		
1989–1992	R40L18 (COLA)	250 km	OI
1992–2007	T80L18 (NMC) T170L28	150 km 75 km	3D-VAR (SSI)
2007	NGFS (T254L64)	50 km	3D-VAR (SSI)
2009	NGFS (T254L64)	50 km	3D-VAR (GSI)
2010	NGFS (T382L64)	35 km	3D-VAR (GSI)
2011–2015	NGFS (T574L64)	22 km	Hybrid 3D-VAR (GSI)
2012–2015	NCUM (N512L70)	25 km	4D-VAR
2015–	NCUM (N768L70)	17 km	Hybrid 4D-VAR
2016–	NGFS (T1534L64)	12 km	Hybrid 3D-VAR (GSI)

2 NCMRWF Unified Model

Unified Model implemented at NCMRWF has been used for numerical weather prediction (NWP) since 2012 (Rajagopal et al. 2012). The NCMRWF UM (named NCUM) system is upgraded periodically to adapt the new developments for improving the forecasts.

Horizontal resolution of the current NCUM global model is 17 km and it has 70 levels in vertical (N768L70), extends from surface to 80 km height above sea level. The heart of the models is the solution of the momentum equation, which is called the 'dynamical core'. The introduction of 'ENDGame' dynamics in the recent versions of the model was a major step forward as it solves unprecedentedly accurate set of equations, which helps the seamless modelling approach. This model has an atmosphere and land configuration of physical process: the Global Atmosphere version 6 (GA6) and Global Land version 6 (GL6) (details are available in Walters et al. 2017). It is mentioned in Walters et al. (2017) that 'Alongside developments of the model's physical parameterizations, ENDGame dynamics also increases variability in the tropics, which leads to an improved representation of tropical cyclones and other tropical phenomena. Further developments of the atmospheric and land surface parameterizations improve other aspects of model performance, including the forecasting of surface weather phenomena (Fig. 1 and Table 2)'.

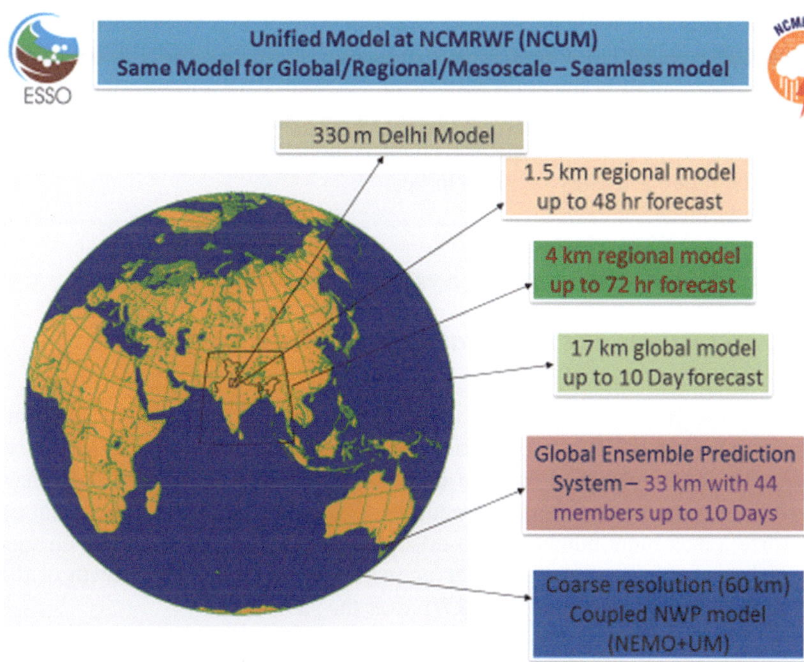

Fig. 1 NCMRWF unified modelling system

Table 2 NCMRWF Unified Model configuration details

Model	Application and domain	Configuration	Forecasts
NCUM	Global NWP	N768L70 (17 km)	00UTC: day 1–day 10 12UTC: day 1–day 5
NEPS	Global ensemble prediction	N400L70 (33 km; Control + 44 member)	00UTC: day 1–day 10
NCUM-R	Regional high-resolution over Indian region (5–40N and 65–100E)	4 km resolution Explicit convection	00UTC: day 1–day 3
NCUM-1.5 km	Regional convective-scale modelling	1.5 km (experimental)	00UTC: day 1–day 3 (experimental/hindcast)
DM-330 m	Delhi fog model	330 m (experimental)	00UTC: up to 36 h
NCUM-Coupled	Global seamless prediction (days to season)	UM (1.875 × 1.25; L85), NEMO (1 × 1), CICE with OASIS coupler	(Experimental/hindcast)

In Sect. 3, a brief description of NCUM Data Assimilation System is given along with details on observation processing. The high-resolution models used operationally and experimentally are discussed in Sect. 4. Section 5 discusses the NCMRWF ensemble prediction system (NEPS) and sensitivity studies involving the NEPS. In Sect. 6 the rainfall verification results for JJAS 2017 are discussed. Section 7 describes the high-resolution regional reanalysis (IMDAA) work carried out jointly with UK Met Office. The NCMRWF coupled modelling system is discussed in Sect. 8. The future activities are given in Sect. 9.

3 NCUM Data Assimilation System

3.1 Observations and Observation Processing

NCMRWF receives global meteorological observations through Global Telecommunication System (GTS) via Regional Telecommunication Hub (RTH), IMD, New Delhi and a large volume of satellite observations through Internet data services directly from various satellite data producers. Continuous efforts are on to acquire and utilize a maximum number of observations from various platforms, with special emphasis on Indian satellite observations. Figure 2 depicts the growth of the observational data reception at NCMRWF over the past decade.

NCMRWF has implemented a robust data monitoring system for both conventional and satellite observations with special emphasis on monitoring of Indian observations. Global observations received in near real time are processed, validated and

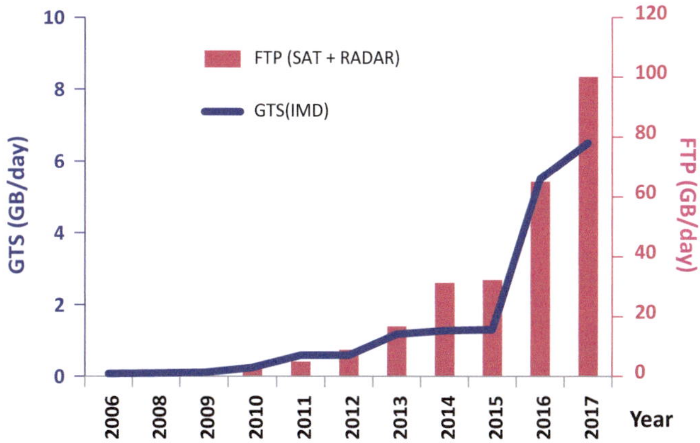

Fig. 2 Growth of the data reception at NCMRWF over the years

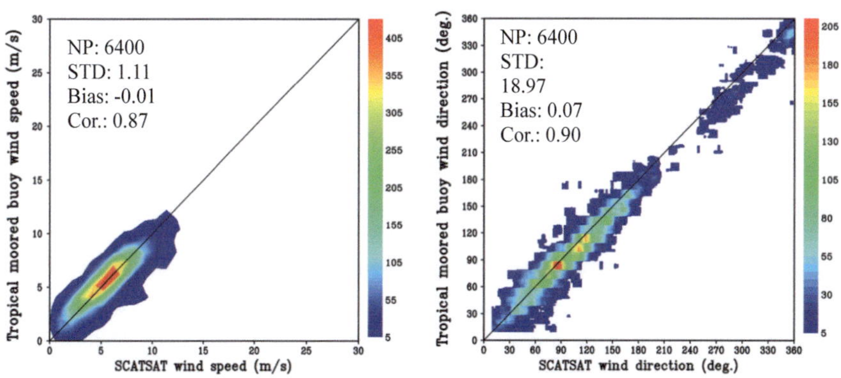

Fig. 3 Validation of SCATSAT ocean winds against buoy observations

monitored for their further utilization in the data assimilation systems. NCMRWF has validated recent SCATSAT-1 ocean surface wind observations against buoy observations (Fig. 3) and a good correlation between these two datasets. Based on this study, it is decided to include this dataset in the operational data assimilation system.

The preprocessing system selects the decoded observations based on observation type and time window. The preprocessed observations are used in the observation processing system (OPS), which does the quality control and thinning of observations.

3.2 NCUM Data Assimilation System

Numerical weather prediction (NWP) is an initial value problem in which prediction of future state of the atmosphere largely depends upon the initial condition (analysis) used by the forecast model. The process of preparation of the 'analysis' is known

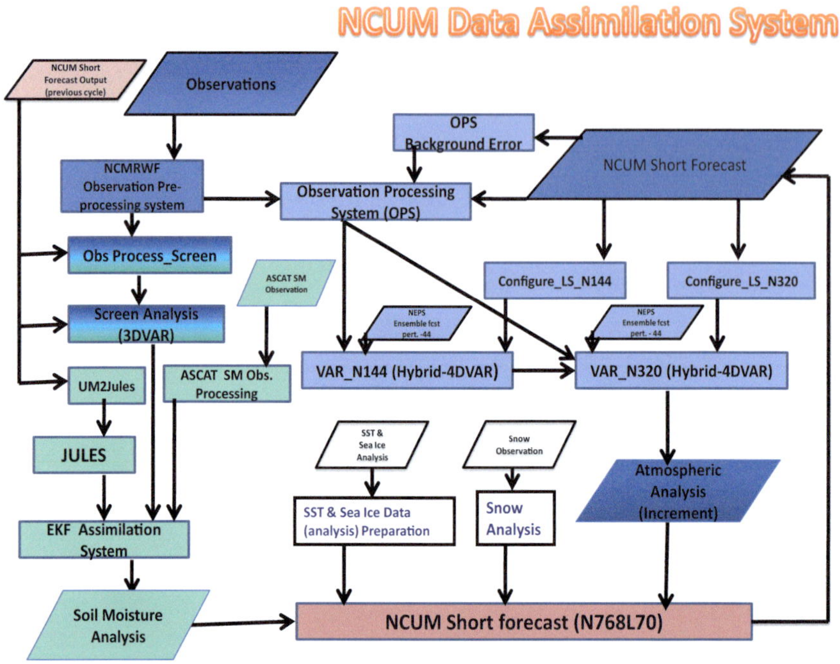

Fig. 4 Flow diagram of hybrid 4D-Var NCUM data assimilation system

as data assimilation (DA). Quality controlled (and thinned, if required) observations from the observation processing system is used for the preparation of analysis. Till October 2016, four-dimensional variational data assimilation (4D-Var) method was being used for the preparation of global analysis. A detailed description of the 4D-Var method can be seen in Rawlins et al. (2007) and Rajagopal et al. (2012). 4D-Var tries to find the best estimate of the state of the atmospheric (or 'analysis') using 'background' or 'first guess' (a short forecast from a previous analysis), observations in a time window as well as error statistics of background and observations. The 4D-Var system produces the analysis by minimizing a cost function (penalty function) which describes the departure of the analysis from background and observations which distributed in the time window. Detailed descriptions can be seen in George et al. (2016).

The data assimilation system was upgraded to Hybrid 4D-Var in October 2016. A flow chart of the NCUM Hybrid 4D-Var data assimilation system is given in Fig. 4. The Hybrid 4D-Var assimilation system combines the advantages of the traditional 4D-Var system and the ensemble data assimilation. The Hybrid 4D-Var system implemented at NCMRWF blends the 'climatological' background error with day-to-day varying flow-dependent background errors from the ensemble prediction system. The ensemble transform Kalman filter (ETKF) based ensemble prediction system (44 +1 members) at NCMRWF provides the flow-dependent background errors to the hybrid 4D-Var data assimilation system. The use of Hybrid 4D-Var at NCMRWF is a major step towards the transition to the use of emerging paradigm of ensemble data assimilation methods in operational NWP in India.

Assimilation of more observations is utmost important for improving the forecast and various efforts are being made to ingest more observational data into the data assimilation system. The volume of data reception, both through Global Telecommunication System (GTS) and Internet data service, has increased manifolds in the recent years. This is achieved with the help IMD as well as National and International Satellite Agencies. It is our endeavour to use all available information related to weather parameters, especially from Indian satellites, radars and in situ observation network, in the data assimilation system for improving the forecast. Table 3 provides the details of the observations currently used in the NCUM data global assimilation system.

NCMRWF is playing a key role in the development of Indian radiance assimilation capability in the data assimilation system. INSAT-3D Sounder and Megha-Tropiques SAPHIR radiances are operationally used in the NCUM-DA system. Ocean surface winds from SCATSAT-1 and Atmospheric Motion Vectors from INSAT-3D are also being used operationally.

To improve the initial position of the tropical cyclones (TC) in NCUM global and regional system using synthetic observation, a new vortex specification (VS) method is also introduced.

UM surface data preparation system (SURF) implemented at NCMRWF is used for the preparation of snow and soil moisture analysis. Operational Sea Surface Temperature and Sea Ice Analysis (OSTIA) SST and Sea Ice analysis is also interpolated to model resolution by SURF system. Extended Kalman filter (EKF) algorithm is used for the preparation of soil moisture analysis (details can be found in Lodh et al. 2016).

A high-resolution data assimilation system is also available for the preparation of initial condition for 4-km resolution regional NCUM model (details are given in Sect. 4). Radial wind observations from Indian Doppler Weather Radar network are also included in this experimental high-resolution regional data assimilation system.

3.3 Impact of Observations on Model Forecasts

It is important to quantify the impact of assimilated observations on the model forecasts. To evaluate the relative impact of each observation in an observing system, adjoint-based forecast sensitivity to observation tool is a good choice. The FSO system is used to evaluate the impact of various observations in the NCUM assimilation-forecast system. Observation impact on the forecast (with 24 h lead time) was studied during different seasons. Impact of the Indian satellite observations is studied using the FSO system in detail, including the channel wise impact of sounder observations. Results of the study indicated that MT-SAPHIR observations have a significant beneficial impact on the forecast of humidity in comparison with other similar

Table 3 Observations assimilated in NCUM global data assimilation system

Observation type	Observation description	Assimilated variables
Aircraft	Upper-air wind and temperature from aircraft	u, v, T
AIRS	Atmospheric infrared sounder of MODIS	*Brightness Temperature* (T_b)
ATOVS	AMSU-A, AMSU-B/MHS, HIRS from NOAA-18 & 19, MetOp-A&B	T_b
ATMS	Advanced Technology Microwave Sounder in NPP satellite	T_b
CrIS	Cross-track infrared sensor observations in NPP satellite	T_b
GOESClear	Cloud clear imager radiances from GOES E & W	T_b
GPSRO	Global positioning system radio occultation observations from various satellites (including MT-ROSA)	Bending angle
GroundGPS	Ground-based GPS observations from various locations	Zenith total delay
IASI	Infrared atmospheric sounding interferometer from MetOp-A&B	T_b
IN3DSndr	INSAT-3D sounder radiances	T_b
MT-SAPHIR	SAPHIR microwave radiances from Megha-Tropiques	T_b
Satwind	Atmospheric motion vectors from various geostationary and polar orbiting satellites (including INSAT-3D)	u, v
Scatwind	Advanced Scatterometer in MetOp-A&B, SCATSAT-1	u, v
SEVIRIClear	Cloud clear observations from SEVIRI of METEOSAT 8 & 10	T_b
Sonde	Radiosonde observations, upper-air wind profile from pilot balloons, wind profiles, VAD wind observation from Indian DWR	u, v, T, q
Surface	Surface observations from land and ocean	u, v, T, q, P_s
SSMIS	SSMIS radiances	T_b

observations, especially over lower levels of the troposphere (Fig. 5). OSEs were also conducted to understand the impact of the SAPHIR observation in the forecasts of longer lead times during the same time. OSE studies also showed similar results which supports the FSO findings.

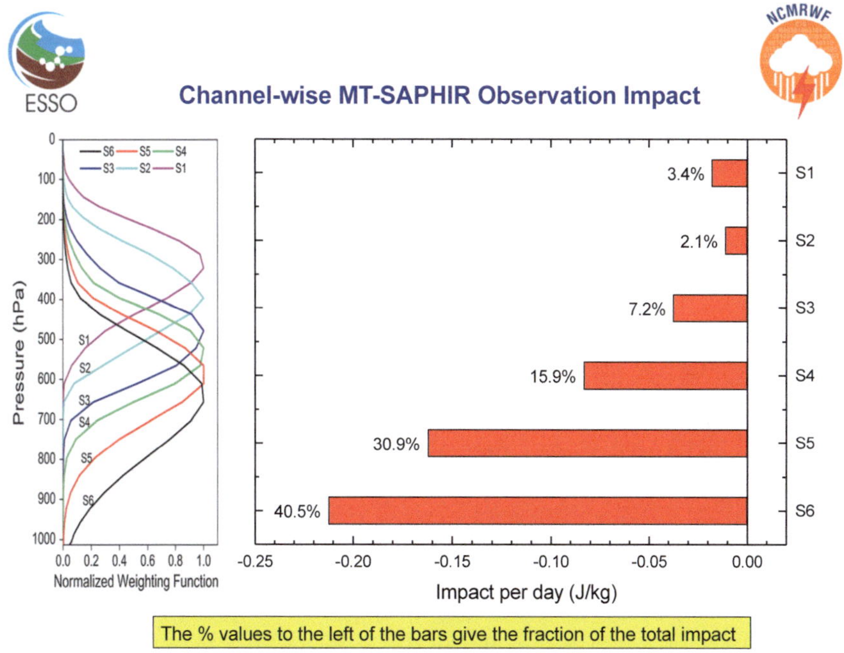

Fig. 5 Channel-wise MT-SAPHIR observation impact

4 High-Resolution Models at NCMRWF

4.1 Operational 4 km and Experimental 1.5 km Models

The operational 4 km model is used to produce high-resolution forecasts up to 75 h (http://www.ncmrwf.gov.in/product_main_ind.php). This model run produces improved forecasts of localized features of severe weather such as floods, thunderstorms, fog, etc. In addition to that forecast, products are used for solar and wind energy applications. This model's domain is 5°–40°N and 65°–100°E and the model's orography is derived from Shuttle Radar Topography Mission (SRTM) DEM. General science configuration of the convection permitted models used are given below.

The initial and lateral boundary conditions are derived from the operational 17 km NCUM global model (Rakhi et al. 2016). The lateral boundary conditions are updated at 3 hourly frequency. The details of the data assimilation employed in the parent global model (NCUM-GL) are given in George et al. (2016). The model uses the ENDGame dynamical core (Wood et al. 2014) and a full suite of parameterized physical processes, such as mixed-phase microphysics (based on Wilson and Ballard 1999), radiation (based on Edwards and Slingo 1996) and The Joint UK Land Environment Simulator (JULES) land surface (Best et al. 2011) scheme.

Table 4 Model physics formulation changes between NCUM-R and DM

	NCUM-R (1.5 km)	DM (330 m)
LBC updation	1 h	15 min
Model time step	30 s	12 s
Radiation time step	15 and 5 min	4 and 1 min
Critical relative humidity	0.96 at surface, decreasing to 0.8 at 1 km, constant above	0.97 at surface, decreasing to 0.9 at 3.5 km, constant above

The experimental 1.5 km model uses the latest tropical science configuration (RA1-T), and it is capable of predicting the local spatio-temporal variation of rainfall, snow, 10 m wind, visibility, 2 m RH, and skin temperature within the selected domain. Sensitivity studies were carried out using high-resolution ISRO's Cartosat-1 digital elevation model (DEM) Version-3R1 (30 m) against SRTM (90 m) orography.

4.2 Experimental 330 m Delhi Model

Delhi model (DM) is of 330 m resolution (Jayakumar et al. 2018) which is set up at multiple nested frameworks. DM is nested inside the NCUM-R of 1.5 km resolution. The DM covers a domain of 100×100 km^2 and is run for 36 h based on 00UTC initial conditions. NCUM-R (1.5 km) provides initial and boundary conditions to the DM, and the boundary conditions are updated every 15 min. All atmospheric prognostic variables are provided by the parent model to the nested models through the lateral boundary conditions. The model uses the ENDGame dynamical core (Wood et al. 2014) and a full suite of parameterized physical processes, such as mixed-phase microphysics (based on Wilson and Ballard 1999), cloud (Smith 1990), radiation (based on Edwards and Slingo 1996) and land surface scheme (Best et al. 2011). Critical relative humidity in DM is higher than NCUM-R (Table 4) following the approach used in the London Model (Boutle et al. 2016), to simulate less sub-grid cloud cover variability in higher resolution models with respect to the coarse resolution model. The sub-grid turbulence scheme used is the blended scheme (Boutle et al. 2014), which dynamically combines the 1D boundary-layer scheme of Lock et al. (2000) with a 3D Smagorinsky scheme using a mixing factor of 0.5. The model time step of the DM is 12 s, and the model has 80 vertical levels with model top at 38.5 km with 14 model levels below 1 km. The DM and NCUM-R setups are compared in Table 4.

4.3 Sensitivity of 4 km, 1.5 km and 330 m Model Horizontal Resolutions

The UM high-resolution models use science configuration of SINGV3 of UM Regional Atmospheric version. The same has been adopted for NCUM-R (4 km). The major differences from the earlier configuration are the moisture conservation and stochastic perturbations. The excessive peak rain rates in the earlier version were mainly due to non-conservation of moisture in the semi-Lagrangian advection scheme. In cases with poorly resolved (grid-scale) updrafts, it is found that the semi-lagrangian scheme does not cope with the stagnation point at the bottom of the column. In the new configuration, the pseudo-lateral boundary flux (PLF) scheme by Aranami et al. (2015) acts as a mass restoration scheme forced to conserve this poorly resolved convection.

In convective-scale models, the uncertainties are related to microphysics or turbulent eddies. The differences in surface-driven heating of boundary layer (BL) that would arise from undersampling of the boundary-layer eddies, or variability in the surface fluxes need to be represented. Hence, modification in NCUM-R includes the perturbation applied to the potential temperature in the convective BL.

The other major changes in the specification of Critical relative humidity (RH_{Crit}). RH_{Crit} is the threshold value at which the cloud will start to form. The uniform probability distribution function with a width related to saturation specific humidity to q_s can be formulated in terms of a threshold in relative humidity—RH_{Crit}. Fractional cloudiness, f, in such a formulation is expressed in terms of the grid-box mean relative humidity, $\bar{r}$, as

$$f = 1 - \sqrt{\frac{1 - \bar{r}}{1 - r_c}}$$

with $f = 0$ for $\bar{r} \leq rc$ and $f = 1$ for $\bar{r} \geq 1$ (Sundqvist et al. 1989).

RH_{Crit} for the DM is higher than NCUM-R (Table 4) follows the approach used as in UK Met Office's London Model (Boutle et al. 2016), to compensate lesser sub-grid cloud cover variability found over higher resolution models with respect to the coarse resolution model. PC2 uses the prognostic large-scale cloud scheme, together with the prognostic RH_{crit} scheme instead of the diagnostic Smith scheme together with the fixed RH_{crit} profile (Fig. 6).

It should be noted that the NCUM-R (4 km) used here is SINGV2, where moisture conservation was not applied, that difference is evident as larger excessive rain cells.

5 NCMRWF Ensemble Prediction System (NEPS)

The operational NCMRWF ensemble prediction system (NEPS) has 44 ensemble members. The resolution of NEPS is N400L70 (horizontal resolution ~33 km). The

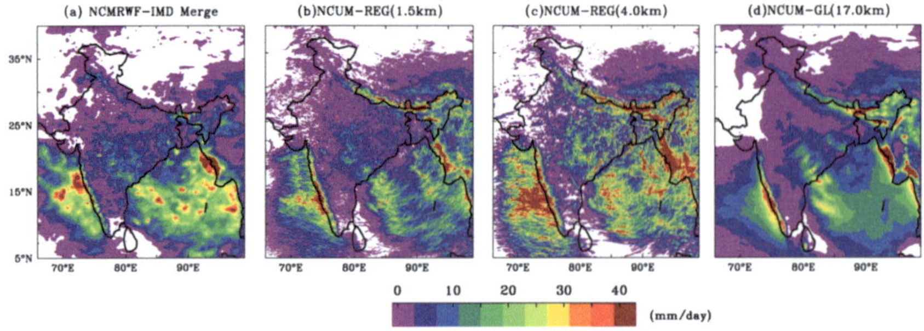

Fig. 6 Mean rainfall amount during the 10–30 June 2016: **a** NCMRWF-IMD merge rainfall and **b** NCUM-R (4 km), NCUM-R (1.5 km) and NCUM-GL (17 km)

global deterministic analysis at 17 km resolution are re-gridded to 33 km resolution for the control model and the perturbed initial conditions for the 44 members are created using Ensemble Transform Kalman Filter (ETKF). ETKF generates global perturbations for wind, temperature, humidity and pressure fields for the 44 ensemble members. In NEPS system, the perturbations generated by ETKF are combined with the operational 4D-Var analysis. The perturbations are added to the re-gridded analysis using the Incremental Analysis Update (IAU) scheme (Clayton 2012) within the NCUM. Model predictions are based on the initial conditions of 00UTC. Additional details about the NCUM and NEPS models can be found in Rajagopal et al. (2012) and Sarkar et al. (2016).

5.1 *Experiments with NCMRWF Ensemble Prediction System (NEPS)*

In designing an effective operational ensemble prediction system under the constraint of limited computational resources it is essential to optimize the ensemble size and trade-off between model resolution and ensemble size (Du et al. 1997; Buizza and Palmer 1998; Buizza et al. 1998; Mullen and Buizza 2002). Atger (1999) noted that in ECMWF EPS, an increase in ensemble size improves ensemble spread but improvement in accuracy of ensemble mean is insignificant. Wang and Bishop (2003) showed that a 16-member ensemble gives a better estimate of analysis error variance and reduces spurious long-distance correlations than an 8-member ensemble.

Juhui et al. (2012) compared the model results of two representative configurations of the NCEP GEFS having equivalent computing costs and found that in a short forecast range increasing model resolution is more beneficial than increasing ensemble size while in long forecast range the reverse is true. The results obtained by Juhui et al. (2012) indicated that ensemble sizes of 20–30 members provide the most effective configuration. Buizza (2010) also noted that the effect of model resolution is strong in the short forecast range. Kay et al. (2013) investigated forecast capabil-

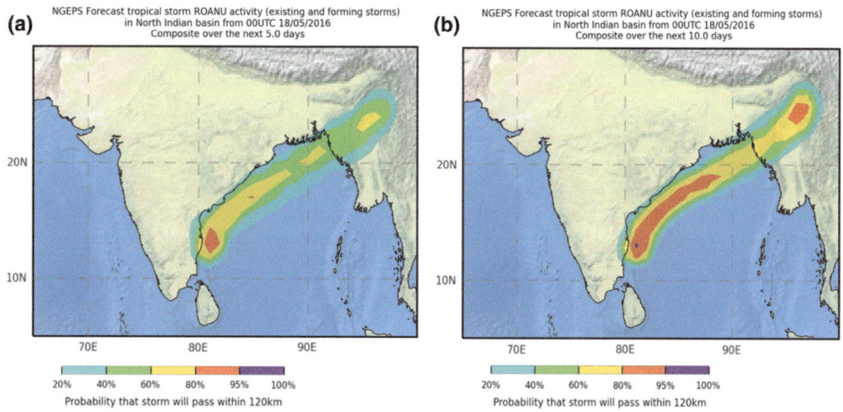

Fig. 7 Strike probability forecast of TC Roanu from NEPS (33 km) for the next 5 days starting from 00 UTC of 18.05.2016 with ensemble sizes 22 (**a**) and 44 (**b**)

ity of the UM-based global ensemble prediction system implemented at KMA by doubling the ensemble size from 24 to 48 and found that the overall performance of the EPS improves only slightly in terms of accuracy, resolution and reliability, especially for longer forecast time.

(a) Sensitivity of Ensemble Size

To investigate the impact of ensemble size and model resolution on the performance of ensemble prediction system, a tropical cyclone case was selected and the forecasts of strike probability for the next 5 days using the NEPS of ensemble sizes 22 and 44 have been compared. The strike probability forecast at a given model grid point gives the probability that the cyclone centre will pass within a distance of 120 km from that point in the next 5 days.

Figure 7 shows the strike probability forecast of Tropical Cyclone 'Roanu' from NEPS at 33 km horizontal resolution with ensemble sizes 22 and 44 members (a and b respectively). The figure clearly demonstrates that the NEPS of ensemble size of 44 could predict higher strike probability of 80–90% over a longer lead time. Similarly, Fig. 8 shows the strike probability forecast from NEPS at 17 km resolution. This figure also shows that NEPS of larger ensemble size gives higher strike probability (80–90%) forecast over a longer lead time.

(b) Sensitivity to Horizontal Resolution

A comparison of the strike probability forecasts from NEPS of same ensemble size (22) but different horizontal resolutions (33 and 17 km) has also been carried out. Figures 8b and 9a shows that the performance of higher resolution NEPS is better as it is able to predict a higher strike probability over a longer range. Figure 9 clearly shows that the strike probability forecasts of a 22 member NEPS at horizontal resolution of 17 km and a 44 member NEPS at horizontal resolution of 33 km are nearly equivalent to each other. So it is required to quantify the trade-off between ensemble performance and cost of computational resources to optimize EPS configuration in

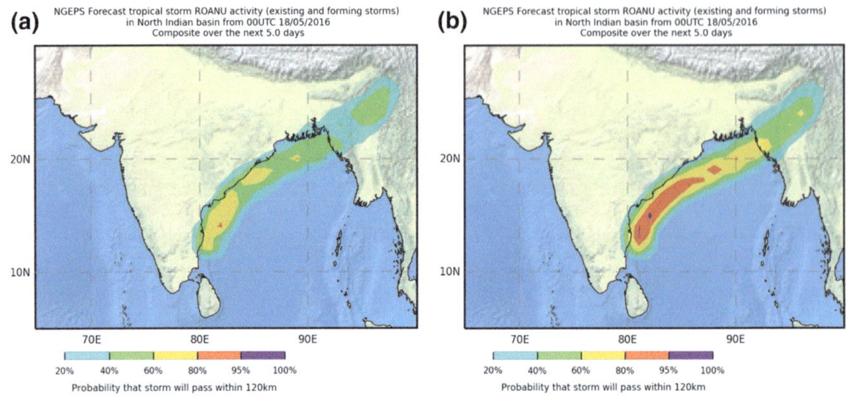

Fig. 8 Strike probability forecast of TC Roanu from NEPS (17 km) for the next 5 days starting from 00 UTC of 18.05.2016 with ensemble sizes 11 (**a**) and 22 (**b**)

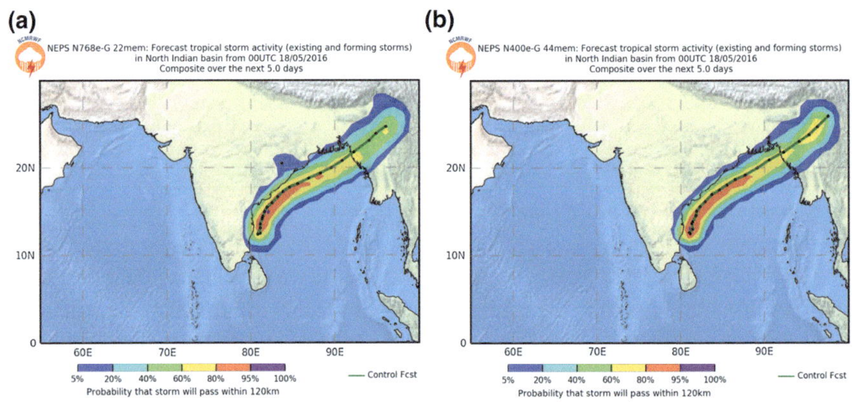

Fig. 9 Strike probability forecast of TC Roanu from NEPS of **a** horizontal resolution 17 km and ensemble size 22 and **b** horizontal resolution 33 km and ensemble size 44 for the next 5 days starting from 00 UTC of 18.05.2016

terms of ensemble size and resolution. Here it may be mentioned that experiments over an extended period of time (about 2–3 months) and for a large number of cases are required to be carried out and performance of the EPS should be evaluated by measuring both the probabilistic and deterministic metrics.

6 Rainfall Forecast Verification over India During JJAS 2017

This section provides brief evaluation results for NCUM and NEPS rainfall forecasts during JJAS 2017. The IMD-NCMRWF daily high-resolution (0.25°) rainfall analysis (Mitra et al. 2009, 2013a, b) is used for the verification. The evaluation

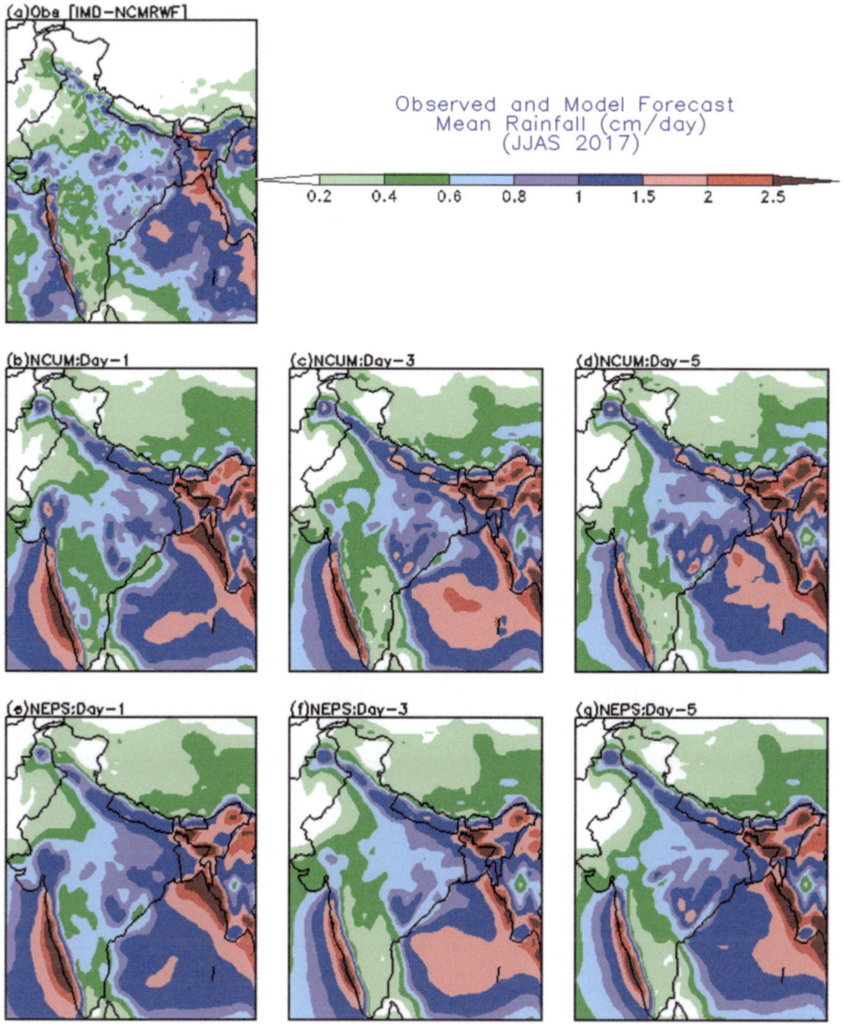

Fig. 10 Observed and forecast mean monsoon rainfall over India during JJAS 2017. Comparison of **a** observed (IMD-NCMRWF merged) and the NCUM forecast, **b** day 1, **c** day 3 and **d** day 5 forecast rainfall (cm/day). Similarly, panels **e**, **f** and **g** correspond to NEPS (ensemble mean) forecast rainfall over India

has been carried out for all days of prediction (day 1–day 10); however, in this report the results of only day 1, day 3 and day 5 forecasts are discussed. In Fig. 10, observed and forecast mean rainfall is presented. The observed peak rainfall amounts (>1.5 cm/day) along the Western Ghats and along the Arakan coast are correctly predicted except over the southern tip of the west coast. However, with the increase in forecast lead time, this amount of rainfall decreases in both the models. Over central and eastern India, the observed mean rain exceeds 0.6 cm/day (core monsoon region) covering a large area with two isolated pockets (M.P., Telangana and A.P.) of higher (>1 cm/day) mean rainfall amounts.

This broad structure of the core monsoon rainfall is fairly well captured by both the models. However, in both the models the area over which rainfall exceeds 1 cm/day is larger than in the observations. Also, in the forecasts from all three models, there are isolated pockets of rainfall exceeding 1.5 cm/day, over Madhya Pradesh and surrounding areas. Rainfall along the foothills of Himalayas is well captured by the two models for all lead times. The dry areas in the observations over parts of Rajasthan and peninsular India are also well reproduced by all the models at all lead times. Rainfall is also over-predicted over northeast India on all days.

Figure 11 shows the Performance Diagram for all the rainfall thresholds for day 5 forecast from both the models. This diagram shows the POD versus the success ratio (SR) [SR = 1 − FAR]. The dashed lines represent the bias scores with labels on the outward extensions of the lines; labelled solid contours are the CSI. For good forecasts, POD, SR, bias and CSI approach unity, such that a perfect forecast lies in the upper right of the diagram. Deviations in a particular direction will indicate the relative differences in POD and SR, and consequently bias and CSI. The figure shows that the lower rainfall thresholds are better predicted by NEPS (relative to NCUM). Both models overestimate the daily rainfall over central and eastern parts of India. Whereas over the west coast, the models are underestimating the daily rainfall. It can be seen that the rainfall peak amounts are overestimated over central India and underestimated over the Bay of Bengal. The NEPS (ensemble mean) shows higher POD and lower FAR as compared to NCUM, especially for lower rainfall thresholds.

7 IMDAA Reanalysis

Indian Monsoon Data Assimilation and Analysis (IMDAA) project under National Monsoon Mission of Ministry of Earth Sciences (Govt. of India) is a collaborate project between UK Met Office, NCMRWF and IMD. This project aims to develop regional reanalysis for Indian Monsoon Region for the period 1979–2018.

This satellite-era atmospheric regional reanalysis at 12 km horizontal resolution aims to improve the understanding of the Asian monsoon. The IMDAA reanalysis is being produced using the advanced data assimilation technique of four-dimensional variational data assimilation (4D-Var). IMDAA project aims to create a multi-decadal dataset of high-resolution gridded, three-dimensional atmospheric analyses of Asian monsoon region, using a state-of-the-art modelling system consistent with the NCUM used at NCMRWF. The IMDAA reanalysis domain is given in Fig. 12.

As a part of IMDAA project, UK Met Office developed the reanalysis system. This reanalysis system has been implemented at NCMRWF for generation of reanalysis for the period 1979–2018. The production runs for the initial 10-year period (1979–1988) were over last year, and the runs for 1988–2018 period were being carried out at NCMRWF. Observations from ECMWF archives as well as observations retrieved

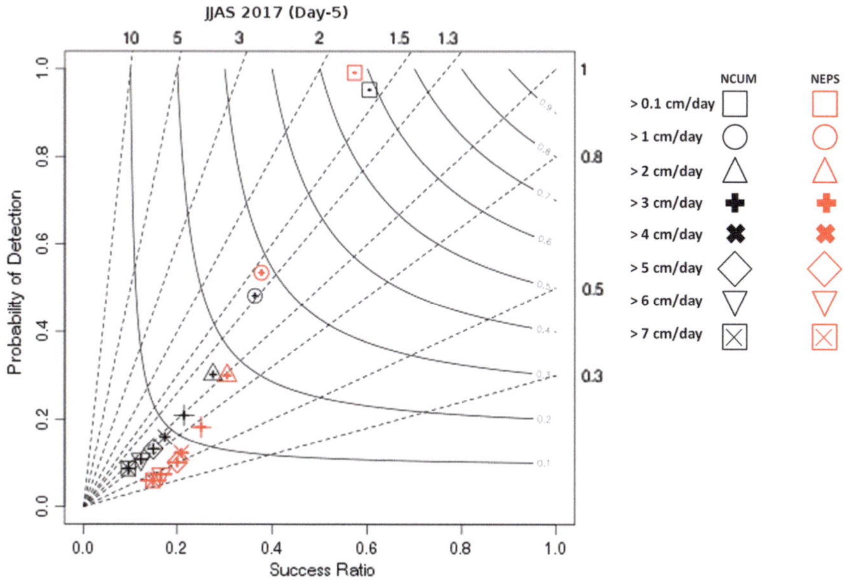

Fig. 11 Performance diagram showing the skill of the models NCUM and NEPS (ensemble mean) rainfall forecast

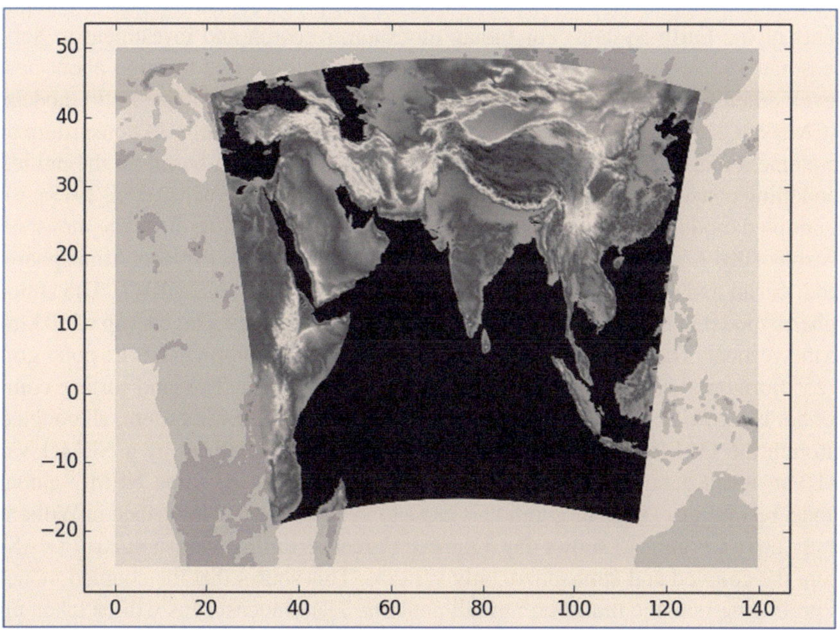

Fig. 12 IMDAA reanalysis domain (highlighted domain)

from IMD and NCMRWF achieves are being used in the IMDAA reanalysis. The reanalysis dataset will be made available for studies by the general Indian research community for further study of the mechanics of the Indian monsoon. The broader impacts envisaged from the IMDAA project include:

- Increased understanding of the details of the onset, development and decay of the Indian monsoon through the period 1979–2018.
- The provision of a 40-year dataset of the Indian climate for further research by Indian scientists.
- Leveraging by NCMRWF scientists working on the development of an operational regional NWP system based on the same data assimilation configuration as IMDAA.
- A vehicle for broader UK-India data assimilation collaboration in future, potentially involving respective university sectors.
- Training of additional Indian scientists through yearly project meetings at the collaborating institutes.

8 NCMRWF Coupled Modelling System

Indian summer monsoon is well known for its complex coupled interactive components of the Earth System. For Indian monsoon, research and investment in Sub-Seasonal to Seasonal (S2S) scale is the most challenging one to bring about new developments in modelling system, which in turn will produce more skilful models. NCMRWF is working in close collaboration with the partners of UM consortium to implement and improve upon the seamless modelling system based on the unified modelling concepts for prediction of day-to-season timescale (Mitra et al. 2013a, b). A coupled model has been implemented and running in real time since the monsoon season of 2017. The end-to-end assimilation-forecast system is in place. Atmosphere, land, ocean and sea ice initial conditions are all produced at NCMRWF. The atmospheric model is UM at N216 (65 km in horizontal) and has 85 layers (up to 80 km) in the vertical. The ocean component is global NEMO3.4 with ORCA tri-polar grid 0.25° horizontal resolution and has 75 layers in the vertical. The land surface component is the JULES and the Sea Ice component is the CICE4.3 system, all coupled through the OASIS coupler. The ocean component is initialized by a NEMO-Var 3-DVar ocean data assimilation system on daily basis at the same NEMO global model resolution. The configuration is close to the GC2 system described in William et al. (2015). Figure 13 shows day 3 rainfall Forecasts during Monsoon 2017 (JJAS) from the coupled and atmosphere only systems. This shows that the coupled model is producing realistic monsoon rainfall patterns. The hindcast runs will be taken up to see the skill in extended/seasonal scale. It is planned that during 2018–19 coupled system will be tested and implemented at all S2S scales with real-time forecasting purpose, namely, the NWP at up to 15 days (17 km), sub-seasonal multi-week (at

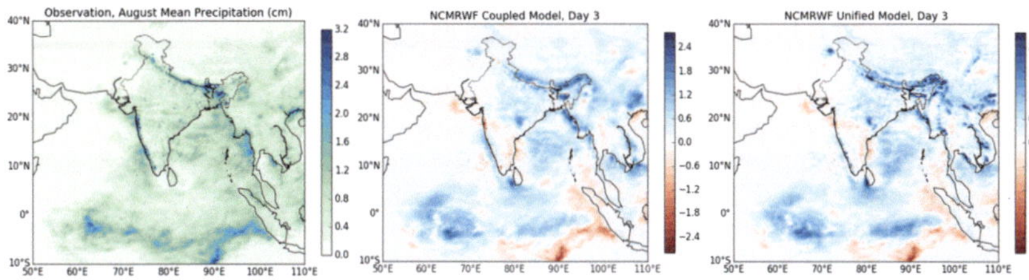

Fig. 13 Day 3 rainfall forecasts during monsoon 2017 (JJAS) from the coupled and atmosphere only systems

25 km), and seasonal runs up to 6 months (at 60 km). The multi-week and seasonal runs will have a 20-member ensemble runs.

9 Future Activities

On the next HPC, there are plans to increase the horizontal resolution of both deterministic and ensemble prediction, including the coupled modelling system. The details of the future activities are the following:

- GFS-DA will be upgraded with EnKF perturbation at T1534 resolution.
- Hybrid 4D-Var for NCUM-G will upgraded with a 20 km ETKF perturbed (44 members) ensemble system.
- The global deterministic model (NCUM-G) will be upgraded to 10 km/L70.
- NCUM EPS will be upgraded to 12 km resolution with 22 members.
- 4D-Var DA for NCUM-R (4 km) will be implemented which will have the capability to assimilate DWR radial winds and radar reflectivities.
- The regional model (NCUM-R) will be upgraded to 1.5 km/L70.
- The ocean DA (NEMO-Var) at a resolution of 25 km/L75 will continue to run once daily.
- Coupled NWP model will be upgraded to 12 km/L85 for 15-day prediction.
- A coupled model at 25 km/L85 will be run once a week for 1-month predictions.
- A seasonal prediction at a resolution of 60 km/L85 will be run once a month for generating 4-month predictions.

References

Aranami, K., T. Davies, and N. Wood. 2015. A mass restoration scheme for limited-area models with semi-Lagrangian advection. *Quarterly Journal of the Royal Meteorological Society* 141: 1795–1803.

Atger, F. 1999. The skill of ensemble prediction systems. *Monthly Weather Review* 127: 1941–1953.

Best, M.J., M. Pryor, D.B. Clark, G.G. Rooney, R.L.H. Essery, C.B. Ménard, J.M. Edwards, M.A. Hendry, A. Porson, N. Gedney, L.M. Mercado, S. Sitch, E. Blyth, O. Boucher, P.M. Cox, C.S.B. Grimmond, and , R.J. Harding. 2011. The Joint UK Land Environment Simulator (JULES), model description – part 1: Energy and water fluxes. *Geoscientific Model Development* 4 (1): 677–699.

Boutle, I.A., J.E.J. Eyre, and A.P. Lock. 2014. Seamless stratocumulus simulation across the turbulent gray zone. *Monthly Weather Review* 142: 1655–1668.

Boutle, I.A., A. Finnenkoetter, A.P. Lock, and H. Wells. 2016. The London model: Forecasting fog at 333 m resolution. *Quarterly Journal of the Royal Meteorological Society* 142: 360–371.

Buizza, R. 2010. Horizontal resolution impact on short and long range forecast error. *Quarterly Journal of the Royal Meteorological Society* 136: 1020–1035.

Buizza, R., and T. Palmer. 1998. Impact of ensemble size on ensemble prediction. *Monthly Weather Review* 126: 2503–2518.

Buizza, R., T. Petroliagis, T.N. Palmer, J. Barkmeijer, M. Hamrud, A. Hollingsworth, A. Simmons, and N. Wedi. 1998. Impact of model resolution and ensemble size on the performance of an ensemble prediction system. *Quarterly Journal of the Royal Meteorological Society* 124: 1935–1960.

Clayton, A. 2012. Incremental Analysis Update (IAU) Scheme. Unified Model Documentation Paper No. 31.

Du, J., S.L. Mullen, and F. Sanders. 1997. Short-range ensemble forecasting of quantitative precipitation. *Monthly Weather Review* 125: 2427–2459.

Edwards, J.M., and A. Slingo. 1996. Studies with a flexible new radiation code. I: Choosing a configuration for a large-scale model. *Quarterly Journal of the Royal Meteorological Society* 122: 689–719.

George, J.P., S. Indira Rani, A. Jayakumar, Saji Mohandas, Swapan Mallick, R. Rakhi, M.N.R. Sreevathsa, and E.N. Rajagopal. 2016. NCUM Data Assimilation System, NMRF/TR/01/2016, March 2016.

Jayakumar, A., E.N. Rajagopal, I.A. Boutle, J.P. George, S. Mohandas, S. Webster, S. Aditi. 2018. An operational fog prediction system for Delhi using the 330 m Unified Model. *Atmospheric Science Letters* 19: e796.

Juhui, M.A., Y. Zhu, R. Wobus, and P. Wang. 2012. An effective configuration of ensemble size and horizontal resolution for the NCEP GEFS. *Advances in Atmospheric Sciences* 29: 782–794.

Kay, J.K., H.M. Kim, Y.Y. Park, and J. Son. 2013. Effect of doubling the ensemble size on the performance of ensemble prediction in the warm season using MOGREPS implemented at the KMA. *Advances in Atmospheric Sciences* 30: 1287–1302.

Lock, A.P., A.R. Brown, M.R. Bush, G.M. Martin, and R.N.B. Smith. 2000. A new boundary layer mixing scheme. Part I: Scheme description and single-column model tests. *Monthly Weather Review* 128: 3187–3199.

Lodh, A., John P. George, and E.N. Rajagopal. 2016. Extended Kalman filter based land data assimilation system for soil moisture analysis at NCMRWF. Technical Report, NMRF/TR/06/2016.

Mitra, A.K., A.K. Bohra, M.N. Rajeevan, and T.N. Krishnamurti. 2009. Daily Indian precipitation analysis from a merge of rain-gauge data with the TRMM TMPA satellite-derived rainfall estimates. *Journal of the Meteorological Society of Japan* 87A: 265–279.

Mitra, A.K., and coauthors. 2013a. Prediction of monsoon using a seamless coupled modelling system. *Current Science* 104 (10): 1369–1379.

Mitra, A.K., I.M. Momin, E.N. Rajagopal, S. Basu, M.N. Rajeevan, and T.N. Krishnamurti. 2013b. Gridded daily Indian monsoon rainfall for 14 seasons: Merged TRMM and IMD gauge analyzed values. *Journal of Earth System Science* 122 (5): 1173–1182.

Mullen, S.L., and R. Buizza. 2002. The impact of horizontal resolution and ensemble size on probabilistic forecasts of precipitation by the ECMWF ensemble prediction system. *Weather Forecasting* 17: 173–191.

Rajagopal, E.N., G.R. Iyengar, John P. George, M. Das Gupta, Saji Mohandas, R. Siddharth, A. Gupta, M. Chourasia, V.S. Prasad, Aditi, K. Sharma, and A. Ashish. 2012. Implementation of Unified Model based analysis-forecast system at NCMRWF. NMRF/TR/2/2012, 45p.

Rakhi, R., A. Jayakumar, M.N.R. Sreevathsa, and E.N. Rajagopal. 2016. Implementation and upgradation of NCUM in Bhaskara HPC. NMRF/TR/03/2016, May 2016.

Rawlins, F., S.P. Ballard, K.J. Bovis, A.M. Clayton, D. Li, G.W. Inverarity, A.C. Lorenc, and T.J. Payne. 2007. The Met Office global four-dimensional variational data assimilation scheme. *Quarterly Journal of the Royal Meteorological Society* 133: 347–362.

Sarkar, A., P. Chakraborty, J.P. George, and E.N. Rajagopal. 2016. Implementation of Unified Model based ensemble prediction system at NCMRWF (NEPS). NMRF/TR/02/2016.

Sundqvist, H., E. Berge, and J.E. Kristjansson. 1989. Condensation and cloud parameterization studies with a mesoscale numerical weather prediction model. *Monthly Weather Review* 117: 1641–1657.

Walters, D., I. Boutle, M. Brooks, T. Melvin, R. Stratton, S. Vosper, H. Wells, K. Williams, N. Wood, T. Allen, A. Bushell, D. Copsey, P. Earnshaw, J. Edwards, M. Gross, S. Hardiman, C. Harris, J. Heming, N. Klingaman, R. Levine, J. Manners, G. Martin, S. Milton, M. Mittermaier, C. Morcrette, T. Riddick, M. Roberts, C. Sanchez, P. Selwood, A. Stirling, C. Smith, D. Suri, W. Tennant, P.L. Vidale, J. Wilkinson, M. Willett, S. Woolnough, and P. Xavier. 2017. The Met Office Unified Model global atmosphere 6.0/6.1 and JULES global land 6.0/6.1 configurations. *Geoscientific Model Development* 10: 1487–1520. https://doi.org/10.5194/gmd-10-1487-2017.

Wang, X., and C.H. Bishop. 2003. A comparison of breeding and ensemble transform Kalman Filter ensemble forecast schemes. *Journal of the Atmospheric Sciences* 60: 1140–1157.

Wilson, D.R. and S.P. Ballard. 1999. A microphysically based precipitation scheme for the UK meteorological office unified model. *Quarterly Journal of the Royal Meteorological Society* 125: 1607–1636.

Williams, K.D., and coauthors. 2015. The Met Office Global Coupled model 2.0 (GC2) configuration. *Geoscientific Model Development* 8: 1509–1524.

Wood, N., A. Staniforth, A. White, T. Allen, M. Diamantakis, M. Gross, T. Melvin, C. Smith, S. Vosper, M. Zerroukat, and J. Thuburn. 2014. An inherently mass-conserving semi-implicit semi-Lagrangian discretization of the deep-atmosphere global non-hydrostatic equations. *Quarterly Journal of the Royal Meteorological Society* 140: 1505–1520.

Microphysics in Goddard Multi-scale Modeling Systems: A Review

W.-K. Tao, J. Chern, T. Iguchi, S. Lang, M.-I. Lee, X. Li, A. Loftus, T. Matsui, K. Mohr, S. Nicholls, C. Peters-Lidard, D. J. Posselt and G. Skofronick-Jackson

Abstract Advances in computing power allow atmospheric prediction and general circulation models to be run at progressively finer scales of resolution, using increasingly more sophisticated physical parameterizations. The representation of cloud microphysical processes is one of the key components of these models. In

---

W.-K. Tao (✉) · J. Chern · T. Iguchi · S. Lang · X. Li · T. Matsui · S. Nicholls
G. Skofronick-Jackson
Mesoscale Atmospheric Processes Laboratory, NASA/Goddard Space Flight Center, Greenbelt, MD 20771, USA
e-mail: Wei-Kuo.Tao-1@nasa.gov

J. Chern · T. Iguchi · A. Loftus · T. Matsui
Earth System Science Interdisciplinary Center, University of Maryland, College Park, MD 20742, USA

S. Lang
Science Systems and Applications Inc., Lanham, MD 20706, USA

M.-I. Lee
School of Urban and Environmental Engineering, Ulsan National Institute of Science & Technology (UNIST), Ulsan 44919, Republic of Korea

X. Li
Goddard Earth Sciences Technology and Research, Morgan State University, Baltimore, MD 21250, USA

A. Loftus
Climate and Radiation Laboratory, NASA/Goddard Space Flight Center, Greenbelt, MD 20771, USA

K. Mohr · C. Peters-Lidard
Earth Sciences Division, NASA/Goddard Space Flight Center, Greenbelt, MD 20771, USA

S. Nicholls
Joint Center for Earth Systems Technology, University of Maryland, Baltimore, MD 21250, USA

D. J. Posselt
Jet Propulsion Laboratory, Pasadena, CA 91109, USA

D. A. Randall et al. (eds.), *Current Trends in the Representation of Physical Processes in Weather and Climate Models*, Springer Atmospheric Sciences,
https://doi.org/10.1007/978-981-13-3396-5_14

addition, over the past decade, both research and operational numerical weather prediction models have started using more complex microphysical schemes that were originally developed for high-resolution cloud-resolving models (CRMs). In the paper, we describe different microphysics schemes that are used in the Goddard multi-scale modeling system, the three major models of which are the Goddard Cumulus Ensemble (GCE), NASA-Unified Weather Research and Forecasting (NU-WRF), and Multi-scale Modeling Framework (MMF) models. The microphysics schemes are the Goddard three class ice (3ICE) and four class ice (4ICE), Morrison two moment (2M), Colorado State University Regional Atmospheric Modeling System (RAMS) 2M five class ice, and spectral bin microphysics schemes. The performance of these schemes is examined and compared with radar and satellite observations. In addition, the intercomparison of different microphysics schemes is conducted. Current and future observations needed for microphysics schemes evaluation as well as major characteristics of current microphysics are discussed.

Keywords Microphysics · Cloud-Resolving Model
Mesoscale Convective System

1 Introduction

Cloud microphysics represents one of the key physical processes in Earth system science (see Fig. 1). For example, its associated latent heat is released or absorbed by the atmosphere as a result of phase changes in water (e.g., condensation or evaporation of cloud droplets and raindrops, freezing of raindrops, melting of snow and graupel/hail, and the deposition or sublimation of ice particles). Cloud microphysics affects the vertical distribution of cloud substances (or hydrometeors) and size distributions (i.e., from small cloud water droplets and ice particles, to medium-sized snow, to large precipitating raindrops and graupel/hail), aspects of which affect active (i.e., radar reflectivity) and passive (i.e., brightness temperature) remote sensing measurements. Since precipitation can be in the form of light rainfall, heavy rainfall, snow, or mixed phase, it influences surface properties (i.e., soil moisture, runoff, albedo, and emissivity) and the energy and water cycles. Convective transport affects the vertical redistribution of chemical species and, in turn, radiative forcing and atmospheric electrification [see a review by Cotton et al. (1995) and Thompson et al. (1997)].

Cloud-resolving models (CRMs) are a type of numerical model wherein mathematical equations are applied at discrete points to simulate the evolution of physical processes over a spatial area. Many CRMs have been developed over the past five decades. They have been applied to improve our understanding of microscale to cloud-scale and mesoscale as well as their interactions with radiation, aerosol, and surface processes. The basic characteristic of CRMs is that their governing equations are non-hydrostatic since the vertical and horizontal scales of atmospheric motion are similar. The CRMs use sophisticated and physically realistic cloud microphysical processes with very fine spatial and temporal resolutions. They represent the inter-

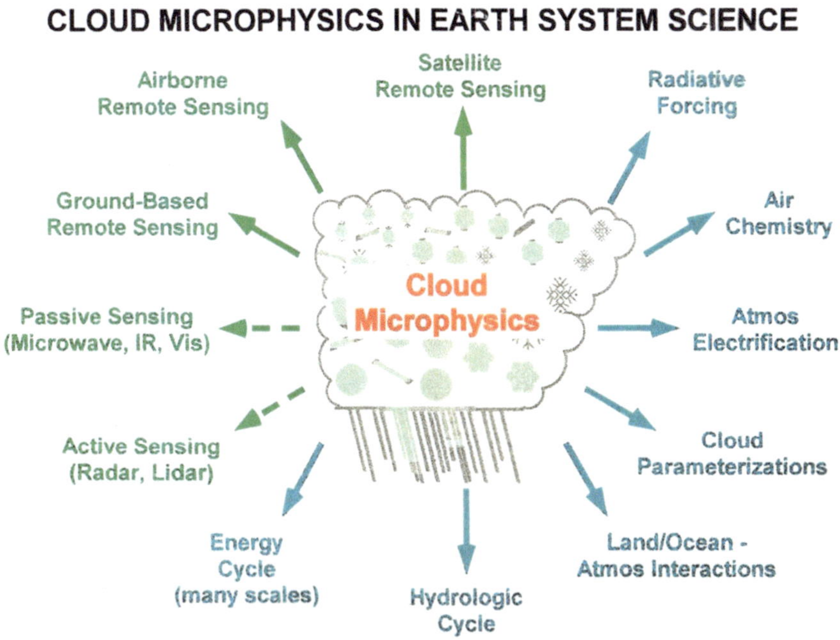

Fig. 1 Schematic diagram showing the interactions between microphysics and other Earth system sciences. Adapted from Tao and Moncrieff (2009)

action between clouds and radiation (and aerosols) with greater fidelity than global models since the spatial and temporal distributions of water substances (vapor, liquid, and ice) are explicitly coupled to the atmosphere circulation at cloud system scale. Other advantages of using CRMs include their ability to quantify the effects of each physical process by means of sensitivity tests [e.g., eliminating a specific process such as evaporative cooling, ice formation, and its associated processes, planetary boundary layer (PBL)], and their detailed dynamic and thermodynamic budget calculations. Figure 2 shows a schematic of the main characteristics of typical CRMs. A review of CRMs including their history and applications can be found in Tao (2003, 2007) and Tao and Moncrieff (2009).

The Goddard Cumulus Ensemble (GCE) model is a CRM that has been developed and improved at NASA Goddard Space Flight Center (GSFC) over the past three and a half decades. It has been used for studying precipitation processes and their impact on rainfall as well as to support NASA satellite missions [i.e., Tropical Rainfall Measurement Mission (TRMM) and Global Precipitation Measurement (GPM); see Simpson et al. (1996), Tao et al. (2016a)]. One of the key developments of the GCE model is the cloud microphysics (microphysical schemes). However, the cloud microphysical processes (nucleation, diffusion growth and collision among cloud, and precipitation particles) still need to be parameterized in GCE (and other CRMs as well). Note that all cloud microphysical schemes have their own set of unique

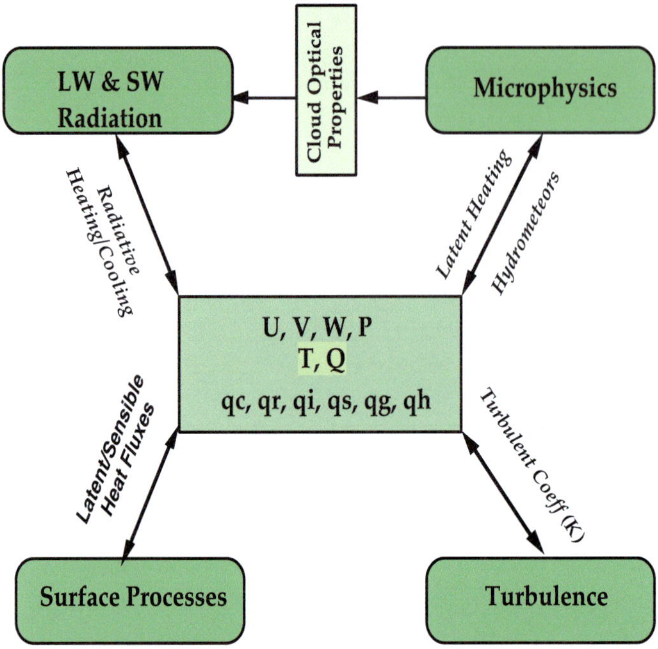

Fig. 2 Schematic showing the characteristics of cloud-resolving models. Solid lines capped by arrows indicate two-way interactions between different physical processes. U, V, W, and P stand for horizontal wind components, vertical wind, and pressure, respectively. T and Q stand for atmospheric temperature and water vapor mixing ratio, respectively, qc, qr, qi, qs, qg, and qh stands for mixing ratio of cloud water, rain, cloud ice, snow, graupel, and hail, respectively, and LW and SW stand for the atmospheric long-wave and short-wave radiative cooling and heating, respectively. Adapted from Tao (2007)

assumptions and capabilities. It is critical, therefore, to sample and evaluate model performance for a comprehensive range of precipitation systems. Observations are crucial to verify model results and improve the initial and boundary conditions as well as the aforementioned physics processes.

The GCE model was recently enhanced to simulate the impact of atmospheric aerosol concentrations on precipitation processes and the impact of land and ocean surface processes on convective systems in different geographic locations (Tao et al. 2007, 2014; Li et al. 2009a; Zeng et al. 2007, 2009a, b; Tao and Li 2016). The GCE model has also been coupled with the Goddard Satellite Data Simulator Unit (G-SDSU), which allows us to scrutinize the performance of the microphysics by analyzing discrepancies between the simulated and observed radiances from remote sensing measurements (Matsui et al. 2009; Li et al. 2010).

Recently, the GCE model has been coupled with a general circulation model (GCM) by replacing the one-dimensional cumulus parameterization scheme with a two-dimensional (2D) GCE model [called super-parameterization or Multi-scale Modeling Framework (MMF)]. In addition, the GCE microphysical schemes and their interactions with radiation and surface processes have also been implemented into NASA-Unified WRF (NU-WRF). The performance of Goddard microphysics schemes can be tested from local, regional to global scales, and for different types of cloud/cloud systems development in different environments by using these three modeling systems (GCE, NU-WRF, and Goddard MMF or GMMF). Together these models are called the Goddard multi-scale modeling system with unified physics (Tao et al. 2009).

The objectives of this paper are to provide a review of developments, improvements, and applications of Goddard microphysics schemes. The Goddard multi-scale system with unified physics and the Goddard microphysics schemes will be described, respectively, in Sects. 2 and 3. The results will be presented in Sect. 4. Summary and future model developments will be presented in Sect. 5.

2 Multi-scale Modeling Systems with Unified Physics

Recently, a multi-scale modeling system with unified physics was developed at NASA Goddard. It consists of (1) the GCE model, a CRM; (2) the NU-WRF, a regional-scale model; and (3) the coupled GCM-GCE, the GCE coupled to a general circulation model (or known as the Goddard MMF or GMMF). The same cloud microphysical processes, long- and short-wave radiative transfer, and land surface processes are applied in all of the models to study precipitation processes, cloud radiation, and cloud surface (i.e., land information system or LIS) interactive processes in this multi-scale modeling system. This modeling system has been coupled with a multi-satellite simulator for comparison and validation with NASA high-resolution satellite data. Figure 3 shows the multi-scale modeling system with unified physics. The same GCE physics will also be utilized in the GMMF. The GCE model and NU-WRF share the same Goddard microphysical and radiative transfer processes (including the cloud interaction) as well as Land Information System (LIS). The same GCE physics is utilized in the GMMF. The idea behind having a multi-scale modeling system with unified physics is to be able to propagate improvements made to a physical process in one component into other components smoothly and efficiently (Tao et al. 2009, 2014). The following will provide descriptions of the GCE model, NU-WRF, Goddard GCM, and GMMF.

2.1 The Goddard Cumulus Ensemble Model (GCE)

The GCE has been developed and improved at NASA-GSFC over the past three and a half decades. A review on GCE model applications to better understand precipitation processes can be found in Simpson and Tao (1993) and Tao (2003). Its development and main features were published in Tao and Simpson (1993) and Tao et al. (2003, 2014). The three-dimension (3D) version of the GCE is typically run using 256×256

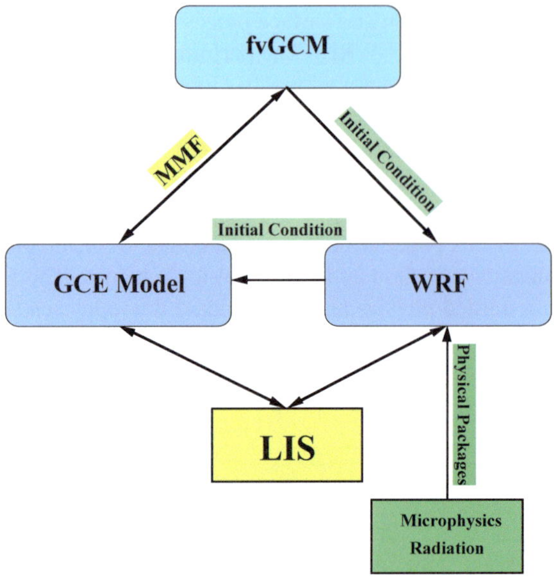

Fig. 3 Schematic diagram of the Goddard multi-scale modeling system with unified physics. The coupling between the fvGCM and GCE is two way, while the coupling between the fvGCM and WRF and WRF and the GCE is only one way. Land Information System (LIS) was developed in the Goddard Hydrological Sciences Branch and has been coupled interactively with both WRF and the GCE. Additionally, WRF has been enhanced by the addition of several of the GCE's physical packages (i.e., microphysical scheme with four different options and short- and long-wave radiative transfer processes with explicit cloud–radiation interactive processes). Adapted from Tao et al. (2014)

up to 4096 × 4096 horizontal grid points at 1-km resolution or better (i.e., 250 m). In typical multi-day to multi-week integrations, the model has performed reasonably well in terms of rainfall, latent heating (LH) profiles, and moisture budget structure compared to observations when driven with observed large-scale forcing derived from sounding networks.

The GCE model's advection scheme uses a multidimensional positive definite advection transport algorithm (Smolarkiewicz and Grabowski 1990). The positive definite advection scheme also produces more light precipitation, which is in better agreement with observations (Johnson et al. 2002). Solar and infrared radiative transfer processes (Chou and Suarez 1999; Chou et al. 1999) have been included (Tao et al. 1996). A sophisticated seven-layer soil/vegetation land process model has also been implemented into the GCE model (Lynn et al. 1998, 2001; Lynn and Tao 2001). Subgrid-scale (turbulent) processes in the GCE model are parameterized using a scheme based on Klemp and Wilhelmson (1978), and the effects of both dry and moist processes on the generation of subgrid-scale kinetic energy have been incorporated (Soong and Ogura 1980). Table 1 shows the major characteristics of the GCE model.

Table 1 Main characteristics of GCE model. Note that many other CRMs have similar physical processes but with different dynamic, vertical coordinate, and microphysics, radiative transfer, turbulence, and surface processes

Parameters/Processes	GCE Model
Dynamics	Anelastic or compressible 2D (slab- and axis-symmetric) and 3D
Vertical coordinate	Z
Microphysics	2-class water and 3-class ice 2-class water and 4-class ice Spectral bin microphysics 2M Morrison, 2M RAMS
Numerical methods	Positive definite advection for scalar variables; fourth order for dynamic variables
Initialization	Initial conditions with forcing from observations/large-scale models
FDDA	Nudging
Radiation	k-distribution and four-stream discrete-ordinate scattering (8 bands) Explicit cloud–radiation interaction
Subgrid diffusion	TKE (1.5 order)
Surface energy budget	Force-restore method Seven-layer soil model (PLACE) TOGA COARE flux module
Parallelization	MPI and MPI-IO

The GCE model has been used to understand the following (see Table 2 for more information on GCE model developments and applications):

- The role of the water and energy cycles in the tropical climate system,
- The redistribution of ozone and trace constituents by individual clouds and well-organized convective systems over various spatial scales,
- The relationship between the vertical distribution of latent heating (phase changes of water), surface rainfall, and the large-scale (pre-storm) environment,
- Climate hypotheses of deep convection related to global warming,
- Precipitation processes (i.e., precipitation efficiency),
- Aerosol impact on precipitation and rainfall in different environments,
- Impact of the surface process on precipitation and rainfall,
- The assumptions used in the representation of cloud and convective processes in climate and global circulation models, and
- The representation of cloud microphysical processes and their interaction with radiative forcing over tropical and midlatitude regions.

Recently, the GCE was adapted to interface with the single- (1M) and two-moment (2M) versions of Colorado State University's Regional Atmospheric Modeling System's (RAMS's) bulk microphysical scheme (Meyers et al. 1997; Saleeby and Cotton 2004), the Morrison 2M scheme (Morrison et al. 2005, 2009) as well as a spectral bin

Table 2 Main developments and applications of GCE model in the past 35 years

Topics	Model characteristics	Descriptions	References
Cloud ensemble model development	2D/3D	Model development and improvements Model descriptions	Soong and Tao (1980), Tao and Simpson (1993), Tao and Soong (1986), Tao et al. (2003, 2007, 2014), Tao and Moncrieff (2009, a review paper)
Cloud–cloud interactions/cloud merger	2D/3D	Cloud downdrafts and their associated cold outflows play a major role in cloud mergers	Tao and Simpson (1984, 1989)
Q_1 and Q_2 budget	2D/3D Warm rain and ice processes	Importance of evaporative cooling and vertical moisture transport by convection in Q_1 and Q_2 budget, respectively	Soong and Tao (1980), Tao and Soong (1986), Tao et al. (2003)
Cloud properties	2D/3D Warm rain	Active convective updrafts cover a small area but are major contributors for mass, Q_1 and Q_2 budgets Model results are in excellent agreement with aircraft observation	Tao and Soong (1986), Tao et al. (1987)
Convective momentum transport	2D/3D Small domain in 3D	Identified the role of horizontal pressure gradient force in upgradient transport of momentum	Soong and Tao (1984), Tao and Soong (1986), Tao et al. (1995)
Saturation for both liquid/ice	2D/3D	Developed an ice–liquid water saturation technique that has been used by many cloud-resolving models	Tao et al. (1989)
Ice processes	2D/3D	Importance of ice processes in stratiform rain formation as well as mass, Q_1 and Q_2 budgets	Tao and Simpson (1989), McCumber et al. (1991), Tao et al. (1993b), Ferrier et al. (1995), Lang et al. (2004)
Convective–stratiform interactions	2D/3D	The horizontal transport of hydrometeors and water vapor from convective towers into the stratiform region are quantified	Tao et al. (1993b, 1995), Sui et al. (1994), Lang et al. (2003), Zeng et al. (2013)

(continued)

Table 2 (continued)

Topics	Model characteristics	Descriptions	References
Cloud–radiation interactions/diurnal variations of precipitation	2D Short- and long-term integration	Long-wave cooling can enhance precipitation significantly for one tropical cloud system, but only slightly for a midlatitude system Modulation of relative humidity by radiative processes is a major reason for the diurnal variation of precipitation	Tao et al. (1993b, 1996), Sui et al. (1998)
Cloud–chemistry interactions	2D/3D	Significant redistribution of trace gases by convection. Enhancement of O3 production related to deep convection in the Tropics	Scala et al. (1990), Pickering et al. (1992a, b), Thompson et al. (1997, a review)
Air–sea interactions	2D/3D	TOGA COARE flux algorithm compared better with observations than other flux algorithms Surface fluxes are important for precipitation processes and maintaining CAPE and boundary layer structure	Wang et al. (1996, 2003)
Precipitation efficiency (PE)	2D	Examined different definitions of PE Identified key atmospheric parameters that determined PE	Ferrier et al. (1996), Tao et al. (2004)
Land processes	2D/3D	Importance of mesoscale circulations induced by soil gradient on precipitation. Identified atmospheric parameters that trigger convection	Lynn et al. (1998, 2001), Lynn and Tao (2001), Baker et al. (2001)
Idealized climate variations in the tropics	2D Long-term integration	Examined several hypotheses associated with climate variations and global warming Identified physical processes that caused two different statistical equilibrium states (warm/humid and cold/dry) in idealized climate	Sui et al. (1994), Lau et al. (1993, 1994), Tao et al. (1999, 2001), Shie et al. (2003)

(continued)

Table 2 (continued)

Topics	Model characteristics	Descriptions	References
TRMM rainfall retrieval	3D	Improved performance of TRMM rainfall retrieval algorithms by providing realistic cloud profiles	Simpson et al. (1996, a review)
Latent heating profiles retrieval	2D	Developed algorithms for retrieving four-dimensional structure of latent heating profiles over global tropics	Tao et al. (1990, 1993a, 2001, 2006, 2010, 2016b review papers), Lang and Tao (2018)
Microphysics developments	2D/3D	Developing, improving, and evaluating Goddard microphysical schemes to better simulating tropical and midlatitude precipitation systems	Ferrier et al. (1995), Lang et al. (2007, 2011, 2014), Li et al. (2009a, b, 2010), Tao et al. (2016a)
Cloud–aerosol interactions	2D	Identified the evaporative cooling and its associated cold pool are the key process for enhancing or suppressing rainfall in three different precipitating systems	Tao et al. (2007 and 2012—a review paper), Zeng et al. (2009a, b, 2010), Li et al. (2013), Tao and Li (2016)
Multi-scale Modeling Framework (MMF)	3D	Developing a coupled global- and cloud-resolving modeling system to study cloud and precipitation processes	Tao et al. (2009), Chern et al. (2016), Mohr et al. (2013), Matsui et al. (2015) Kidd et al. (2016), Tao and Chern (2017)

microphysics (SBM) scheme (Khain et al. 2004; Tao et al. 2007; Li et al. 2009a, b). The GCE's own 1M bulk microphysics, especially, ice processes have been significantly improved starting with the reduction of excessive graupel (Lang et al. 2007) and unrealistically high dBZs aloft (Lang et al. 2011) and culminating in the new 4ICE scheme (Lang et al. 2014; Tao et al. 2016b) capable of simulating a wide range of precipitation systems better than previous generations of the Goddard bulk microphysics. These schemes will be described in Sect. 3.

2.2 The NASA-Unified Weather Research and Forecasting Model (NU-WRF)

The NASA-Unified Weather Research and Forecasting (NU-WRF) model combines the capabilities of the Advanced Research WRF (ARW, Michalakes et al. 2001) with various modules developed at NASA-GSFC: the Land Information System (LIS, Kumar et al. 2006), the Goddard Chemistry Aerosol Radiation and Transport (GOCART) model (Chin et al. 2000), the Goddard microphysics (Lang et al. 2014; Tao et al. 2016b) and Goddard radiation (Chou and Suarez 1999; Chou et al. 2001; Matsui and Jacob 2014), and Ensemble Data Assimilation (EDA) system (Zhang et al. 2017). In addition to traditional reanalysis and global forecasting data, NU-WRF supports high-resolution initial and boundary conditions from the Modern-Era Retrospective analysis for Research and Applications 2 (MERRA2, Bosilovich et al. 2015), LIS land surface model (LSM) spin-up, and various aerosol emissions databases, including dynamic 1-km dust erosion maps (Kim et al. 2017). These packages enable fully coupled aerosol–cloud–precipitation–land surface simulations and satellite-based model evaluation at satellite-resolvable scales (Peters-Lidard et al. 2015). Figure 4 shows the physical processes and applications of NU-WRF.

NU-WRF has been used to provide real-time forecasts for GPM field campaigns [i.e., Midlatitude Continental Convective Clouds Experiment (MC3E), Iowa Flood Studies (IFloodS), and Integrated Precipitation and Hydrology Experiment (IPHEX)]. It has also been used to simulate a variety of precipitation systems, e.g., Canadian CloudSat/CALIPSO Validation Project (C3VP) (Shi et al. 2010; Iguchi et al. 2012a), Light Precipitation Validation Experiment (LPVEx) (Iguchi et al. 2014), NASA African Monsoon Multidisciplinary Analyses (NAMMA) (Shi et al. 2014), MC3E (Iguchi et al. 2012b; Tao et al. 2013, 2016b), and IFloodS (Wu et al. 2016).

2.3 Goddard Finite-Volume GCM (fvGCM)

The fvGCM has been constructed by combining the finite-volume dynamic core developed at Goddard (Lin 2004) with the physics package of the NCAR Community Climate Model CCM3, which represents a well-balanced set of processes

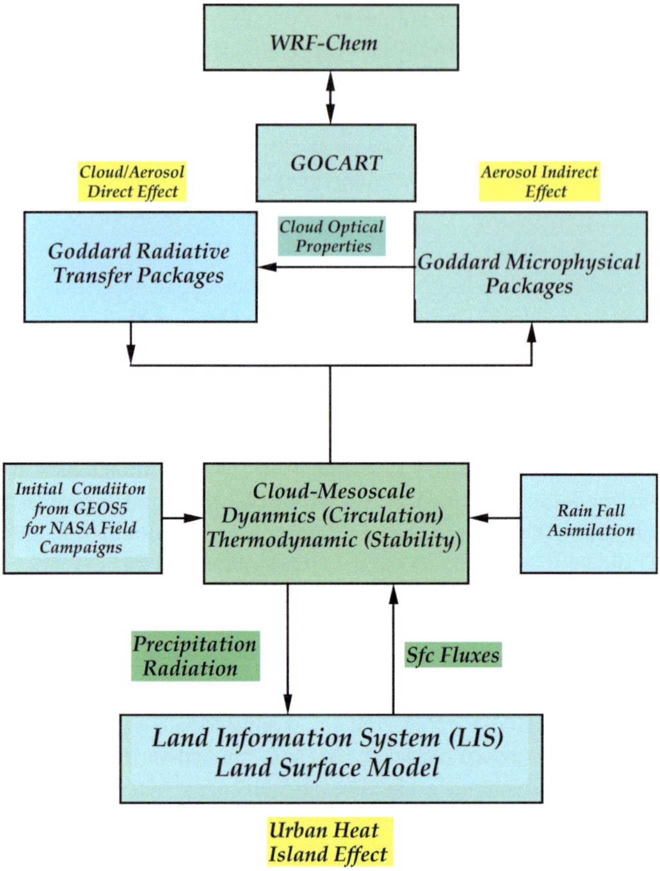

Fig. 4 Schematic diagram showing main characteristics (physical processes) and applications of NASA-Unified WRF (NU-WRF)

with a long history of development and documentation (Kiehl et al. 1998). The unique features of the finite-volume dynamical core include an accurate conservative Flux-Form Semi-Lagrangian transport algorithm (FFSL) with a monotonicity constraint on subgrid distributions that are free of Gibbs oscillation (Lin and Rood 1996, 1997), a physically consistent integration of the pressure gradient force for a terrain-following Lagrangian control-volume vertical coordinate (Lin 1997), and a mass-, momentum-, and total-energy-conserving vertical remapping algorithm. The physical parameterizations of the fvGCM have been upgraded by incorporating the gravity-wave drag scheme of the NCAR Whole Atmosphere Community Climate Model (WACCM) and the Community Land Model version 2 (CLM-2; Bonan et al. 2002).

2.4 Goddard Multi-scale Modeling Framework (GMMF)

The Goddard MMF is based on the Goddard fvGCM and the GCE model. The fvGCM provides global coverage while the GCE allows for the explicit simulation of cloud and microphysical processes, providing simulated cloud property profiles to the fvGCM for radiation calculations.

Goddard MMF typically is conducted using fvGCM with 2.5° × 2° horizontal grid spacing with 32 layers from the surface to 0.4 hPa, and the two-dimensional (2D) GCE using 32 horizontal grids (oriented east–west) and 32 levels with 4-km horizontal grid spacing and cyclic lateral boundaries.[1] The time step for the GCE is 10 s, and the fvGCM–GCE coupling interval is 1 h (which is the fvGCM physical time step). Because the vertical coordinate of the fvGCM (a terrain-following coordinate) is different from that of the GCE [a height (z) coordinate], vertical interpolations are needed in the coupling interface. An interpolation scheme, based on a finite-volume Piecewise Parabolic Mapping (PPM) algorithm, has been developed to conserve mass, momentum, and moist static energy between the two coordinates. The coupling between fvGCM and GCE is shown in Fig. 5.

2.5 Goddard Satellite Data Simulator Unit (G-SDSU)

Modern multi-sensor satellite observations provide a more complete view of land, cloud, precipitation, and aerosols processes from space; meanwhile, it is becoming a challenge for remote sensing and modeling communities to harness these observations simultaneously due to inconsistent physics assumptions and spatial scales between satellite retrievals and CRM physics. To this end, a unified system of multi-sensor simulators, the G-SDSU, has been developed through multi-institutional collaborations (Matsui et al. 2013, 2014). The G-SDSU is an end-to-end satellite simulator, which computes satellite-consistent level (L1) measurements (e.g., radiance/brightness temperature or backscatter), from outputs of GCE, NU-WRF, and GMMF simulations through radiative transfer, antenna gain patterns, and satellite orbit/scan simulators for passive microwave and visible-IR sensors, radar, lidar, and broadband and hyperspectral sensors (see Fig. 6). In addition to the satellite sensors, a polarimetric radar simulator was recently included for supporting ground-based polarimetric weather radars (Matsui et al. 2017).

All radiative transfer modules consistently treat the CRM's microphysics assumptions (phase, size, and effective density) to calculate single-scattering properties and backscatter/radiance. G-SDSU-simulated L1 signals can be directly compared with the satellite-observed L1 signals; therefore, G-SDSU bridges model and satellite remote sensing through the following paths: (i) radiance-based model evaluation and development (Matsui et al. 2009; Li et al. 2010), (ii) an operator of radiance-based data assimilation system (Zhang et al. 2017), and (iii) development of synthetic

[1]Please see Table 1 in Tao and Chern (2017) for other MMFs' configurations.

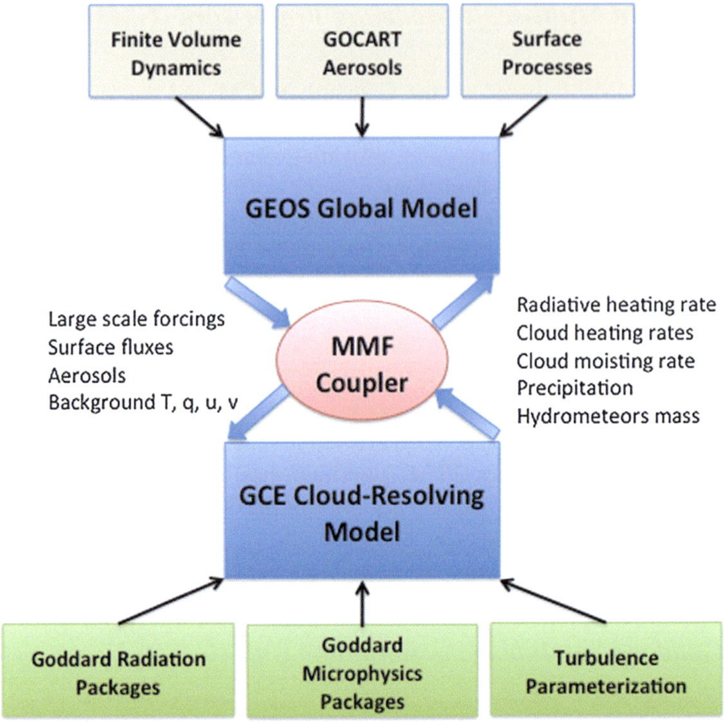

Fig. 5 Schematic diagram showing the coupled GEOS and GCE model (Goddard MMF or GMMF). The physical processes exchanged between the two modeling systems is also shown

satellite observations for future satellite missions (Matsui et al. 2014) and retrieval algorithm database (Kidd et al. 2016).

3 Microphysics

3.1 Goddard One-Moment (1M) 3-Class Ice Scheme (3ICE)

The Goddard Cumulus Ensemble (GCE) model's 1M bulk microphysical scheme is mainly based on Lin et al. (1983) with additional processes from Rutledge and Hobbs (1984). However, the Goddard microphysics scheme, known as 3ICE, contains several configuration options. One option allows the user to choose either graupel or hail as the third class of ice (McCumber et al. 1991). Graupel has a relatively low density and a high intercept value (i.e., more numerous small particles). In contrast, hail has a relatively high density and a low intercept value (i.e., more numerous large particles). These differences can affect not only the description of the hydrometeor population and formation of the anvil–stratiform region, but also the relative importance of the

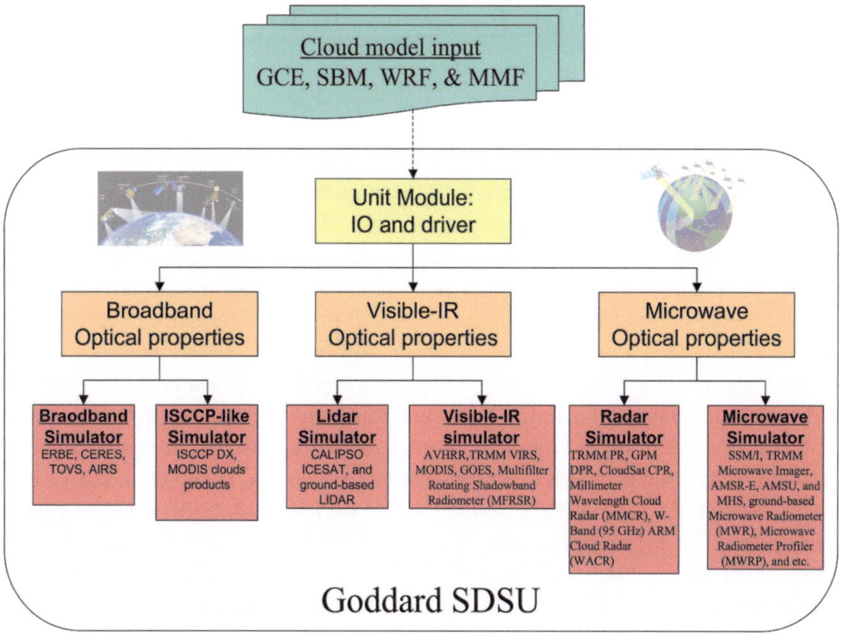

Fig. 6 Schematic diagram showing the major characteristics of the G-SDSU and their respective roles in satellite missions. Adapted from Tao et al. (2014)

microphysical–dynamical–radiative processes. The Goddard microphysics scheme has been modified to reduce unrealistic cloud water in the stratiform region (Tao and Simpson (1993)). In addition, the GCE model's own microphysics—especially its ice processes—has been improved to reduce unrealistically high dBZs aloft (Lang et al. 2007, 2011). Also, ice crystal concentration can be introduced as an independent factor into the scheme to increase the cloud ice (or anvil) significantly so that the modeled clouds are close to observations (e.g., Zeng et al. 2009a, b, 2010).

The GCE's Rutledge and Hobbs (1983, 1984)-based 3-class ice scheme was further improved via the following processes: a snow size/density mapping was added, whereby snow densities are increased with decreasing size in a more realistic fashion, a rain evaporation correction based on bin model results was used to remap the rain sizes below cloud base where the fixed intercept in the bulk scheme leads to excessively small rain sizes at smaller rain mixing ratios, and hence excessive evaporation, the saturation adjustment scheme was further modified to allow for cloud ice to persist in subsaturated conditions and for a small amount of ice supersaturation to exist even at extremely cold temperatures, a simple two-tier graupel density scheme was added, wherein graupel densities are increased to a higher value at higher mixing ratios to mimic the effect of prolonged riming, and finally the snow/graupel size mapping scheme was adjusted to reduce particle sizes for higher mixing ratios as the addition of hail reduced the need for snow/graupel to produce peak reflectivity values aloft.

3.2 Goddard 1M 4-Class Ice Scheme (4ICE)

Almost all microphysics schemes are 3ICE (cloud ice, snow, and graupel). Very few 3ICE schemes have the option to alternatively have hail processes (cloud ice, snow, graupel, or hail); see Table 1 in Tao et al. (2016b) for key papers using high-resolution numerical cloud models with these microphysics schemes to study the impact of microphysics schemes on precipitation. Hail and graupel can occur in real weather events simultaneously; therefore, a 4ICE scheme (cloud ice, snow, graupel, and hail) is required for real-time forecasts (especially, for high-resolution prediction of severe local thunderstorms, midlatitude squall lines, and tornadoes). In addition, current and future global high-resolution cloud-resolving models need the ability to predict/simulate a variety of weather systems from weak to intense (i.e., tropical cyclones, thunderstorms) over the globe, and this requires the use of a 4ICE scheme.

A new 1M 4-class ice (cloud ice, snow, graupel, and frozen drops/hail) microphysics scheme (4ICE) was recently developed for the GCE (Lang et al. 2014). Hail processes from the GCE's 3ICE scheme based on Lin et al. (1983) were added to the improved 3ICE graupel scheme (Lang et al. 2007, 2011) and further refined to create a 1M 4ICE scheme (Lang et al. 2014) capable of simulating a wide range of convective systems, from weak to intense. Its key features include no dry collection of ice species by hail, resulting in realistic narrow hail cores as well as peak echoes that monotonically decrease with height, a spectral bin microphysics (SBM) based rain evaporation correction (Li et al. 2009a, b), which reduces excessive evaporation and upshear-tilted convective cores, and a refined snow size (and density) mapping scheme with an enhanced aggregation effect. The scheme was further modified to include the effects of snow break up by graupel/hail while further improving the snow aggregation effect, resulting in improved radar structures (i.e., a transition region and more horizontally stratified stratiform features), and a simple hail size mapping, which eliminates the need to select the hail intercept a priori (Tao et al. 2016b). The Goddard 4ICE scheme was first validated in the GCE where it outperformed 3ICE graupel schemes in terms of peak intensity and overall echo distributions versus observations for both moderate and intense convection (Lang et al. 2014). It was then added to NU-WRF with the additional modifications where it was similarly shown to be superior to 3ICE schemes, including a hail scheme (Tao et al. 2016b). The scheme was also tested in 2-year-long global GMMF simulations, where it produced improved global cloud ice distributions (Chern et al. 2016) and reflectivity/TB relations over land and ocean (Matsui et al. 2015).

Table 3 shows the evolution of the Goddard microphysics during the last three decades. The performance of Goddard 3ICE and 4ICE microphysics schemes will be discussed in Sect. 4.

Table 3 Evolution of Goddard microphysical scheme in the past 30 years. *Psfi is Bergeron processes for snow

Major characteristics	References
Implemented/coded a 3ICE based on Lin et al. (1983) with modifications	Tao and Simpson (1989, 1993)
Saturation adjustment Adapted by many other 3ICE schemes	Tao et al. (1987)
Implemented/coded Rutledge and Hobbs (1984) 3ICE with hail or 3ICE with graupel Available in MM5 and ARP	McCumber et al. (1991)
Modified saturation adjustment and reduced cloud ice near melting layer in the stratiform region	Tao et al. (2003)
Reduced excessive graupel (eliminated dry collection of snow/ice by graupel), reduced excessive snow by reducing snow collection efficiency of cloud water Available in NCAR WRF	Lang et al. (2007)
Added snow and graupel mappings (both are functions of temperature and mixing ratio, smaller sizes at colder temperatures), corrected the *Psfi term by accounting for RH and cloud ice size, added Hallett–Mossop rime splintering, allowed for ice supersaturation, added contact nucleation and immersion freezing, replaced Fletcher curve with Meyers curve for IN, added cloud ice fall speeds, and allowed for graupel/snow sublimation	Lang et al. (2011)
Added frozen drops/hail category, added bin-based rain evaporation correction (made physical by adjusting raindrop sizes), and many others	Lang et al. (2014)
Added hail mapping, ice supersaturation (w), and snow breakup, switched from Meyers curve (f(SSI)) to Cooper curve (f(T) Available in NU-WRF and GMMF	Tao et al. (2016b)

3.3 *Spectral Bin Microphysics (SBM)*[2]

The SBM scheme in the Hebrew University Cloud Model (HUCM; Khain et al. 2011, 2012) was implemented into GCE model to study the impact of aerosols on deep convective precipitation systems (Tao et al. 2007). The SBM includes the following processes: (1) nucleation of droplets and ice particles (Pruppacher and Klett 1997; Meyers et al. 1992), (2) immersion freezing (Bigg 1953), (3) contact freezing (Meyers et al. 1992), (4) ice multiplication (Hallett and Mossop 1974; Mossop and Hallett 1974), (5) detailed melting (Khain et al. 2004), (6) condensation/evaporation of liquid drops (Pruppacher and Klett 1997; Khain et al. 2000), (7) deposition/sublimation of ice particles (Pruppacher and Klett 1997; Khain et al. 2000), (8) drop/drop, drop/ice,

[2]The same SBM has been implemented into the GCE model (Tao et al. 2007; Li et al. 2009a, b and 2010).

and ice/ice collision/coalescence (Pruppacher and Klett 1997; Pinsky et al. 2001), (9) turbulence effects on liquid drop collisions (Pinsky et al. 2000), and (10) collisional breakup (Seifert and Beheng 2001; Seifert et al. 2005). In the first process, ice nucleation includes both condensation–freezing and homogeneous nucleation. The Meyers formula is applied in a semi-Lagrangian approach (see Khain et al. 2000), and the concentration of newly nucleated ice crystals at each time step is calculated by the increase in the value of supersaturation. Sedimentation of liquid and ice particles is also considered. SBMs are specially designed to take into account the effect of atmospheric aerosols on cloud development and precipitation formation. The activation of aerosols in each size bin is explicitly calculated in this scheme (Khain et al. 2000). This added level of sophistication will improve our understanding of microphysical processes and positively influence the development of TRMM and GPM rain/snowfall retrieval algorithms. Note that bulk schemes can also represent aerosol impacts on cloud processes [see a review on aerosol impact on convective precipitation processes by Tao et al. (2012)].

The bulk microphysics drop size distribution (DSD) assumptions have been evaluated against explicitly simulated DSDs using the SBM scheme, because SBM simulations yield much more realistic radar echo profiles than the bulk one-moment microphysics (e.g., Fig. 6, Li et al. 2009a). The results suggest that SBM-simulated DSDs are more dependent on temperature than mass mixing ratio. In addition, the numerical results are in good agreement with observations, indicating the microstructure of clouds depends strongly on cloud–aerosol interactions (Tao et al. 2007; Li et al. 2009a, b, 2010).

3.4 The Morrison 2-Moment (2M) Microphysics Scheme

The Morrison 2M microphysical scheme (Morrison et al. 2005, 2009) predicts number concentrations and mass mixing ratios of five hydrometeor types (cloud droplets, ice crystals, raindrops, snow particles, and graupel particles). The particle size distributions are assumed to be gamma distributions. The precipitating hydrometeor types (rain, snow, and graupel) are fully prognostic in the CRM. Droplet activation is calculated at each CRM grid cell based on the parameterization of Abdul-Razzak and Ghan (2000) and related to the resolved vertical velocity and the turbulent kinetic energy with a minimum vertical velocity of 0.1 m s^{-1}. A resolved droplet activation scheme with prognostic aerosol (serving as CCN) activation and advection is also included in the Morrison scheme and tested using both GCE and WRF models (e.g., Li et al. 2017; Fridlind et al. 2017). The scheme also includes prognostic equations for both the mass mixing ratio and number concentration for ice nucleation (both homogeneous and heterogeneous), ice multiplication, auto-conversion of droplets to form rain and ice to form snow, accretion of droplets by rain and snow, of rain by snow, and of ice by snow, freezing, melting, self-collection, condensation/deposition, and evaporation/sublimation.

The Morrison microphysics scheme has the option to choose the third precipitating ice type, either graupel or hail, similar to the Goddard 3ICE microphysics scheme, which also has the option to choose graupel or hail (McCumber et al. 1991). In addition, the Goddard scheme has another option to use just two classes of ice (cloud ice and snow or cloud ice and graupel) (Shi et al. 2010).

3.5 The Colorado State University Regional Atmospheric Modeling System

The Colorado State University Regional Atmospheric Modeling System (RAMS) two-moment (2M) bulk cloud microphysical scheme (Meyers et al. 1997; Cotton et al. 2003; Saleeby and Cotton 2004, 2008; Lee et al. 2009) has been implemented in the GCE model. The RAMS 2M scheme assumes gamma particle size distributions for three species of liquid (small and large cloud droplets and rain) and five species of ice (small and large vapor-grown crystals, aggregates, graupel, and hail). Consistent with observations of bimodal cloud droplet size distributions, the cloud droplet spectrum is decomposed into two modes; one for droplets up to about 50 μm in diameter and the second for droplets 50 to roughly 100 μm in diameter. Ice crystal habit is allowed to vary as a function of temperature and humidity. The scheme accounts for mass and number changes of each hydrometeor specie owing to cloud and ice nucleation, vapor diffusion, evaporation, droplet self-collection (autoconversion), collision–coalescence, freezing, melting, sedimentation, and secondary ice production. In addition, the RAMS bulk aerosol module (Saleeby and van den Heever 2013) is incorporated for explicit simulation of cloud droplet activation from sub- and super-micron sulfate, sea salt and dust aerosols, wet/dry deposition, and aerosol regeneration upon hydrometeor evaporation. Ice nucleation follows either Meyers et al. (1992) or Demott et al. (2010) based on user specification. A species-dependent soluble fraction parameter, ε, accounts for aerosol hygroscopicity in a manner analogous to the κ-parameter (Petters and Kreidenweis 2007, 2008; Sullivan et al. 2009). User-specified aerosol profiles are initialized horizontally homogeneously within the model domain, although aerosol mass is tracked within hydrometeors, and non-activated aerosols are advected with the model-predicted flow fields. Collection is simulated using stochastic collection equation solutions, facilitated by bin-emulating look-up tables that incorporate size-dependent collection kernels for liquid species, rather than by continuous accretion approximations. The philosophy of bin representation for collection is extended to calculations of hydrometeor sedimentation (Feingold et al. 1998; Loftus et al. 2014) and riming (Saleeby and Cotton 2008) melting, and shedding (hail only) (Meyers et al. 1997). The scheme also includes explicit prediction of supersaturation, a critical consideration for conducting aerosol–cloud interaction studies, and has the ability to keep track of microphysical budgets. The Goddard radiation scheme (Chou and Suarez 1999; Chou et al. 2001) fully interacts with all eight hydrometeor species and accounts for changes in particle size distributions as cloud systems develop and evolve.

4 Results

4.1 GCE Model Results

The GCE model has been extensively used for two types of cloud and precipitation process studies. The first uses large-scale forcing in temperature and water vapor derived from a sounding network to drive the model. In this mode, the model is typically integrated for multiple weeks to sample and obtain cloud statistics for many systems. The GCE model simulated apparent heating (Q_1) and apparent moisture sink (Q_2), as defined in Yanai et al. (1973), and surface rainfall are typically in excellent agreement with the observed (sounding derived) values. Results using this approach have been used in support of both the TRMM and GPM missions (e.g., Tao et al. 2001, 2010; Lang and Tao 2018). The second type of GCE study is the case study, which is used to study cloud processes as well as cloud–radiation, cloud–aerosol, and cloud–surface interactions for specific cloud systems, and the GCE model is usually only integrated for a short term (12–24 h). In the following subsection, the performance of the Goddard microphysics for these two types of simulations is presented.

4.1.1 Long-Term Integration and Diurnal Variation

The diurnal variation of tropical oceanic convection is one of the most important components in tropical variability and plays a crucial role in regulating tropical hydrological and energy cycles. A successful simulation of the diurnal variability of the hydrologic cycle and radiative energy budget provides a robust test of physical processes represented in atmospheric models (e.g., Slingo et al. 1987; Randall et al. 1991; Lin et al. 2000). Simulations were performed to examine the sensitivity to cloud microphysics (3ICE versus 4ICE) using the 2D GCE model with 512 horizontal grids at 1-km spacing and 43 vertical levels. Large-scale advective tendencies for dry static energy and moisture were obtained from the sounding network during the Dynamics of MJO (Madden–Julian Oscillation) (DYNAMO) over Gan Island (Yoneyama et al. 2013) to specify the 3-hourly forcing during the model integration. The surface fluxes were specified uniformly over the model domain.

Figure 7 shows the observed and the GCE-simulated precipitation rate for the DYNAMO case. Overall, both 3ICE and 4ICE cases captured the observed active period of deep convection associated with the MJO event during 17–30 November, and the suppression periods before and after the MJO event. The result also shows that both 3ICE and 4ICE cases simulate very similar temporal variation and rainfall intensity. This feature is better illustrated with the scatter plot of precipitation rate between the observations and the two GCE runs with the 3ICE and the 4ICE microphysics schemes (Fig. 8). The correlation with observed precipitation rate is as high as 0.92 in both simulations, suggesting that the 4ICE scheme can be successfully applied for tropical oceanic convective events that rarely contain hail.

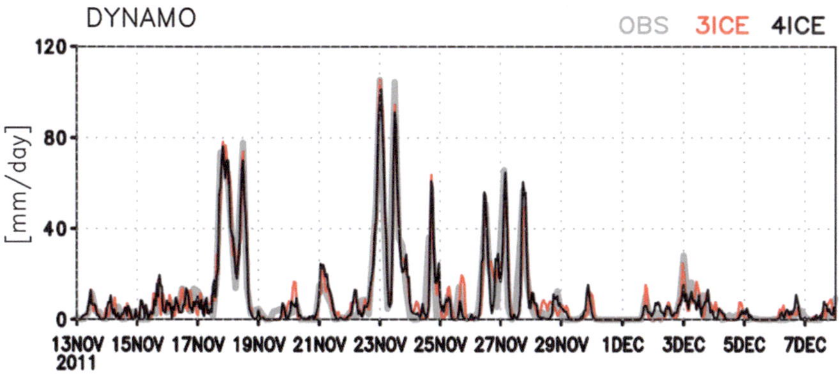

Fig. 7 Time series of observed precipitation rate (gray) and two GCE runs with the 3ICE (red) and the 4ICE (black) microphysics during DYNAMO. The unit is mm day^{-1}

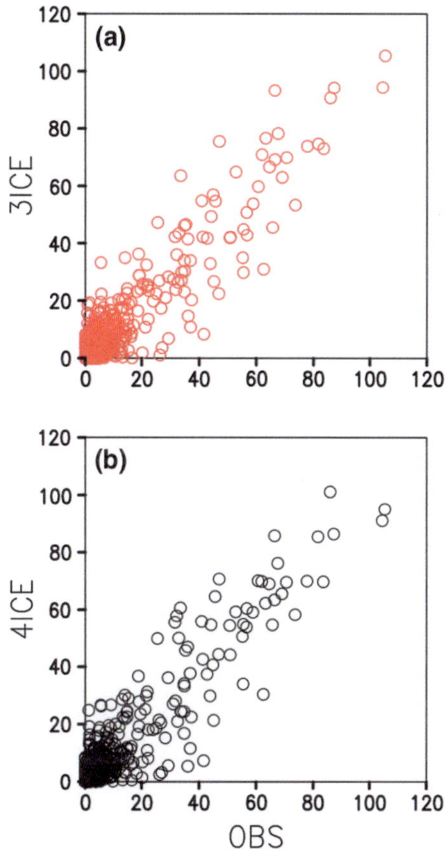

Fig. 8 Scatter plots of precipitation rate between observation and the two GCE runs with the 3ICE (**a**) and the 4ICE (**b**) microphysics. The unit is mm day^{-1}

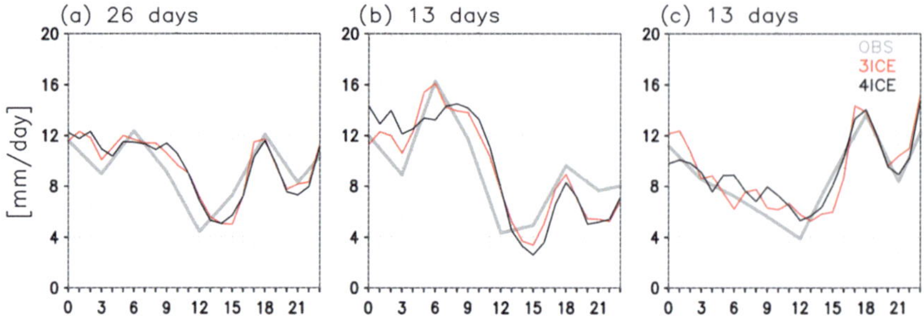

Fig. 9 Diurnal variation of precipitation rate averaged for **a** all days during, **b** the days with morning peak (0000–1100 LST), and **c** the days with evening peak (1200–2300 LST). Observed variations (gray) are compared and the two GCE runs with the 3ICE (red) and the 4ICE (black) microphysics. The unit is mm day^{-1}

The sounding estimated and GCE-simulated diurnal variation of precipitation is also examined. Figure 9a shows the observed and simulated diurnal cycle of precipitation, where the hourly rainfall data were averaged each hour. The observation shows double peaks, one from midnight to morning (0000–0900 LST) and the other in late afternoon (~1800 LST). The nighttime peak is distributed more widely in time, whereas the afternoon peak is centered at 1800 LST. Overall, the nighttime precipitation is more dominant rather than the daytime, consistent with the previous observational studies (i.e., Kraus 1963; Gray and Jacobsen 1977; Randall et al. 1991; Sui et al. 1997).

Both 3ICE and 4ICE cases capture the observed diurnal cycle of rainfall fairy well, with the double peaks in the nighttime and late afternoon. There is discrepancy at the timing of minimum precipitation rate between observation (~1200 LST) and the model simulations (~1500 LST), but this may be partly caused by less frequent data sampling of observed precipitation. This study further separated the cases of nighttime and the late afternoon rainfall (Fig. 9b, c). Both GCE cases also reproduce the observed variation quite well.

The prevalence of nocturnal rainfall in tropical oceans has been suggested by the cloud–radiation interaction mechanism that emphasizes the dominant role of convective clouds in the nighttime through enhanced IR cooling (Kraus 1963; Randall et al. 1991) and a large-scale dynamic response to the radiational differences between cloudy and clear regions (Gray and Jacobson 1977). The simulated cloud–radiation feedback can be affected by the implemented microphysics schemes, particularly by the differences in vertical distribution of ice clouds. Both GCE cases show almost no sensitivity in the simulated diurnal cycle of precipitation (c.f. Fig. 9a). However, the simulation difference becomes relatively larger in the nighttime, suggesting the connection between the ice clouds and long-wave radiation. This aspect should be tested more rigorously for the GCE model.

4.1.2 Short-Term Integration—Case Studies

The Goddard 4ICE scheme was used to simulate an intense continental squall line observed on May 20, 2011 during the Midlatitude Continental Convective Clouds Experiment (MC3E) to evaluate its ability to simulate intense convection with significant hail. Additionally, a loosely organized transient line of moderate convection from TRMM Large-Scale Biosphere–Atmosphere Experiment in Amazonia (LBA, February 23, 1999) was also simulated to ensure the scheme does not overpredict less intense convection. The MC3E May 20, 2011 case featured an intense squall line that formed over central Oklahoma as a deep, upper-level low over the central Great Basin moved through the central and southern Rockies before lifting into the central and northern plains. 3D GCE model simulations were conducted using 1-km horizontal grid spacing and a 256-km × 256-km horizontal domain with a stretched vertical grid having 70 levels and a top near 23 km. The LBA case was characteristic of the widespread, weaker monsoon-like convection observed within the westerly wind regime during TRMM LBA. For the LBA case, a horizontal domain of 128 km × 128 km with 200 m horizontal grid resolution was used with 70 stretched vertical grids with a top near 23 km. For each case, seven numerical experiments were conducted. Three experiments were made using previous versions of the 3ICE-graupel scheme: the original Tao et al. (2003, named 3ice0), Lang et al. (2007, named 3ice1), and Lang et al. (2011, named 3ice3). Four variations of the new 4ICE scheme were tested with smaller, medium-, and larger sized hail with a bin rain evaporation correction (4iceb sml, 4iceb med, and 4iceb lrg, respectively) and smaller sized hail without the evaporation correction (4ice sml). Smaller, medium-, and larger sized hail use fixed hail distribution intercepts of 0.020, 0.0020, and 0.0002 cm^{-4}, respectively.

Figure 10 shows vertical profiles of maximum radar reflectivity for the MC3E and TRMM LBA cases. For the MC3E case, all three 3ICE simulations have a pronounced low bias that ranges from about 5 dBZ below the freezing level to as much as 15 dBZ above the freezing level (Fig. 10a). The 4ICE simulations show a marked improvement in the bias at almost all levels except for 4iceb lrg, which produces excessively large reflectivities (~15 dBZ) near the melting level. The medium hail profile has the smallest overall bias and agrees best with the observed. Though not quite as good, the smaller hail runs are significantly improved over the 3ICE with a consistent low bias of just 5 dBZ at all levels. Peak reflectivity profiles from the LBA case are shown in Fig. 10b; the 4ICE simulations with smaller hail clearly perform the best and show almost no bias (less than ~4 dBZ) through nearly the entire depth of the storm. Remarkably, none of the 4ICE runs produced the over bias evident in runs 3ice0 and 3ice1 in the top part of the storm; furthermore, all of the 4ICE runs produced monotonically decreasing profiles with height in agreement with the observations. However, the medium to larger hail sizes in runs 4iceb med and 4iceb lrg are obviously too large, producing high biases of up to ~10–15 dBZ around the melting level. These results suggest that the new 4ICE scheme is quite capable of responding appropriately to the intensity of the convective environment and can outperform the 3ICE-graupel scheme in terms of peak reflectivities even in a moderate intensity environment when the hail parameters are selected properly.

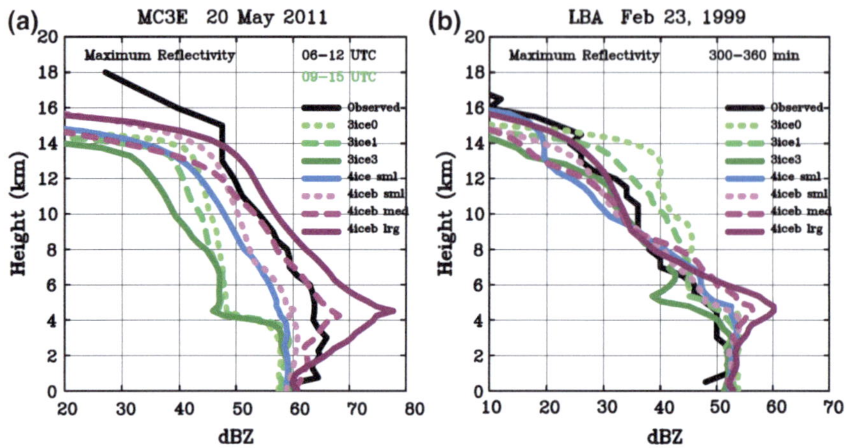

Fig. 10 Vertical profiles of maximum radar reflectivity for the 20 May 2011 MC3E case extracted between 06 and 12 UTC from Doppler radar observations and between 09 and 15 UTC from the three Goddard 3ICE simulations and four Goddard 4ICE simulations (**a**, see Lang et al. 2014 for details). **b** Shows similar results for the moderate convective intensity 23 February 1999 TRMM LBA case over Amazonia. Adapted from Lang et al. (2014)

In addition, contoured frequency with altitude diagrams (CFADS, Yuter and Houze 1995) are used to evaluate the overall reflectivity distributions. For the 20 May MC3E squall line case (Fig. 11a), the observed CFAD has high frequencies of 0–40 dBZ below the melting level, whereas aloft, a coherent core of even higher echo probabilities (i.e., more concentrated dBZ values) increases from ~10 dBZ near 200 mb to ~25 dBZ just above the freezing level, a signature of increased particle size due mainly to aggregation in the stratiform region. Infrequent but much more intense echoes associated with the convective cores extend out to near ~65 dBZ above and below the freezing level, 50 dBZ at 12 km and 40 dBZ at 16 km. The improved 3ICE graupel scheme (3ice3, Fig. 11b) cannot reproduce the strong reflectivities over 50 dBZ above the freezing level, and while there is some evidence of an aggregation effect, the highest probabilities (i.e., most abundant echoes) occur at dBZ values that are too weak compared to the observations. In contrast, the 4ICE simulations (Fig. 11c, represented by the medium hail run) can much more realistically capture the infrequent but intense echoes that arise from hail and, although the proportion of weak echoes (i.e., below 10 dBZ) is still too high, have better distributions of weaker echoes with a better aggregation signature relative to 3ice3. CFADs for the weaker, less-organized February 23, 1999 TRMM LBA case (Fig. 11d) show that the 4ICE scheme even with medium-sized hail can match the performance of the improved 3ICE graupel scheme (3ice3) for weak echoes. The 4ICE scheme with medium hail is comparable in its ability to replicate the infrequent but more moderate echoes, tending to be slightly too strong versus slightly too weak, but it does eliminate the tendency of the graupel scheme to produce elevated reflectivity maxima above the freezing level. The 4ICE scheme with smaller hail performed the best overall for

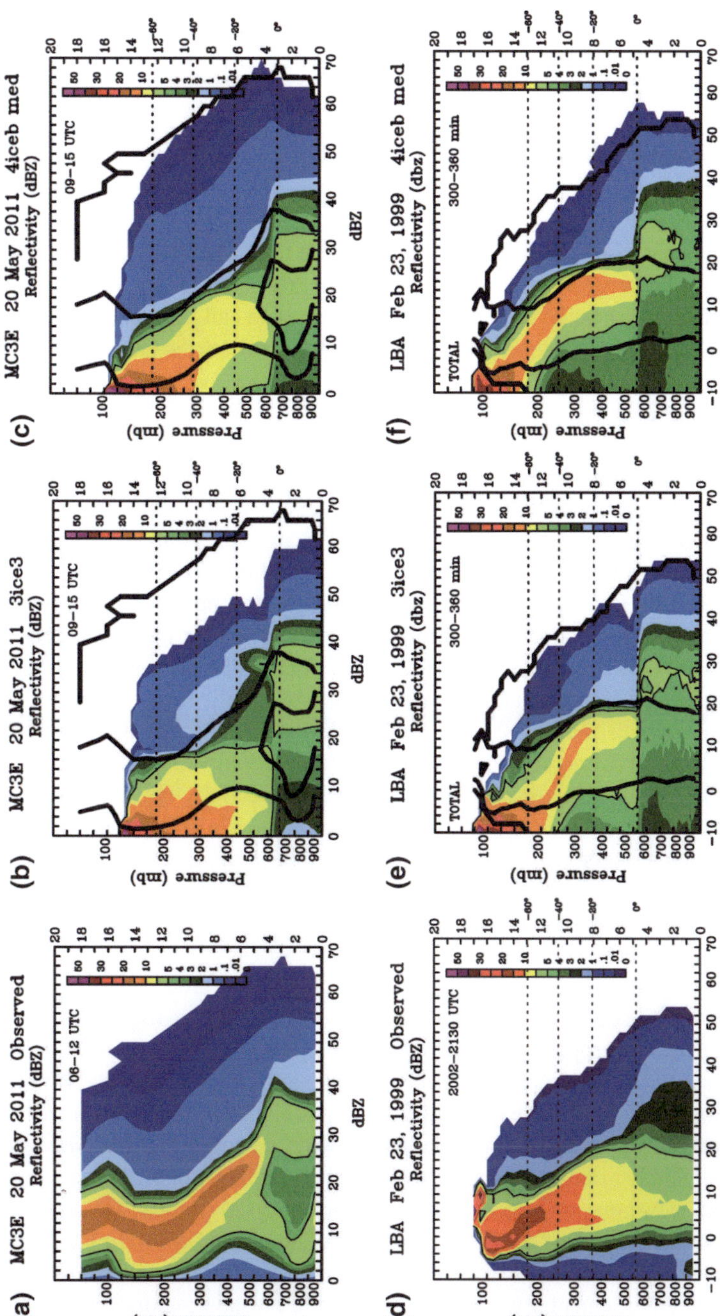
(a) MC3E 20 May 2011 Observed
Reflectivity (dBZ)
06–12 UTC
(b) MC3E 20 May 2011 3ice3
Reflectivity (dBZ)
09–15 UTC
(c) MC3E 20 May 2011 4iceb med
Reflectivity (dBZ)
09–15 UTC
(d) LBA Feb 23, 1999 Observed
Reflectivity (dBZ)
2002–2130 UTC
(e) LBA Feb 23, 1999 3ice3
Reflectivity (dbz)
300–360 min
TOTAL
(f) LBA Feb 23, 1999 4iceb med
Reflectivity (dbz)
300–360 min
TOTAL
Pressure (mb)
dBZ
(dBZ)

◀**Fig. 11** Radar reflectivity CFADs for the 20 May 2011 MC3E case constructed from **a** NEXRAD Doppler radar observations and GCE simulations using Lang et al. (2011) improved 3ICE (**b**) and new 4ICE scheme with moderate hail and bin rain evaporation correction (**c**). **d–f** Show corresponding results for the moderate February 23, 1999 TRMM LBA convective case. Heavy thick lines in the simulation CFADs show the edges of the core observed frequency probabilities (i.e., the 5% contours shown in a/d) and the outer limits of the observed frequency distributions (i.e., the 0% contours shown in a/d). Right axes are heights in km, while horizontal dashed lines show the level of indicated environmental temperatures in degrees C. Adapted from Lang et al. (2014)

the 23 February case (see Fig. 12 in Lang et al. 2014). The 4ICE scheme was further improved by implementing, among other modifications, a simple hail-mapping scheme that eliminates the need to select an appropriate hail intercept value for each case. The improved 4ICE scheme was implemented into both the GCE and NU-WRF (Tao et al. 2016b).

4.1.3 Bin Microphysics

The 2D GCE model with the SBM scheme was used to simulate a summertime midlatitude squall line case that occurred during the PRESTORM campaign in the central US. This is a case study using the open lateral boundary condition with observed atmospheric conditions ahead of the squall line as the initial condition. There are 1024 horizontal and 33 vertical grid points. The horizontal resolution is 1 km at the center of the domain, and is stretched toward the lateral boundaries. The vertical grids are also stretched, with finer resolution (~0.2 km) near the ground and coarser resolution (2 km) at the top. Figure 12 shows a comparison of observed radar reflectivity (Fig. 12a) along with simulated radar reflectivity by the SBM scheme (Fig. 12b) and the Goddard 3ICE bulk scheme (Fig. 12c). The SBM simulation compares much better with the observation in that it has an extensive stratiform region that has a horizontally uniform structure. The 3ICE bulk scheme produced a narrower stratiform region that consists of previous convective cells propagating from the leading edge.

The differences between the SBM and the bulk scheme are also used to improve the Goddard 3ICE bulk scheme. For example, the raindrop size distribution assumption in the bulk microphysical scheme artificially enhances rain evaporation rate. This is because bulk microphysical schemes have to make assumptions on raindrop size distributions. When the exponential assumption is used, rain mass is artificially redistributed to the small size tail compared with the more realistic SBM model simulation. Smaller raindrops evaporate faster than the larger ones. The cooling produced by rain evaporation largely determines the cool pool strength, which is crucial for storm regeneration and propagation (e.g., Rotunno et al. 1988). In the bulk simulation shown in Li et al. (2009a), the strong cool pool circulation dominates the near-surface environmental wind shear, producing pulsating updraft cores that tilt toward the rear and propagate into the stratiform region, as shown in Fig. 12c. This is in contrast to the near balance between the cool pool and the wind shear simulated

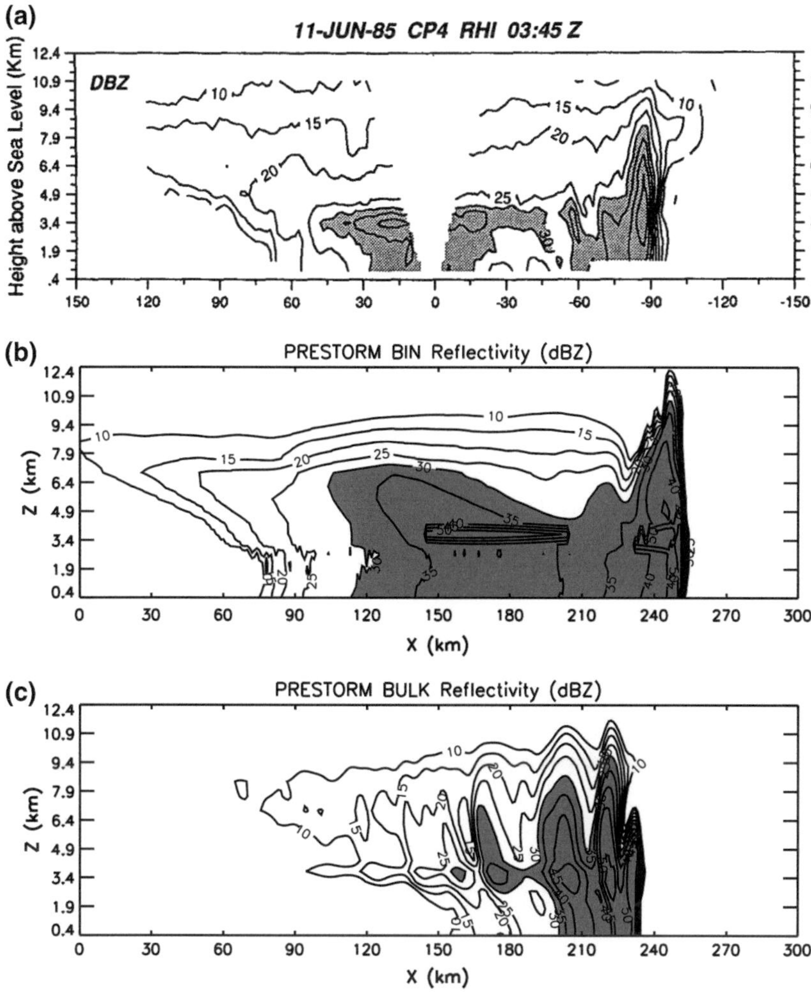

Fig. 12 Comparisons of the radar reflectivity simulated by the SBM scheme (**b**) and Goddard 3ICE bulk scheme (**c**) with surface radar observation (**a**), for a squall line simulation during the PRESTORM field campaign. Adapted from Li et al. (2009a)

in the bin scheme (Li et al. 2009b), which results in upright and steady updraft cores and a homogeneous stratiform region (Fig. 12b). In order to improve 3ICE bulk scheme, rain evaporation rate is reduced according to an empirical formula derived from the SBM scheme (Li et al. 2009b). This resulted in a better agreement in the radar reflectivity comparisons.

The SBM scheme has helped to improve the Goddard bulk scheme (Lang et al. 2014). However, the SBM scheme itself is not perfect. For example, 9 years of TRMM Precipitation Radar (PR) and 85 GHz TRMM Microwave Imager (TMI) data during the late spring and early summer over central US were compiled and compared against an SBM simulation of the PRESTORM (Fig. 12) squall line (Li et al. 2010). Figure 13 shows the observations and the SBM simulations. Comparisons against a surface C-band radar (Fig. 13a) and the TRMM PR radar (Fig. 13b) show an overestimation of radar reflectivity in the original scheme, especially, between the height of 5–8 km (Fig. 13d, e). To improve the simulated radar reflectivity profiles, the temperature dependence of the collection efficiency between ice-phase particles, especially, the plate type was modified. This modification reduced the coalescence of various ice-phase particles and produced smaller aggregates, resulting in better radar CFAD comparisons in the stratiform region for both C-band radar and TRMM PR, as shown

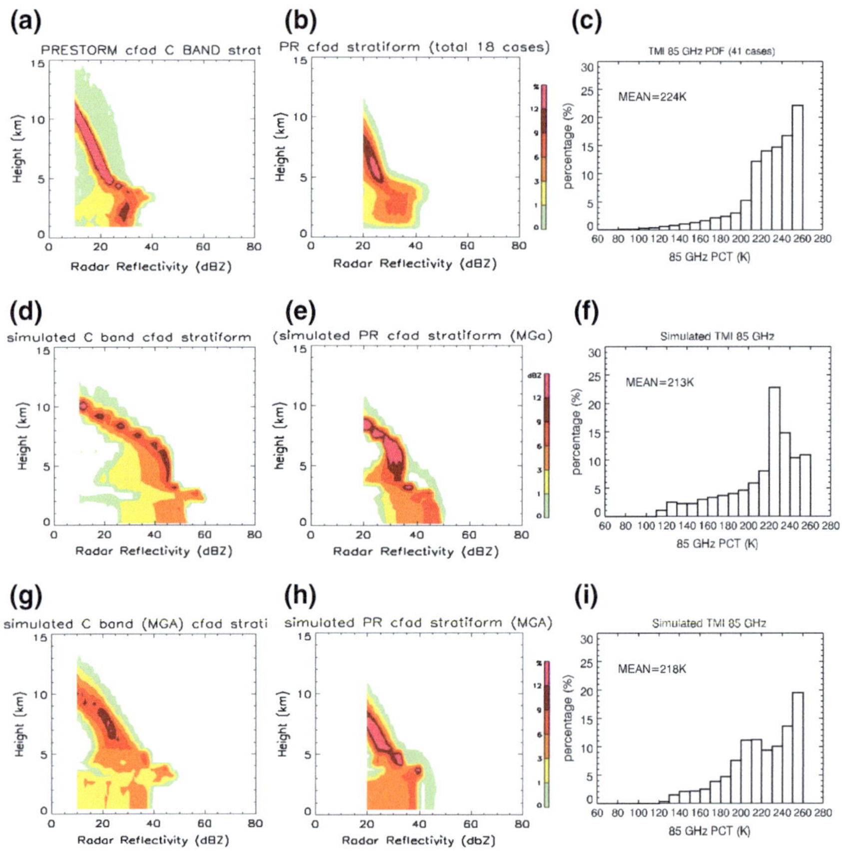

Fig. 13 Comparisons between observations (first row) and SBM scheme model simulations (second and third rows). The second row uses the original SBM scheme; the third row uses the improved SBM scheme. The first column represents surface C-band radar CFADs for the stratiform region. 9 years of TRMM PR observed radar reflectivity CFADs for the stratiform region over the US Southern Great Plain are shown in the second column. The third column compares distributions of TRMM TMI 85 GHz polarization corrected brightness temperature. Adapted from Li et al. (2010)

in Fig. 13g, h. In addition, the simulated 85 GHz brightness temperature distributions compare reasonably well with the TMI observation (Fig. 13i). We will continue testing and improving the SBM scheme for other precipitating events (especially, for convective systems and snow events observed at the aforementioned GPM-related GV sites).

4.1.4 GCE-Morrison 2M Microphysics Scheme

Li et al. (2017) have conducted a model intercomparison study to examine the differences and similarities of precipitation processes between the Goddard 3ICE (graupel version) and Morrison microphysics. The DYNAMO case (Sect. 4.1.1) was used and a long-term 3D GCE model integration is performed, with the domain size of 256 × 256 km and 60 vertical levels. Figure 14a and 14b show the simulated echo-top height distributions from GCE 3ICE and Morrison scheme, respectively. Generally speaking, both schemes captured the variations of the radar echo-top height variations during the MJO event. The clouds transitioned from mainly shallow convection during the MJO suppressed phase (November 9–15) to a mixture of convection with different heights during the developing phase (November 16–22) to the deep convection dominant during the mature phase (November 23–29). The Morrison scheme simulated slightly lower echo-top heights compared to GCE 3ICE (i.e., at the height of 5 km and below, and around November 4, 10, and 14). This is consistent with its simulated rainfall intensity (more light rainfall compared to 3ICE scheme simulated). The Morrison scheme also simulated more frequent and higher echo-top heights compared to GCE 3ICE scheme prior to November 24 during the developing stage. This could explain why the Morrison scheme simulated more heavy rainfall compared to 3ICE scheme.

Lastly, the TRMM surface rainfall product 3B42 within the DYNAMO sounding array is used to compare simulated surface rainfall by both Goddard 3ICE and Morrison schemes. 3B42 is a merged multi-satellite, near real-time product with 3-hourly temporal resolution and 0.25° spatial resolution. Model simulated surface rainfall rates are sampled every 3 h and averaged over 28-km × 28-km grids to match the 3B42 resolution. Figure 14c shows the comparison of probability distributions of simulated surface rainfall for the 3ICE scheme (blue) and Morrison scheme (orange). The satellite observations and model simulations compare reasonably well for most of the rainfall rates. However, the model simulations overestimate light surface rainfall below 1 mm/h and are also missing the extremely strong events of higher than 20 mm h^{-1} rainfall. Generally, the Morrison scheme simulated more light rainfall (<5 mm h^{-1}) and more heavy rainfall (>30 mm h^{-1}) compared to TRMM rainfall product. On the other hand, GCE 3ICE scheme simulated more moderate rainfall (between 5 and 30 mm h^{-1}) compared to TRMM rainfall.

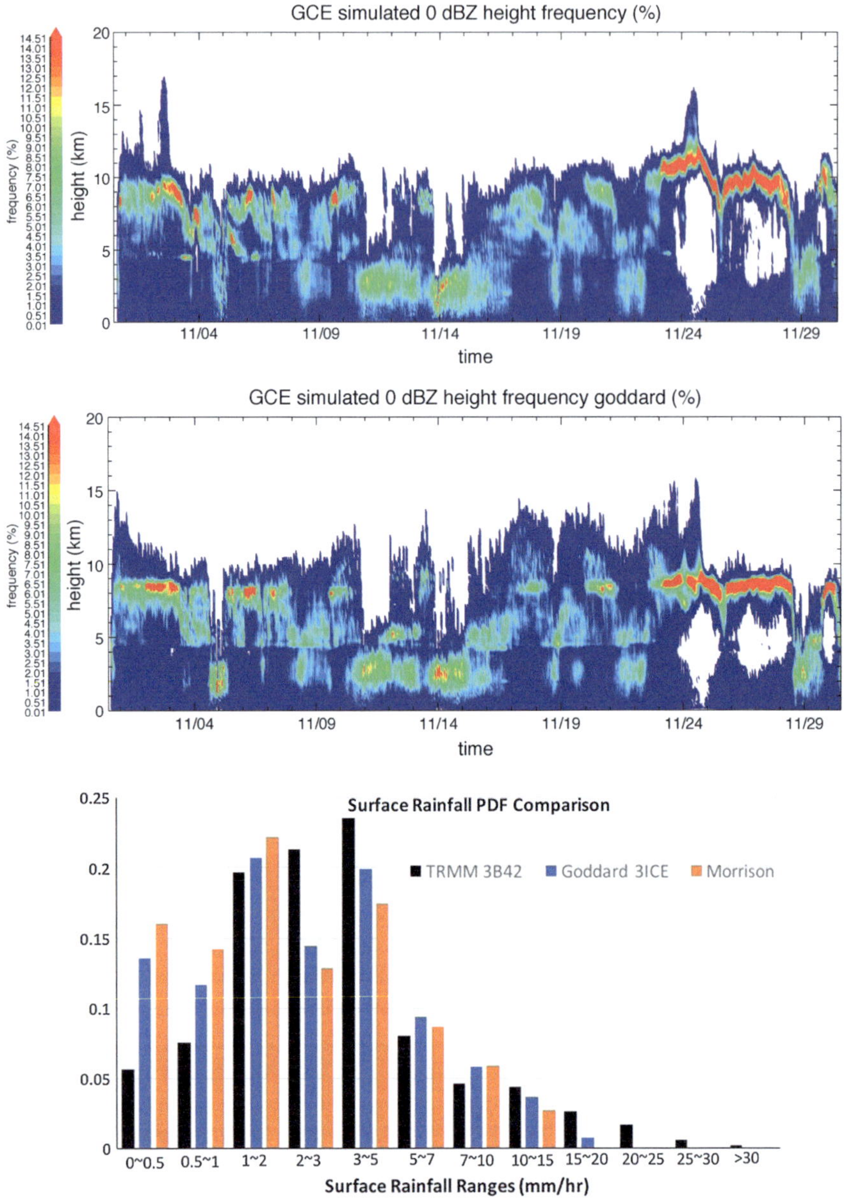

Fig. 14 Comparison of GCE simulations using Goddard 3ICE scheme and Morrison scheme. Upper panel: echo-top height distributions simulated by Goddard 3ICE scheme; middle panel: echo-top height distributions simulated by Morrison scheme; lower panel: probability distributions of surface rainfall for Goddard 3ICE scheme (blue) and Morrison scheme (orange)

4.1.5 GCE-RAMS

Sample results from a 3D GCE simulation using RAMS 2M microphysics with the aerosol module of Saleeby and van den Heever (2013) depict the nighttime development stage of a low-level stratocumulus deck over northern Vietnam (Fig. 15) that was observed on April 7–8, 2013 during the 7-SEAS/BASELInE field campaign (Loftus et al. 2016; Tsay et al. 2016). The model domain was 14 × 14 × 13.6 km in the horizontal and vertical directions, respectively, with horizontal grid spacing of 200 m, and vertical grid spacing stretched from 30 m at the lowest model level to 300 m above 13 km. The large and small time steps were 1 and 0.25 s, respectively. For this particular simulation, only sulfate aerosols served as potential CCN, and the model was initialized with an exponentially decreasing aerosol concentration profile with a maximum of 300 mg^{-1} at the lowest model level. The model was initialized horizontally homogeneously using the 12 UTC atmospheric sounding from Hanoi, Vietnam, and a slightly supersaturated (0.05%) layer in the initial sounding forced the cloud layer to form shortly after model start-up. With time, the cloud layer thickened along with gradual increases in LWC (Fig. 15a) and simulated W-band radar reflectivity (Fig. 15c), similar to the observations (Fig. 15f). Peak N_c values exceeded

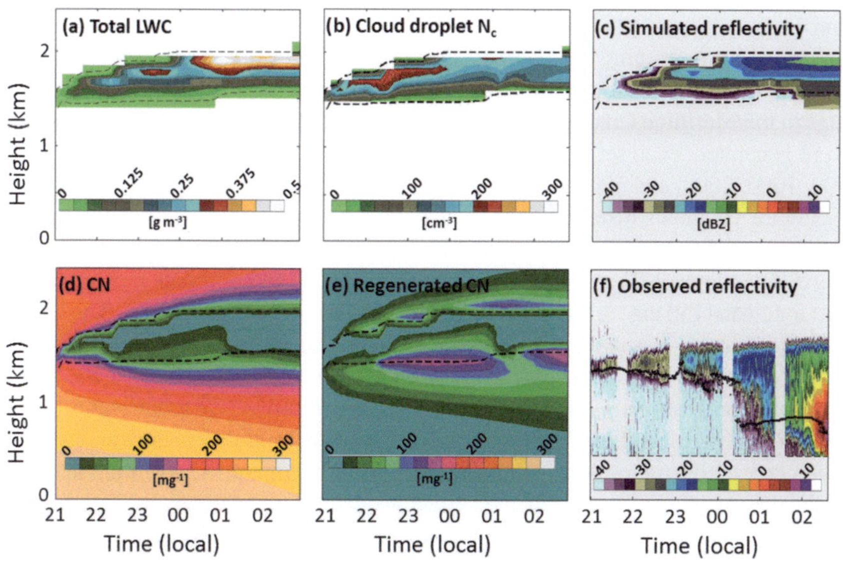

Fig. 15 Time–height plots of **a** total liquid water content LWC (g m^{-3}), **b** cloud droplet number concentration N_c (cm^{-3}), **c** G-SDSU attenuated W-band equivalent reflectivity Z_e (dBZ), **d** cloud nuclei (CN) aerosol concentration (mg^{-1}), and **e** regenerated CN aerosol concentration (mg^{-1}) for a GCE simulation of **f** a low-level stratocumulus development event observed by ground-based W-band radar over northern Vietnam on April 7–8, 2013. Dashed lines in panels (**a**–**e**) denote approximate cloud boundaries (cloud water mixing ratio ≥ 0.01 g m^{-3}), and black dots in panel (**f**) represent measured cloud base height from a ground-based ceilometer. Adapted from Loftus et al. (2016)

200 cm^{-3} during the initial cloud development stage, followed by lower N_c values owing to droplet self-collection (Fig. 15b). However, because no additional forcing (e.g., moisture advection) was applied beyond model initialization, the simulation did not develop a secondary lower cloud deck and eventual drizzle that were evident in the observations after roughly 00:30 UTC. Notably, depletion of CN within the cloud (Fig. 15d) and the regeneration of CN owing to evaporation along the cloud boundaries (Fig. 15e) are well captured by the model.

In addition, Lee et al. (2009) used the GCE with RAMS 2M microphysics to examine the effects of enhanced aerosol loading on thin marine stratocumulus clouds in environments characterized by different cloud top relative humidity (RH) values. For greater aerosol loading, the authors noted an increase in cloud droplet number concentrations (CNDC), along with increased reflection and absorption of downward SW radiation owing to smaller droplet sizes, although changes in cloud Liquid Water Path (LWP) were found to depend on RH near cloud top as well due to feedbacks among CDNC, condensation, and dynamics (i.e., vertical velocities, cloud top entrainment, and evaporative cooling below cloud base).

Recently, experiments have been performed in which large ensembles of GCE-RAMS simulations of convection are run, and used to test the spectrum of convection—aerosol responses in different environments. Simulations of the aforementioned TRMM LBA February 23, 1999 case were performed with 2-km grid spacing and 72 model levels [c.f. Posselt (2016) for details]. A 9,900 member ensemble was generated by perturbing the February 23 sounding with temperature, water vapor, and wind Empirical Orthogonal Functions (EOFs) generated from 8 years of soundings in the Maritime Continent (Bukowski et al. 2017). Six sets of 1665 simulations were generated, each set having different sulfate aerosol concentrations: [150, 300, 500, 1000, 2000, and 5000 cm^{-3}].

Figure 16 shows histograms of GCE output precipitation rate (Fig. 16a–c) and mean upward vertical velocity (Fig. 16d–f) averaged over times corresponding to storm development (Fig. 16a, d), maturity (Fig. 16b, e) and dissipation (Fig. 16c, f), and for clean (150 cm^{-3}; blue) and polluted (5000 cm^{-3}; red) conditions. The largest effect of increase in aerosol is in the precipitation rates, which exhibit the well-known shift in precipitation rate that occurs during squall line evolution. Early in the development (warm rain only) phase, larger CCN concentrations lead to rain suppression (Fig. 16a). There is little difference in the distribution of rainfall at maturity (Fig. 16b), when convective updrafts are strongest (Fig. 16e). Postmaturity, when the majority of the precipitation is stratiform and vertical velocities are relatively weak, rain rates are larger in the more heavily polluted environment. The simulations exhibit little evidence of vertical velocity enhancement in polluted conditions, though the variance in updraft speeds is larger at maturity (Fig. 16e), and mean vertical velocities are larger for the polluted cases in the dissipating phase (Fig. 16f).

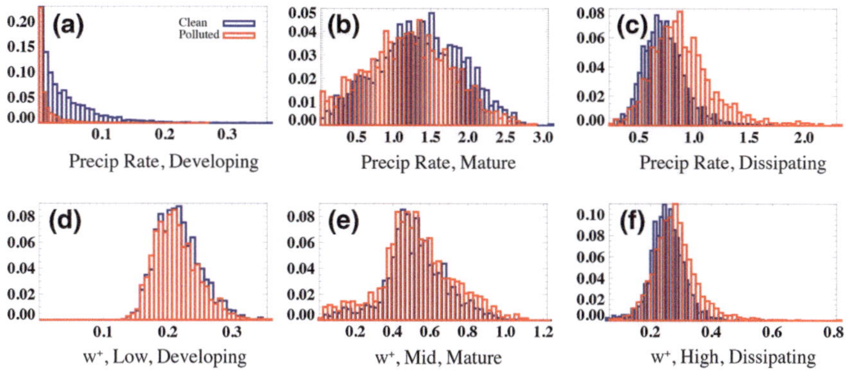

Fig. 16 Histograms of GCE output precipitation rate (**a**–**c**) and mean upward vertical velocity (**d**–**f**) averaged over times corresponding to storm development (**a**, **d**), maturity (**b**, **e**), and dissipation (**c**, **f**), and for clean (150 cm^{-3}; blue) and polluted (5000 cm^{-3}; red) conditions. Note that vertical velocities are shown for different levels at each time, being averaged over 0–3 km in the developing phase (**d**), 3–8 km in the mature phase (**e**), and 8–16 km in the dissipating phase (**f**)

4.2 NU-WRF

Both GCE and NU-WRF have been utilized to improve Goddard latent heating retrieval and surface rainfall/snowfall retrieval for TRMM and GPM (Tao et al. 2006, 2016a; Simpson et al. 1996). This section will present the performance of microphysics schemes associated with different weather events simulated by NU-WRF.

4.2.1 3ICE Versus 4ICE—MC3E

NU-WRF was used at a relatively high horizontal resolution (i.e., 1 km for the innermost domain[3]) to examine the performance of the Goddard 3ICE and 4ICE microphysics schemes. The strong, well-organized MC3E MCS (20 May 2011) with intense leading edge convection and a well-developed trailing stratiform region (the same GCE case presented in Sect. 4.1.2) was simulated and the different schemes evaluated in terms of their radar reflectivity structures and distributions, propagation, rainfall, and surface rain rate histograms versus NMQ NEXRAD radar data [see details in Tao et al. (2016b)].

Figure 17 shows the observed and NU-WRF-simulated CFADs. The NU-WRF-simulated CFADs with several different microphysics schemes (Goddard 3ICE, Goddard 4ICE, and Morrison scheme with two different third class ice options, i.e., graupel or hail) are also shown for comparison. For the observed CFAD (Fig. 17a), the highest probabilities follow a coherent pattern with the peak density steadily decreas-

[3]A triple nested domain with 9, 3, and 1 km resolution, respectively, was used.

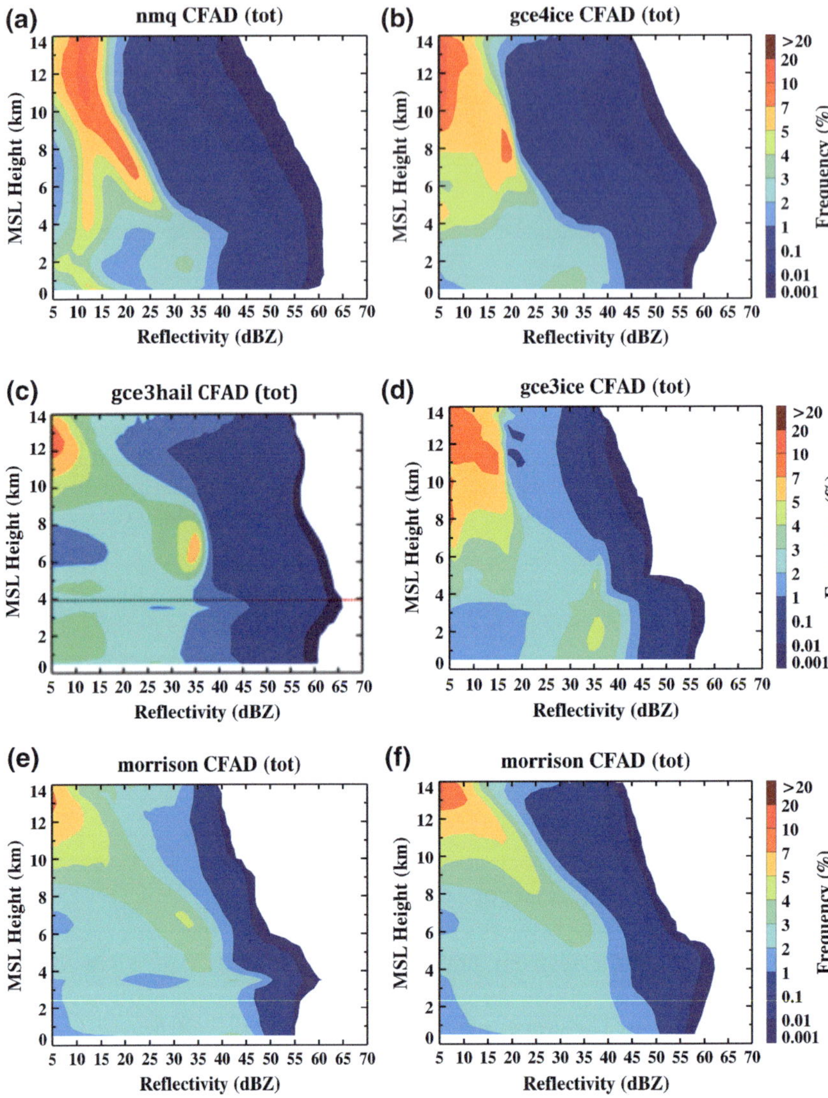

Fig. 17 Comparison of radar reflectivity CFADs both **a** observed and simulated with NU-WRF using the Goddard **b** 4ICE, **c** 3ICE-hail and **d** 3ICE-graupel options and the Morrison **e** graupel and **f** hail options microphysics for the May 20, 2011 MC3E case study. The Goddard 4ICE and Morrison 3ICE scheme with hail are in better agreement with observations than the other 3ICE schemes

ing with height from between 20 and 35 dBZ near the melting level to between 5 and 15 dBZ above 12 km, indicative of a robust sedimentation/aggregation effect. Maximum reflectivities at the lowest frequency contour of 0.001% are just over 60 dBZ from the surface up to 6 km and drop-off steadily aloft to around 45 dBZ at 14 km. The Goddard 3ICE-graupel scheme simulated CFAD (Fig. 17d) has some notable discrepancies with the observed. Similar to the GCE results, it lacks all of the reflectivity values higher than 45 dBZ above the freezing level. Second, although it captures some of the aggregation effect evident in the observed CFAD, it is too weak with too few echoes in the 20–25 dBZ range between 4 and 8 km. In contrast, the 4ICE scheme (Fig. 17b) can simulate the rare high reflectivity values above the freezing level as was observed. It produces a very realistic radar reflectivity CFAD with a more robust and coherent aggregation signature than the 3ICE graupel scheme that much more closely resembles the observed as well as peak reflectivities similar to the observed and which realistically monotonically decrease with height as observed. Although the Goddard 3ICE-hail scheme was able to simulate the more intense dBZ values like the 4ICE scheme, its proportion of intense echoes is too high, indicating its hail cores are too wide relative to the other echoes, and its most intense echoes do not decrease monotonically with height above the freezing level (Fig. 17c). Finally, its modal dBZ values have a very poor aggregation signature that is not in good agreement with the observations. The Morrison scheme CFAD with the hail option (Fig. 17f) is in better agreement with the observed than when using the graupel option (Fig. 17e). These results (3ICE versus 4ICE and graupel versus hail) suggested that hail processes are essential for this particular case. In addition, the results suggested that the predetermined hail option in the Morrison scheme performed better than the graupel option for this case (Fig. 17e, f).

Vertical cross sections of the observed NEXRAD- and NU-WRF-simulated radar reflectivity are shown in Fig. 18 for comparison. This MC3E case shows a classic continental unicellular squall line structure (Rutledge et al. 1998; Johnson and Hamilton 1988; see review by Houze 1997) with deep, erect leading convective cell(s) followed by a wide trailing stratiform region, featuring a distinct high radar reflectivity bright band near the melting level separated from the convective core(s) by a transition area with a less prominent bright band. The Goddard 3ICE graupel scheme produced a wide trailing stratiform region as observed but with too much moderate precipitation and leading edge reflectivities that were too weak (Fig. 18b). Without hail, the graupel scheme simply cannot match the intense radar returns associated with such large solid ice particles while too much moderately falling graupel is transported rearward into the stratiform region. The Goddard 3ICE hail scheme (Fig. 18c) is able to replicate the intense echoes in the convective cores, but the cores are too broad, and the structure of the stratiform region is quite different from the observed with dBZs maximized well above the freezing level and their distribution very nonuniform. In contrast, the 4ICE scheme, especially, the modified version, can reproduce the narrow, but intense convective cores, a broad, uniform stratiform area with radar echoes that are strongly vertically stratified, and a more well-defined transition region separating it from the leading convection (Fig. 18d). All of these features with 4ICE are in good or better agreement with the observations compared with the other two schemes.

Figure 19 shows PDFs of the total simulated and radar-derived surface rain rate intensities. Both the hail and 4ICE schemes have a higher proportion of heavy precip-

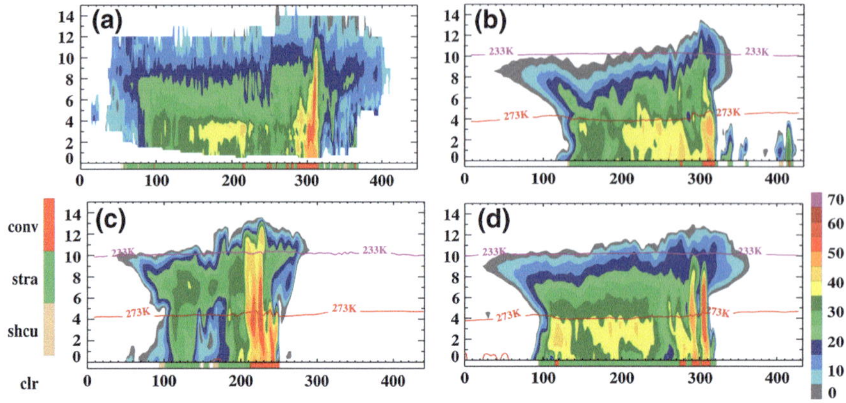

Fig. 18 Vertical cross sections of NEXRAD radar observed (**a**) and NU-WRF-simulated (**b–d**) radar reflectivity at 10 UTC on May 20, 2011, (**b**) is from 3ICE-graupel, (**c**) is from 3ICE-hail, and (**d**) is from 4ICE scheme. Adapted from Tao et al. (2016b)

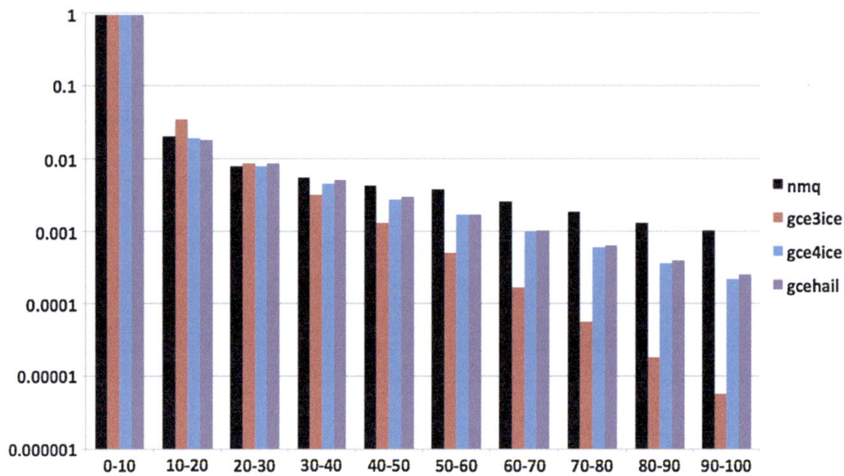

Fig. 19 PDFs of observed and NU-WRF simulated rainfall intensity in mm h^{-1} from three different variations of the Goddard microphysical schemes (3ICE with graupel, 3ICE with hail and 4ICE). The observed rain rates are estimated from Stage IV bias-corrected Q2 radar estimates. PDFs were calculated every 10 min from both the observed and simulated datasets from 06 to 12 UTC on May 20, 2011 within the analysis domain shown in Fig. 17. Adapted from Tao et al. (2016a)

itation (i.e., >30 mm h^{-1}) as well as less moderate precipitation (i.e., 10–20 mm h^{-1}) than does the graupel scheme, placing them in better agreement with the bias-corrected Q2 radar estimates in both situations. Overall, the hail scheme produces the best agreement with the observed rain intensity frequencies despite its unrealistic anvil radar structure (see Fig. 12 in Tao et al. 2016a).

4.2.2 Bin—C3VP, MC3E, LPVex

The spectral bin microphysics in the Hebrew University Cloud Model (HUCM; Khain et al. 2011, 2012) was coupled with the WRF (Skamarock et al. 2008) model. This coupling enabled cloud-resolving simulations using the spectral bin microphysics for hindcast simulations under realistic conditions beyond conventional idealized simulations. It is beneficial particularly to studies based on comparison with various types of measurement for the validation of the model performance and the development of the discussion on the atmospheric processes. The versions of WRF coupled with the spectral bin microphysics (WRF-SBM) model developed at NASA-GSFC employ advanced parameterizations with the following functions, compared to the community versions of WRF model released by National Center for Atmospheric Research (NCAR). The particle size distributions (PSDs) of atmospheric hydrometeors are represented by 43 doubling mass bins (33 bins in the community versions). The SBM traces changes of bulk density of snow aggregates through explicit prediction of rimed mass fraction on snow. In addition, a time-dependent melting scheme (Phillips et al. 2007) in the SBM to calculate liquid water fractions of ice hydrometeors was replaced with an outdated instantaneous melting scheme. These advanced functions were included to conduct cloud-resolving hindcast simulations for specific precipitation events to support the GPM mission. Several precedence publications showed how WRF-SBM demonstrated the characteristics of cloud microphysics captured in remotely sensed measurements as well as in situ ground-based and aircraft measurements. The following provide brief reviews of studies based on WRF-SBM simulations.

Iguchi et al. (2012a) investigated two distinct snowfall events observed during the field campaign of the Canadian CloudSat/CALIPSO Validation Project (C3VP) conducted near Toronto, Canada. The first snow event was local intense lake-effect snowfall that developed from Georgian Bay of Lake Huron, and the second event was a widespread moderate snowfall caused by the passage of a synoptic low-pressure system. The cloud microphysics of these two events was observed by in situ measurements. Their characteristics were distinguished by different bulk density of solid-phase hydrometeor particles, which were attributable to the presence or absence of the interaction with super-cooled droplets. The WRF-SBM simulations were conducted to analyze these two snowfall events. A double-nesting domain with 3- and 1-km horizontal grid spacing was used, and the vertical domain extending to a height of approximately 20 km was divided into 60 layers with intervals increasing with altitude. The simulations successfully reproduced these distinct characteristics in the two snowfall events. In particular, riming of snow caused by super-cooled droplets was a key factor in the microphysical characteristics observed in the lake-effect snowstorm. Sensitivity experiments with different planetary boundary layer (PBL) schemes showed that PBL process had a large impact on the cloud microphysics of the lake-effect snowstorm through the change in the generation of super-cooled water in the system.

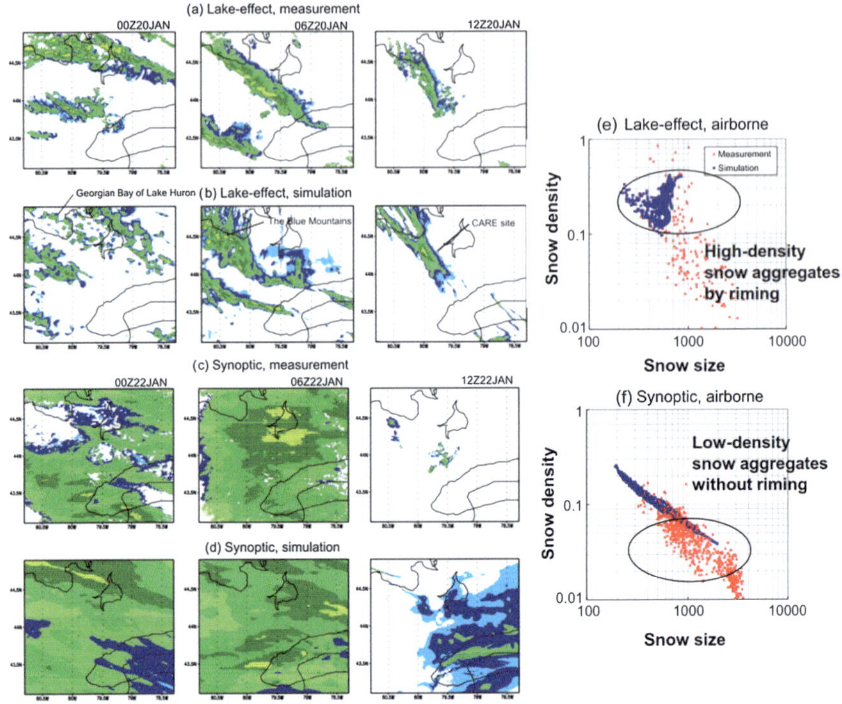

Fig. 20 Left panels show the horizontal distributions of vertically maximum C-band reflectivity (dB) in two specific snowfall events in January 2007 in a region near Toronto, Canada **a**, **c** derived from King City ground radar measurements and **b**, **d** simulated by WRF-SBM. The right panels show scatter diagrams between bulk snow density (g cm^{-3}) and bulk snow effective radius (μm) that are compatible with quantities derived from airborne cloud/precipitation imaging probes. Red and blue marks denote values from the aircraft measurements and the WRF-SBM simulations, respectively. Adapted from Iguchi et al. (2012a) with minor changes

Figure 20 shows a comparison among the ground-based C-band radar, aircraft in situ measurements, and the WRF-SBM simulation results. The observed maximum radar reflectivity distribution characterized the lake-effect snowstorm by narrow and straight reflectivity bands, and the synoptic-system snowfall by a wide and uniform reflectivity pattern covering the entire domain. The WRF-SBM reproduced the overall reflectivity distribution through the radar reflectivity simulation using the G-SDSU, except for some forecast errors in snowfall spatial distribution and timing. The aircraft in situ measurements showed the mixture of high-density rimed snow and low-density snow in the lake-effect snowstorm, whereas the measurements for the synoptic-system snowfall exhibited the presence of low-density snow only. The WRF-SBM simulated roughly the difference of the snow bulk density between the two events, but the mixture of high- and low-density snow in the lake-effect snowstorm was not well reproduced. Employment of different PBL schemes could change the variability of the bulk snow density in the lake-effect snowstorm simulation.

The WRF-SBM was also employed to simulate a midlatitude continental convective and shallow cloud system observed at the Atmospheric Radiation Measurement (ARM) Southern Great Plains (SGP) site in central Oklahoma during the MC3E field campaign (Iguchi et al. 2012b). The configuration of the model grid resolutions is the same as in the C3VP case simulations. Ground-based disdrometer measurements revealed two distinct modes in the observed rainfall variables caused by precipitation from deep convective clouds and from shallow clouds. The WRF-SBM simulation successfully reproduced two distinct modes in calculated rainfall variables that were similar to those from the disdrometer measurements. In addition, the analysis of the simulated atmospheric fields revealed how the cloud physics and the weather conditions associated with the precipitation event closely interacted in forming the unique rainfall characteristics.

Figure 21 presents the observations and the WRF-SBM simulation results for the precipitation event (April 25, 2011) over the ARM SGP site during MC3E. The vertical reflectivity profiles obtained from Ka-band ground zenith radar measurements show that deep convective clouds passed over the site between roughly 9 and 11 UTC and shallow boundary layer clouds existed over the site intermittently during most of the day. The bulk effective radii of sampled raindrops and the rainfall rates derived from the disdrometer measurements (Fig. 21c) exhibit two distinct modes according to the sampling time of the precipitation in the day. A corresponding scatter plot (Fig. 21e) derived from the WRF-SBM simulation similarly shows two distinct modes. The horizontal distributions of cloud top temperature and surface raindrop radius at 09 UTC in the simulation (Fig. 21b, d) exhibit negative correlation between the two quantities. The shallow clouds with warmer cloud top temperature moved over the ARM SGP site after the passage of the deep convective clouds with colder cloud top temperature.

The ability to simulate radar bright bands caused by melting ice particles is an advantage of the WRF-SBM model compared with conventional cloud microphysics parameterizations employed in typical weather or climate models. Two precipitation events with mixed-phase clouds over the southern part of Finland during the Light Precipitation Validation Experiment (LPVEx) field campaign were simulated using the WRF-SBM (Iguchi et al. 2014). 36-hour hindcast simulations using WRF-SBM were conducted in a double-nesting domain with 3- and 1.5-km horizontal grid spacing, and 60 vertically stretched layers that extended up to approximately 20-km height. Two types of WRF-SBM simulations with or without the time-dependent melting scheme were compared to highlight the effects of melting ice particles on the radar reflectivity and Doppler velocity profiles containing bright band structure.

Figure 22 shows observed and simulated K-band (24 GHz) radar reflectivity and Doppler velocity profiles in the form of normalized contoured frequency by temperature diagrams (CFTDs) in the first precipitation event. A CFTD can show the correlation between frequency distribution of a target quantity and temperature that is important for discussion of the melting ice particles, as compared to conventional normalized CFADs. The radar reflectivity CFTD from the observations (Fig. 22a) exhibits relatively large reflectivity between 0 and 3 °C. The WRF-SBM simulation with the time-dependent melting scheme yields a similar CFTD structure (Fig. 22c),

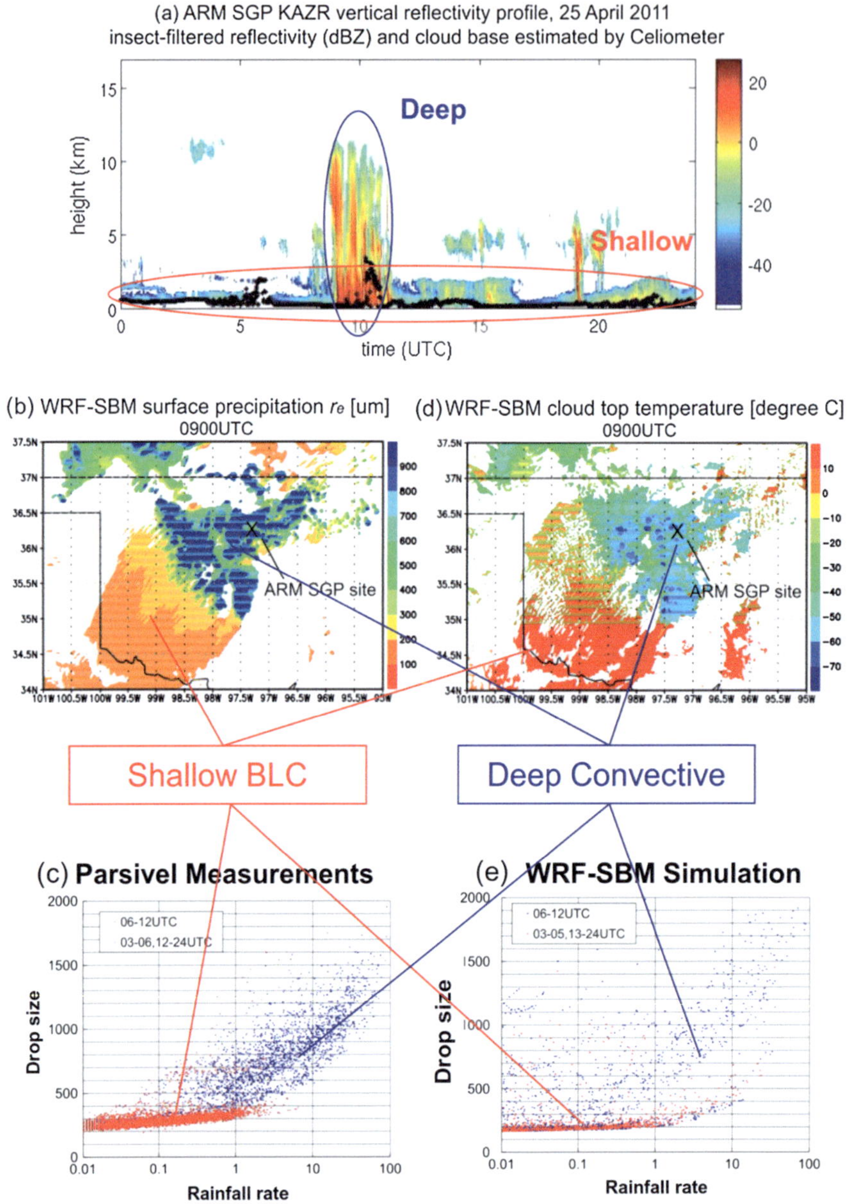

Fig. 21 **a** Time series of the Ka-Band Zenith Radar (KAZR) vertical reflectivity profiles and cloud base heights by the ceilometer deployed at the ARM SGP central facility on April 25, 2011. **b**, **d** Horizontal distributions of raindrop bulk effective radius at the surface and cloud top temperature at 0900 UTC April 25, 2011 derived from WRF-SBM simulation. **c**, **e** Scatter diagrams between surface rainfall rate (mm h^{-1}) and raindrop bulk effective radius (μm) derived from the Parsivel disdrometer measurements and from the WRF-SBM simulation results. Adapted from Iguchi et al. (2012b) with minor changes

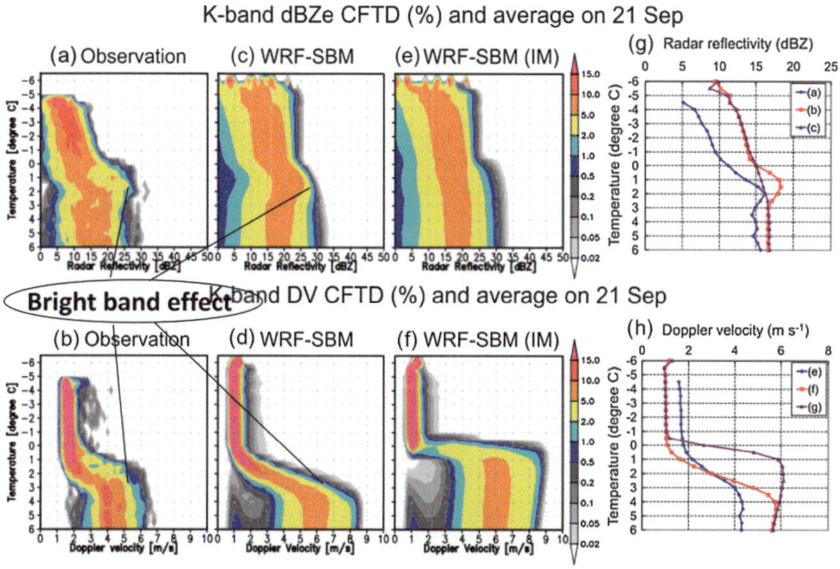

Fig. 22 **a**, **c**, **e** Normalized Contoured Frequency by Temperature Diagrams (CFTDs) of the water-equivalent radar reflectivity factor (dBZ) derived from **a** the micro rain radar measurements deployed at Jarvenpaa, Finland and vertical temperature profiles of sounding data at Jokioinen, Finland on September, 21, 2010 and obtained from **c** WRF-SBM simulations using a time-dependent melting scheme and **e** using an instantaneous melting scheme. **g** Vertical profile of mean water-equivalent radar reflectivity calculated from the CFTDs. **b**, **d**, **f**, **h** Are same as (**a**, **c**, **e**, **g**) but for radar-based mean Doppler velocity (m s^{-1}). Adapted from Iguchi et al. (2014) with minor changes

whereas the corresponding simulation using the instantaneous melting scheme is not able to reproduce the structure (Fig. 22e). These features are also confirmed in the line graph for the averages (Fig. 22g). Figure 22b, d, f, and h show that the Doppler velocity gradually increases with temperature in the range from 0 to 3 °C in both the observation and simulation using the time-dependent melting scheme. In contrast, the CFTD of the simulation using the instantaneous melting scheme exhibits an abrupt increase in velocity.

4.2.3 NE Storm (4ICE Versus Other WRF Microphysical Parameterization Schemes)

Nicholls et al. (2017) investigated how five bulk microphysics parameterization schemes (BMPSs) affected WRF simulations of seven intense wintertime cyclone events ("nor'easters"). The five evaluated BMPSs include the single-moment, six-class Lin (Lin6; Lin et al. 1983), the WRF single-moment, six-class (WSM6; Hong and Lim 2006), the Goddard Cumulus Ensemble (GCE) single-moment, six-class "3 ice" (GCE6; Lang et al. 2007), the GCE single-moment, seven-class "4 ice" (GCE7;

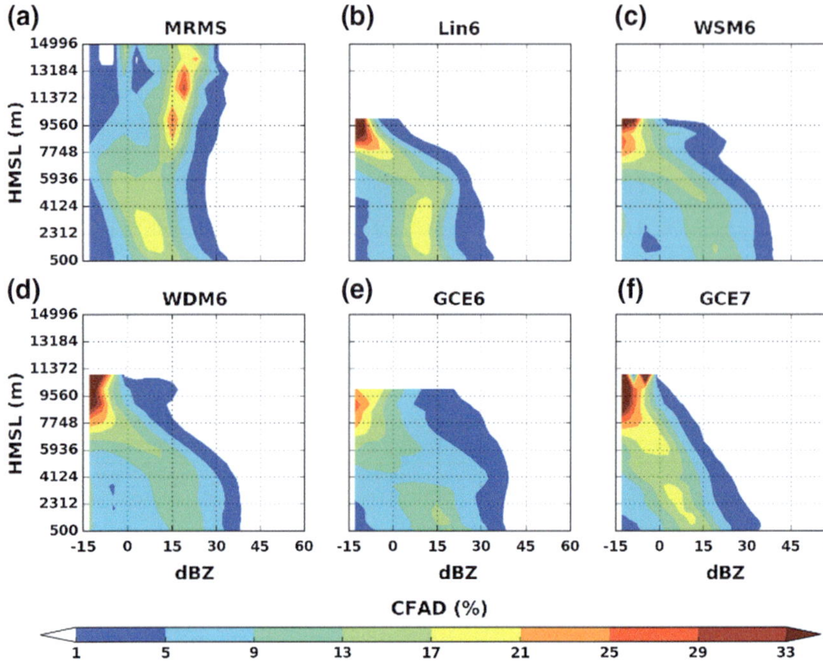

Fig. 23 Case 4 (January 2015), Domain 3 (5-km grid spacing) CFAD of observed and simulated radar reflectivity during the domain 4 transit period (12UTC January, 26–27, 2015). The y-axis shows the height above mean sea level (HMSL, m). Adapted from Nicholls et al. (2017). Panels show CFADs derived from **a** MRMS and **b–f** WRF simulations using the indicated BMPS

Lang et al. 2014), and the WRF double-moment, six-class (WDM6; Lim and Hong 2010) schemes. WRF was configured with 61 vertical levels, four model domains (45-, 15-, 5-, and 1.67-km grid spacing) and was integrated for 7 days starting 24 hours prior the onset of rapid cyclogenesis. Model validation of simulated microphysical properties and radar reflectivity structures focused on the National Severe Storms Laboratory (NSSL) Multi-Radar, Multi-Sensor 3D radar reflectivity product (MRMS).

Figure 23 shows the contoured frequency with altitude diagram (CFAD) for the January 2015 nor'easter event derived from MRMS data and WRF model output, which covers a 24-hour period and is representative of all seven events. The shown CFAD is derived from measured (MRMS) and simulated (WRF) S-band radar reflectivity data and it extends vertically from the surface to the cloud top. Both MRMS- and WRF-simulated radar reflectivity values are based in the S-band. The MRMS CFAD (Fig. 23a) shows two distinct frequency maxima centered around 2,500 m and 12,000 m Above Mean Sea Level (AMSL), respectively. GCE7- and Lin6-based WRF simulations best reproduce the lower frequency maximum value, but only GCE7 correctly produced CFADs of similar frequency and slope as MRMS below 6-km AMSL. All other BMPS shifted toward higher reflectivity values due to comparatively higher graupel mixing ratios. As compared to GCE6, GCE7 mitigates graupel generation by including a new snow size map, including deposition processes, improved aggregation physics, and a general reduction in super-cooled cloud droplets. Approaching cloud top (6,000–9,000 m AMSL), all WRF CFADs unrealistically collapse toward lower reflectivity values due to increased entrainment closer to the cloud top, or scheme-specific underlying aggregation assumptions, or combination of both. Despite, its success at lower levels, GCE7 may curtail graupel production too much above 6,000-m AMSL, where graupel mixing ratios are up to three times lower than in any other WRF simulation; severely lowering its associated radar reflectivity values. No WRF simulation reproduced the MRMS-based CFAD frequency maximum above 10,000-m AMSL due to a lack of any hydrometeors (and thus simulated radar reflectivity values) above the simulated cloud top, subsaturated conditions, and weak vertical motion (mean 0.3 m s^{-1}). MRMS values above 10 km, however, appear suspect because detected echoes generally appear offshore where radar coverage is poor and are subject to weak echo filtering during data post-processing.

Complimenting Fig. 23, Fig. 24 shows CFAD scores in height and time. A CFAD score measures forecast skill by determining the degree of overlap (0 = no overlap, 1 = identical) between radar reflectivity probability density functions calculated for each height and time between MRMS and each WRF simulation. Results from Fig. 24 show that GCE7 and Lin6 have the best forecast skill below 5,000-m AMSL, yet GCE7 forecast skill falls below other BMPSs are higher altitudes due to its dearth of graupel relative to other simulations, a result consistent with Fig. 23. The hourly time frame of Fig. 24 (vs. daily for Fig. 23) does demonstrate these results to be robust and not a product of outlier points or times throughout the event. More discussions can be found in Nicholls et al. (2017).

4.3 GMMF

The performance of the Goddard new 4ICE scheme has been examined by comparison between GMMF and satellite observation (i.e., CloudSat and TRMM). In addition, GMMF with 4ICE scheme results has been compared with a cloud-permitting global circulation model.

4.3.1 MMF Versus CloudSat

The embedded CRMs in an MMF make it possible to apply CRM-based cloud microphysics directly within a GCM. However, most such schemes are typically developed and evaluated using special field campaign datasets or short-term case study simulations. How well these schemes perform in a global environment for long-term climate simulations is still uncertain and requires comprehensive evaluation. Four one-moment Goddard microphysical schemes, including three 3ICE class

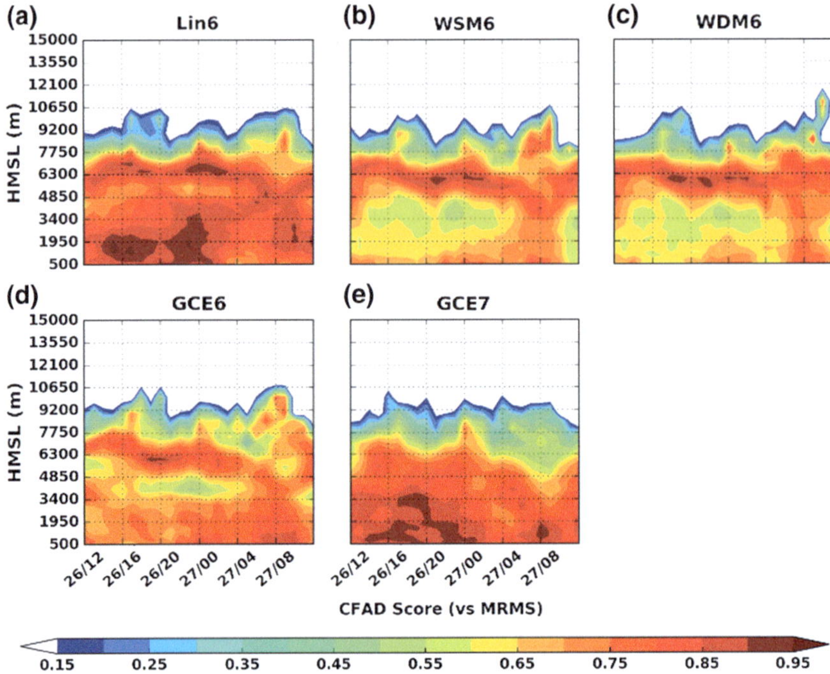

Fig. 24 Case 4, Domain 3, (5-km grid spacing), hourly CFAD scores with height above sea level (HMSL, m) between 12 UTC January, 26–27, 2015. CFAD scores measure the overlap between radar reflectivity PDFs between each WRF simulation and MRMS radar reflectivity product (1 = identical PDFs, 0 = no overlap). Adapted from Nicholls et al. (2017). Panels **a** through **e** show CFAD scores from WRF simulations using the indicated BMPS

(cloud ice, snow, and graupel) and one 4-ice class (cloud ice, snow, graupel, and hail), are implemented into the Goddard MMF and their results validated against CloudSat/CALIPSO cloud ice products and other satellite data. Four GMMF control experiments (named T2003, L2007, L2011, and L2014) are carried out with four different Goddard microphysical schemes (i.e., Tao et al. 2003; Lang et al. 2007, 2011, 2014) from January 1, 2007 to December 31, 2008. The GMMF was configured to run with $2° \times 2.5°$ (latitude $\times$ longitude) GEOS's horizontal grid and 48 vertical layers stretching from the surface to 0.4 hPa. Within each GEOS grid column there is a 64-column 2D GCE with a horizontal spacing of 4 km and a 10 s time step. The 2-year (2007–2008) global total ice water content (TIWC) from CloudSat/CALIPSO 2C-ICE Release 4 product (Deng et al. 2010) and cloud ice water content (CIWC) derived from 2C-ICE (Li et al. 2012) are used as the primary validation. The annual mean zonal height distribution of TIWC and CIWC from the 2C-ICE observations and the four GMMF experiments are shown in Fig. 25. The GMMF TIWC zonal patterns (Fig. 25b–e) are in good agreement with observation with relatively high values over the tropics and midlatitude storm tracks (though the magnitudes tend to be too low) and low values over subtropical subsidence regions. The asymmetric

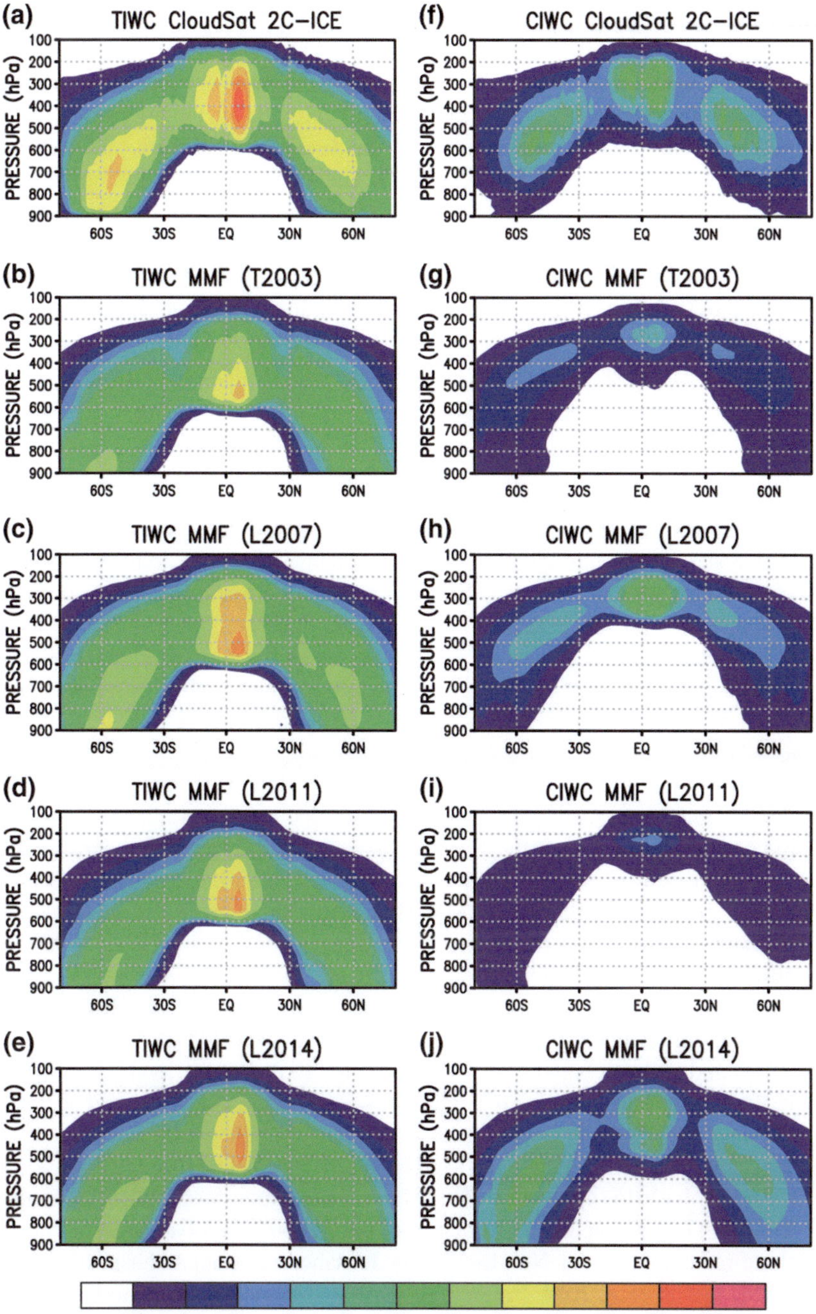

Fig. 25 Annual zonal mean (left) total and (right) cloud ice mixing ratio (10−6 g g^{-1}) from (first row) 2C-ICE, and the GMMF simulations using the (second row) T2003 (Tao et al. 2003), (third row) L2007 (Lang et al. 2007), (fourth row) L2011 (Lang et al. 2011), and (fifth row) L2014 (Lang et al. 2014) microphysics. Adapted from Chern et al. (2016)

patterns with a stronger northern branch in the Tropics are well simulated and the southern midlatitudes have more ocean surface, and hence produce more ice condensate than the northern. The vertical distribution of total ice in the GMMF simulations is similar to observations except the model cloud tops are lower in high latitude. This may indicate that the vertical resolution of the model is inadequate for simulating the thin cirrus clouds observed by CALIPSO. The 4ICE scheme is superior in terms of its mean zonal cloud ice vertical structure and amount among the four schemes (Fig. 25e, j). In contrast to the other GMMF simulations, which have high cloud ice bases near 400–500 hPa in the Tropics, CIWCs extend down to the freezing level near 600 hPa in L2014 (Fig. 25j) in good agreement with the 2C-ICE products. The better simulation in L2014 is mainly due to the improvements in cloud ice depositional growth and snow/graupel size mappings as discussed in Chern et al. (2016).

4.3.2 TRMM Versus MMF

The GMMF results are also compared with multi-sensor radiance composites from TRMM observations through the G-SDSU. Figure 26 shows joint diagrams of infrared brightness temperature (Tb_{IR}) and radar echo-top height (Het) from the TRMM and the GMMF, separated into ocean and land components. It also shows CFADs from the TRMM PR and the GMMF. Based on the joint diagrams, four types of cloud classification can be identified [left panels, a, c, and e in Fig. 26; also see Matsui et al. (2009)]. These are (1) shallow ($Tb_{IR} > 260$ K and $H_{ET} < 4$ km), (2) congestus ($Tb_{IR} > 245$ K and 4 km < $H_{ET} < 7$ km), (3) mid-cold ($Tb_{IR} < 245$ K and 4 km < H_{ET} < 7 km), and (4) deep ($Tb_{IR} < 260$ K and 7 km < H_{ET} b). For the purpose of model evaluation, this separation method is advantageous in that identical radiance-based separation can be applied to both the TRMM observations and simulator-coupled CRM simulations (Masunaga et al. 2008). Two different Goddard microphysics schemes (4ICE and 3ICE) are evaluated. TRMM PR CFADs show three distinct transitions of reflectivity distributions (right panel, b in Fig. 26): solid-phase (i.e., >8000 m), mixed-phase (i.e., 5000–8000 m), and liquid-phase zones (i.e., <5000 m). The 4ICE CFAD simulated three distinct microphysics zones as shown in the TRMM observations. The 4ICE microphysics has improved in CFAD in both land and ocean (Fig. 26f) and is much better than the 3ICE (Fig. 26d). For the joint Tb_{IR}-H_{et} diagram, the performance of the 4ICE microphysics scheme is not as good shown in the CFADs, yet it seems to perform better than the 3ICE scheme for the deep and the shallow cloud categories. Please see Kidd et al. (2016) for more detailed discussions.

Matsui et al. (2015) applied TRMM Triple-sensor Three-Step Evaluation Framework (T3EF) to evaluate the performance of GMMF and a global cloud-permitting (or cloud-resolving) model, Non-hydrostatic Icosahedral Cloud Atmospheric Model (NICAM) (Satoh et al. 2014, Progress in Earth and Planetary Science (PEPS)[4]). A month-long integration (June 2008) was conducted for the GMMF with the embedded GCE at 4-km grid spacing. However, only 1 week (starting 15 June 2008) was conducted at 3.5-km grid spacing due to expensive computation costs and data stor-

[4]Please see http://progearthplanetsci.org/.

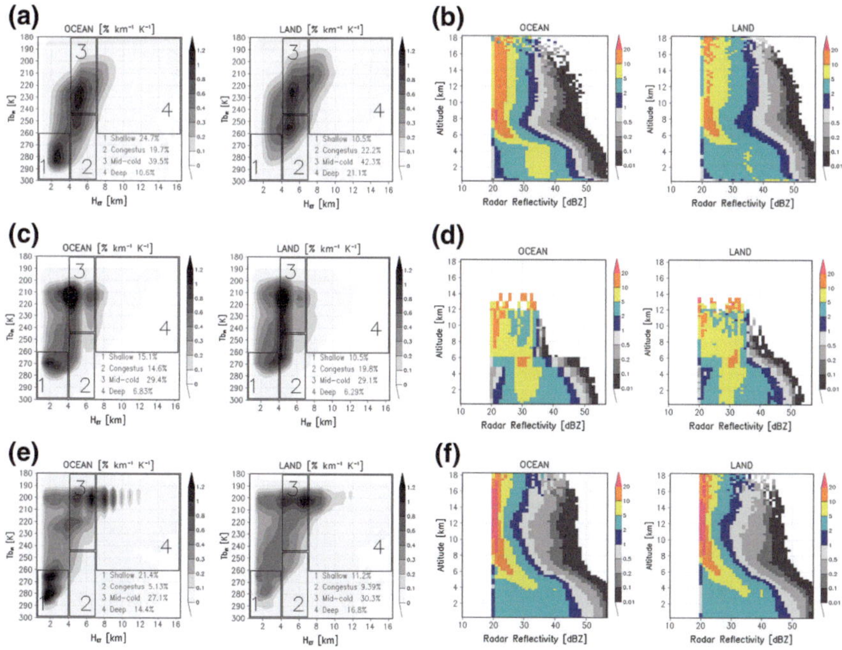

Fig. 26 Joint Tb_{IR}-H_{ET} precipitation cloud classification diagrams and CFAD of deep convection (echo-top height>7 km) over the tropics from TRMM observation (**a**) and (**b**) and two GMMF simulations with different microphysics scheme [3ICE (**c**) and (**d**) and 4ICE (**e**) and (**f**)] over land and ocean

age. The GMMF and NICAM used the Goddard 4ICE and the NICAM 1-M Water 6 (NSW6; Tomita et al. 2008) microphysics scheme, respectively. In addition, Matsui et al. (2015) modified the cloud classifications from four to five types based on joint Tb_{IR}-H_{et} diagram. They are shallow warm, shallow cold, mid-warm, mid-cold, and deep.

Figure 27 shows joint Tb_{IR}-H_{et} diagrams from the TRMM product, and output from the GMMF and NICAM simulations to identify the land–ocean contrast in characteristics of five different precipitating cloud classifications. While both the GMMF and the NICAM simulations capture convective land–ocean contrasts in the warm precipitation to some extent (Fig. 27a–c), they found that near-surface conditions over land are relatively moister in the NICAM than the GMMF, which appears to be the key driver in the divergent warm precipitation results between the two models. However, continental convective vigor is not captured by the GMMF probably because the GCE in the GMMF is driven by the homogeneous surface forcing, which does not have realistic see-breeze-driven convective system over islands. Nevertheless, neither model could reproduce a realistic land–ocean contrast in deep convective precipitation microphysics characterized by the PR CFADs. A realistic contrast between land and ocean remains an issue in global storm-resolving modeling. Please see Matsui et al. (2015) for more details.

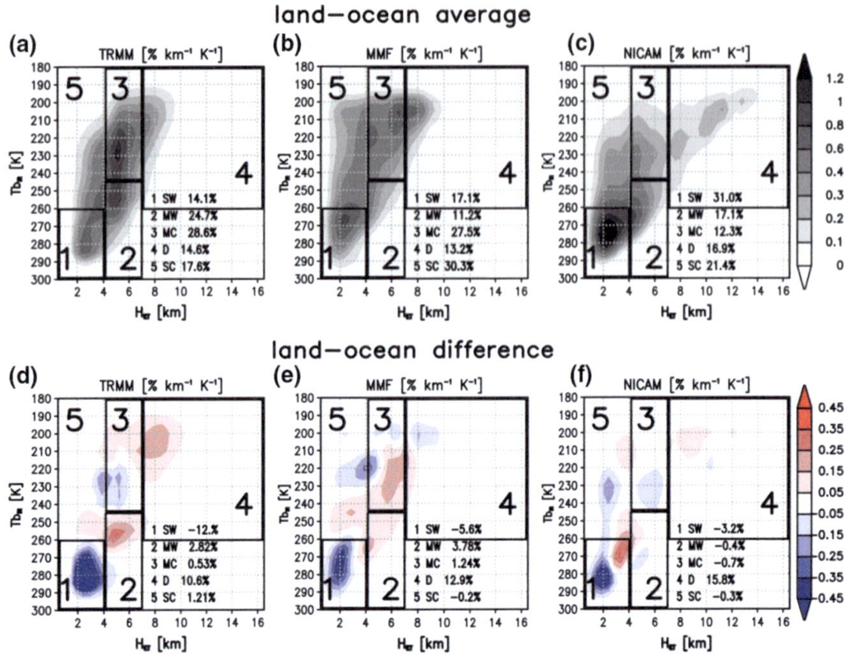

Fig. 27 **a–c** Joint Tb_{IR}-H_{ET} precipitation cloud classification diagrams for June 2008 over the tropics from TRMM observations, the GMMF simulation, and the NICAM simulation over land and ocean. **d–f** Are corresponding land–ocean differences. Adapted from Matsui et al. (2015)

4.3.3 GMMF and MCS

The importance of precipitating mesoscale convective systems (MCSs) has been quantified from TRMM precipitation radar and microwave imager retrievals. MCSs generate more than 50% of the rainfall in most tropical regions. MCSs usually have horizontal scales of a few 100 kilometers (km); therefore, a large domain of several hundred is required for realistic simulations of MCSs in cloud-resolving models (CRMs) (i.e., Ooyama 2001; Johnson et al. 2002; Petch and Gray 2001). Almost all traditional global and climate models do not have adequate parameterizations to represent MCSs. Typical MMFs may also lack the resolution (4-km grid spacing) and domain size (128 km) to realistically simulate MCSs.

The impact of MCSs on precipitation is examined by conducting model simulations using GMMF. Figure 28 shows the 2-year (2007–2008) annual mean precipitation rate from Global Precipitation Climatology Project (GPCP) and three different GMMF simulations with different MMF embedded CRM configurations (M32: 32 grid points with 4-km grid spacing, M64: 64 grid points with 4-km grid spacing, M128: 128 grid points with 2-km grid spacing, and M256: 256 grid points with 1-km grid spacing). Overall, the GMMF simulations show very similar surface rainfall patterns and capture the major weather phenomena, such as a single ITCZ

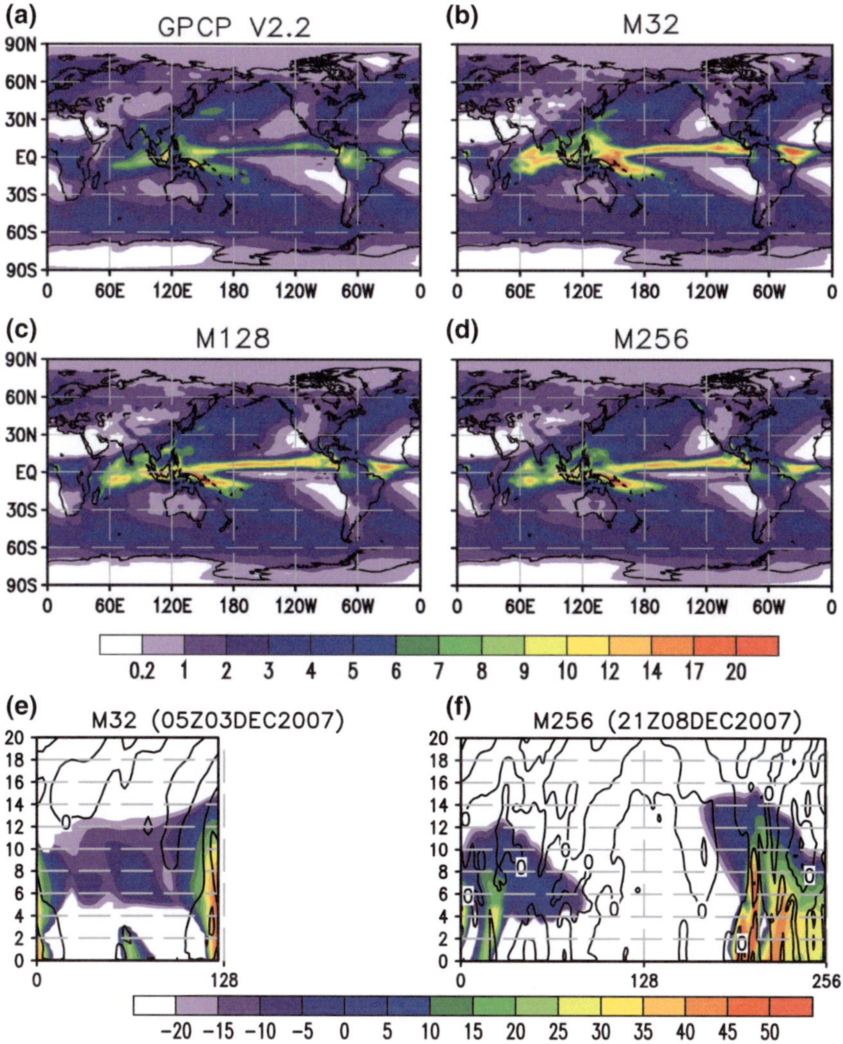

Fig. 28 Annual mean precipitation rates (mm/day) from **a** GPCP and **b** TRMM observations, GMMF simulations using the **c** M32 (embedded CRM with 32 grid points and 4-km grid spacing), and **d** M256 (embedded CRM with 256 grid points and 1-km grid spacing). **e** and **f** Are GMMF-simulated radar reflectivity (dBZ, color shading) and vertical velocity (m s^{-1}, contours) at the maximum precipitation time for a GCM grid point at (76°E, 2°N), which is near the DYNAMO field campaign site (76°E, 2°N), in December 2007 for the M32 and M256 experiments, respectively. Adapted from Tao and Chern (2017)

and SPCZ and large rainfall over the Indian Ocean, S. America, and Eastern Atlantic. Yet, all of the GMMF simulations overestimated the total rainfall amount compared to satellite estimates from GPCP (Fig. 28).

The GMMF with more CRM grid points and higher resolution (M256) has a lower bias, smaller RMSE, and higher correlation versus surface rainfall compared to those with fewer grid points and lower resolution (i.e., M32, M64, and M128). Overall, the M256 and M128 simulations are in better agreement with observations than the M32 and M64 (Fig. 28a, d). The results indicate that models can realistically simulate MCSs with more grid points and higher resolutions compared to those simulations with fewer grid points and low resolution (Fig. 28e, f). The modeling results also reveal that the strengths of the Hadley circulation, mean zonal and regional vertical velocities, surface evaporation, and amount of surface rainfall are weaker or reduced in the GMMF when using more CRM grid points and higher CRM resolution. In addition, the results indicate that large-scale surface evaporation and wind feedback are key processes for determining the surface rainfall amount in the GMMF. Please see Tao and Chern (2017) for more details.

5 Current and Future Research

In this paper, the cloud microphysics schemes used in the Goddard multi-scale modeling systems are reviewed. The performance of these microphysics schemes have been examined by conducting high-resolution model simulations at local (using GCE), regional (using NU-WRF), and global (using GMMF) scales, spanning a wide range of precipitation systems and then validating the results from these model simulations with radar and satellite observations.

Still, uncertainties exist in the cloud and microphysical processes and include the simulated vertical profiles of the cloud/precipitation properties in convective and stratiform regions, mixed-phase processes (melting, riming, and ice processes) as well as the life cycles of cloud and precipitation systems. These uncertainties can impact the numerical models across all spatial scales. As the resolutions in general circulation models increase (i.e., NICAM shown in Sect. 4.3.2), explicit cloud and microphysics schemes developed for CRMs are being used for global models (called convection-permitting global models). In this section, current and future research, including the impact of the microphysical schemes on precipitation processes as well as evaluations of their performances using current and future observations are briefly described.

5.1 Microphysics

All microphysical schemes (including the Goddard 4ICE microphysics schemes) have their own set of unique assumptions and capabilities. Therefore, it is critical

Table 4 Main characteristics in terms of predicted variables for 1M, 2M, 3M, and P3 microphysical scheme. P3 stands for predicted particle properties

	1M (mass)	2M (# of concentration)	3M (reflectivity)	P3 (ice particle)	P3
Cloud water	Qc	Nc	Rc	Qc	Nc
Rain	Qr	Nr	Rr	Qr	Nr
Cloud ice	Qi	Ni	Ri	Qi	Ni
Snow	Qs	Ns	Rs		
Graupel	Qg	Ng	Rg		
Hail	Qh	Nh	Rh		
Ice from riming				Qrime	
Bulk rime volume				Brim	

to evaluate model performance for a comprehensive range of precipitation systems. Table 4 shows the main characteristics of 1M, 2M, 3M, and P3 (Morrison and Milbrandt 2015) microphysical schemes. For 1M schemes, only mass mixing ratio of cloud hydrometeors (i.e., cloud water, rain, cloud ice, snow, graupel, and/or hail) is predicted. In addition to the mass, the total number concentrations of these hydrometeors are also predicted in 2M schemes. For 3M schemes, the reflectivity factors of hydrometeors (related to the sixth moment of the particle size distribution) are also predicted in addition to the mixing ratio and number concentration. Lastly, the P3 scheme does not differentiate precipitating ice particles (i.e., snow, graupel, and hail) and instead predicts these particles based on riming and bulk rime volume. The cloud water, rain, and cloud ice are 2M in the P3 scheme.[5]

The bin microphysical scheme explicitly resolves the hydrometeor size distributions using 43 mass doubling size bins, thus there is no need to assume any predefined particle size distribution. Eight different species, i.e., aerosols serving as CCN, liquid drops, three types of pristine ice crystals (column, plate, and dendrite), snow aggregates, graupel as well as hail are included. Riming fractions for aggregates and graupel, and melting fractions for all ice-phase species are also predicted.

More comprehensive case studies are needed using the different microphysics schemes (i.e., Goddard 3ICE, Goddard 4ICE, Morrison, and RAMS). In order for a fair comparison, all common and predetermined cloud properties and parameters (i.e., densities, intercept, and size distributions) need to be identical (i.e., hail or graupel in the 3ICE scheme as shown in Fig. 17). Such intercomparison studies could identify the "uncertainties" of microphysics schemes by conducting sensitivity tests (i.e., eliminate or reduce/increase some precipitation processes (for example, riming) and examining their impact on cloud and precipitation structures and properties). These

[5]Note that the P3 microphysics scheme will be available in NCAR WRF V3.9. The Goddard 4ICE is currently being implementing into NCAR WRF and will be available to the WRF community in the next version.

comparison studies could also identify the strengths and weaknesses of each scheme by validating their results with observations. The ultimate goal of these comparisons is to reduce uncertainties and improve the performance of each scheme.

5.2 Aerosols and Microphysics

Some microphysics schemes require initial Cloud Condensation Nuclei (CCN) and Ice Nuclei (IN) to activate cloud condensation and ice deposition. These initial CCN and IN distributions can have a major impact on the simulated cloud and precipitation properties and surface rainfall [see a review by Tao et al. (2012)]. Therefore, detailed case studies also need to (1) identify the activation of cloud CCN (and/or giant CCN) and IN in each scheme and set them as close as possible, (2) set the background aerosols identical,[6] (3) have all schemes produce common cloud properties and surface rainfall datasets, and (4) couple the radiation explicitly and consistently with the microphysics assumptions and simulated cloud properties (size distributions and optical properties).

5.3 Microphysics Evaluation

Radar has and will continue to be central to model validation with 3D comprehensive sampling and the ability to profile intense convective cores (see examples shown in Sect. 4). However, simple radar reflectivity (i.e., the sixth moment of particle size distributions) is a necessary but not unique solution, thus comparisons with additional parameters, including polarimetric radar quantities, can be of great benefit for validating and further refining model physics. As such, in addition to regular radar and microwave signatures, it is planned to compare model-simulated cloud microphysical properties with ZDR, LDR, Kdp, and HID (hydrometeor identification) analyses from polarimetric radar for a variety of different cloud systems. Doppler-derived winds and polarimetric-radar-derived surface rain maps are also critical for model validation.

[6]The IN and CCN used in the GCE model will be provided by high-resolution Goddard Chemistry Aerosol Radiation and Transport (GOCART). The CCN is calculated from the 14 aerosol species predicted by GOCART based on the Kohler curve (Andreae and Rosenfeld 2008), while ice nuclei (IN) is obtained following the approach in DeMott et al. (2010).

Specifically, cloud and microphysical schemes can be validated using the following measurements of cloud properties: (1) 3D vertical velocity structures, (2) high temporal resolution aerosol/CCN measurements, (3) vertical hydrometeor particles (ice and liquid droplet spectrum, condensation, size, and density) in situ measurements, and (4) comprehensive polarimetric radar measurements (i.e., S/C-band ground-based for convective cores and air/spaceborne or vertically pointing X/K-band and Ka/W-band for convective/anvil/stratiform characteristics). These measurements can be used to constrain as well as improve the performance of microphysics schemes.

5.4 Clouds, Convection, and Precipitation (CCP, a Future Satellite Mission)

Looking toward the future, global measurements related to cloud and precipitation processes are needed to provide critical data for fundamental improvements in the understanding of cloud processes and cloud models as described herein. Indeed, in 2017, CRMs are at a roadblock and cannot improve without observational constraints to assess the fidelity of existing microphysical schemes and processes (e.g., Hagos et al. 2014; Bassill 2014; Stephens and Ellis 2008). This necessitates a paradigm shift away from our current practices that largely observe states to future observing strategies that can deliver information on both states and the processes that govern model physics and prediction skill. Thus, it becomes essential to understand at both the local and global scale the underlying cloud processes (via measurable proxies such as ice microphysics and vertical velocities) that result in precipitation in order to improve the next generation of climate and Numerical Weather Prediction (NWP) models. As the resolutions of these climate models improve over time to be able to explicitly represent cloud and convective processes, it is equally imperative to plan for timely observations to constrain, evaluate, and define these processes to produce more accurate predictions of the water cycle. Central to this required knowledge are better predictions of atmospheric water across all spatial scales to know where, when, and how clouds form, whether they precipitate or not, and how those patterns may change in a future climate.

As such, the 2017 NASA Earth Science Decadal Survey recommended the Clouds, Convection, and Precipitation (CCP) observing system. The science and application target of the CCP observing system is the improvement of cloud and precipitation processes in Earth system models through focused global spaceborne measurements of cloud and precipitation properties and corresponding vertical air motions. These relate directly to microphysical processes that form the key linchpin of uncertainty in Earth system predictive capacities and would bring fundamental insights and essential improvements to the models described herein.

5.5 *Data Distribution Through a Cloud Library*

The Goddard Mesoscale and Dynamics Modeling Group has generated and made available a multidimensional (space, time, multivariate, and multiple cloud/cloud system type) cloud database representing different geographic locations/climate regimes to the global modeling community to help improve the representation and performance of moist processes in climate models, and to improve our understanding of cloud and precipitation processes globally. This database is available to modelers and other researchers aiming to improve representations of cloud processes in GCMs and climate models. The cloud dataset is available to the public community via ftp access from a website created within NASA Goddard (Goddard Cloud Library, http://cloud.gsfc.nasa.gov/).

Acknowledgements The authors appreciate the inspiring and enthusiastic support by Dr. Joanne Simpson for many years. The author is grateful to Dr. R. Kakar and Dr. D. Consitine at NASA headquarters for their continuous support of Goddard Cumulus Ensemble (GCE), NU-WRF, and GMMF model improvements and applications. The NASA PMM and MAP mainly support the work. Acknowledgment is also made to Dr. T. Lee at NASA headquarters, the NASA Goddard Space Flight Center and the NASA Ames Research Center for computer time used in this research.

References

Abdul-Razzak, H., and S.J. Ghan. 2000. A parameterization of aerosol activation: 2. Multiple aerosol types. *Journal of Geophysical Research* 105: 6837–6844.

Adams-Selin, R.D., S.C. van den Heever, and R.H. Johnson. 2013. Quantitative evaluation of bow echo microphysical sensitivity. *Weather and Forecasting* 28: 1188–1209.

Andreae, M.O., and D. Rosenfeld. 2008. Aerosol–cloud–precipitation interactions. Part 1. The nature and sources of cloud-active aerosols. *Earth-Science Reviews* 89: 13–41. https://doi.org/10.1016/j.earscirev.2008.03.001.

Baker, R.D., B.H. Lynn, A. Boone, W.-K. Tao, and J. Simpson. 2001. The influence of soil moisture, coastline curvature, and the land-breeze circulation on sea-breeze initiated precipitation. *Journal of Hydrometeorology* 2: 193–211.

Bassill, N.P. 2014. Accuracy of early GFS and ECMWF Sandy (2012) track forecasts: Evidence for a dependence on cumulus parameterization. *Geophysical Reseach Letters* 41: 3274–3281. https://doi.org/10.1002/2014gl059839.

Bigg, E.K. 1953. The formation of atmospheric ice crystals by the freezing of droplets. *Quarterly Journal of the Royal Meteorological Society* 79: 510–519. https://doi.org/10.1002/qj.49707934207.

Bonan, G.B., K.W. Oleson, M. Vertenstein, S. Levis, X.B. Zeng, Y.J. Dai, R.E. Dickinson, and Z.L. Yang. 2002. The land surface climatology of the community land model coupled to the NCAR community climate model. *Journal of Climate* 15: 3123–3149.

Bosilovich, M.G., and Co-authors. 2015. MERRA-2: Initial Evaluation of the Climate, NASA/TM-2015-104606/Vol. 43. 145 pp.

Bryan, G.H., and H. Morrison. 2012. Sensitivity of a simulated squall line to horizontal resolution and parameterization of microphysics. *Monthly Weather Review* 140: 202–225. https://doi.org/10.1175/mwr-d-11-00046.1.

Bukowski, J., D.J. Posselt, J.S. Reid, and S.A. Atwood. 2017. Modes of thermodynamic and wind variability over the Maritime Continent. *Atmospheric Chemistry and Physics* 17: 4611–4626. https://doi.org/10.5194/acp-17-4611-2017.

Chern, J.-D., W.-K. Tao, S.E. Lang, T. Matsui, J.-L.F. Li, K.I. Mohr, G.M. Skofronick-Jackson, and C.D. Peters-Lidard. 2016. Performance of the Goddard multiscale modeling framework with Goddard ice microphysical schemes. *Journal of Advances in Modeling Earth Systems* 7. https://doi.org/10.1002/2015ms000469.

Chin, M., R.B. Rood, S.-J. Lin, J.-F. Müller, and A.M. Thompson. 2000. Atmospheric sulfur cycle simulated in the global model GOCART: Model description and global properties. *Journal Geophysical Research* 105 (D20): 24671–24687. https://doi.org/10.1029/2000jd900384.

Chou, M.-D., K.-T. Lee, S.-C. Tsay, and Q. Fu. 1999. Parameterization for cloud longwave scattering for use in atmospheric models. *Journal of Climate* 12: 159–169.

Chou M.-D., and M.J. Suarez. 1999. A solar radiation parameterization for atmospheric studies. NASA Technical Report NASA/TM-1999-10460, vol. 15, 38 pp.

Chou, M.-D., M. J. Suarez, X.-Z. Liang. M.-H. Yan. 2001. A thermal infrared radiation parameterization for atmospheric studies. NASA/TM-2001-104606, vol. 19, 55 pp.

Cotton, W.R., R.D. Alexander, R. Hertenstein, R.L. Walko, R.L. McAnelly, and M. Nicholls. 1995. Cloud venting—A review and some new global annual estimates. *Earth Science Reviews* 39: 169–206.

Cotton, W.R., and Coauthors. 2003. RAMS 2001: Current status and future directions, *Meteorology and Atmospheric Physics* 82: 5–29.

DeMott, P.J., A.J. Prenni, X. Liu, M.D. Petters, C.H. Twohy, M.S. Richardson, T. Eidhammer, S.M. Kreidenweis, and D.C. Rogers. 2010. Predicting global atmospheric ice nuclei distributions and their impacts on climate. *Proceedings of the National Academy of Sciences of the United States of America* 107 (25): 11217–11222. https://doi.org/10.1073/pnas.0910818107.

Deng, M., G.G. Mace, Z. Wang, and H. Okamoto. 2010. Tropical composition, cloud and climate coupling experiment validation for cirrus cloud profiling retrieval using CloudSat radar and CALIPSO lidar. *Journal of Geophysical Research* 115: D00J15. https://doi.org/10.1029/2009jd013104.

Feingold, G., R.L. Walko, B. Stevens, and W.R. Cotton. 1998. Simulations of marine stratocumulus using a new microphysical parameterization scheme. *Atmospheric Research* 47–48: 505–528.

Ferrier, B.S., J. Simpson, and W.-K. Tao. 1996. Factors responsible for precipitation efficiencies in midlatitude and tropical squall simulations. *Monthly Weather Review* 124: 2100–2125.

Ferrier, B.S., W.-K. Tao, and J. Simpson. 1995. A double-moment multiple-phase four-class bulk ice scheme. Part II: Simulations of convective storms in different large-scale environments and comparisons with other bulk parameterizations. *Journal of the Atmospheric Sciences* 52: 1001–1033.

Fridlind, A.M., X. Li, D. Wu, M. van Lier-Walqui, A.S. Ackerman, W.-K. Tao, G.M. McFarquha, W. Wu, X. Dong, J. Wang, A. Ryzhkov, P. Zhang, M.R. Poellot, A. Neumann, and J.M. Tomlinson. 2017. Derivation of aerosol profiles for MC3E convective studies and use in simulations of the 20 May squall line case. *Atmospheric Chemistry and Physics* 17: 5947–5972. https://doi.org/10.5194/acp-17-5947-2017.

Gray, W.M., and R.W. Jacobsen. 1977. Diurnal variation of deep cumulus convection. *Monthly Weather Review* 105: 1171–1188.

Hagos, S., Z. Feng, C.D. Burleyson, K.-S.S. Lim, C.N. Long, D. Wu, and G. Thompson. 2014. Evaluation of convection-permitting model simulations of cloud populations associated with the Madden-Julian Oscillation using data collected during the AMIE/DYNAMO field campaign. *Journal Geophysical Research* 119: 12052–12068.

Hallett, J., and S.C. Mossop. 1974. Production of secondary ice crystals during the riming process. *Nature* 249: 26–28.

Hong, S.-Y., and J.-O.J. Lim. 2006. The WRF single-moment 6-class microphysics scheme (WSM6). *Journal of the Korean Meteorological Society* 42: 129–151.

Houze Jr., R.A. 1997. Stratiform precipitation in regions of convection: A meteorological paradox? *Bulletin of the American Meteorological Society* 78: 2179–2196.

Iguchi, T., T. Matsui, J.J. Shi, W.-K. Tao, A.P. Khain, A. Hou, R. Cifelli, A. Heymsfield, and A. Tokay. 2012a. Numerical analysis using WRF-SBM for the cloud microphysical structures in the C3VP field campaign: Impacts of supercooled droplets and resultant riming on snow microphysics. *Journal Geophysical Research* 117: D23206. https://doi.org/10.1029/2012jd018101.

Iguchi, T., T. Matsui, W.-K. Tao, A.P. Khain, V.T.J. Phillips, C. Kidd, T. L'Ecuyer, S.A. Braun, and A. Hou. 2014. WRF-SBM simulations of melting layer structure in mixed-phase precipitation events observed during LPVEx. *Journal of Applied Meteorology and Climatology* 53: 2710–2731. https://doi.org/10.1175/jamc-d-13-0334.1.

Iguchi, T., T. Matsui, A. Tokay, P. Kollias, and W.-K. Tao. 2012b. Two distinct modes in one-day rainfall event during MC3E field campaign: Analyses of disdrometer observations and WRF-SBM simulation. *Geophysical Reseach Letters* 39: L24805. https://doi.org/10.1029/2012gl053329.

Johnson, R.H., and P.J. Hamilton. 1988. The relationship of surface pressure features to the precipitation and air flow structure of an intense midlatitude squall line. *Monthly Weather Review* 116: 1444–1472.

Johnson, D.E., W.-K. Tao, J. Simpson, and C.-H. Sui. 2002. A study of the response of deep tropical clouds to large-scale thermodynamic forcings. Part I: Modeling strategies and simulations of TOGA COARE convective systems. *Journal of the Atmospheric Sciences* 59: 3492–3518.

Kain, J.S., and Coauthors. 2008. Some practical considerations regarding horizontal resolution in the first generation of operational convection-allowing NWP. *Weather and Forecasting* 23: 931–952.

Kain, J.S., and Coauthors. 2013. A feasibility study for probabilistic convection initiation forecasts based on explicit numerical guidance. *Bulletin of the American Meteorological Society* 94: 1213–1225. https://doi.org/10.1175/bams-d-11-00264.1.

Kain, J.S., S.J. Weiss, J.J. Levit, M.E. Baldwin, and D.R. Bright. 2006. Examination of convection-allowing configurations of the WRF model for the prediction of severe convective weather: The SPC/NSSL Spring Program 2004. *Weather and Forecasting* 21: 167–181.

Khain, A., M. Ovtchinnikov, M. Pinsky, A. Pokrovsky, and H. Krugliak. 2000. Notes on the state-of-the-art numerical modeling of cloud microphysics. *Atmospheric Research* 55: 159–224.

Khain, A.P., V. Phillips, N. Benmoshe, and A. Pokrovsky. 2012. The role of small soluble aerosols in the microphysics of deep maritime clouds. *Journal of the Atmospheric Sciences* 69: 2787–2807. https://doi.org/10.1175/2011JAS3649.1.

Khain, A., A. Pokrovsky, M. Pinsky, A. Seifert, and V. Phillips. 2004. Simulation of effects of atmospheric aerosols on deep turbulent convective clouds using a spectral microphysics mixed-phase cumulus cloud model. Part I: Model description and possible applications. *Journal of the Atmospheric Sciences* 61: 2963–2982.

Khain, A., D. Rosenfeld, A. Pokrovsky, U. Blahak, and A. Ryzhkov. 2011. The role of CCN in precipitation and hail in a mid-latitude storm as seen in simulations using a spectral (bin) microphysics model in a 2D dynamic frame. *Atmospheric Research* 99: 129–146. https://doi.org/10.1016/j.atmosres.2010.09.015.

Kidd, C., T. Matsui, J. Chern, K.I. Mohr, C. Kummerow, and D. Randall. 2016. Global precipitation estimates from cross-track passive microwave observations using a physically-based retrieval scheme. *Journal of Hydrometeorology* 17: 383–400. https://doi.org/10.1175/jhm-d-15-0051.1.

Kiehl, J.T., J.J. Hack, and J.W. Hurrell. 1998. The energy budget of the NCAR Community Model: CCM3. *Journal of Climate* 11: 1151–1178.

Kim, D., M. Chin, E.M. Kemp, Z. Tao, C.D. Peters-Lidard, P. Ginoux. 2017. Development of high-resolution dynamic dust source function—A case study with a strong dust storm in a regional model. *Atmospheric Environment* 159: 11–25. https://doi.org/10.1016/j.atmosenv.2017.03.045.

Klemp, J.B., and R.B. Wilhelmson. 1978. The simulation of three dimensional convective storm dynamics. *Journal of the Atmospheric Sciences* 35: 1070–1096.

Kraus, E.B. 1963. The diurnal precipitation change over the sea. *Journal of the Atmospheric Sciences* 20: 546–551.

Kumar, S.V., C.D. Peters-Lidard, Y. Tian, P.R. Houser, J. Geiger, S. Olden, L. Lighty, J.L. Eastman, B. Doty, P. Dirmeyer, J. Adams, K. Mitchell, E.F. Wood, and J. Sheffield. 2006. Land information system—An interoperable framework for high resolution land surface modeling. *Environmental Modelling and Software* 21: 1402–1415.

Lang, S.E., and W.-K. Tao. 2018. The next-generation Goddard convective-stratiform heating algorithm: New tropical and warm season retrievals for GPM. *Journal of Climate* 31: 5997–6026. https://doi.org/10.1175/JCLI-D-17-0224.1.

Lang, S.E., W.-K. Tao, J.-D. Chern, D. Wu, and X. Li. 2014. Benefits of a fourth ice class in the simulated radar reflectivities of convective systems using a bulk microphysics scheme. *Journal of the Atmospheric Sciences* 71 (10): 3583–3612. https://doi.org/10.1175/jas-d-13-0330.1.

Lang, S., W.-K. Tao, J. Simpson, and B. Ferrier. 2003. Modeling of convective-stratiform precipitation processes: Sensitivity to partitioning methods. *Journal of Applied Meteorology* 42: 505–527.

Lang, S., W.-K. Tao, J. Simpson, and B. Ferrier. 2004. Comment on "Modeling of convective-stratiform precipitation processes: Sensitivity to partitioning methods". *Journal of Applied Meteorology* 43: 962–965.

Lang, S., W.-K. Tao, X. Zeng, and Y. Li. 2011. Reducing the biases in simulated radar reflectivities from a bulk microphysics scheme: Tropical convective systems. *Journal of the Atmospheric Sciences* 68: 2306–2320. https://doi.org/10.1175/jas-d-10-05000.1.

Lang., S.E., et al. 2007. Improving simulations of convective systems from TRMM LBA: Easterly and westerly regimes. *Journal of the Atmospheric Sciences* 64: 1141–1164. https://doi.org/10.1175/jas3879.1.

Lau, W.K., K.M. Kim, J.J. Shi, T. Matsui, M. Chin, Q. Tan, C. Peters-Lidard, and W.-K. Tao. 2016a. Impacts of aerosol-monsoon interaction on rainfall and circulation over Northern India and the Himalayan Foothills. *Climate Dynamics*. https://doi.org/10.1007/s00382-016-3430-y.

Lau, K.-M., C.-H. Sui, M.-D. Chou, and W.-K. Tao. 1994. An inquiry into the cirrus-cloud thermostat effect for tropical sea surface temperature. *Geophysical Research Letters* 21: 1157–1160. https://doi.org/10.1029/94GL00222.

Lau, W.K., J.J. Shi, W.-K. Tao, and K.M. Kim. 2016b. What would happen to superstorm Sandy under the influence of a substantially warmer Atlantic sea surface temperature? *Geophysical Reseach Letters* 43 (2): 802–811. https://doi.org/10.1002/2015gl067050.

Lau, K.-M., C.-H. Sui, and W.-K. Tao. 1993. A preliminary study of the tropical water cycle and its sensitivity to surface warming. *Bulletin of the American Meteorological Society* 74: 1313–1321.

Lee, S.S., J.E. Penner, and S.M. Saleeby. 2009. Aerosol effects on liquid-water path of thin stratocumulus clouds. *Journal Geophysical Research* 114: D07204. https://doi.org/10.1029/2008jd010513.

Li, J.-L.F., et al. 2012. An observationally based evaluation of cloud ice water in CMIP3 and CMIP5 GCMs and contemporary reanalyses using contemporary satellite data. *Journal Geophysical Research* 117: D16105. https://doi.org/10.1029/2012jd017640.

Li, X., M. Janiga, S. Wang, W.-K. Tao, A. Rowe, W. Xu, C. Liu, and T. Matsui. 2017. Evolution of precipitation structure during the November DYNAMO MJO event: Cloud-resolving model inter-comparison and cross-validation using radar observations. *Journal of Geophysical Research* (accepted).

Li, X., W.-K. Tao, A. Khain, J. Simpson, and D. Johnson. 2009a. Sensitivity of a cloud-resolving model to bulk and explicit-bin microphysics schemes: Part I: Comparisons. *Journal of the Atmospheric Sciences* 66: 3–21.

Li, X., W.-K. Tao, A.P. Khain, J. Simpson, and D.E. Johnson. 2009b. Sensitivity of a cloud-resolving model to bulk and explicit bin microphysical schemes. Part II: Cloud microphysics and storm dynamics interactions. *Journal of the Atmospheric Sciences* 66: 22–40. https://doi.org/10.1175/2008jas2647.1.

Li, X., W.-K. Tao, T. Matsui, C. Liu, and H. Masunaga. 2010. Improving a spectral bin microphysical scheme using long-term TRMM satellite observations. *Quarterly Journal of the Royal Meteorological Society* 136 (647): 382–399.

Li, X., W.-K. Tao, H. Masunaga, G. Gu, and X. Zeng. 2013. Aerosol effects on cumulus congestus population over the Tropical Pacific: A cloud-resolving modeling study. *Journal of the Meteorological Society of Japan* 91: 817–883. https://doi.org/10.2151/jmsj.2013-607.

Lim, K.-S., and S.-Y. Hong. 2010. Development of an effective double-moment cloud microphysics scheme with prognostic cloud condensation nuclei (CCN) for weather and climate models. *Monthly Weather Review* 138: 1587–1612.

Lin, S.-J. 1997. A finite-volume integration method for computing pressure gradient forces in general vertical coordinates. *Quarterly Journal of the Royal Meteorological Society* 123: 1749–1762.

Lin, S.-J. 2004. A "vertically-Lagrangian" finite-volume dynamical core for global atmospheric models. *Monthly Weather Review* 132: 2293–2307.

Lin, Y.-L., R.D. Farley, and H.D. Orville. 1983. Bulk parameterization of the snow field in a cloud model. *Journal of Applied Meteorology and Climatology* 22: 1065–1092.

Lin, X., D.A. Randall, and L.D. Fowler. 2000. Diurnal variability of the hydrological cycle and radiative fluxes: Comparisons between observations and a GCM. *Journal of Climate* 13: 4159–4179.

Lin, S.-J., and R.B. Rood. 1996. Multidimensional flux form semi-Lagrangian transport schemes. *Monthly Weather Review* 124: 2046–2070.

Lin, S.-J., and R.B. Rood. 1997. An explicit flux-form semi-Lagrangian shallow water model on the sphere. *Quarterly Journal of the Royal Meteorological Society* 123: 2477–2498.

Loftus, A.M., W.R. Cotton, and G.G. Carrió. 2014. A triple-moment hail bulk microphysics scheme. Part I: Description and initial evaluation. *Atmospheric Research*. https://doi.org/10.1016/j.atmosres.2014.05.013.

Loftus, A.M., S.-C. Tsay, W.-K. Tao, P. Pantina, C. Nguyen, P.M. Gabriel, A.X. Nguyen, A.M. Sayer, and T. Matsui. 2016. Coupled aerosol-cloud systems over Northern Vietnam during 7-SEAS/BASELInE: A radar and modeling perspective. *Aerosol and Air Quality Research*. https://doi.org/10.4209/aaqr/2015.11.0631.

Lynn, B.H., and W.-K. Tao. 2001. A parameterization for the triggering of landscape generated moist convection. Part II: Zero order and first order closure. *Journal of the Atmospheric Sciences* 58: 593–607.

Lynn, B.H., W.-K. Tao, and F. Abramopoulos. 2001. A parameterization for the triggering of landscape generated moist convection. Part I: Analyses of high resolution model results. *Journal of the Atmospheric Sciences* 58: 575–592.

Lynn, B.H., W.-K. Tao, and P. Wetzel. 1998. A study of landscape generated deep moist convection. *Monthly Weather Review* 126: 928–942.

Masunaga, H., M. Satoh, and H. Miura. 2008. A joint satellite and global cloud-resolving model analysis of a Madden-Julian Oscillation event: Model diagnosis. *Journal Geophysical Research* 113: https://doi.org/10.1029/2008JD009986.

Matsui, T., J.-D. Chern, W.-K. Tao, S. Lang, M. Satoh, T. Hashino, and T. Kubota. 2015. On the land-ocean contrast of tropical convection and microphysics statistics derived from TRMM satellite signals and global storm-resolving models. *Journal of Hydrometeorology* 17: 1425–1445. https://doi.org/10.1175/jhm-d-15-0111.1.

Matsui, T., T. Iguchi, X. Li, M. Han, W.-K. Tao, W. Petersen, T. L'Ecuyer, R. Meneghini, W. Olson, C.D. Kummerow, A.Y. Hou, M.R. Schwaller, E.F. Stocker, and J. Kwiatkowski. 2013. GPM satellite simulator over ground validation sites. *Bulletin of the American Meteorological Society* 94: 1653–1660. https://doi.org/10.1175/bams-d-12-00160.1.

Matsui, T., J. Santanello, J.J. Shi, W.-K. Tao, D. Wu, C. Peters-Lidard, E. Kemp, M. Chin, D. Starr, M. Sekiguchi, and F. Aires. 2014. Introducing multisensor satellite radiance-based evaluation for regional Earth system modeling. *Journal Geophysical Research* 119: 8450–8475. https://doi.org/10.1002/2013jd021424.

Matsui, T., X. Zeng, W.-K. Tao, H. Masunaga, W. Olson, and S. Lang. 2009. Evaluation of long-term cloud-resolving model simulations using satellite radiance observations and multifrequency satellite simulators. *Journal of Atmospheric and Oceanic Technology* 26: 1261–1274.

Matsui, T., W.-K. Tao, S.J. Munchak, M. Grecu, and G.J. Huffman. 2015. Satellite view of quasi-equilibrium states in tropical convection and precipitation microphysics. *Geophysical Research Letters* 42: 1959–1968. https://doi.org/10.1002/2015GL063261.

Matsui, T., and J.P. Jacob. 2014. Goddard radiation scheme for NU-WRF version 2014. NASA Goddard Space Flight Center, 4 pp. [available in NU-WRF package].

Matsui, T., B. Dolan, S. Rutledge, W.-K. Tao, T. Iguchi, J. Barnum, and S. Lang. 2017. POLARRIS: Polarimetric radar retrievals and instrumental simulator. *Journal of Geophysical Research* (submitted).

McCumber, M., W.-K. Tao, J. Simpson, R. Penc, and S.-T. Soong. 1991. Comparison of ice-phase microphysical parameterization schemes using numerical simulations of convection. *Journal of Applied Meteorology* 30: 987–1004.

Meyers, M.P., P.J. DeMott, and W.R. Cotton. 1992. New primary ice nucleation parameterizations in an explicit cloud model. *Journal of Applied Meteorology* 31: 708–721.

Meyers, M.P., R.L. Walko, J.Y. Harrington, and W.R. Cotton. 1997. New RAMS cloud microphysics parameterization. II: The two-moment scheme. *Atmospheric Research* 45: 3–39.

Michalakes, J., S.-H. Chen, J. Dudhia, L. Hart, J. Klemp, J. Middlecoff, and W. Skamarock. 2001. Development of a next-generation regional weather research and forecast model, In Developments in Teracomputing: Proceedings of the Ninth ECMWF Workshop on the Use of High Performance Computing in Meteorology, Eds. W. Zwieflhofer and N. Kreitz, World Scientific, Singapore, 269–276.

Mohr, K.I., W.-K. Tao, J. Chern, S.V. Kumar, and C. Peters-Lidard. 2013. The NASA-Goddard Multi-scale Modeling Framework-Land Information System: Global land/atmosphere interaction with resolved convection. *Environmental Modeling & Software* 39: 103–115. https://doi.org/10.1016/j.envsoft.2012.02.023.

Morrison, H., J.A. Curry, and V.I. Khvorostyanov. 2005. A new double-moment microphysics parameterization for application in cloud and climate models. Part I: Description. *Journal of the Atmospheric Sciences* 62: 1665–1677.

Morrison, H., and A. Gettelman. 2008. A new two-moment bulk stratiform cloud microphysics scheme in the community atmosphere model, version 3 (CAM3). Part I: Description and numerical tests. *Journal of Climate* 21: 3642–3658.

Morrison, H., G. Thompson, and V. Tatarskii. 2009. Impact of cloud microphysics on the development of trailing stratiform precipitation in a simulated squall line: Comparison of one- and two-moment schemes. *Monthly Weather Review* 137: 991–1007.

Morrison, H., and J.A. Milbtandt. 2015. Parameterization of cloud microphysics based on the prediction of bulk ice particle properties. Part I: Scheme description and idealized tests. *Journal of the Atmospheric Sciences* 287–311.

Mossop, S.C., and J. Hallett. 1974. Ice crystal concentration in cumulus clouds: Influence of the drop spectrum. *Science* 186: 632–634. https://doi.org/10.1126/science.186.4164.632.

Nicholls, S.D., S.G. Decker, W.-K. Tao, S.E. Lang, J.J. Shi, and K.I. Mohr. 2017. Influence of bulk microphysics schemes upon Weather Research and Forecasting (WRF) version 3.6.1 nor'easter simulations. *Geoscientific Model Development* 10: 1033–1049. https://doi.org/10.5194/gmd-10-1033-2017.

Ooyama, K.V. 2001. A dynamic and thermodynamic foundation for modeling the moist atmosphere with parameterized microphysics. *Journal of the Atmospheric Sciences* 58: 2073–2102.

Petch, J.C., and M.E.B. Gray. 2001. Sensitivity studies using a cloud-resolving model simulation of the tropical west Pacific. *Quarterly Journal Royal Meteorological Society* 127: 2287–2306.

Peters-Lidard, C.D., E.M. Kemp, T. Matsui, J.A. Santanello, Jr., S.V., Kumar, J.P. Jacob, T. Clune, W.-K. Tao, M. Chin, A. Hou, J.L. Case, D. Kim, K.-M. Kim, W. Lau, Y. Liu, J.-J. Shi, D. Starr, Q. Tan, Z. Tao, B. F. Zaitchik, B. Zavodsky, S.Q. Zhang, and M. Zupanski. 2015. Integrated modeling of aerosol, cloud, precipitation and land processes at satellite-resolved scales. *Environmental Modeling & Software* 67: 149–159. https://doi.org/10.1016/j.envsoft.2015.01.007.

Petters, M.D., and S.M. Kreidenweis. 2007. A single parameter representation of hygroscopic growth and cloud condensation nucleus activity. *Atmospheric Chemistry and Physics* 7: 1961–1971.

Petters, M.D., and S.M. Kreidenweis. 2008. A single parameter representation of hygroscopic growth and cloud condensation nucleus activity—Part 2: Including solubility. *Atmospheric Chemistry and Physics* 8: 6273–6279.

Phillips, V.T.J., A. Pokrovsky, and A. Khain. 2007. The influence of time-dependent melting on the dynamics and precipitation production in maritime and continental storm clouds. *Journal of the Atmospheric Sciences* 64: 338–359. https://doi.org/10.1175/JAS3832.1.

Pickering, K.E., A.M. Thompson, J.R. Scala, W.-K. Tao, and J. Simpson. 1992a. Ozone production potential following convective redistribution of biomass emissions. *Journal of Atmospheric Chemistry* 14: 297–313.

Pickering, K.E., J.R. Scala, A.M. Thompson, W.-K. Tao, and J. Simpson. 1992b. A regional estimate of convective transport of CO from biomass burning. *Geophysical Research Letters* 19: 289–292. https://doi.org/10.1029/92GL00036.

Pinsky, M., A. Khain, and M. Shapiro. 2000. Stochastic effects of cloud droplet hydrodynamic interaction in a turbulent flow. *Atmospheric Research* 53: 131–169.

Pinsky, M., A. Khain, and M. Shapiro. 2001. Collision efficiency of drops in a wide range of Reynolds numbers: Effects of pressure on spectrum evolution. *Journal of the Atmospheric Sciences* 58: 742–764.

Posselt, D.J. 2016. A Bayesian examination of deep convective squall line sensitivity to changes in cloud microphysical parameters. *Journal of the Atmospheric Sciences* 73: 637–665.

Pruppacher, Hans R., and James D. Klett. 1997. *Microphysics of clouds and precipitation*. Dordrecht: Kluwer Academic Publishers.

Randall, D.A., Harshvardhan, and D.A., Dazlich. 1991. Diurnal variability of the hydrologic cycle in a general circulation model, *Journal of the Atmospheric Sciences* 48: 40–62.

Rotunno, R., J.B. Klemp, and M.L. Weisman. 1988. A theory for strong, long-lived squall lines. *Journal of the Atmospheric Sciences* 45: 463–485.

Rutledge, S.A., R.A. Houze Jr., and M.I. Biggerstaff. 1988. The Oklahoma-Kansas mesoscale convective system of 10–11 June 1985: Precipitation structure and single-doppler radar analysis. *Monthly Weather Review* 116: 1409–1430.

Rutledge, S.A., and P.V. Hobbs. 1983. The mesoscalc and microscale structure of organization of clouds and precipitation in midlatitude cyclones. VIII: A model for the "seeder-feeder" process in warm-frontal rainbands. *Journal of the Atmospheric Sciences* 40: 1185–1206. https://doi.org/10.1175/1520-0469(1983)040,1185:TMAMSA.2.0.CO;2.

Rutledge, S.A., and P.V. Hobbs. 1984. The mesoscale and microscale structure and organization of clouds and precipitation in midlatitude cyclones. Part XII: A diagnostic modeling study of precipitation development in narrow cold-frontal rainbands. *Journal of the Atmospheric Sciences* 41: 2949–2972. https://doi.org/10.1175/1520-0469(1984)041,2949:TMAMSA.2.0.CO;2.

Saleeby, S.M., and W.R. Cotton. 2008. A binned approach to cloud-droplet riming implemented in a bulk microphysics model. *Journal of Applied Meteorology and Climatology* 47: 694–703.

Saleeby, S.M., and S.C. van den Heever. 2013. Developments in the CSU-RAMS aerosol model: Emissions, nucleation, regeneration, deposition, and radiation. *Journal of Applied Meteorology and Climatology* 52: 2601–2622.

Saleeby, S.M., and W.R. Cotton. 2004. A large-droplet mode and prognostic number concentration of cloud droplets in the Colorado State University Regional Atmospheric Modeling System (RAMS). Part I: Module descriptions and supercell test simulations. *Journal of Applied Meteorology* 43 (1): 182–195.

Satoh, M., H. Tomita, H. Yashiro, H. Miura, C. Kodama, T. Seiki, A.T. Noda, Y. Yamada, D. Goto, M. Sawada, T. Miyoshi, Y. Niwa, M. Hara, Y. Ohno, S. Iga, T. Arakawa, T. Inoue, and H. Kubokawa. 2014. The non-hydrostatic icosahedral atmospheric model: Description and development. *Progress in Earth and Planetary Science* 1: 18. https://doi.org/10.1186/s40645-014-0018-1.

Scala, J.R., M. Garstang, W.-K. Tao, K.E. Pickering, A.M. Thompson, J. Simpson, V. Kirchhoff, E.V. Browell, G.W. Sachse, A.L. Torres, G.L. Gregory, R.A. Rasmussen, and M.A. Khalil. 1990. Cloud draft structure and trace gas transport. *Journal Geophysical Research* 95: 17015–17030.

Seifert, A. 2005. On the shape-slope relation of drop size distribution in convective rain. *Journal of Applied Meteorology and Climatology* 44: 1146–1151.

Seifert, A., and K.D. Beheng. 2001. A double-moment parameterization for simulating autoconversion, accretion and selfcollection. *Atmospheric Research* 59–60: 265–281.

Seifert, A., A. Khain, U. Blahak, and K.D. Beheng. 2005. Possible effects of collisional breakup on mixed-phase deep convection simulated by a spectral (bin) cloud model. *Journal of the Atmospheric Sciences* 62: 1917–1931. https://doi.org/10.1175/jas3432.1.

Shi, J.J., W.-K. Tao, T. Matsui, A. Hou, S. Lang, C. Peters-Lidard, G. Jackson, R. Cifelli, S. Rutledge, and W. Petersen. 2010. Microphysical properties of the January 20–22 2007 snow events over Canada: Comparison with in-situ and satellite observations. *Journal of Applied Meteorology and Climatology* 49 (11): 2246–2266.

Shi, J.J., T. Matsui, W.-K. Tao, Q. Tan, C. Peters-Lidard, M. Chin, K. Pickering, N. Guy, S. Lang, and E.M. Kemp. 2014. Implementation of an aerosol-cloud-microphysics-radiation coupling into the NASA Uni ed WRF: Simulation results for the 6–7 August 2006 AMMA special observing period. *Quarterly Journal of the Royal Meteorological Society*. https://doi.org/10.1002/qj.2286.

Shie, C.-L., W.-K. Tao, J. Simpson, and C.-H. Sui. 2003. Quasi-equilibrium states in the tropics simulated by a cloud-resolving model. Part I: Specific features and budget analysis. *Journal of Climate* 16: 817–833. https://doi.org/10.1175/1520-0442(2003)016<0817:QESITT>2.0.CO;2.

Simpson, J., C. Kummerow, W.-K. Tao, and R.F. Adler. 1996. On the Tropical Rainfall Measuring Mission (TRMM). *Meteorology and Atmospheric Physics* 60: 19–36.

Simpson, J., and W.-K. Tao. 1993. The Goddard cumulus ensemble model. Part II: applications for studying cloud and precipitating processes and for NASA TRMM. *Terrestrial Atmospheric and Oceanic Sciences* 4: 73–116.

Skamarock, W.C., J.B. Klemp, J. Dudhia, D.O. Gill, D.M. Barker, M.G. Duda, X.-Y. Huang, W. Wang, and J.G. Powers. 2008. A description of the advanced research WRF version 3, NCAR Technical Note, NCAR/TN-475flSTR (2008), p. 113.

Slingo, A., R.C. Wilderspin, and S.J. Brentnall. 1987. Simulation of the diurnal cycle of outgoing longwave radiation with an atmospheric GCM. *Monthly Weather Review* 115: 1451–1457.

Smolarkiewicz, P.K., and W.W. Grabowski. 1990. The multidimensional positive advection transport algorithm: Nonoscillatory option. *Journal of Computational Physics* 86: 355–375. https://doi.org/10.1016/0021-9991(90)90105-a.

Soong, S.-T., and Y. Ogura. 1980. Response of trade wind cumuli to large-scale processes. *Journal of the Atmospheric Sciences* 37: 2035–2050. https://doi.org/10.1175/1520-0469(1980)037,2035:rotctl.2.0.co;2.

Soong, S.-T., and W.-K. Tao. 1980. Response of deep tropical clouds to mesoscale processes. *Journal of the Atmospheric Sciences* 37: 2016–2036.

Soong, S.-T., and W.-K. Tao. 1984. A numerical study of the vertical transport of momentum in a tropical rainband. *Journal of the Atmospheric Sciences* 41: 1049–1061.

Stephens, G.L., and T.D. Ellis. 2008. Controls of global-mean precipitation increases in global warming GCM experiments. *Journal of Climate* 21: 6141–6155.

Sui, C.H., K.M. Lau, Y.N. Takayabu, and D.A. Short. 1997. Diurnal variations in tropical oceanic cumulus convection during TOGA COARE. *Journal of the Atmospheric Sciences* 54: 639–655.

Sui, C.-H., K.M. Lau, W.-K. Tao, and J. Simpson. 1994. The tropical water and energy cycles in a cumulus ensemble model. Part I: Equilibrium climate. *Journal of the Atmospheric Sciences* 51: 711–728.

Sui, C.-H., K.M. Lau, and X. Li. 1998. Convective-radiative interaction in simulated diurnal variations of tropical cumulus ensemble. *Journal of the Atmospheric Sciences* 55: 2345–2357.

Sullivan, R.C., M.J.K. Moore, M.D. Petters, S.M. Kreidenweis, G.C. Roberts, and K.A. Prather. 2009. Effect of chemical mixing state on the hygroscopicity and cloud nucleation properties of calcium mineral dust particles. *Atmospheric Chemistry and Physics* 9: 3303–3316. https://doi.org/10.5194/acp-9-3303-2009.

Tao, W.-K., et al. 2003. Microphysics, radiation and surface processes in the Goddard Cumulus Ensemble (GEC) model. *Meteorology and Atmospheric Physics* 82: 97–137.

Tao, W.-K. 2007. Cloud resolving modeling. *Journal of the Meteorological Society of Japan* 85B: 305–330.

Tao, W.-K., et al. 2009. A multiscale modeling system: Developments, applications, and critical Issues. *Bulletin of the American Meteorological Society* 90: 515–534. https://doi.org/10.1175/2008bams2542.1.
Tao, W.-K., et al. 2014. The Goddard cumulus ensemble model (GCE): Improvements and applications for studying precipitation processes. *Atmospheric Research* 143: 392–424. https://doi.org/10.1016/j.atmosres.2014.03.005.
Tao, W.-K., and J. Simpson. 1989. Modeling study of a tropical squall-type convective line. *Journal of the Atmospheric Sciences* 46: 177–202.
Tao, W.-K., and J. Simpson. 1993. The Goddard cumulus ensemble model. Part I: Model description. *Terrestrial Atmospheric and Oceanic Sciences* 4: 35–72.
Tao, W.-K., J. Simpson, S. Lang, J. Simpson, and R. Adler. 1993a. Retrieval Algorithms for estimating the vertical profiles of latent heat release: Their applications for TRMM. *Journal of the Meteorological Society of Japan* 71: 685–700.
Tao, W.-K., D. Wu, S. Lang, J.-D. Chern, C. Peters-Lidard, A. Fridlind, and T. Matsui. 2016b. High-resolution NU-WRF simulations of a deep convective-precipitation system during MC3E: Further improvements and comparisons between Goddard microphysics schemes and observations. *Journal Geophysical Research* 121: 1278–1305. https://doi.org/10.1002/2015jd023986.
Tao, W.-K., D. Wu, T. Matsui, C. Peters-Lidard, S. Lang, A. Hou, M. Rienecker, W. Petersen, and M. Jensen. 2013. Precipitation intensity and variation during MC3E: A numerical modeling study. *Journal Geophysical Research* 118: 7199–7218. https://doi.org/10.1002/jgrd.50410.
Tao, W.-K., J. Simpson, and M. McCumber. 1989. An ice-water saturation adjustment. *Monthly Weather Review* 117: 231–235.
Tao, W.-K. 2003. Goddard cumulus ensemble (GCE) model: Application for understanding precipitation processes, in Cloud Systems, Hurricanes and TRMM. *AMS Meteorological Monographs* pp. 107–138, Boston, Mass.
Tao, W.-K., and J.-D. Chern. 2017. The impact of mesoscale convective systems on global precipitation: A modeling study. *Journal of Advances in Modeling Earth Systems*. https://doi.org/10.1002/2016ms000836
Tao, W.-K., and X. Li. 2016. The relationship between latent heating, vertical velocity, and precipitation processes: The impact of aerosols on deep convective systems. *Journal of Geophysical Research* 121. https://doi.org/10.1002/2015jd024267.
Tao, W.-K., and M.W. Moncrieff. 2009. Multi-scale cloud-system modeling. *Reviews of Geophysics* 47: RG4002. https://doi.org/10.1029/2008rg000276.
Tao, W.-K., S. Lang, J. Simpson, C.-H. Sui, B. Ferrier, and M.-D. Chou. 1996. Mechanisms of cloud–radiation interaction in the tropics and midlatitudes. *Journal of the Atmospheric Sciences* 53: 2624–2651.
Tao, W.-K., X. Li, A. Khain, T. Matsui, S. Lang, and J. Simpson. 2007. The role of atmospheric aerosol concentration on deep convective precipitation: Cloud-resolving model simulations. *Journal of Geophysical Research* 112: D24S18. https://doi.org/10.1029/2007jd008728.
Tao, W.-K., et al. 2016a. TRMM latent heating retrieval: Applications and comparisons with field campaigns and large-scale analyses. *Amer. Meteor. Soc. Meteor. Mono. - Multi-scale Convection-Coupled Systems in the Tropics, 56, Chapter 2,* https://doi.org/10.1175/amsmonographs-d-15-0013.1.
Tao, W.-K., and J. Simpson. 1984. Cloud interactions and merging: Numerical simulations. *Journal of the Atmospheric Sciences* 41: 2901–2917.
Tao, W.-K., J. Simpson, S. Lang, M. McCumber, R. Adler, and R. Penc. 1990. An algorithm to estimate the heating budget from vertical hydrometeor profiles. *Journal of Applied Meteorology* 29: 1232–1244.
Tao, W.-K., and S.-T. Soong. 1986. A study of the response of deep tropical clouds to mesoscale processes: Three-dimensional numerical experiments. *Journal of the Atmospheric Sciences* 43: 2653–2676.

Tao, W.-K., J. Simpson, C.-H. Sui, B. Ferrier, S. Lang, J. Scala, M.-D. Chou, and K. Pickering. 1993b. Heating, moisture and water budgets of tropical and midlatitude squall lines: Comparisons and sensitivity to long wave radiation. *Journal of the Atmospheric Sciences* 50: 673–690.

Tao, W.-K., J. Simpson, and S.-T. Soong. 1987. Statistical properties of a cloud ensemble: A numerical study. *Journal of the Atmospheric Sciences* 44: 3175–3187.

Tao, W.-K., J.-P. Chen, Z. Li, C. Wang, and C. Zhang. 2012. Impact of aerosols on convective clouds and precipitation. *Reviews of Geophysics* 50: RG2001. https://doi.org/10.1029/2011RG000369.

Tao, W.-K., J. Scala, and J. Simpson. 1995. The effects of melting processes on the development of a tropical and a midlatitude squall line. *Journal of the Atmospheric Sciences* 52: 1934–1948. https://doi.org/10.1175/1520-0469(1995)052<1934:TEOMPO>2.0.CO;2.

Tao, W.-K., J. Simpson, C.-H. Sui, C.-L. Shie, B. Zhou, K.M. Lau, and M.W. Moncrieff. 1999. On equilibrium states simulated by cloud-resolving models. *Journal of the Atmospheric Sciences* 56: 3128–3139. https://doi.org/10.1175/1520-0469(1999)056<3128:ESSBCR>2.0.CO;2.

Tao, W.-K., S. Lang, W.S. Olson, S. Yang, R. Meneghini, J. Simpson, C. Kummerow, E. Smith, and J. Halverson. 2001. Retrieved vertical profiles of latent heat release using TRMM products for February 1998. *Journal of Applied Meteorology* 40: 957–982. https://doi.org/10.1175/1520-0450(2001)040<0957:RVPOLH>2.0.CO;2.

Tao, W.-K., D. Johnson, C.-L. Shie, and J. Simpson. 2004. The atmospheric energy budget and large-scale precipitation efficiency of convective systems during TOGA COARE, GATE, SCSMEX, and ARM: Cloud-resolving model simulations. *Journal of the Atmospheric Sciences* 61: 2405–2423. https://doi.org/10.1175/1520-0469(2004)061<2405:TAEBAL>2.0.CO;2.

Tao, W.-K., et al. 2006. Retrieval of latent heating from TRMM measurements. *Bulletin of the American Meteorological Society* 87: 1555–1572. https://doi.org/10.1175/BAMS-87-11-1555.

Tao, W.-K., S. Lang, X. Zeng, S. Shige, and Y. Takayabu. 2010. Relating convective and stratiform rain to latent heating. *Journal of Climate* 23: 1874–1893. https://doi.org/10.1175/2009JCLI3278.1.

Thompson, A.M., W.-K. Tao, K.E. Pickering, J.R. Scala, and J. Simpson. 1997. Tropical deep convection and Ozone formation. *Bulletin of the American Meteorological Society* 78: 1043–1054.

Tomita, H., K. Goto, and M. Satoh. 2008. A new approach of atmospheric general circulation model: Global cloud resolving model NICAM and its computational performance. *SIAM Journal on Scientific Computing* 2008 (30): 2755–2776.

Tsay, S.-C., et al. 2016. Satellite-surface perspectives of air quality and aerosol-cloud effects on the environment: An overview of 7-SEAS/BASELInE. *Aerosol and Air Quality Research* 16: 2581–2602. https://doi.org/10.4209/aaqr.2016.08.0350.

Varble, A., E.J. Zipser, A.M. Fridlind, P. Zhu, A.S. Ackerman, J.-P. Chaboureau, S. Collis, J. Fan, A. Hill, and B. Shipway. 2014. Evaluation of cloud-resolving and limited area model intercomparison simulations using TWP-ICE observations: 1. Deep convective updraft properties. *Journal of Geophysical Research* 119: 13,891–13,918. https://doi.org/10.1002/2013jd021371.

Wang, Y., W.-K. Tao, and J. Simpson. 1996. The impact of ocean surface fluxes on a TOGA COARE convective system. *Monthly Weather Review* 124: 2753–2763. 10.1175/1520-0493(1996)124< 2753:TIOOSF>2.0.CO;2.

Wang, Y., W.-K. Tao, J. Simpson, and S. Lang. 2003. The sensitivity of tropical squall lines (GATE and TOGA COARE) to surface fluxes: 3-D cloud resolving model simulations. *Quarterly Journal of the Royal Meteorological Society* 129: 987–1007. https://doi.org/10.1256/qj.00.131.

Wu, D., C. Peters-Lidard, W.-K. Tao, and W. Petersen. 2016. Evaluation of NU-WRF real-time rainfall forecast for IFloodS. *Journal of Hydrometeorology*. https://doi.org/10.1175/jhm-d-15-0134.1.

Yanai, M., S. Esbensen, and J. Chu. 1973. Determination of bulk properties of tropical cloud clusters from large-scale heat and moisture budgets. *Journal of the Atmospheric Sciences* 30: 611–627.

Yoneyama, K., C. Zhang, and C.N. Long. 2013. Tracking pulses of the Madden-Julian oscillation. *Bulletin of the American Meteorological Society* 94: 1871–1891. https://doi.org/10.1175/bamsd-12-00157.1.

Yuter, S.E., and R.A. Houze Jr. 1995. Three-dimensional kinematic and microphysical evolution of Florida cumulo- nimbus. Part II: Frequency distributions of vertical velocity, reflectivity, and differential reflectivity. *Monthly Weather Review* 123: 1941–1963. https://doi.org/10.1175/1520-0493(1995)1231941:tdkame.2.0.CO;2.

Zeng, X., W.-K. Tao, T. Matsui, S. Xie, S. Lang, M. Zhang, D. Starr, and X. Li. 2010. Estimating the ice crystal enhancement factor in the tropics. *Journal of the Atmospheric Sciences* 68: 1424–1434.

Zeng, X., W.-K. Tao, M. Zhang, A.Y. Hou, S. Xie, S. Lang, X. Li, D. Starr, X. Li, and J. Simpson. 2009a. The indirect effect of ice nuclei on atmospheric radiation. *Journal of the Atmospheric Sciences* 66: 41–61.

Zeng, X., W.-K. Tao, M. Zhang, S. Lang, C.D. Peters-Lidard, J. Simpson, S. Xie, S. Kumar, J.V. Geiger, C.-L. Shie, and J.L. Eastman. 2007. Evaluation of long-term cloud-resolving modeling with observational cloud data. *Journal of the Atmospheric Sciences* 64: 4153–4177.

Zeng, X., W.-K. Tao, M.-H. Zhang, A. Y. Hou, S. Xie, S. Lang, X. Li, D. O'C Starr, and X. Li. 2009b. A contribution by ice nuclei to global warming. *Quarterly Journal of the Royal Meteorological Society* 135: 1614–1629.

Zeng, X., W.-K. Tao, S. Powell, R.A. Houze, P.E. Ciesielski, N. Guy, H. Piece, and T. Matsui. 2013. A comparison of the water budgets between clouds from AMMA and TWP-ICE. *Journal of the Atmospheric Sciences* 70: 487–503. https://doi.org/10.1175/JAS-D-12-050.1.

Zhang, J., K. Howard, C. Langston, B. Kaney, Y. Qi, L. Tang, H. Grams, Y. Wang, S. Cocks, S. Martinaitis, and A. Arthur. 2016. Multi-radar multi-sensor (MRMS) quantitative precipitation estimation: Initial operating capabilities. *Bulletin of the American Meteorological Society* 97: 621–638.

Zhang, S.Q., T. Matsui, S. Cheung, M. Zupanski, and C. Peters-Lidard. 2017. Impact of assimilated precipitation-sensitive radiances on the NU-WRF simulation of the West African Monsoon. *Monthly Weather Review* (in revision).

Zhou, Y., D. Wu, K.-M. Lau, and W.-K. Tao. 2016. Scale dependence of land atmospheric interactions in wet and dry regions as simulated with NU-WRF over Southwest and Southeast US. *Journal of Hydrometeorology*. https://doi.org/10.1175/jhm-d-16-0024.1.

Stochastic Representation of NCEP GEFS to Improve Sub-seasonal Forecast

Yuejian Zhu, Wei Li, Xiaqiong Zhou and Dingchen Hou

Abstract The National Centers for Environmental Prediction (NCEP) Global Ensemble Forecast System (GEFS) has been in daily operation to provide probabilistic guidance for public since December 1992. Since July 2017, the GEFS was extended from 16 days to 35 days forecast to support NCEP Climate Prediction Center (CPC)'s sub-seasonal forecast. The latest GEFS version was upgraded in three areas to improve sub-seasonal forecast: (1) introducing a new set of stochastic physical perturbations to improve model uncertainty representation for the tropics; (2) a 2-tiered SST approach to consider ocean impact; and (3) a new scale-aware convection scheme to improve model physics for tropical convection and MJO forecasts. The new set of stochastic physical perturbations include stochastic kinetic energy backscatter to make up subscale energy lost during model integration; stochastic physics perturbation tendency with five different spatial and temporal scales to perturb physical tendency; and stochastic perturbed humidity on the model lower level. After upgraded to new set of stochastic physical perturbations, the MJO forecast skill has been improved from 12.5 days of a 25-month period to nearly 22 days by combining all three modifications include stochastic physics. In the extratropics, the 500-hPa geopotential height; surface temperature and precipitation are improved for sub-seasonal timescale as well. However, the raw forecast skills of surface temperature and precipitation are extremely low, and the results imply that calibration may be important and necessary for surface temperature and precipitation forecast for the sub-seasonal timescale due to the large systematic model errors.

Keywords NCEP · GEFS · Stochastic representation · Sub-seasonal forecast

---

Y. Zhu (✉) · W. Li · X. Zhou · D. Hou
Environmental Modeling Center, NCEP/NWS/NOAA, 5830 University Research Ct., College Park, MD 20740, USA
e-mail: Yuejian.Zhu@noaa.gov

D. A. Randall et al. (eds.), *Current Trends in the Representation of Physical Processes in Weather and Climate Models*, Springer Atmospheric Sciences,
https://doi.org/10.1007/978-981-13-3396-5_15

1 Introduction

With the improvement of accuracy of weather forecasting and the increasing computational capacity, a seamless forecast that ranges from weather to seasonal timescale is in growing interest and demanding in general public and service sectors in order to protect life and properties. Extending the weather forecast to cover sub-seasonal timescale clearly has great socioeconomic significance. However, in scientific aspect, improving the forecast skill on this timescale is quite challenging. This gap in the forecast skill between weather and climate is partially due to the limitation of forecast predictability (Lorenz 1969) and less sensitivity to the initial condition which benefits the weather scale yet insufficient sensitivity to the boundary and external forcing which benefits the seasonal and longer lead time (Vitart 2014; Johnson et al. 2014; Liu et al. 2016; Troccoli 2010; Tian et al. 2017). Imperfectness of the representation of the model dynamics and physics, however, should be considered as the major source of uncertainties and errors for all lead time (Buizza et al. 1999). The approaches that aim to reasonably represent the model uncertainty thus become a practical method to reduce the model errors in recent years. The efforts in this regards include a multi-model ensemble method (Shin and Krishnamurti 2003; Palmer et al. 2004; Kirtman et al. 2014) that represents the overall uncertainty from different models; a stochastic total tendency perturbation method (STTP, Hou et al. 2008) that represents the uncertainty related to both dynamic and physics in single model; a stochastic physics perturbation tendency scheme (SPPT, Buizza et al. 1999; Palmer et al. 2009) that represents the uncertainty related to total model physical process; In addition to the stochastic perturbation on the tendency, Stochastic Kinetic Energy Backscatter (SKEB, Shutts and Palmer 2004; Shutts 2005; Berner et al. 2009; Shutts et al. 2015) is another way to present forecast uncertainty through considering the energy at non-resolved scales which cannot cascade to larger scales due to the model's finite resolution. All these methods have been used in operational centers and research community (Palmer et al. 2009).

Since it was implemented into operation in 1992, the NCEP GEFS has been widely used as probabilistic forecast guidance for the forecast within 2 weeks (Zhou et al. 2017). Regardless of the initial perturbation, the operational version of GEFS uses STTP to represent the model uncertainty. With the contribution of both initial uncertainties and perturbation in total tendency, the ensemble mean forecast of GEFS has outperformed the GFS deterministic forecast on the anomaly correlation of forecast lead at Day-8 of Northern Hemisphere 500 hPa geopotential height to represent mid-level general circulation for the past few years (Fig. 1).

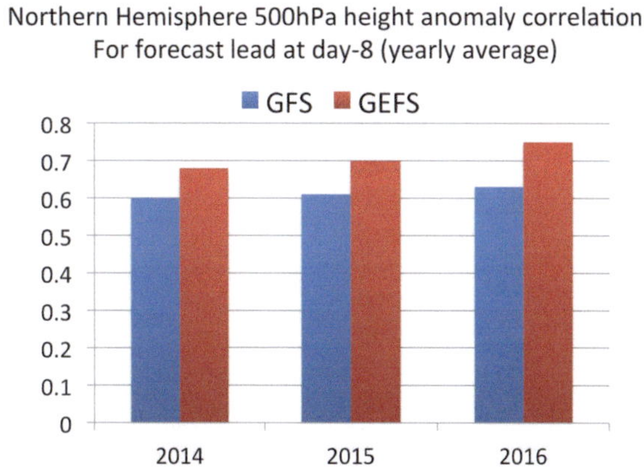

Fig. 1 Northern hemisphere 500 hPa geopotential height anomaly correlation for forecast lead at day-8 of GFS forecast (blue) and ensemble mean forecast (red) during years 2014–2016

2 Stochastic Physics Perturbation Schemes Tested in NCEP GEFS 35-Day Forecast

To be aligned with NOAA's mission of generating a unified coupled forecast system to cover the timescale from weather to seasonal, GEFS has carried out investigations on the strategy to potentially improve the forecast skill on week 3 and 4 time range (sub-seasonal timescale), and further to cover monthly forecast. A recent investigation is testing the impact of different stochastic perturbation schemes that represent the model uncertainty on the performance of sub-seasonal forecast (Zhu et al. 2017, 2018; Li et al. 2018). The motivation for this work came from the concerns of the under-dispersion (or overconfidence) of the current operational version of GEFS (GEFS v11 with EnKF initial perturbation + STTP) on medium range forecast especially over the tropics (Hou et al. 2008; Zhou et al. 2016, 2017).

Although STTP scheme compensates the less error growth from initial perturbations to some degree, the impact of the STTP is mainly over extratropics during boreal winter season with less impact on the spread over tropical region. It is well known that MJO is a major source of the predictability on sub-seasonal timescale. Therefore, improving the representation of the model uncertainty over tropics is a possible pathway to potentially improve this source of sub-seasonal predictability. A suite of three widely accepted stochastic perturbation methods (SPs hereafter) is thus applied to GEFS to represent the model uncertainties instead of STTP more efficiently (Table 1, second row). The scheme of SPs are: SKEB from expectation of making up subscale energy lost due to imperfect computation algorithms; SPPT with five different spatial and temporal scales (Figs. 2 and 3); Stochastic Perturbed Humidity (SHUM; Tompkins and Berner 2008) with single spatial-temporal scale, and near model surface layers. These schemes have already been implemented in the National Center Environmental Prediction (NCEP) Global Forecast System (GFS) model in the hybrid-EnKF data assimilation system, then basically available for use

Table 1 The configuration differences for four experiments

Experiments	Stochastic schemes	Boundary (SST)	Convection
CTL	STTP	Default	Default
SPs	SKEB + SPPT + SHUM	Default	Default
SPs + SST_bc	SKEB + SPPT + SHUM	2-Tiered SST	Default
SPs + SST_bc + SA_CV	SKEB + SPPT + SHUM	2-Tiered SST	Scale-aware convection

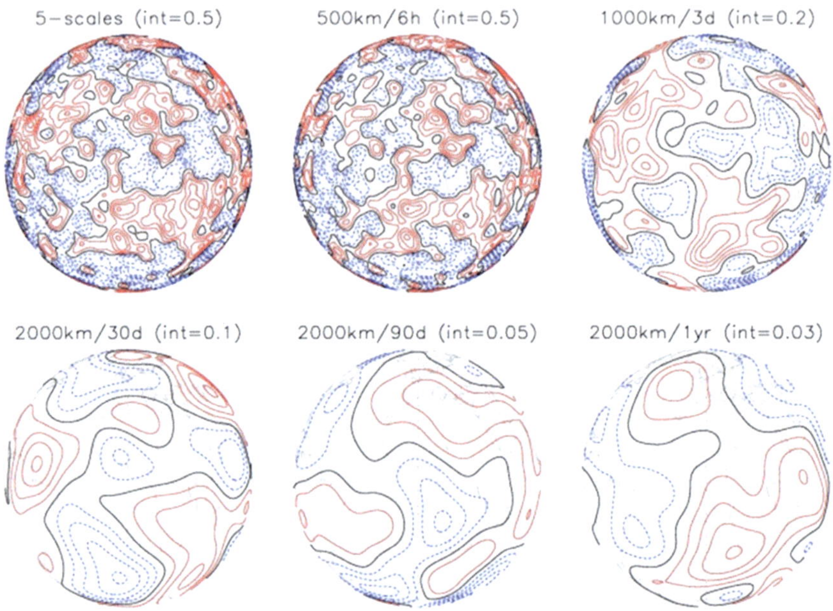

Fig. 2 5-scale random patterns used in stochastic perturbed physics tendencies (SPPT). On the top of each plot, the numbers (except for upper left) represent the scales of spatial and temporal perturbations with contour intervals in the bracket. The upper left is for combined total 5-scales

in the GEFS for testing and modification. Detailed descriptions of these schemes are as follows.

The SKEB scheme has been used to represent dynamical uncertainty through subgrid-scale processes that propagate upscale. A stream function forcing from the total dissipation has been applied to SKEB. Depending on numerical model design, the numerical dissipation (i.e., the diffusions) is only part to be considered in current GFS version. The generations of such perturbations on each vertical level are independently to provide some vertical coherence through vertical smoothing. Overall, the SKEB scheme should improve the global power spectrum and increase forecast spread.

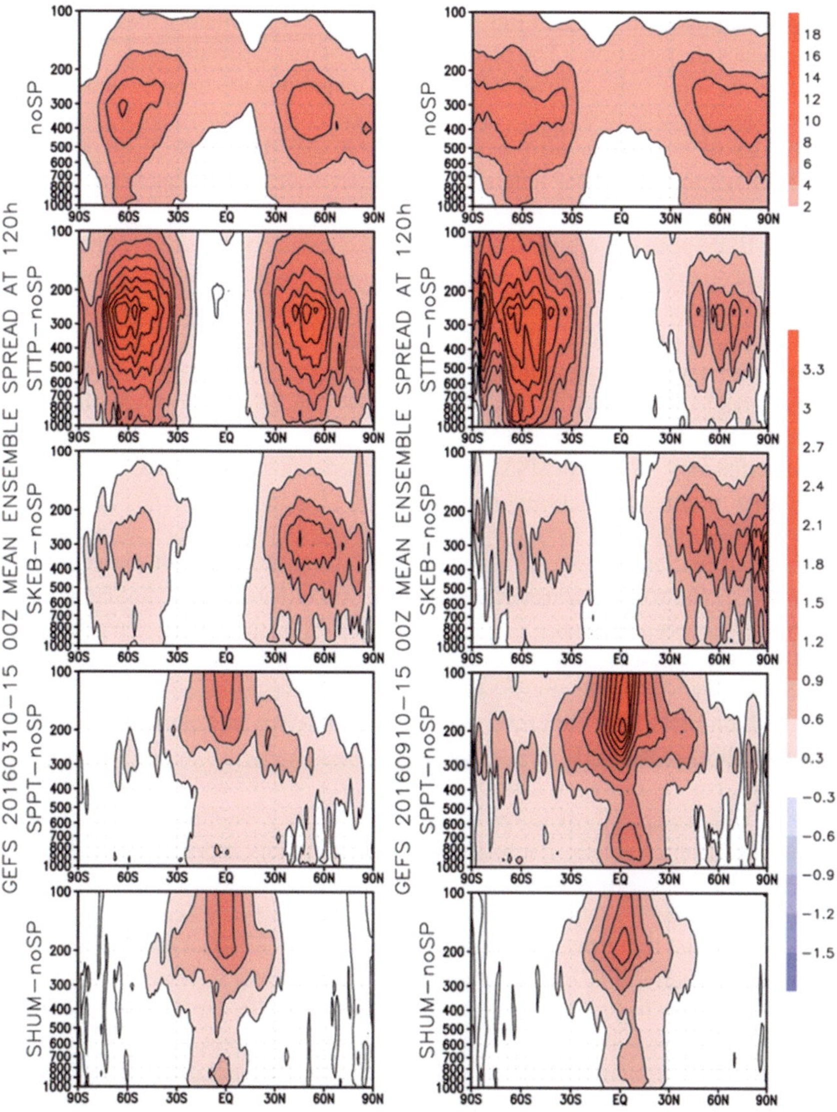

Fig. 3 Global meridional cross section showing the impact of stochastic perturbations for the atmosphere (cross section) for 120 h forecasts from six spring initializations (left) and six fall initializations (right). Paneled are the differences of zonal wind spread from CTL for (top) no stochastic physical perturbations, and the difference of STTP (upper middle); SKEB (middle); SPPT; (lower middle) and SHUM (bottom)

The SPPT scheme perturbs the total tendencies of temperature, wind, and water vapor during numerical integration generated by the GFS physics parameterizations (after all physics processes). The current version of SPPT implies five different random patterns with different timescales and correlation length scales to generate the tendency perturbations. The patterns of the stochastic perturbations, in general, are uniform in the vertical levels, except their magnitude are reduced and taped to zero gradually for both of near surface and above the tropopause. The maximum amplitudes of five scales are 0.8, 0.4, 0.2, 0.08, and 0.04 respectively. Figure 2 demonstrates the individual independent random scale patterns and combined 5-scale random pattern.

The SHUM scheme perturbs the near-surface humidity only; based on the concept that the uncertainty in humidity can have nonlinear impacts as thresholds in physical parameterizations are crossed (e.g., convective initiation). SHUM uses the same random pattern generator as SPPT scheme but only a single spatial-temporal scale is used with maximum amplitude of 0.006. The perturbation is a maximum in the lowest model level and decreases exponentially with height. However, its impact transports rapidly to upper level of troposphere.

Since the new schemes (SPs), which is the combination of three schemes are introduced to replace the current operational STTP scheme in GEFS, an averaged spread at 120-hr forecast in two seasons are demonstrated in Fig. 3 to indicate the relative contribution and effect of each individual stochastic scheme. Compared to the ensemble spread without considering stochastic perturbations (i.e., noSP, top row), the control experiment with STTP (operational GEFS, the second row) produced extra spread in the extratropics area without major impact in the tropics. For the package of new stochastic physics schemes (SPs), however, an additional spread is produced through SKEB (3rd row), which has similar spatial contribution of STTP (2nd row). Both of the SPPT (row 4) and SHUM (last row) generate additional spread for tropics, but the evolution characteristics are slightly different (not shown here) from SHUM which only perturbs humidity in the near boundary. An increasing spread over tropics greatly improves forecast uncertainty representation and also enhances the tropical forecast skill.

Apparently, the SPs (combined three stochastic schemes) present forecast spread globally, especial for tropical area when compared to NCEP operational GEFS solo stochastic scheme (STTP). The statistical scores for tropical zonal winds of 850 and 250 hPa show huge improvements from introduced SPs for 2 years experiment periods (Fig. 4). For both of upper and lower atmosphere levels, increased forecast spread does also result in the reduced forecast error (root mean square error—RMSE; left plots of Fig. 4); and higher continuous ranked probability skill scores (CRPSS; right plots of Fig. 4). Moreover, the spreads are more closed to forecast errors (left plots of Fig. 4) indicates the representation of forecast uncertainty is more realistic than current operational GEFS (STTP).

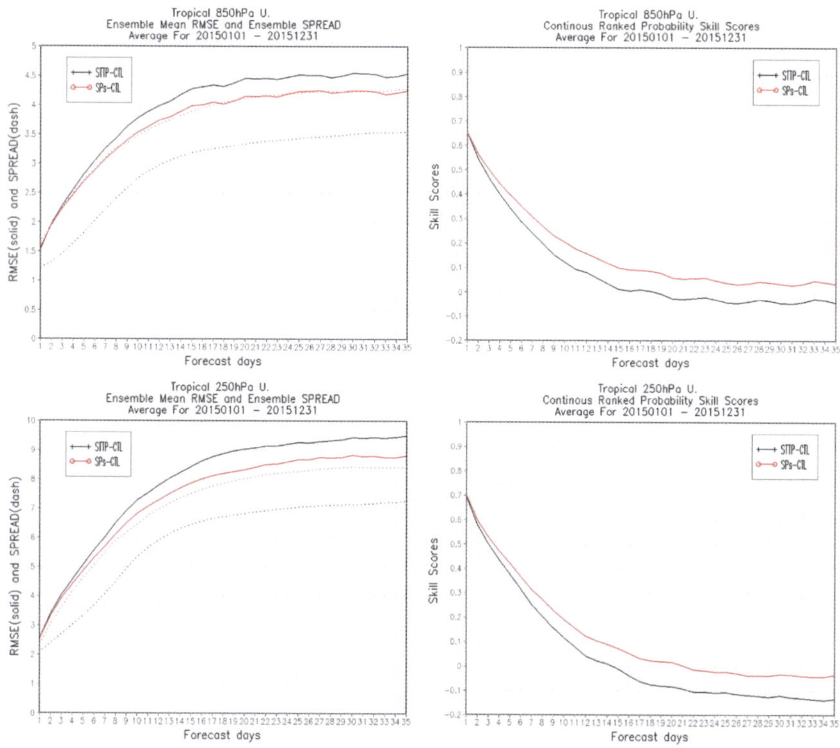

Fig. 4 RMS error of the ensemble mean (solid) and the ensemble spread (dash) (left), and CRPSS (right) are plotted every 24 h out to 35 days for 850-hPa (top) tropical (20°N–20°S) zonal wind during the Jan. 2015 to Dec. 2015 period comparing CTL (black) and SPs (red)

3 Other Strategies on Improving Ensemble Forecast on Sub-seasonal Timescale

The sub-seasonal forecast has different dependences from the short-term forecast. While the short-term forecast largely relies on the initial condition, the sub-seasonal forecast more and more relies on the boundary and external forcing. As such, for an uncoupled forecast system on sub-seasonal timescale, an accurate representation of the prescribed Sea Surface Temperature (SST) is of great importance (Li et al. 2001; Ling et al. 2015). The operational version of GEFS uses a prescribed SST that is initiated from analysis data and damps to climatology. Taking into account the day-to-day variability of the SST and as an intermediate stage between uncoupled and coupled forecast system, the underlying SST is updated using the bias-corrected SST from coupled model forecast (i.e., two-tiered SST, Table 1, third row).

As for the forecast system, an accurate representation of the physical process is critical to the forecast skill, the last strategy (or configuration) (Table 1, fourth row)

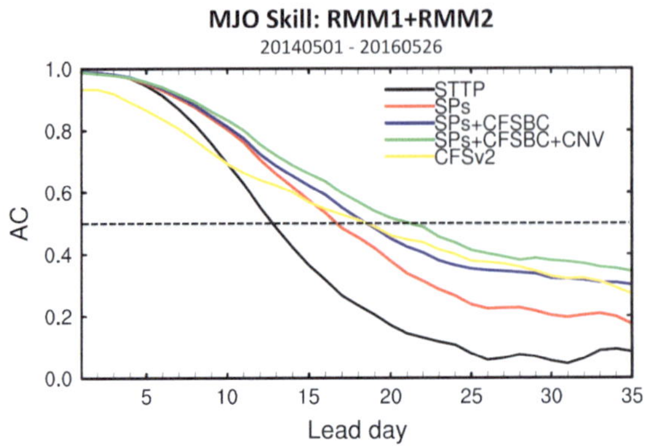

Fig. 5 MJO skills of the four different configurations of GEFS and CFSv2

that was tested is combining new SPs; two-tiered SST; and an upgraded Simplified Arakawa-Schubert (SAS) cumulus parameterization scheme that is both scale- and aerosol-aware (Han et al. 2017). The highlights of this upgraded convective parameterization scheme include: (1) the change of the rain conversion rate; (2) the modification of convective adjustment time in deep convection; (3) the cloud base mass flux in the shallow convection scheme becomes a function of mean updraft velocity; (4) convective inhibition (CIN) in the sub-cloud layer is an additional trigger condition to suppress unrealistic spotty rainfall; and (5) convective cloudiness is enhanced by suspended cloud condensate in an updraft.

The performance of the different GEFS configurations is demonstrated in Figs. 5 and 6. Since the Madden Julian Oscillation (MJO) is the dominant mode on the sub-seasonal predictability, MJO and its associated components are one of the emphases to evaluate the capability of the forecast system on sub-seasonal timescale. Compared to STTP scheme, the performance of the 850 hPa zonal wind over the tropics indicated a significant improvement associated with the increase of the spread in SPs (Fig. 4). The skill of the upper level zonal wind showed similar improvement (Figure not shown. please confirm), indicating a positive impact of the SPs on the MJO associated circulation. The RMM MJO skill increased from ~12.5 days in STTP scheme to 16.8 days in SPs. Combing SPs and updated SST further result in the increased MJO skill to 18.5 days. Combing SPs with updated SST and updated convection scheme lead to increase the MJO skill to 22 days (Fig. 5). The impact of the different configurations on the Northern Hemisphere large-scale circulation indicated the consistent result as the MJO (Fig. 6), with the improvement from STTP to SPs. The statistics, in terms of NH 500 hPa geopotential height anomaly correlations for average period of week-2 (days 8–14) and weeks 3 and 4 (days 15–28), include NCEP Climate Forecast System version 2 (CFSv2). The results indicate (1) All three new configurations show similar or better score than GEFS operation (ctl) for week-2, but much better than CFSv2; (2) All there new configurations demonstrate the very valuable skills for weeks 3 and 4 than GEFS operation (ctl), and much better than CFSv2.

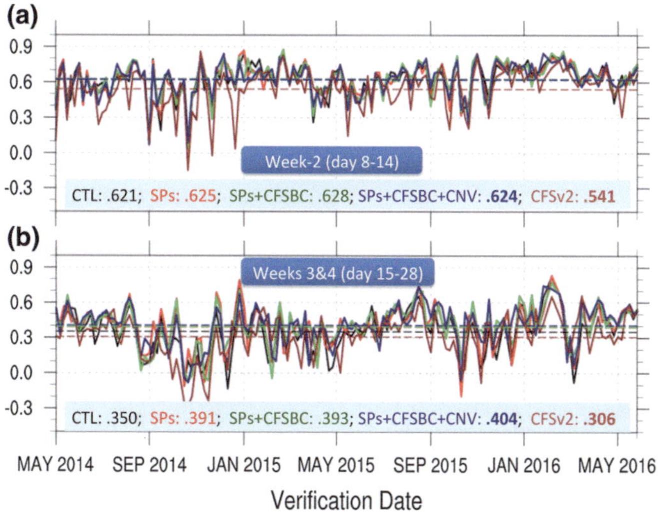

Fig. 6 The time series of ensemble mean anomaly correlation for Northern Hemisphere (20°N–80°N) 500-hPa geopotential height from May 2014 to May 2016 for different configurations (CTL-black; SPs-red; SPs + CFSBS-green and SPs + CFSBC + CNV-purple) and CFSv2 (orange) for lead week-2 (**a**) and weeks 3 and 4 (**b**). Days 15–28 (weeks 3 and 4 average). Average scores are shown in the bottom of each plot

4 Towards Physically Based Stochastic Parameterization

As we demonstrated in Sect. 2 for various stochastic perturbation schemes, most of them are in current operational ensemble forecast system that is still preliminary approach to assimilate model based uncertainties. There are many limitations in the application of the stochastic schemes. For example, the SKEB scheme highly depends on the accumulation of dissipation in the numerical integration, the horizontal and vertical diffusions scheme, gravity wave drag and mountain blocking parameterization et al. A contribution of SKEB will be greatly reduced when model resolutions are increased and when numerical schemes are improved. In addition, perturbation in SPPT varies with model physics thus varies with total physics tendency. The spatial and temporal de-correlation of the stochastic patterns thus does not really reflect uncertainty associated with individual physical process.

Figure 7 is a schematic diagram which demonstrates the current status of the stochastic perturbations, and the approach that represents the model uncertainties through realistic stochastic parameterization which is most possibly to be applied in the future. In the same time, two valuable studies have been done based on the operational ECMWF ensemble forecast system to apply (1) Independent random patterns to perturb different physical processes (or iSPPT). It is a similar procedure to current stochastic schemes (SPPT) but accounting stochastics for each individual physical process (Christensen et al. 2017); (2) Stochastic perturbed selected 20 phys-

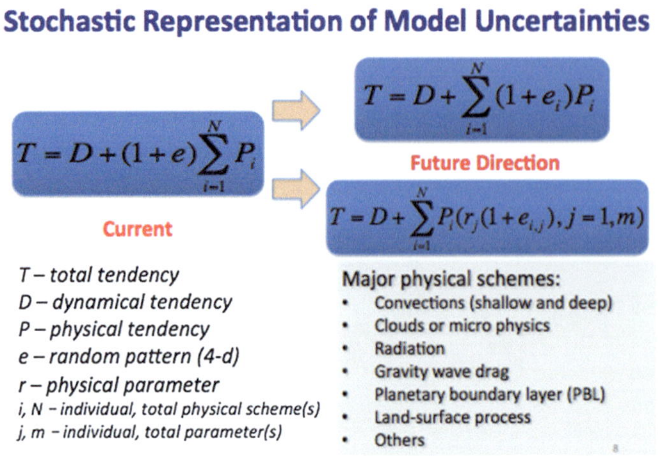

Fig. 7 Schematic diagram to present current status and future direction for stochastic representation of model uncertainties

ical parameters (SPP). Both of them increase ensemble spreads in general. The later one could change vertical distributions of forecast uncertainties significantly and thus may represent model uncertainties through the interaction of physical processes more realistically.

With the rapid progress in ensemble forecast system development and better understanding on the model physical process, the representation of forecast uncertainties from model dynamics and physics should be more approach to realistic atmosphere. Following this progress, many other sources of uncertainties, such as soil moisture and soil temperature from land model, sea surface temperature from ocean model, reflectivity of snow, and sea ice from sea ice model will be considered to improve weather forecast and sub-seasonal-climate prediction.

5 Summary

Stochastic perturbation is important processes that can help to improve sub-seasonal prediction after it succeeds for weather forecast. It advanced MJO skills significantly and associated tropical atmospheric circulation (850 and 200 hPa zonal winds, Li et al. 2018). It also enhanced extratropical prediction skills for weeks 3 and 4 average. In contrast to NCEP CFS v2 that is a coupling system with lower model resolution and older model physics, latest GEFS configuration has taken great advantage with new SPs, two-tiered SST and new convective parameterization in terms of tropical and extratropical, weather and sub-seasonal prediction. There are two areas we should focus on in near future: (1) Improve current physical tendency perturbation scheme to represent physical processes more realistically; (2) Consider other sources of uncertainties from land, sea, and other surface boundary.

Acknowledgements The authors would like to thank all of the helps from EMC ensemble team members, and Dr. Bing Fu helped to provide Figs. 2 and 3; Mr. Eric Sinsky provided Figs. 4 and 6 in particular. This study is partially supported through NWS's Office of Science and Technology Integration (OSTI) and NOAA's Climate Program Office (CPO)'s Modeling, Analysis, Predictions, and Projections (MAPP) program.

References

Berner, J., G.J. Shutts, M. Leutbecher, and T.N. Palmer. 2009. A spectral stochastic kinetic energy backscatter scheme and its impact on flow-dependent predictability in the ECMWF ensemble prediction system. *Journal of the Atmospheric Sciences* 66 (3): 603–626.

Buizza, R., M. Miller, and T. Palmer. 1999. Stochastic representation of model uncertainties in the ECMWF ensemble prediction system. *Quarterly Journal Royal Meteorological Society* 125 (560): 2887–2908.

Christensen, H.M., S.-J. Lock, I.M. Moroz, and T.M. Palmer. 2017. Introducing independent patterns into the stochastically perturbed parametrization tendencies (SPPT) scheme. *Quarterly Journal of the Royal Meteorological Society, Part A* 143 (706): 2168–2181.

Han, J., W. Wang, Y.C. Kwon, S.-Y. Hong, V. Tallapragada, and F. Yang. 2017. Updates in the NCEP GFS cumulus convection schemes with scale and aerosol awareness. *Weather and Forecasting*. https://doi.org/10.1175/WAF-D-17-0046.1.

Hou, D., Z. Toth, Y. Zhu, and W. Yang. 2008. Evaluation of the impact of the stochastic perturbation schemes on global ensemble forecast. In *Proceedings of the 19th conference on probability and statistics*, New Orleans, LA, American Meteor Society. https://ams.confex.com/ams/88Annual/webprogram/Paper134165.html.

Johnson, N.C., D. Collins, S. Feldstein, M. L'Heureux, and E. Riddle. 2014. Skillful wintertime North American temperature forecasts out to 4 weeks based on the state of ENSO and the MJO. *Weather and Forecasting* 29: 23–38. https://doi.org/10.1175/WAF-D-13-00102.1.

Kirtman, B.P., D. Min, and J.M. Infanti. 2014. The North American multimodel ensemble: Phase-1 seasonal-to-interannual prediction; Phase-2 toward developing intraseasonal prediction. *Bulletin of the American Meteorological Society* 95: 585–601.

Li, W., R. Yu, H. Liu, and Y. Yu. 2001. Impacts of diurnal cycle of SST on the intraseasonal variation of surface heat flux over the western Pacific warm pool. *Advances in Atmospheric Sciences* 18 (5): 793–806.

Li, W., Y. Zhu, X. Zhou, D. Hou, E. Sinsky, C. Melhauser, M. Pena, H. Guan, and R. Wobus. 2018. Evaluating the MJO prediction skill from different configurations of NCEP GEFS extended forecast. *Climate Dynamics*, https://doi.org/10.1007/s00382-018-4423-9.

Ling, T., M. Xu, X.-Z. Liang, J.X.L. Wang, and Y. Noh. 2015. A multilevel ocean mixed layer model resolving the diurnal cycle: Development and validation. *Journal of Advances in Modeling Earth Systems* 07. https://doi.org/10.1002/2015ms000476.

Liu, X., T. Wu, S. Yang, T. Li, W. Jie, L. Zhang, Z. Wang, X. Liang, Q. Li, Y. Cheng, H. Ren, Y. Fang, and S. Nie. 2016. MJO prediction using the sub-seasonal to seasonal forecast model of Beijing Climate Center. *Climate Dynamics*. https://doi.org/10.1007/s00382-016-3264-7.

Lorenz, E. 1969. The predictability of a flow which possesses many scales of motion. *Tellus* 21: 289–307. https://doi.org/10.1111/j.2153-3490.1969.tb00444.x.

Palmer, T.N., et al. 2004. Development of a European multi-model ensemble system for seasonal to inter-annual prediction (DEMETER). *Bulletin of the American Meteorological Society* 85: 853–872.
Palmer, T.N., R. Buizza, F. Doblas-Reyes, T. Jung, M. Leutbecher, G. Shutts, M. Steinheimer, and A. Weisheimer. 2009. Stochastic parametrization and model uncertainty. Technical Report ECMWF RD Tech. Memo. 598, 42 pp. http://www.ecmwf.int/publications/.
Shin, D.W., and T.N. Krishnamurti. 2003. Short- to medium-range superensemble precipitation forecasts using satellite products: 1. Deterministic forecasting. *Journal of Geophysical Research* 108 (D8): 8383. https://doi.org/10.1029/2001jd001511.
Shutts, G. 2005. A kinetic energy backscatter algorithm for use in ensemble prediction systems. *Quarterly Journal Royal Meteorological Society* 131: 3079–3102.
Shutts, G. 2015. A stochastic convective backscatter scheme for use in ensemble prediction systems. *Quarterly Journal of the Royal Meteorological Society: Part A* 141 (692): 2602–2616.
Shutts, G., and T.N. Palmer. 2004. The use of high-resolution numerical simulations of tropical circulation to calibrate stochastic physics schemes. In *Proceedings of the ECMWF/CLIVAR simulation and prediction of intra-seasonal variability with emphasis on the MJO*, Reading, United Kingdom, European Centre for Medium-Range Weather Forecasts, 83–102.
Tian, D., Eric F. Wood, and X. Yuan. 2017. CFSv2-based sub-seasonal precipitation and temperature forecast skill over the contiguous United States. *Hydrology and Earth System Sciences* 21: 1477–1490.
Tompkins, A.M., and J. Berner. 2008. A stochastic convective approach to account for model uncertainty due to unresolved humidity variability. *Journal Geophysical Research* 113: D18101.
Troccoli, A. 2010. Seasonal climate forecasting. *Meteorological Applications* 17: 251–268. https://doi.org/10.1002/met.184.
Vitart, F. 2014. Evolution of ECMWF sub-seasonal forecast skill scores. *Quarterly Journal of the Royal Meteorological Society* 140: 1889–1899. https://doi.org/10.1002/qj.2256.
Zhou, X., Y. Zhu, D. Hou, and D. Kleist. 2016. Comparison of the ensemble transform and the ensemble Kalman filter in the NCEP global ensemble forecast system. *Weather and Forecasting* 31: 2058–2074.
Zhou, X., Y. Zhu, D. Hou, Y. Luo, J. Peng, and D. Wobus. 2017. The NCEP global ensemble forecast system with the EnKF initialization. *Weather and Forecasting* 32: 1989–2004.
Zhu, Y.X., M. Zhou, W. Pena, C.Melhauser Li, and D. Hou. 2017. Impact of sea surface temperature forcing on weeks 3 and 4 forecast skill in the NCEP global ensemble forecasting system. *Weather and Forecasting* 32: 2159–2173. https://doi.org/10.1175/WAF-D-17-0093.1.
Zhu, Y., X. Zhou, W. Li, D. Hou, C. Melhauser, E. Sinsky, M. Pena, B. Fu, H. Guan, W. Kolczynski, R. Wobus, and V. Tallapragada. 2018. An assessment of subseasonal forecast skill using an extended global ensemble forecast system (GEFS). *Journal of Geophysical Research* 6732–6745. https://doi.org/10.1029/2018JD028506.

New Pathways for Moist Convection Parameterisation

A. Pier Siebesma and Jesse Dorrestijn

Abstract This chapter starts with an introduction to standard mass-flux parameterisations and closures for moist convection which are used in most large-scale models. The shortcomings of this approach will be discussed, especially for higher resolutions where standard assumptions such as quasi-equilibrium start to break down. New pathways that use a stochastic approach and are scale aware will be discussed. These will also allow to incorporate the effect of mesoscale organisation into parameterisations for moist convection.

Keywords Clouds · Moist convection · Stochastic parameterisation · Grey zone Conditional Markov Chains

1 Large-Scale Budget Equations

The grid averaged prognostic equations for the dry static energy $s_\mathrm{d} \equiv c_\mathrm{p}T + gz$, specific humidity q_v and condensed water q_c in large-scale models can be written as

$$\overline{\mathrm{D}}_t\overline{s}_\mathrm{d} = -\frac{1}{\rho}\partial_z\rho\,\overline{w's_\mathrm{d}'} + \ell_\mathrm{v}(c-e) + Q_\mathrm{rad} \tag{1a}$$

$$\overline{\mathrm{D}}_t\overline{q}_\mathrm{v} = -\frac{1}{\rho}\partial_z\rho\,\overline{w'q_\mathrm{v}'} - (c-e) \tag{1b}$$

$$\overline{\mathrm{D}}_t\overline{q}_\mathrm{c} = -\frac{1}{\rho}\partial_z\rho\,\overline{w'q_\mathrm{c}'} + (c-e) - G \tag{1c}$$

where ρ is the average density, w the vertical velocity, ℓ_v the latent heat, $c - e$, the net condensation rate, i.e. condensation minus evaporation, G the autoconversion rate from condensed cloud water to rain water and Q_rad the radiative heating tendency.

A. P. Siebesma (✉) · J. Dorrestijn
Faculty of Civil Engineering and Geoscience, Delft University of Technology, Stevinweg 1, 2628CN Delft, The Netherlands
e-mail: a.p.siebesma@tudelft.nl

D. A. Randall et al. (eds.), *Current Trends in the Representation of Physical Processes in Weather and Climate Models*, Springer Atmospheric Sciences,
https://doi.org/10.1007/978-981-13-3396-5_16

The primes denote deviations from the grid mean values. The operator D_t represent a total time derivative while the overbars denote a averaging filtering operator over the horizontal size of the grid mesh of the large-scale model, i.e. for an arbritrary variable ϕ.

$$\overline{\phi}(x, y) = \frac{1}{\Delta x \Delta y} \int_x^{x+\Delta x} \int_y^{y+\Delta y} \phi(x', y')\mathrm{d}x'\mathrm{d}y' \tag{2}$$

where Δx and ΔY represents the size of the grid mesh. The terms on the right-hand side of Eq. (1) compromise the subgrid terms that require parameterisation. Cumulus parameterisation aims to represent the turbulent fluxes, the thermodynamic effects of phase changes of the water and the conversion from cloud condensate to rain.

In the absence of evaporation of rain, the surface precipitation rate P can be diagnosed by vertically integrating the autoconversion rate

$$P = \int_0^{\infty} G\rho dz \tag{3}$$

Inspection of (1) promotes the use the liquid water static energy $s_\ell \equiv s_\mathrm{d} - \ell_\mathrm{v} q_\mathrm{c}$ and the total water specific humidity $q_\mathrm{t} \equiv q_\mathrm{v} + q_\mathrm{c}$, leading to

$$\overline{\mathrm{D}}_t \overline{s}_\ell = -\frac{1}{\rho}\partial_z \rho \overline{w' s'_\ell} + Q_\mathrm{rad} \tag{4a}$$

$$\overline{\mathrm{D}}_t \overline{q}_\mathrm{t} = -\frac{1}{\rho}\partial_z \rho \overline{w' q'_\mathrm{t}} - G \tag{4b}$$

Note that, the condensation terms now have been absorbed in the moist conserved variables s_ℓ and q_t so that in the absence of precipitation and radiative processes, the cumulus parameterisation challenge is reduced into a turbulent mixing problem. Alternatively, the moist static energy, defined as $s_\mathrm{e} \equiv s_\mathrm{d} + \ell_\mathrm{v} q_\mathrm{v}$ can also be used as a conserved energy variable.

2 Traditional Mass-Flux Parameterisation for Moist Convection

2.1 Mass-Flux Approximation

Most conventional cumulus parameterisations are based on the assumption that the turbulent flux of cumulus convection can be well approximated in terms of one or multiple saturated convective updrafts and downdrafts. In its simplest form, a grid box of a large- scale model can be decomposed into two parts: a convective updraft part with the fractional area a_u consisting of all saturated convective updrafts and

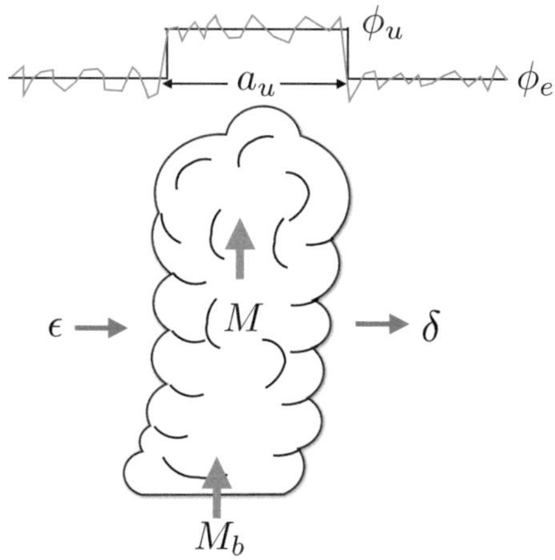

Fig. 1 Upper panel: illustration of the top-hat approximation. Lower panel: illustration of the entraining plume model. The model is feed from below by a cloud base mass flux. The updraft is diluted by entrained air ϵ from the environment while detrainment δ describes the outflow of updraft air into the environment

a complementary environmental fraction with a fractional area $1 - a_\mathrm{u}$. The domain averaged value of any variable $\overline{\phi}$ can be expressed in terms of its updraft average as

$$\overline{\phi} = a_\mathrm{u}\phi_\mathrm{u} + (1 - a_\mathrm{u})\,\phi_\mathrm{e} \tag{5}$$

where the subscripts u and e indicate, respectively, the updraft and the environmental average. For cumulus convection, the differences between the updraft and environmental averages are in general larger than the typical fluctuations within the updrafts and the environment. This supports the use of a top-hat approximation, i.e. neglecting the fluctuations within the updrafts and the environment, as schematically indicated in Fig. 1. In doing so, the turbulent flux can be simplified to

$$\rho\overline{w'\phi'} = \rho\frac{a_\mathrm{u}}{1 - a_\mathrm{u}}\left(w_u - \overline{w}\right)\left(\phi_u - \overline{\phi}\right). \tag{6}$$

Since cumulus updrafts are organised in narrow updrafts with updraft velocities, much larger than the large-scale vertical velocity $\overline{w}$, Eq. (6) is usually further approximated as

$$\overline{w'\phi'} \approx \mathcal{M}_\mathrm{u}\left(\phi_u - \overline{\phi}\right), \quad \mathcal{M}_\mathrm{u} = \rho a_\mathrm{u} w_\mathrm{u} \tag{7}$$

where the convective mass flux $\mathcal{M}_\mathrm{u}$ has been introduced. Equation (7) constitutes the simplest bulk mass-flux approximation. More complex spectral mass-flux models can be defined by subdividing the cloud updraft ensemble into subensembles of updrafts that have a similar cloud top height labelled by a subscript i

$$\rho\overline{w'\phi'} \approx \sum_i \mathcal{M}_{\mathrm{u},i}\left(\phi_{\mathrm{u},i} - \overline{\phi}\right). \tag{8}$$

Extensions to downdrafts can be included in a similar manner. The advantage of the mass-flux approximation is that the difficult task of parameterising the turbulent fluxes for heat, moisture and momentum, as displayed by Eqs. (1a) and (1b) has been reduced to finding updraft values for ϕ_u and mass flux profiles $\mathcal{M}_u$.

2.2 Cloud Updraft Equations

The standard way of parameterising the mass flux and the updraft fields are through the use of a steady-state entraining plume model. Applying the same top-hat approximation for the budget updraft equations for heat and moisture gives (Tiedtke 1989; Siebesma 1998)

$$\partial_z s_{\ell,u} = -\varepsilon\left(s_{\ell,u} - \overline{s}_\ell\right) \tag{9a}$$

$$\partial_z q_{t,u} = -\varepsilon\left(q_{t,u} - \overline{q}_t\right) - G_u/\mathcal{M}_u \tag{9b}$$

$$\partial_z \ln \mathcal{M}_u = \varepsilon - \delta \tag{9c}$$

where ϵ represents the fractional entrainment of environmental air into the updrafts and δ the outflow by detrainment of updraft air into the environment. The first terms on the right-hand side of Eqs. (9a) and (9b) represent the dilution of the updrafts by entrainment of environmental air. The autoconversion rate G_u in the updraft is related to the grid averaged autoconversion rate G using the tophat approximation as $G = a_u G_u$, where it is implicitly assumed that no autoconversion takes place outside the updrafts. Finding the appropriate parameterisations for the ϵ and δ is an active field of research as these parameters are the largest source of uncertainty in climate model projections (de Rooy et al. 2013).

Once the conserved variables $q_{t,u}$ and $s_{\ell,u}$ of the updraft equations have been determined from Eq. (9), the non-conserved variables T_u, q_v, u and $q_{c,u}$ can be easily derived, because the updrafts can be assumed to be saturated so that $q_v = q_s(p, T)$, where q_s is the saturation specific humidity.

Often, a vertical velocity equation is added to the entraining plume model which can be used to find the cloud top height, defined as the height where the updraft vertical velocity vanishes. The updraft velocity equation can be derived in a similar fashion as the entraining plume equations for q_t and s_ℓ as

$$\frac{1}{2}\partial_z {w_u}^2 = a B_u - \varepsilon\, {w_u}^2, \tag{10}$$

where the constant $a \simeq 0.75$ is a tunable prefactor that aims to correct for the fact that pressure gradient terms are not taken into account explicitly (de Roode et al. 2012). The buoyancy term B_u is expressed as

$$B_u = \frac{g}{T_0}\left(T_{\rho,u} - \overline{T}_\rho\right) \tag{11}$$

where T_0 refers to a reference temperature and where a density temperature $T_\rho = T\,(1 - 0.622q_\mathrm{v} + q_\mathrm{c})$ has been introduced. In the absence of a vertical velocity equation, cloud top height can be estimated to be at the level of neutral buoyancy (LNB), defined as the height where the updraft buoyancy B_u becomes negative.

Cloud updrafts are assumed to be initiated at the lifting condensation level (LCL) defined as the height where an adiabatic parcel initiated from near the surface becomes saturated. The adiabatic parcel, that mimics a dry thermal in the subcloud layer, can also be used to find the updraft temperature and humidity at cloud base. Moist convection is triggered if the adiabatic parcel has enough kinetic energy to reach its level of free convection, the height where the saturated updraft becomes positively buoyant.

The main remaining task of the convection parameterisation is to find the value for the mass flux at cloud base $\mathcal{M}_{\mathrm{u},b}$ which determines the overall intensity of the convection and is referred to as the principal closure of the convection scheme. Two standard closures will be reviewed in the next subsection.

2.3 Mass-Flux Closures

Parameterisation of the cloud base mass flux essentially requires and understanding of the relation between the intensity of convection and the properties of the environment. Two classes of closures can be distinguished: one based on convective quasi-equilibrium and another based on boundary layer quasi-equilibrium. Both will be briefly reviewed in this subsection.

Convective quasi-equilibrium assumes that cumulus convection stabilises the atmosphere by convective mixing at a rate much faster than the destabilisation through radiative cooling and other non-convective processes (Arakawa and Schubert 1974). Quasi-equilibrium can be phrased most conveniently in terms of the conditional available potential energy (CAPE) $\mathcal{A}$

$$\mathcal{A} \equiv \int_{z_\mathrm{f}}^{z_\mathrm{n}} B_\mathrm{u}\, dz = \frac{\beta}{c_p} \int_{z_\mathrm{f}}^{z_\mathrm{n}} (s_{\rho,\mathrm{u}} - \overline{s}_\rho)\, dz \qquad (12)$$

where β represents the buoyancy parameter $\beta \equiv g/T_0$. The integral extents from the level of free convection z_f to the level of neutral buoyancy z_n. In the second step, the buoyancy has been re-expressed in terms of the density dry static energy, $s_\rho \equiv c_p T_\rho + gz$, a variable conserved for dry adiabatic displacements and a measure for buoyancy at the same time.

$\mathcal{A}$ essentially measures the amount of potential energy in the atmosphere that can be released by moist convection. The CAPE Quasi-equilibrium implies that $\mathcal{A}$ is consumed by moist convection in a typical convective adjustment time τ_adj

$$(\partial_t \mathcal{A})_\mathrm{conv} = -\frac{\mathcal{A}}{\tau_\mathrm{adj}}, \qquad (13)$$

much shorter than the time it takes by radiative cooling and other non-convective processes to produce $\mathcal{A}$.

Assuming that the compensating subsidence in the environment due to moist convection is the dominant effect that moist convection exerts on the the environment, the density dry static energy tendency due to convection can be written as

$$\left(\partial_t \overline{s}_\rho\right)_{\text{conv}} \simeq \frac{\mathcal{M}_{\text{u}}}{\rho} \partial_z \overline{s}_\rho, \tag{14}$$

Rewriting Eq. (14) in terms of $\mathcal{A}$ and combining it with the adjustment closure Eq. (13) gives the following expression for the cloud base mass flux (Fritsch and Chappell 1980):

$$\mathcal{M}_{u,b} = \frac{\mathcal{A}}{\tau_{\text{adj}}} \left(\frac{\beta}{c_p} \int_{z_b}^{z_t} m \, \partial_z \overline{s}_\rho dz \right)^{-1} \tag{15}$$

where m is a normalised mass flux

$$m(z) \equiv \frac{\mathcal{M}_{\text{u}}}{\mathcal{M}_{\text{u,b}}} \tag{16}$$

In summary, the CAPE closure Eq. (15) determines a cloud base mass flux that is required to consume the existing $\mathcal{A}$ in a convective adjustment time τ_{adj} by moist convection.

On the contrary, boundary layer quasi-equilibrium is based on the observation that the subcloud layer, especially over the oceans, is in an approximate steady state for the moist static energy $s_{\text{e}} \equiv c_{\text{p}} T + gz + l_{\text{v}} q_{\text{v}}$. Such an equilibrium can be established if the sources of the moist static energy in the subcloud layer such as surface fluxes and low level convergence of moist static energy are compensated by the venting of heat and moisture out of the boundary layer by moist convection. Such a boundary layer equilibrium for the subcloud layer can be expressed as

$$\mathcal{M}_{\text{u}} \left(s_{\text{e,u}} - \overline{s}_{\text{e}} \right) \Big|_{\text{b}} = \int_0^{z_{\text{b}}} \left(\rho \, \partial_t \overline{s}_{\text{e,adv}} - \partial_z \rho \overline{w' s_e'} \right) dz \tag{17}$$

where the left-hand side represents the turbulent flux of moist static energy at cloud base while the right-hand side represents the surface flux and the resolved, large-scale advective tendencies of $\overline{s}_{\text{e}}$ (Tiedtke 1989).

Note that, the CAPE closure (15) is based on the instability of the atmosphere above cloud base while the boundary equilibrium is based on the convergence of moist static energy below cloud base.

Both approaches provide a closure in terms of the mass flux. Therefore, mass-flux schemes for moist convection do not require separate information of the updraft vertical velocity w_{u} nor on the updraft fraction a_{u}, the two components that make up the mass flux $M \equiv \rho w_{\text{u}} a_{\text{u}}$. In a sense, this is the elegance of mass-flux schemes but in the next section, the drawbacks of this approach will be discussed.

3 Large-Scale Controls of Moist Convection

The prime challenge of any parameterisation is to find and utilise relationships between the resolved, large-scale forcing and its response on the smaller unresolved scale. This can be illustrated by expanding the total time derivative in the prognostic equations (4a) and (4b) in an Eulerian form. This way the prognostic equations for $\phi \in \{s_\ell, q_t\}$ can be written in a schematic form as

$$\partial_t\overline{\phi} = \left(\partial_t\overline{\phi}\right)_{\text{res}} + \left(\partial_t\overline{\phi}\right)_{\text{subgrid}} \tag{18}$$

where the resolved part is given by

$$\left(\partial_t\overline{\phi}\right)_{\text{res}} = -\partial_x\overline{u}\overline{\phi} - \partial_y\overline{v}\overline{\phi} - \frac{1}{\rho}\partial_z\rho\overline{w}\overline{\phi} \tag{19}$$

while the subgrid or parameterised part is provided by the terms on the right-hand side of Eqs. (4a) and (4b).

Cumulus parameterisations need to be formulated in terms of the resolved variables, either through the mean state $\overline{\phi}$ or through the large-scale forcing $\left(\partial_t\overline{\phi}\right)_{\text{res}}$. Two examples are the CAPE closure (15) where the intensity of cumulus convection is related to the atmospheric instability, as expressed by the mean state $\overline{\phi}$, and the boundary equilibrium closure where convergence provided by the large-scale forcing $(\partial_t\overline{s_e})_{\text{adv}}$ is used as a closure. A relevant question for moist convection parameterisations is therefore: how strong are the relative contributions of CAPE and convergence of $\overline{s_e}$ to the intensity of cumulus convection?

To address this question, Loriaux et al. (2017) (hereafter referred to as L17) analysed a large number of large eddy simulations (LES) of strong moist convection in which the atmospheric stability, the atmospheric relative humidity and the strength of the moist convergence were systematically varied. The simulations were performed on a 100 by 100 km domain, indicative for the size of a large-scale model grid cell, at a resolution of 200 m on a time scale of a day. Large-scale convergence was prescribed as well as the surface fluxes. The relation between convergence, surface evaporation and precipitation can be made explicit by integrating Eq. (4b) over the depth of the atmosphere

$$\int_0^\infty \partial_t\overline{q}_t\rho dz = -\int_0^\infty \left(\partial_x\overline{u}\,\overline{q}_t + \partial_y\overline{v}\,\overline{q}_t\right)\rho\, dz + \rho\overline{w'q'_{t0}} - P, \tag{20}$$

or, symbolically

$$S_q = MC + E - P, \tag{21}$$

where S_q denotes the storage of moisture in the atmosphere, MC indicates the moisture convergence and E the surface evaporation. Equation (21) shows that the precipitation amount is determined by moist convergence and surface evaporation on longer

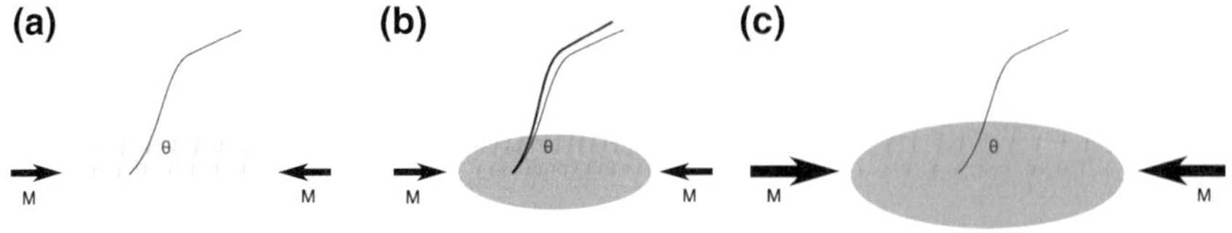

Fig. 2 Conceptual drawing showing the effects of stability (**b**) and moisture convergence perturbations (**c**) on precipitation, with respect to the reference experiment (**a**). Destabilisation and increased moisture convergence both lead to an increase in precipitation. This occurs through intensification (**b**), or through an increase in the precipitation area fraction (**c**), respectively, (figure reproduced from L17)

timescales when the storage term becomes irrelavant. On subdiurnal timescales, S_q can strongly modify the precipitation amounts.

The large eddy simulation results in L17 show an increase of the precipitation rate P with increasing moist convergence, but increased precipitation is also observed for larger values of CAPE, i.e. for a more unstable atmosphere. While this dependency of the precipitation itself might not be that surprising, a more interesting question is how the change in precipitation is precisely realised. In general, the precipitation rate can be written as the product of intensity P_r and the fractional precipitation area a_r[1]

$$P = a_r P_r \tag{22}$$

so that a change in precipitation can be written as

$$\delta P = P_r da_r + a_r d P_r \tag{23}$$

In L17, it is demonstrated that the mean precipitation for stronger moisture convergence is mainly caused by an increase of the precipitation area fraction a_r, with minor changes in the precipitation intensity P_r. The opposite is true when increasing the atmospheric instability. In this case, the precipitation intensity increases while the convective area remains relatively constant. These effects of the large-scale controls on precipitation are schematically visualised in Fig. 2.

Since precipitation can be interpreted as a measure of the intensity of convection, the results can also be used to interpret the mass flux response to large-scale controls. Changes in the large-scale convergence control changes in the fractional updraft area of the mass flux. It regulates boundary layer equilibrium through the convective area fraction. Atmospheric instability affects the updrafts strength but leaves the convective area relatively constant. This finding makes sense in terms of Eq. (10) which indeed couples updraft strength directly to CAPE. So, both convergence and stability do influence the strength of convection as measured by the mass flux but in different

[1]The precipitation area is not equal to the updraft area but for precipitating cumulus convection it is reasonable to assume that they are proportional to each other.

manners. Whereas, increase of instability is manifested in the mass flux through more intense updraft velocities, increase in convergence and induces a stronger mass flux through a larger convective fractional area a_u.

4 Resolution, Grey Zone and Stochastic Parameterisations

Traditionally, convection parameterisations are deterministic. In deterministic parameterisations, the tendency of each model variable ϕ is a deterministic function of the prognostic variables and large-scale forcings $\mathbf{x} \in \{u, v, w, s_\ell, q_u, \ldots\}$

$$\partial_t \phi = f(\mathbf{x}). \tag{24}$$

The initial state $\mathbf{x}_0$ determines the future time state $\mathbf{x}_T$ in a deterministic way (Fig. 3). In stochastic models, random numbers affect the model variable tendencies:

$$\partial_t \phi = f(\mathbf{x}, \alpha), \tag{25}$$

in which $\alpha(x, y, z, t)$ is a stochastic process, producing random numbers that depend on time t and grid point location (x, y, z). In the schematic illustration in the right panel of Fig. 3, the model variable trajectories of ϕ are not uniquely determined by the initial state $\mathbf{x}_0$, instead several trajectories are possible and consequently, several outcomes for the future time state $\mathbf{x}_T$. These random numbers can affect the time derivatives every time step, thereby possibly changing model behaviour.

The development of stochastic schemes have been pioneered by the European Centre for Medium-Range Weather Forecasts (ECMWF) and have been adopted by many modelling centres worldwide. One widely used approach is the stochastically perturbed parametrisation tendencies scheme (SPPT), whereby a spatially and temporally correlated random number field is used to perturb the parametrisation tendencies from the model's physics package (Buizza et al. 2007). This pragmatic approach to represent uncertainty was initially proposed to improve the reliability

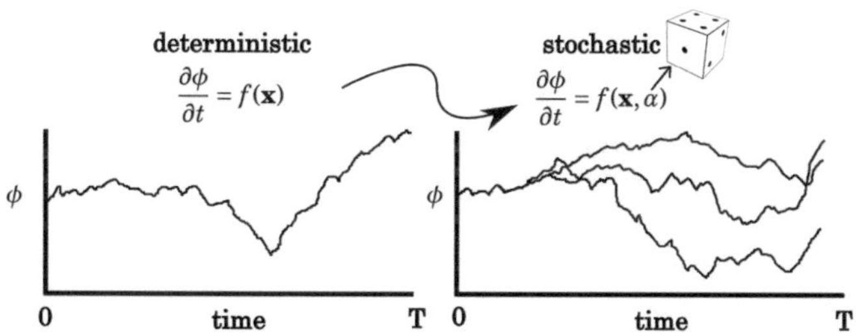

Fig. 3 In deterministic models (left), the initial state uniquely determines the trajectories of the model variables while in stochastic models (right), several trajectories are possible

of the Centre's medium-range forecasts by correcting the under-dispersive nature of their forecast ensemble. The imposed spatial and temporal correlation scales have no physical basis, though coarse-graining studies have been used to retrospectively justify the multiplicative nature of the noise and its magnitude. Despite the ad hoc choices in the formulation of the scheme, SPPT is surprisingly effective at improving ensemble reliability, reducing forecast error, and improving biases in the mean and variability of the model's climate.

In the next subsection, it will be discussed under which conditions stochastic convection parameterisations become desirable and how convection parameterisations can be made stochastic in a physical consistent manner.

4.1 The Grey Zone

Historically, convection parameterisation has been developed under the assumption of quasi-equilibrium. If the grid size Δx is much larger than the individual size of the convective updrafts ℓ_u, it implies that the number of convective updrafts in a grid box of a large-scale model will be large, i.e. $N_u \gg 1$. Under these conditions, a cumulus parameterisation does not aim to predict the behaviour of individual updrafts but rather the overall effect of an ensemble of many of these updrafts in a deterministic statistical sense. If furthermore, the rate of change of the large scale state is slow enough to allow the cloud ensemble with sufficient time to adjust accordingly, quasi-equilibrium can be assumed. This quasi-equilibrium assumption forms the basis of the CAPE closure expressed in Eq. (15). More generally, only in the case of quasi-equilibrium, it is reasonable to expect a one-to-one correspondence between the resolved processes and the parameterised response of the subgrid processes, which would justify a deterministic approach. This situation is schematically sketched in the right panel of Fig. 4 where $\Delta x \gg l_u$ so that $N_u \gg 1$. Because the horizontal size of convective updrafts is in the range of 1–10 km, this requires a grid sizes of at least several hundreds of kilometres for deep convection.

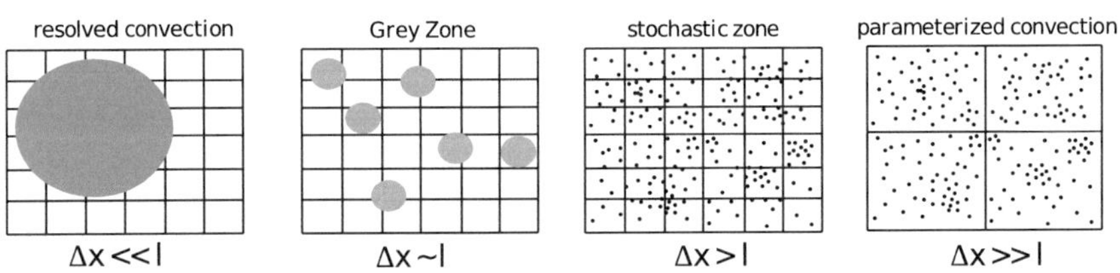

Fig. 4 Depending on the grid size, convection can be fully resolved (left panel) or parameterised deterministically (right panel). In between these two extreme situations (middle panels), scale-aware and stochastic parameterisations become more appropriate

At the other end of the spectrum, where $\Delta x \ll l_u$, we have the situation where the resolution is so fine that simulations become cloud resolving and no explicit parameterisation of the updrafts is required anymore.

In between these two extremes, there is the situation where $\Delta x \simeq l_u$. In this case, the resolution is such that it partially resolves effects of cumulus convection but where an additional parameterisation is still required. Resolutions around this value, are referred to as the grey zone. Within the grey zone, a partial parameterisation of the convection process is required but the relative contribution depends on the used resolution. It is for this reason that in this grey zone, parameterisations need to be scale-aware.

4.2 *Stochastic Parameterisations*

The intermittent and random character of moist convection vanishes for the case $\Delta x \gg l_u$ when only statistical effects of a whole ensemble is relevant. For resolutions $\Delta x > l_u$, close or in the grey zone as depicted in the second panel from the right in Fig. 4, it can be seen that the number of updrafts present in a model grid box varies significantly from grid box to grid box, even when they are subjected to the same large-scale forcing. Therefore, for these resolutions, stochastic parameterisations have more potential to adequately represent convection, because a deterministic convection parameterisation will provide a too strict and too hard-wired relationship with the corresponding large-scale state.

Inspection of Fig. 4 suggests that the level of stochasticity should increase with finer resolution and should be manifested in the convective area fraction. At coarse resolutions, a constant convective area fraction $a_u \ll 1$ is sufficient and could be absorbed in the mass-flux formulation as explained in the previous section. At finer resolutions, however, a stochastic treatment of a_u is required. Note also that for finer resolutions, the convective area fraction can take any value between 0 and 1, so that also the assumption $a_u \ll 1$ breaks down when approaching the grey zone (Arakawa et al. 2011). This requires an explicit stochastic parameterisation for a_u. Along with an appropriate parameterisation for the updraft velocity w_u this could replace the necessity to have a separate parameterisation of the mass flux as expressed by Eq. (9c). This will also have the advantage that an explicit parameterisation of detrainment will become obsolete.

Recently, several approaches have been proposed to parameterise convection stochastically: e.g. Plant and Craig (2008) introduced a stochastic multiplume model to obtain the cloud base mass flux in a deep convection scheme; Sakradzija et al. (2015) used this framework to parameterise shallow convection stochastically; Teixeira and Reynolds (2008) directly affected the convective tendencies; Bengtsson et al. (2013) used cellular automata to influence convective parameterisation, while Sušelj et al. (2013) introduced a stochastic eddy-diffusivity– mass-flux approach. In all these papers, it has been shown that the natural variability of convection can be better captured with a stochastic rather than with a deterministic parameterisation.

In the next section, a new method will be described that uses conditional Markov chains to parameterise the convective updraft fraction stochastically in a scale-aware manner and consistent with the relations between the large-scale forcing and the convective response.

5 Stochastic Convection Parameterisation Using Conditional Markov Chains

5.1 Markov Chains

One of the simplest ways to construct a stochastic model is through the use of Markov chains. To this purpose, we subdivide a grid box of a large-scale model into a microgrid that consists of N nodes (see Fig. 5). Such a grid box is populated with different cloud types in the various stages of their life cycle. Inspired by Khouider et al. (2010) and Dorrestijn et al. (2015), it is assumed that each node is occupied by one of the following five possible states

1. clear sky
2. moderate cumulus congestus
3. strong cumulus congestus
4. deep convective cumulus
5. stratiform cloud (anvil)

Of course, different choices are possible with more or fewer states. Once the state of each node is known the fraction of the sky covered by a certain cloud type, a_m can be simply found by adding the number of nodes that belong to the same cloud type

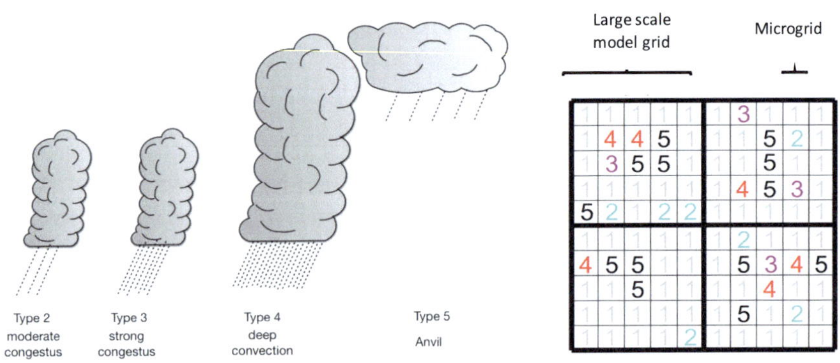

Fig. 5 Each node is occupied by one of the five states: (1) clear sky, (2) moderate congestus, (3) strong congestus, (4) deep convective cloud and (5) stratiform cloud (anvil). Nodes form a microgrid that is contained inside the large-scale model grid

and dividing this by N, the total number of nodes

$$a_{\mathrm{m}} = \frac{1}{N}\sum_{n=1}^{N} \mathbf{1}\,[Y_n = m] \tag{26}$$

in which $\mathbf{1}[\cdot]$ is the indicator function ($\mathbf{1}(A) = 1$ if A is true and $\mathbf{1}(A) = 0$ if A is false). At the next time step, each node can be in a different state according to a transition probability matrix $P(m, n)$ that denotes the probability that a node which is at time t in state m will be in state n at time $t + \Delta t$. The transition probabilities can be inferred from physical considerations (Khouider et al. 2010), from observational data (Dorrestijn et al. 2015) or from high resolution modelling data (Dorrestijn et al. 2013). In Dorrestijn et al. (2015) (hereafter referred to as D15), the transition probabilities were inferred from observational data of precipitation intensity and cloud top height from the Darwin C-band polametric (CPOL) Radar in Darwin (Kumar et al. 2013). By defining a low and a high precipitation intensity (less or more than $12\,\mathrm{mm\,h^{-1}}$) and high and low cloud top (less or more than 6 km), the four cloud categories can be identified.

In this case, the transition probability matrix $P(m, n)$ can be estimated as

$$P(m, n) = \frac{T(m, n)}{\sum_n T(m, n)}, \tag{27}$$

where $T(m, n)$ counts the number of transitions from m to n observed in the training data set.

Each node of the microgrid can be considered as a finite state Markov chain as indicated in Fig. 5. Since in this case, the transition probability only depends on the present state, all the N Markov chains defined on the microgrid can be considered as N independent processes that do not interact with each other.

5.2 *Conditional Markov Chains*

The finite state Markov chains cannot be expected to provide a realistic representation of the cloud dynamics as long as the transition probabilities only depend on the present state. The occurrence and strength of moist convection depends in reality on the large-scale physical state of the atmosphere. For instance, divergence, a stable atmosphere and a low relative humidity are all factors that will favour transitions to a cloud-free state while convergence, unstable conditions and a more humid atmosphere will enhance the probability towards states with congestus or deep convection.

Such dependencies can be taken into account by making the transition probabilities conditional to a relevant aspect of the large-scale state $X(t)$ of the atmosphere. In that case, we can define the conditional probability

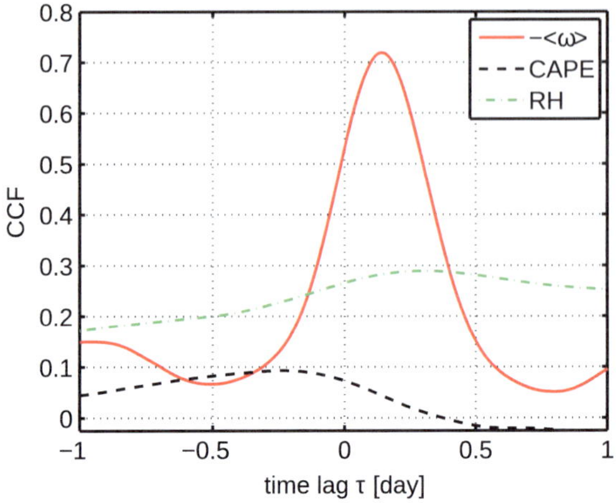

Fig. 6 Time-lagged cross-correlation between the deep convective area fraction and the large-scale vertical velocity $-\omega$, CAPE and relative humidity (RH) at 640 hPa

$$P(m,n)_\gamma \equiv \text{Prob}(Y(t+\Delta t) = n | Y(t) = m, X(t) = \gamma) \tag{28}$$

where $P(m,n)_\gamma$ can be obtained from observations by counting the number of transitions from m to n, given that the large-scale state is equal to γ at time t.

The question is therefore: what is a good indicator for states displayed in Fig. 5 that can be used for γ, or phrased differently: which atmospheric variable or process correlates strongly with the occurrence of convection. In D15, three possible indicators have been investigated: the thermal instability $\mathcal{A}$ of the atmosphere as measured by CAPE, the large-scale vertical velocity $\overline{w}$ in the lowest 5 km and the atmospheric relative humidity at 640 hPa. All these indicators have been derived from the ERA reanalysis (Dee et al. 2011) co-located with the CPOL radar measurements in Darwin.

The time-lagged cross-correlation function (CCF), derived in D15, between the indicators and the deep convective area fraction is displayed in Fig. 6. It can be seen that the vertical velocity has by far the strongest correlation at $t = 0$ (i.e. no time lag) with the deep convective area fraction a_4. Note that, the maximum correlation of the vertical velocity (or convergence) occurs for a positive time lag which indicates that deep convection reinforces convergence, which might suggest that convection is the cause rather than the result of large-scale convergence. But even at $t = 0$, the time at which deep convection is diagnosed, the large scale vertical velocity has still the strongest correlation with deep convection. Therefore, in D15, convergence is used as an indicator for constructing the Conditional Markov Chains (CMC) by calculating 25 matrices $P(m,n)_\gamma$ for 15 different binned values of $\overline{w}$. Not surprisingly, the bins with the largest values of $\overline{w}$ coincide with the largest transition probabilities to deep convection. Extensions to the use of multiple indicators are possible. For instance, in Peters et al. (2017), the transition probability to deep convective states is made dependent on both the vertical velocity and the mid-tropospheric relative humidity.

5.3 Scale Adaptivity

The degree of stochasticity is controlled by N, the number of conditional Markov chains (CMCs) used within one grid box (see Fig. 5). The parameter N in Eq. 26 controls the number of clouds that can be contained inside a large-scale model grid column. By increasing N, the cloud type area fractions a_m become less stochastic and converge towards their expectation values. These expectation values are determined by the large-scale variables that are used to condition the Markov chains.

The number of CMCs is related to the used resolution. The resolution of the microgrid is determined by the training data set. For instance, in D15, the CPOL rain radar was used for the training of the transition probabilities. Since the radar data were retrieved at a resolution of $\Delta x_{\text{node}} = 2\,\text{km}$, it is naturally to take this as the resolution for the microgrid displayed in Fig. 5. Therefore, the number of CMCs used in a parameterisation for a large-scale model with a resolution Δx can be expressed as

$$N \simeq \left(\frac{\Delta x}{\Delta x_{\text{node}}} \right)^2 \tag{29}$$

which indeed implies that finer resolutions (smaller Δx) require a smaller number N of CMCs per gridbox, and consequently more stochastic intermittancy. As an illustration, Fig. 7 shows a time series of the area fraction of deep convective nodes a_4 observed in Darwin along with two simulations of the CMCs for $N = 100$ and $N \to \infty$. The latter simulation can be regarded as the deterministic limit of the stochastic model. The timing of the deep convective events produced by the CMC's is almost correct for both cases. However, the deterministic parameterisation underestimates

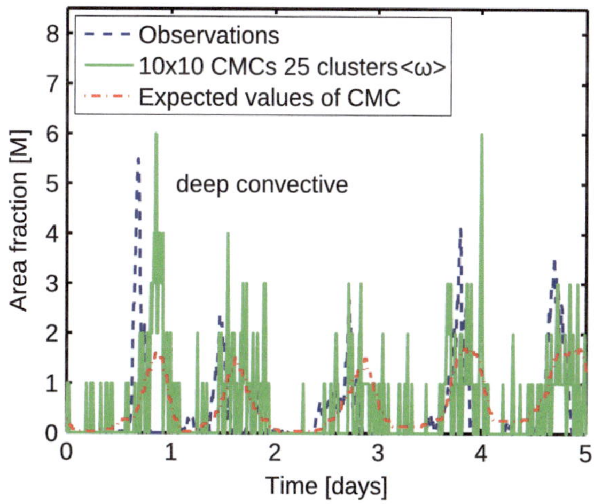

Fig. 7 Time series of the area fraction of deep convective nodes a_4 observed in Darwin (blue dashed line) produced by 100 CMCs (solid green line) and the corresponding expected deep convective area fractions for the deterministic limit, corresponding to the case $N \to \infty$ (red dashed line)

the maximum values of the precipitation events. This demonstrates how the stochastic fluctuations of the multicloud model fractions are needed in order to produce the correct maximum values of the deep convective area fraction.

6 Synthesis and Outlook

Standard convection parameterisations are traditionally formulated in terms of a mass flux. This approach has the advantage that knowledge about the updraft area fraction and updraft velocity, which are the components that constitute the mass flux, is not required. Although this has been proven to be a powerful and elegant approach, there is an increasing number of arguments suggesting that estimating the mass-flux explicitly as a product of the cloud updraft fraction and the updraft velocity could be more advantageous

1. The dependence of the mass-flux profile is traditionally determined by the entrainment and detrainment profiles as expressed by Eq. (9c). Many studies (de Rooy et al. 2013; Neggers 2015) indicate that entrainment profiles depend on the updraft size distribution and on the updraft vertical velocity strength while detrainment profiles are closely connected to the vertical gradient of the cloud fraction a_u. So although Eq. (9c) allows a direct formulation of M in terms of fractional entrainment and detrainment, information on the w_u and a_u is required. A direct formulation of the mass flux in terms of a_u and w_u would be physically more consistent and would also make an explicit parameterisation of detrainment obsolete.
2. L17 suggests that both CAPE and convergence influence the intensity of convection, but in different ways. Whereas, moist convergence influences the convective fractional area, the atmospheric instability is coupled to the updraft velocity. This dependency can only be incorporated if the mass flux is expressed in terms of a_u and w_u.
3. The assumption that $a_u \ll 1$ fails for resolutions less than 10 km. An explicit parameterisation of a_u allows for a realistic parameterisation also at these finer resolutions in the vicinity of the grey zone.
4. The quasi-equilibrium assumption breaks down at finer resolutions and requires a consistent stochastic approach in terms of a_u and w_u.

The stochastic approach described in Sect. 5 provides a way to determine values for the convective area fractions. Even more so, it provides information on area fractions of the different updraft types that together form the cumulus ensemble. The question is how to use this stochastic approach in standard mass-flux approaches.

The simplest implementation of this approach in a global large-scale model has been explored in Dorrestijn et al. (2016) and Peters et al. (2017). In Dorrestijn et al. (2016) N, CMCs are implemented in each grid box of the large-scale model that work with conditional transition probabilities $P(m, n)_\omega$ depending on the value of the vertical velocity of the large-scale model. So, at each time step, the CMCs provide information on the convective area fractions a_i, as described in Sect. 5. The CMCs are coupled to the convection scheme, simply by replacing the existing cloud base mass-flux closure of the deep convection parameterisation by

$$M_\mathrm{b} = \rho\, w_{\mathrm{u,b}}\, a_{\mathrm{u,b}} \qquad \text{with} \qquad a_{\mathrm{u,b}} = a_4 + a_3, \tag{30}$$

while choosing a typical value of the vertical updraft velocity at cloud base of $w_{\mathrm{u,b}} = 1\,\mathrm{ms}^{-1}$. So, the intensity of the cloud base mass-flux closure is proportional to the area fraction of the deep convective and strong congestus states of the CMC. The implementation in Peters et al. (2017) is similar in spirit, but uses only the deep convective states (a_4) to estimate the area fraction, while the transition probabilities not only depend on vertical velocity but also on tropospheric relative humidity.

Despite its simplicity, the coupling of the stochastic CMCs to the cloud base mass flux influences the nature of the mass-flux scheme in a number of ways. The strength of the convection scheme becomes stochastic but not in a purely random fashion as in standard stochastic approaches (Buizza et al. 2007) but proportional to the convective area fraction. Since the transition probabilities of the CMCs are conditional to the large-scale convergence, stronger convergence will lead to larger convective area fractions and hence to stronger convection and more precipitation, in line with L17. Note also that through the CMCs, memory of the cloud base mass flux is introduced.

Global simulations with the new stochastic closure Eq. (30) show an improved representation of the tropical intraseasonal variability of precipitation, including the Madden–Julian Oscillation (MJO) and a more continuous deep convection on sub-daily times scales, due to the memory effects of the cloud base mass flux (Peters et al. 2017). In Dorrestijn et al. (2016), positive impact on the MJO is less clear, and improvements of the MJO and Kelvin waves are mainly achieved by simply reducing the prescribed vertical updraft velocity at cloud base $w_{\mathrm{u,b}}$.

The current implementations Eq. (30) obviously does not take full advantage of the information of the CMCs. This choice was deliberate as these studies should be regarded as a proof of concept to motivate further efforts in improving the representation of convection through the use of CMCs. An obvious next step is to use CMSs that allow to incorporate spatial organisation. The transition probabilities of the CMCs introduced in Sect. (5.2) were only conditioned on atmospheric conditions, which make them essentially independent of each other so that no spatial coherence is taken into account. Alternatively, CMSs can also be conditioned on the states of the neighbouring nodes which introduces short-range spatial interactions and allows for spatial organisation (Dorrestijn et al. 2013). In doing so, CMSs essentially become stochastic cellular automata (SCA) that not only provide information on the area fractions of the different states, but also how they are spatially arranged. Such stochastic cellular automata can be coupled to multiplume approach such as proposed by Neggers (2015) and Plant and Craig (2008). In this approach, closely related to the original spectral approach of Arakawa and Schubert (1974), multiple plumes with different horizontal sizes ℓ are being released. The plumes differ by their entrainment strength, assumed to be inverse proportional to their horizontal size

$$\epsilon(\ell_i) = \frac{\eta}{\ell_i}, \tag{31}$$

following from simple area–perimeter considerations. So, plume sizes but also their spatial organisation affect the strength of the entrainment (Mapes and Neale 2011). Such information can be obtained from stochastic cellular automata, while multi-plume models can subsequently be used to calculate how deep each plume can rise. Such an approach also allows for parameterising the simultaneous occurrence of shallow and deep convection, a phenomenon, often observed in nature but rarely parameterised in models.

References

Arakawa, A., and H. Schubert. 1974. Interaction of a cumulus cloud ensemble with the large-scale environment. Part I: Theoretical formulation and sensitivity tests. *Journal of the Atmospheric Sciences* 31: 674–701.

Arakawa, A., J.-H. Jung, and C.-M. Wu. 2011. Toward unification of the multiscale modeling of the atmosphere. *Atmospheric Chemistry and Physics* 11 (8): 3731–3742.

Bengtsson, L., M. Steinheimer, P. Bechtold, and J.-F. Geleyn. 2013. A stochastic parametrization for deep convection using cellular automata. *Quarterly Journal of the Royal Meteorological Society* 139 (675): 1533–1543.

Buizza, R., M. Miller, and T.N. Palmer. 2007. Stochastic representation of model uncertainties in the ECMWF ensemble prediction system. *Quarterly Journal of the Royal Meteorological Society* 125 (560): 2887–2908.

de Roode, S.R., A.P. Siebesma, H.J.J. Jonker, and Y. de Voogd. 2012. Parameterization of the vertical velocity equation for shallow cumulus clouds. *Monthly Weather Review* 140: 2424–2436.

de Rooy, W.C., P. Bechtold, K. Frohlich, C. Hohenegger, H.H.J. Jonker, D. Mironov, A.P. Siebesma, J. Teixeira, and J.-I. Yano. 2013. Entrainment and detrainment in cumulus convection: an overview. *QJRMS* 139 (670): 1–19.

Dee, D.P., S.J. Uppala, A.P. Simmons, P. Berrisford, S. Poli, U. Kobayashi, A. Andrae, M.G. Balmaseda, P. Balsamo, P. Bauer, M. Bechtold, A.C. Beljaars, L.J. van de Berg, N. Bidlot, C. Bormann, R. Delsol, M. Dragani, J. Fuentes, A.L. Geer, B. Haimberger, S.H. Healy, V. Hersbach, E.L. Hlm, P. Isaksen, M. Kllberg, M. Khler, P. Matricardi, A.M. McNally, B. MongeSanz, J.J. Morcrette, B.K. Park, C. Peubey, P. de Rosnay, C. Tavolato, J.N. Thpaut, and F. Vitart. 2011. The ERAInterim reanalysis: Configuration and performance of the data assimilation system. *Quarterly Journal of the Royal Meteorological Society* 137 (656): 553–597.

Dorrestijn, J., D.T. Crommelin, J.A. Biello, and S.J. Böing. 2013. A data-driven multi-cloud model for stochastic parametrization of deep convection. *Philosophical Transactions of the Royal Society A* 371 (1991): 20120374.

Dorrestijn, J., D.T. Crommelin, A.P. Siebesma, H.J.J. Jonker, and C. Jakob. 2015. Stochastic parameterization of convective area fractions with a multicloud model inferred from observational data. *Journal of the Atmospheric Sciences* 72 (2): 854–869.

Dorrestijn, J., D.T. Crommelin, A.P. Siebesma, H.J.J. Jonker, and F. Selten. 2016. Stochastic convection parameterization with Markov chains in an intermediate-complexity GCM. *JAS* 73: 1367–1382.

Fritsch, J.M., and C.F. Chappell. 1980. Numerical prediction of convectively driven mesoscale pressure systems. Part I: Convective parameterization. *Journal of the Atmospheric Sciences* 37 (8): 1722–1733.

Khouider, B., J. Biello, A.J. Majda, et al. 2010. A stochastic multicloud model for tropical convection. *Communications in Mathematical Sciences* 8 (1): 187–216.

Kumar, V.V., C. Jakob, A. Protat, P.T. May, and L. Davies. 2013. The four cumulus cloud modes and their progression during rainfall events: AC-band polarimetric radar perspective. *Journal of Geophysical Research: Atmospheres* 118 (15): 8375–8389.

Loriaux, J.M., G. Lenderink, and A.P. Siebesma. 2017. Large-scale controls on extreme precipitation. *Journal of Climate* 30 (3): 955–968.

Mapes, B.E., and R.B. Neale. 2011. Parameterizing convective organization to escape the entrainment dilemma. *Journal of Advances in Modeling Earth Systems* 3 (06).

Neggers, R.A.J. 2015. Exploring bin-macrophysics models for moist convective transport and clouds. *Journal of Advances in Modeling Earth Systems* 7 (4): 2079–2104.

Peters, K., T. Crueger, C. Jakob, and B. Möbis. 2017. Improved MJO-simulation in ECHAM6.3 by coupling a stochastic multicloud model to the convection scheme. *Journal of Advances in Modeling Earth Systems* 9 (1): 193–219.

Plant, R.S., and G.C. Craig. 2008. A stochastic parameterization for deep convection based on equilibrium statistics. *Journal of the Atmospheric Sciences* 65 (1): 87–105.

Sakradzija, M., A. Seifert, and T. Heus. 2015. Fluctuations in a quasi-stationary shallow cumulus cloud ensemble. *Nonlinear Processes in Geophysics* 22 (1): 65–85.

Siebesma, A.P. 1998. Shallow cumulus convection. In *Buoyant convection in geophysical flows*, vol. 513, ed. E.J. Plate, E.E. Fedorovich, D.X. Viegas, and J.C. Wyngaard, 441–486. New York: Kluwer Academic Publishers.

Sušelj, K., J. Teixeira, and D. Chung. 2013. A unified model for moist convective boundary layers based on a stochastic eddy-diffusivity/mass-flux parameterization. *Journal of the Atmospheric Sciences* 70 (7): 1929–1953.

Teixeira, J., and C.A. Reynolds. 2008. Stochastic nature of physical parameterizations in ensemble prediction: A stochastic convection approach. *Monthly Weather Review* 136 (2): 483–496.

Tiedtke, M. 1989. A comprehensive mass flux scheme for cumulus parameterization in large-scale models. *Monthly Weather Review* 117 (8): 1779–1800.

Land Surface Processes

Dev Niyogi

Abstract The role of land surface processes on monsoonal mesoscale convection and rainfall is discussed. The chapter initially provides a foundational framework for the land surface and land–atmosphere coupling processes. This is followed by a discussion regarding the role and the need for improving land feedbacks within multiscale models and processes. In particular, the role of land surface models in providing surface boundary conditions within numerical weather prediction models is highlighted. Building on the observational understanding related to the role of land surface processes and changes on the regional weather and climate, the land surface feedbacks over the monsoon region are summarized. The chapter concludes by highlighting the need for work on three challenges to advance the high-impact monsoon weather prediction. These challenges include (i) more realistic inclusion of land processes and model parameterizations that account for the land–atmosphere feedbacks; (ii) development of datasets including data fusion products that can be used for input conditions including initialization of the land models as well as validation of newer schemes; and (iii) more coordinated model calibration/validation efforts with focus on the monsoon region by community of users (such as Indian land model intercomparison project—using both offline and coupled studies). Studies are especially needed for improving the representation of human-managed landscapes such as urban areas and agricultural croplands in the monsoon region. A more concerted effort linking land model with boundary layer coupling, and convection / microphysical parameterizations is needed as they intimately impact the performance of the coupled numerical weather prediction system.

Keywords Land atmosphere interactions · Surface energy balance
Mesoscale convection · Heavy rainfall · Land use land cover change
Soil moisture · Atmospheric boundary layer

D. Niyogi (✉)
Purdue University, West Lafayette, IN 47907, USA
e-mail: niyogi@gmail.com; climate@purdue.edu
URL: http://www.landsurface.org

D. A. Randall et al. (eds.), *Current Trends in the Representation of Physical Processes in Weather and Climate Models*, Springer Atmospheric Sciences,
https://doi.org/10.1007/978-981-13-3396-5_17

1 Introduction

This chapter provides a framework for improving land surface representation as a means to enhance the convection and rainfall simulations (predictions) with a particular focus on the Indian Monsoon Region (IMR). The framework and the need are presented via the following couplets:

Land basks in the Sun's radiation and like a Yogi balances the energies
the moist latent, the potent sensible, and the ground
and through these energies, the turbulence and thermal rises
amidst the flux gradients and the ensuing wind convergences
lays the drama for multiscale convection and even rains
but alas, the land lays quiet.. looking up to the mighty clouds
often ignored as the cause for those heavy rains untold.
So, here we say, know the land, know the couplings
and let the monsoonal vagaries unfold ....

The understanding of land surface feedbacks on the meteorological features over the IMR is important. The Indian monsoon rainfall is at the cynosure of multi-scale feedbacks. A variety of processes actively interact to create the optimal environmental setting that causes the convective precursors and the resulting rains. The large-scale variables such as those forced from sea surface temperature (SST) settings *vis a vis* the El Nino Southern Oscillation (ENSO), Indian Ocean Dipole (IOD) and synoptic atmospheric oscillations such as Madden–Julian Oscillations (MJO) are important and have been extensively studied (see review by Webster et al. 1998). Figure 1

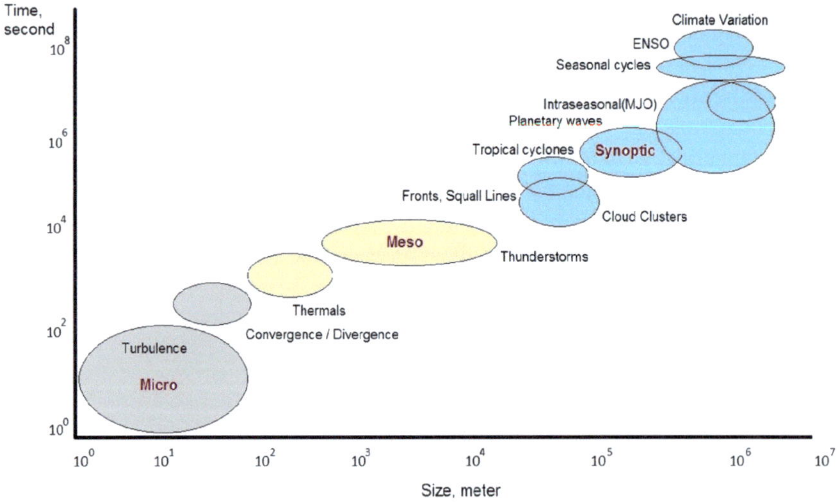

Fig. 1 Multi-scale atmospheric forcings affecting Indian Monsoon Region (*Source* Gadgil 2003)

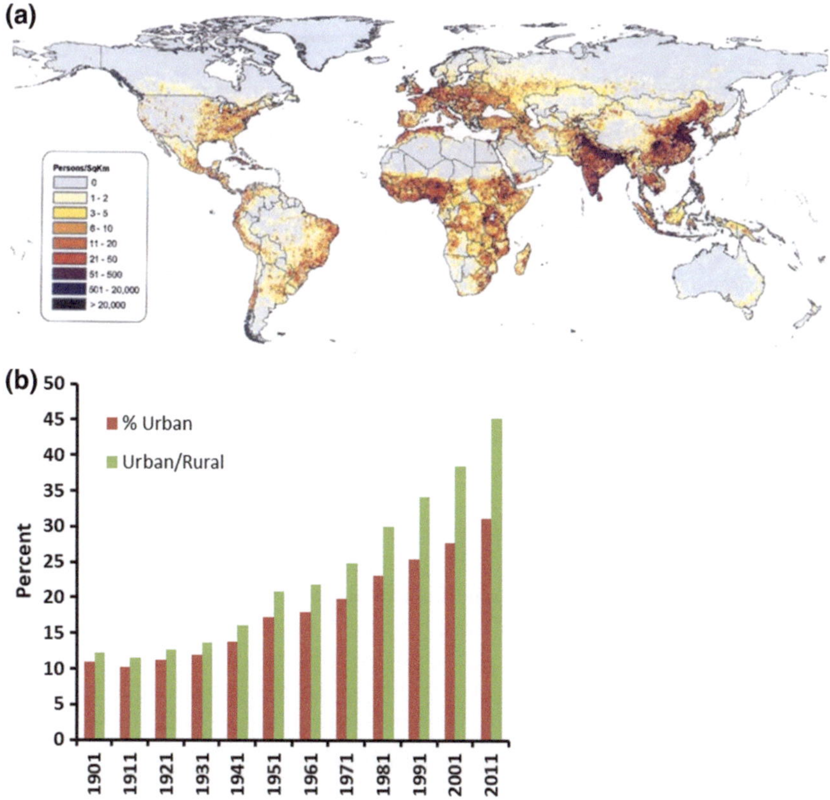

Fig. 2 **a** Population density map showing the concentration over the Indian Monsoon Region (*Source* FAO). **b** Percentage of urban population and the urban versus rural population ratio for India

adapted from a review by Gadgil (2003) provides a scale perspective of these different atmospheric settings.

With the growth in the available computational power and the finer grid spacing (resolutions) at which the models are being run, the role of fine-scale land surface processes is expected to become increasingly important. Indeed, another factor which highlights the importance of land surface representation is the rapid growth in population and regional economies both of which are leading to pressure on agricultural landscapes and development of more cities as well as sprawl in the existing cities.

Figure 2a from the Food and Agriculture Organization (FAO) shows the population density map and the notable population concentrations over the Indian and South Asia, China region. As discussed in Kishtawal et al. (2010), India's population growth and migration in the twentieth century, for example has increased nearly sixfold with the percentage of urban dwellers changing from nearly 5% at the start of twentieth century to nearly 30% at the start of the twenty-first century (Fig. 2b). To support

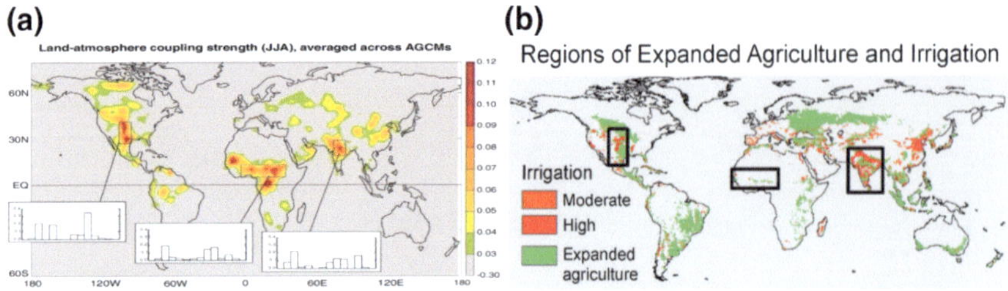

Fig. 3 **a** Land–atmosphere coupling hotspots found in the GLACE study (figure from Koster et al. 1999), **b** Regions of intensive agricultural management/irrigation

the food and fibre requirements of the growing population and the consumer demand in burgeoning economies, there has been a concurrent intensification of agricultural production and land transformation.

The rapid growth in population and the resulting seats of economic development such as urban areas and agricultural landscapes show up dramatically in the intensity of land surface forcings. What is being presented for India here is true for other emerging economies and indeed economic centres across the globe. Therefore, land cover change due to agricultural intensification and urbanization can be considered as the new global change underway and it is perhaps best manifested in the IMR. This chapter focuses on the IMR, and builds on the recent works that show that land–surface feedbacks are causing a significant and detectable change in the weather and climate through temperature and rainfall modifications (see a review by Pielke et al. 2011).

The effect of the land surface is thus an important emergent forcing that needs to be considered in prediction/synthesis studies. Another reason the effect of land surface representation and improvements in land surface important for the IMR is due to the strong land–atmosphere coupling noted in the global modelling studies and synthesis reported in studies such as Koster et al. (2004) and the Global Land–Atmosphere Coupling Experiment—GLACE), and Guo et al. (2006). An interesting aspect of the GLACE hotspots is how they align with the regions of human activities such as agricultural intensification/irrigation (and shown in Fig. 3).

2 Why Does the Land Surface Matter?

Land–surface feedbacks provide surface forcings to the atmosphere. This forcing is via energy balance and the surface roughness as well as surface heating that contribute to shear and buoyancy/convection, which, in turn, aids turbulence that drives the boundary layer evolution. The local land–atmosphere interactions are manifested through energy and water exchanges through the surface vegetation, soil moisture and the soil texture. These, in turn, lead to complex feedback loops and cycles. Examples include surface radiation partitioning with a part being used for evaporation and vegetation transpiration, while partly being used for soil and subsurface heating and sensible heating of the surface. The increase in the soil and surface heat storage leads to increased temperature, which, in turn, further modifies the transpiration/evaporation rates and depletes the soil moisture from the surface transferring it in atmospheric humidity. The reduced soil moisture on the surface leads to a change in the soil albedo and emissivity, and affects the partitioning (Bowen ratio), and could lead to further changes in the surface temperature and convection potential.

The incoming solar radiation (insolation) is a combination of the shortwave flux (predominantly) and longwave radiation. Through the atmosphere, it is impacted by aerosols, clouds and atmospheric gases which reflect, absorb and transmit the insolation. Roughly, half of the radiation reaches the earth's surface and is partitioned into radiative surface energy balance terms that can be expressed as

$$\begin{aligned}\text{Net Radiative Flux} &= \text{Sensible Heat Flux} + \text{Latent Heat Flux}\\ &\quad + \text{Ground Heat Flux} + \text{Storage.}\end{aligned}$$

The net radiative flux is prescribed as a forcing when the land surface models are run in an 'offline' mode or are provided as an input from the atmospheric (radiative) component of the model. A component of the shortwave radiation up to 700 nm is typically taken as the photosynthetically active radiation (PAR). When PAR information is not available, typically a fraction (e.g. 70%) of the net radiative flux is assumed to be PAR. The net radiation is then used for computing the evaporation and transpiration terms. The bare ground evaporation follows classical hydrological formulations, which are dependent on the soil moisture availability and soil characteristics which are often prescribed via lookup tables for soil texture. The soil textural information (available from global datasets in the land models), provides values for soil hydraulic conductivity, thermal conductivity, retention coefficient terms, etc. An example summary and information regarding values typically used in land models can be found in Noilhan and Planton (1989) as well as Sellers et al. (1996).

Indeed, while highlighting the critical role of land surface to help improve convection and rainfall simulation, it is important to recognize that the land surface is the driving force but the resulting effects are intimately tied to the role of the boundary layer, the convective parameterization and the aerosol-microphysical processes. Therefore, when reviewing the impact of land surface and the potential for improvement, the results need to be reviewed in the context of the coupled components. The increased temperature at the surface and the availability of moisture can lead to growth in surface convective thermals and boundary layer heating which in turn can lead to the evolution and growth of the surface/atmospheric boundary layer.

For representing the vegetation feedback, the transpiration effect is important and is often computed as a variance of Penman–Monteith formulation. This requires a surface (stomatal) resistance term that helps link the surface–atmosphere feedback. This term is central to considering the vegetation feedbacks in the model and is computed from the so-called Jarvis model which modulates a minimum resistance (Rcmin) constant prescribed as a function of vegetation/land use type in a lookup table to the model. This Rcmin term is considered in conjunction with PAR, soil moisture, air temperature and atmospheric humidity and simple empirical functional forms are prescribed to develop the stress functions which increase the resistance value. Higher the resistance lesser is the transpiration potential for the land surface. The Jarvis-type approach is typically used in most Numerical Weather Prediction (NWP) linked land models, while climate models and the so-called second- and third-generation land models consider a photosynthesis–transpiration relation (e.g. Niyogi et al. 2009; Collatz et al. 1991; Niu et al. 2011; Lawrence et al. 2007). An advantage of using the photosynthesis models is outlined in Niyogi et al. (2009). For example, the classical Noah land model uses the Jarvis-type transpiration/vegetation formulation, while the Noah-MP model uses the photosynthesis (so-called Ball–Berry–) model. Most land models in climate studies use the Ball–Berry formulation with Collatz–Farquhar formulation and some variance (see Niu et al. 2011 for the model formulations and the citations).

The evaporation–transpiration formulation is used to compute the latent heat flux component of energy balance. The ground heat flux, on the other hand, is computed iteratively as a function of the initial soil temperature forcing and the energy balance for the soil layers for each time steps. A positive radiative energy flux term is translated into equivalent temperature change using thermal diffusion equation and the thermal conductivity information (represented as a function of soil texture in a lookup table within land models). The sensible heat flux is computed using a force restore approach in which the difference between soil temperature and air temperature is used with an atmospheric thermal conductivity (diffusion) exchange coefficient. The numerical time steps in the prognostic equations allow for the computation of soil temperatures at different layers, i.e. air temperature information. The air temperature and the evaporative–transpirative flux are used to compute the atmospheric humidity for different time steps. Interestingly, the changes in the evaporative flux lead to the changes in the surface soil moisture availability. Note that in the simpler slab model only the soil temperature is a prognostic variable and the soil moisture availability is taken as a soil texture-based constant. Precipitation and dew formation (estimated as negative latent heat flux) leads to an increase in the initial (or prior time step) soil moisture, and there is an exchange in the soil–moisture flux between the different soil layers as a function of soil moisture in the adjacent layers and the hydraulic conductivity. Realistic representation of the energy balance and the surface boundary condition has been the principal reason for incorporating land surface processes.

A summary developed regarding the relationship between the energy balance and boundary layer/convection observations over the agricultural landscapes and forest sites highlights the need for accurate representation of the land state. Over the agricultural landscape, net radiation was found to be 65% of the global radiation, while 25% was reflected back to the atmosphere. Nearly 80% of the net radiation reaching the surface is partitioned into latent heat flux, 15% is used in storage and about 5% is utilized for sensible heating. The resulting thermals and convection cause the afternoon boundary layer to be about 1 km deep. Contrasting this with the data from the forest site, nearly 85% of the global radiation is available as net radiation with 15% being reflected back to the atmosphere. The surface latent heat flux is about 60% of the net radiation, while the sensible heat flux is half of that at 30% of the net radiation, and an 8% is stored in the forest, while 2% is used as ground heat flux. The resulting sensible and momentum flux lead to a deeper boundary layer up to about 1.5 km. These estimates are a gross generalization but provide a perspective on the coupled dynamics of the land surface affecting the boundary layer and the resulting cloud–convection processes.

As the prime objective of the land–surface representation is to provide a realistic and representative surface energy balance terms for energy and water fluxes, the developments in the land surface models have evolved to include more realism in recent years. There has been a growing focus on considering more process-based approaches rather than empiricism. An example is the evolution from the constant soil–moisture-based slab model to multilayer soil moisture prognostic terms in the land model. Another example is replacing the empirical Jarvis-type transpiration model to a more realistic and physiologically correct photosynthesis–transpiration model. As a result, land models can provide detailed surface information for land state related to vegetation, as well as, bare ground temperatures, soil temperature and moisture at different layers within the soil and in some cases even within the canopy. There are snow algorithms and groundwater–soil aquifer interactions available in some formulations. There are also modules and equations now available for dynamic vegetation for forests and more recently for crops. The main advantage and motivation for such developments have been to help improve forecasts at multiple timescales especially the regional/seasonal climate simulations. This is possible because just like large-scale atmospheric motion, the sea surface temperature provides the requisite memory for aiding predictability, the land models provide memory through soil moisture information. This memory in terms of the evolution of the deeper layer and surface soil moisture, as well as snow and vegetation state, helps provide the right boundary conditions needed for the energy balance and the association boundary layer/mesoscale forcing. Thus, the efforts associated with improved regional and subseasonal forecasts need to include better land model physics as part of the coupled modelling system.

Pitman (2003) provides a overview of the evolution of the land models and their role in terms of the coupling between atmospheric models—land surface—and the full terrestrial energy and water balance and dynamics. From the NWP perspective, the evolution of the land boundary conditions from Slab, to Noah model to Noah-MP (multi-parameterization options) is of interest as it is part of number of modelling systems such as the NOAA's Eta model, NAM modelling system, GFS, CFS, MM5 and the WRF suite including the Land Data Assimilation System (LDAS) and Global LDAS (GLDAS). More details regarding the model equations and evolution can be found in Niu et al. (2011), Yang et al. (2014), and Barlage et al. (2015). A more recent update by Liu et al. (2016) shows the improvements associated with the dynamic crops as part of the Noah-MP-Crops.

3 Detecting the Land Feedbacks from Observations and Models

Much of the model development has its roots in the ability to identify the manner in which land has affected regional weather and climate. Some of the best-known examples of the land use/cover change based climatic impact have been the Urban Heat Islands (UHIs) created due to growing cities. Urban areas have higher temperatures as compared to the surrounding rural regions because of the surface characteristics that allow for more heat storage and temperature increase. This feature about UHI had been reported nearly a century ago. This has been typically observed by comparing the air temperature from a rural versus a nearby urban location. A number of review papers (e.g. Oke 1973; Santamouris 2015) show a distinct 3°–5° warming in the urban regions. The 'island' feature of the UHI is better depicted in satellite imagery, where the urban regions are seen are temperature hotspots against the cooler rural background (see an example figure from Kishtawal et al. 2010 for a satellite temperature image for UHI over city of Ahmedabad in Western India) (Fig. 4).

An important extension of the UHI findings is that the multitude and magnitude of this heating are not just limited to isolated select cities, but is prevalent enough to cause a climatic signature in terms of the regional temperature patterns. In a study over China, Zhou et al. (2004) have shown that urbanization can cause up to 0.05 C warming per decade. An interesting question to consider then is if the landscape were to be greener, such as due to afforestation or agriculture intensification, would

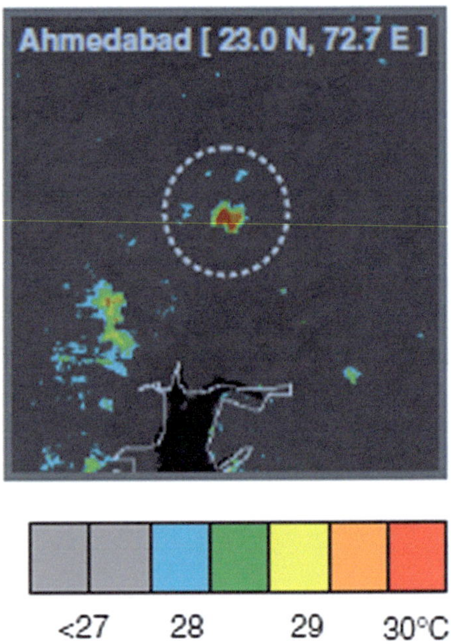

Fig. 4 Urban Heat Island (UHI) (shown by the thermal imagery within the dotted circle) for the city of Ahmedabad in Western India (*Source* Kishtawal et al. 2010)

a cooling signature be detected? Indeed, looking at the temperature data for pre- and post- 'green' revolution or the agricultural intensification, Roy et al. (2007) showed that during the growing season agriculture in India for the post-green revolution has likely led to 0.34 C cooling. These results are not isolated and a careful synthesis of the landscape change versus temperature change analysis shows it is possible to attribute specific warming or cooling signatures to different vegetation types and landscape conversions. Fall et al. (2010) have reported such an analysis for the continental U.S. region, again highlighting the cooling effect due to agricultural intensification and warming in other land conversions. Thus, the scale of landscape interactions—particularly due to agricultural intensification, and urbanization—are becoming significant and spatially notable enough to cause atmospheric feedbacks that are becoming spatiotemporally organized and detectable in climatic teleconnection or attribution studies. Pielke et al. (2002) also present a perspective on landscape energy balance forcing as a climatic driver beyond greenhouse gas emissions.

Land feedbacks can also have a profound impact on the regional rainfall and associated precursors. A cascade of impacts has been outlined in Pielke et al. (2011). Landscape changes such as deforestation, a transformation of grasslands to agricultural regions, urbanization and practices such as irrigation, harvesting or multi-cropping are detectable from the biophysical mapping of different datasets. These physical changes have a detectable impact on the surface and regional energy balance including the partitioning of surface radiative flux (different Bowen's ratio), leading to changes in the surface roughness, albedo, surface temperatures and humidity values. These changes in turn influence boundary layer, thermodynamic and dynamic variables including profiles of temperature, humidity and wind shear; which in turn manifest changes in the mesoscale convergence patterns and resulting impact on the mesoscale convection and precipitation. These changes can directly or via interactions affect regional hydrology, basin-scale water budgets, snow cover and regional CO_2 source/sinks.

Detecting the impact on the regional rainfall patterns due to landscape changes is more challenging than identifying the impact on the regional temperature changes. This is because, temperature changes are typically coincidental to the regions of landscape changes, while the rainfall changes are often at a region that is advected away from the seat of surface heterogeneity and along locales of mesoscale flux gradients. To identify such changes in the rainfall effects due to landscape changes, it is possible to review the change in regional urbanization and rainfall (Changnon et al. 1971). The UHI, for instance, can create zones of localized surface convergence and upper level divergence, due to surface heating and resulting instability. The urban areas also typically have higher aerosols/pollution that can cause differential heating rates over the city and surrounding region. The instability over the urban regions, if fuelled with sufficient moisture and aerosols can become preferential zones of convection that often get advected downwind of the urban regions. When a thunderstorm or rain-bearing system approaches the urban area, it is likely to be more invigorated on the periphery or downwind of the urban region. This feature was noted in a synthesis of different case studies for select cities such as Chicago, St. Louis, New York City, Atlanta, Mexico City, Beijing, etc. Niyogi et al. (2011) provided a decade-

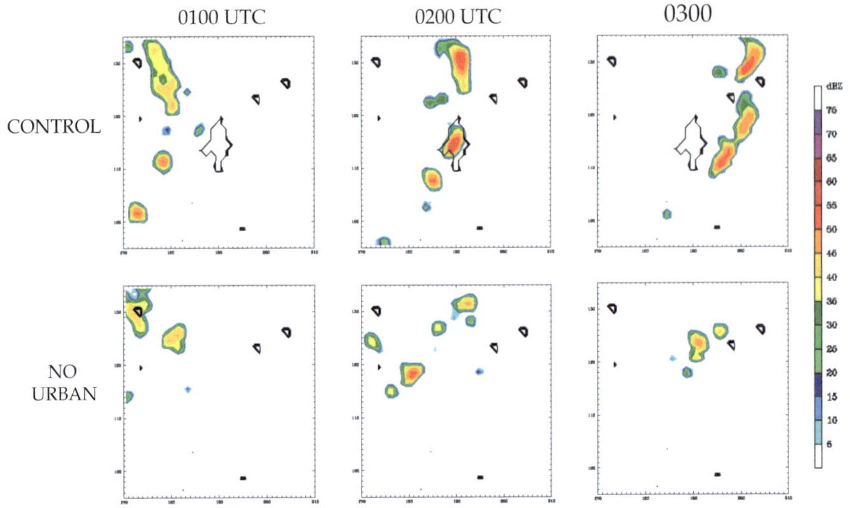

Fig. 5 Model simulations showing storm and rainfall changes with and without urban region (shown by the solid outline in the top panel) for the same meteorological setting (*Source* Niyogi et al. 2011)

long analysis using in situ and radar data over Indianapolis region over the Midwest U.S. to conclude that indeed 60% of daytime storms during summer passing over the city can have a downwind intensification and enhanced rainfall potential. In fact, the climatological assessment of the historical rainfall dataset in the vicinity of the city also showed a zone of increased rainfall occurrence. This feature was also reported and considered possible for downwind of Chicago in a classical study by Changnon (1980). This precipitation modification was referred to as 'LaPorte Anomaly' in consideration of the increased rainfall over LaPorte, IN downwind of the Chicago urban locale. Modelling analysis and if-then process-scale assessments indeed reveal that the mesoscale heterogeneity caused by the urban region can trigger the storms and even cause potential splitting. When the urban area is not present in the model setup, similar mesoscale atmospheric conditions did not lead to the development of the storm (Fig. 5 from Niyogi et al. 2011).

Synthesis of the urban—rainfall modification studies highlight these changes are due to gradients in the surface properties such as albedo and thermal storage potential, and also surface roughness enhancements which lead to wind shear and directional changes. The cities also need to be of a sufficient size so as to cause an impact on the mesoscale atmosphere, which a study by Schmid and Niyogi (2013) suggests is about 25 km. Indeed, urban aerosols also cause a notable impact and may in some instance be more dominant than the city size itself (Schmid and Niyogi 2017). The flux gradients can create mesoscale convergence/diverge zones and the resulting rainfall bands. Note that what was seen for cities was also applicable to similar heterogeneity in landscape characteristics such as due to forest and agriculture. In

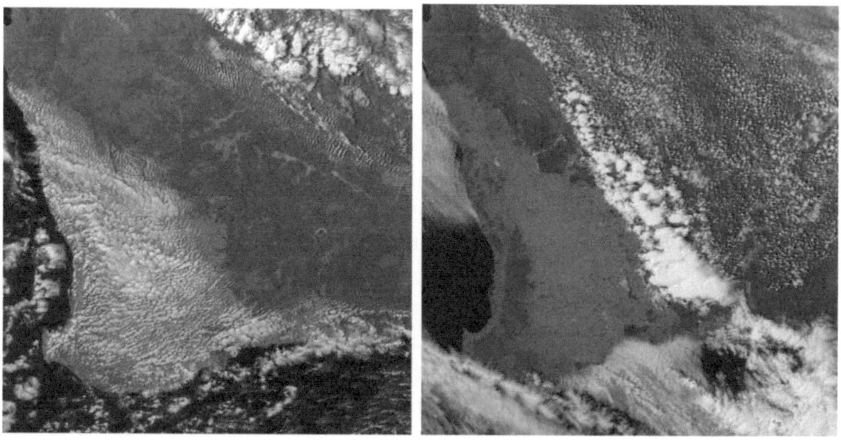

Fig. 6 Satellite image showing cloud–convection occurrence only over the agricultural region (left) and no clouds over the forest and in other instance clouds occurring only over the forested region (right) and no clouds over the agricultural region, illustrating the coupling between land–atmosphere convection (*Source* Pielke et al. 2011)

Pielke et al. (2011), an example is illustrated for the region of Southwest Australia with a distinct agriculture and forest boundary shown in Fig. 6.

Satellite imagery shows distinct regimes that have cloud cover either only over the agricultural landscape or over the forested land, illustrating the effect of the land–atmosphere interactions showing the creation of mesoscale convective environment that is conducive for convection over one landscape compared to the other. A number of such studies now exist which provide a convincing summary that land–surface feedback and heterogeneity has a significant impact on the timing, location, intensity, magnitude of regional convection and rainfall.

4 Land Feedbacks for Improved Predictions

There have been notable improvements in representing the landscape changes especially with the advent of satellite data products. An enduring challenge is to develop the model capabilities so that the predictions can benefit from these observations. The lag that remains in the model parameterization development despite the tremendous improvements in the land surface measurements has been one of the reasons that a broader impact of the benefits that can accrue from the land surface features is still lacking. As an example, a number of soil moisture products are now available either through direct measurements such as from the Soil Moisture Active Passive (SMAP) sensor or the blended products from the Land Data Assimilation Systems (LDAS). For the NWP or climate models to benefit from these products the land model needs to be beyond the Slab land model—which has a constant soil moisture consideration and only has soil temperature as a prognostic variable. When the model is updated from the Slab to say a Noah land model which has both soil temperature as well as soil moisture prognostic equations, then the impact of soil moisture observations can be better realized. In the same vein, if vegetation canopy temperature or morphological information is available such as from a MODIS or some other biophysical mapping product then having a more detailed land model such as Noah-MP would be of value in bringing out the benefits of the products and likely improving the model performance. Such a feature is highlighted in this section taking a number of examples.

In an early demonstration of the value of improved vegetation and the soil moisture initial conditions, Holt et al. (2006) analysed a severe thunderstorm case over U.S. Southern Great Plains from the International H2O Project (IHOP_2002). In their study, improved model performance was noted when the model was updated from the Slab land scheme to the Noah land surface model. Further, the performance of the Noah model was shown to improve when there was better initial soil moisture/temperature information available. This had a clear impact on the model Quantitative Precipitation Forecast (QPF), which was a result of the improved land surface heterogeneity and the resulting convergence zones as well as convection potential.

The value of improving land is not just for inland storm cases. In a series of studies conducted over regions with land-sea boundary and tropical rainfall events, it was shown that the model QPF was intimately related to the vegetation/transpiration model used within the WRF model. A similar conclusion was made for rains from tropical storms not just near the coast but after there has been notable inland track. This indicates that the vegetation and land representation has a significant impact on the QPF not just for mesoscale convective events, but also for synoptic events such as large tropical storms. An example of this was seen in the post-landfall rains from Tropical Storm Alison (2001) both near the coasts as well as inland, and more recently for TS Faye over Florida (Bozeman et al. 2012), and for TS Erin which reintensified over Oklahoma (Kellner et al. 2012).

Building off these results, a series of collaborative projects between the academia and operational community have shown the value of land representation in post-landfall characteristics of landfalling tropical systems using the Hurricane WRF (HWRF) model. For example, Bozeman et al. (2012) showed that the track prediction improved when the Noah model was used in place of Slab, as it allowed for an appropriate regional convergence zone to be simulated. The improvement in the storm track, as a response to a more realistic simulation of the mesoscale and regional atmospheric setting also meant that the model was able to produce more accurate rainfall in terms of the location and intensity. For the IMR, Osuri et al. (2012) have shown a potential benefit of using Noah in place of Slab for simulating Bay of Bengal storm in the WRF model (Fig. 7).

Fig. 7 Improved simulation of observed monsoon depression track (back) using Noah land model (blue) as compared to the Slab (red) (*Source* Osuri et al. 2012)

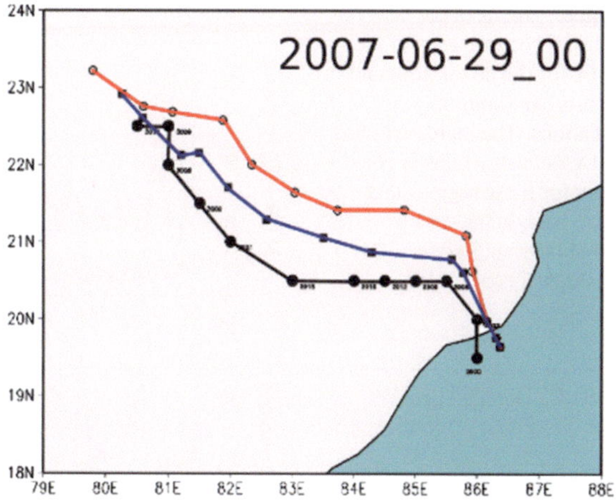

5 Land Surface Feedbacks for Indian Monsoon Region

As highlighted in the introduction, it is well known that the IMR is affected by large-scale oceanic and atmospheric feedbacks, and therefore an interesting question arises whether the land surface feedback would be significant in the backdrop of these synoptic forcings? A comprehensive summary of possible impacts and evidence has been outlined in Niyogi et al. (2018). The perspective being highlighted is to recognize monsoon is not a giant sea breeze and the rainfall during the course of the monsoon season is due to a myriad of coherent clusters that aid seasonal and interseasonal rainfall variability and amounts (as well as trends). As an example, rains from thunderstorms and monsoon depressions can contribute to about third to half of the heavy rain amounts during a season. Also, there is significant spatiotemporal variability in actual rainfall that needs to be captured in order to get the rainfall variations and amounts right.

Niyogi et al. (2010) showed an example of how agriculture intensification can affect monsoon rainfall and could play a role in reducing the rainfall in Northern India. The setting for the landscape change in Northern India is well documented with extensive ground water depletion to aid intensive agricultural activities. The Northern India/Punjab ground water extraction (order of 2.5 m per decade) has been recorded from GRACE satellite products in studies such as Rodell et al. (2009) and Rodell et al. (2018). This availability of irrigation water has moved the agricultural activities to become lesser dependent on the monsoonal rains. As a result, the peak greenness in the region has been recorded 1–2 weeks earlier than prior decades. This greening leads to important changes in the regional meteorological setup—most notably a weaker monsoon low-pressure centre over the Northern India region and a slower advancement of monsoon. The resulting impact of such cascade of feedbacks is that the region with more agriculture due to irrigation shows lesser rainfall than climatology—and further leads the agricultural community to move towards irrigation.

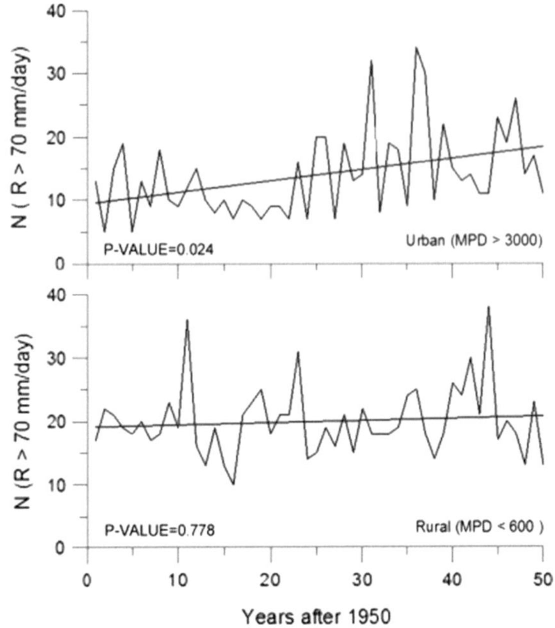

Fig. 8 Trends in the heavy rainfall occurrence over the IMR for the urban (top) and non-urban (bottom) locations. The increase in the extreme/heavy rains is noted only for those regions that have been impacted by urbanization (*Source* Kishtawal et al. 2010)

This complex feedback shows the link between the human modification of landscape and human-management decisions becoming one important component of the earth system change for something as significant and large-scale driven as the monsoon rains. Similar results of agriculture intensification contributing to the potential for reduced soil moisture and drought potential has been reported in Liu et al. (2016) for China. In their study, it was shown that the intense agricultural activities are causing an increase in the transpiration loss leading to decreased soil moisture availability, which in turn is leading to altered hydroclimatology of the region. Contrary to the reduction in the rainfall or increase drought recorded in India and China, the effect of irrigation and agricultural intensification in midlatitudes is that of likely increased rainfall (such as in the U.S. Great Plains). This again highlights that the impacts from land–surface feedbacks are intimately linked to other factors and should not be considered singularly.

Urbanization is the other landscape change underway over the Indian region. A question of interest then is whether the increased urbanization contributes to increased rainfall intensity that has been reported for the monsoon region (Goswami et al. 2006)? This question was addressed in a study by Kishtawal et al. (2010) looking at the gridded rainfall products and satellite data rain rates. The study shows clear evidence that the increase in the extreme rainfall events reported over the Indian monsoon region in studies such as Goswami et al. (2006) is only noted for 'urban' grids and the non-urban or grids over rural region show no change in the monsoon rainfall, as shown in Fig. 8.

A similar conclusion was reached from the TRMM-based rainfall rate with higher rain rates for urban grids versus rural. This finding was also noted for northeastern monsoon rainfall indicating that urbanization is an important ingredient of the rainfall intensification and increased extreme rains over the Indian monsoon region. Consequently, it is important to consider the urban landscape in the models that seek to simulate heavy rains over the Indian monsoon region.

Follow-up studies building off GFS modelling, GLDAS, WRF run as well as observations by a number of research teams across India (e.g. Saha et al. 2011; Shastri et al. 2015; Unnikrishnan et al. 2013; Roxy et al. 2015; Osuri et al. 2017) have shown that land feedbacks over the Indian monsoon region coherently organize themselves to modify regional/multi-scale temperature, rainfall and regional climate. The implication of this finding is that improved representation of land state can lead to improved prediction of weather and climate over the IMR.

6 The Approach Ahead

Land model applications over IMR need a combination of twofold strategy: (i) model development which has realistic physics, and (ii) development of data to create model parameters/initialization (and validation). Building off this premise, as part of a National Monsoon Mission project a 4 km grid spacing, sub-daily scale, soil moisture/soil temperature data product has been developed. The data product has been created using downscaling of MERRA and GLDAS fields using a local LDAS domain enhanced with vegetation cover, radar and satellite rainfall information. The calibration/validation of the LDAS fields has been undertaken by comparing with available in situ and satellite estimates and other reanalyses products. Figure 9 shows one such comparison for the LDAS fields against in situ observations from Nayak et al. (2018).

One of the goals of the developed data product is to improve rainfall predictions as part of the Indian National Monsoon Mission project. Osuri et al. (2017) used these soil moisture fields to assess the value of simulate thunderstorms. The soil

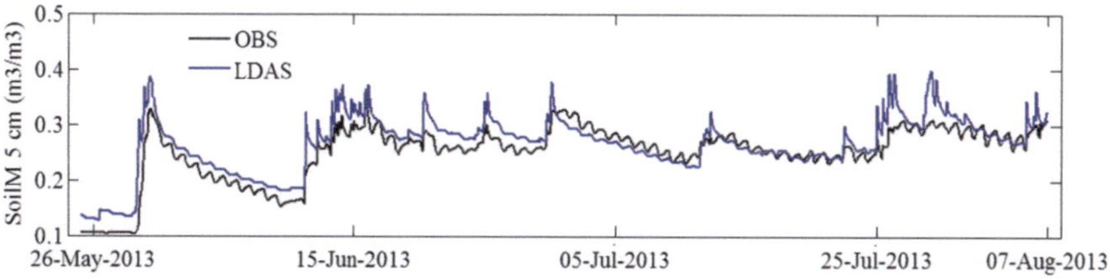

Fig. 9 Comparison of the LDAS products developed as a part of a National Monsoon Mission project and available for Indian region at 4 km grid spacing, at a field site in Kharagpur, Eastern India

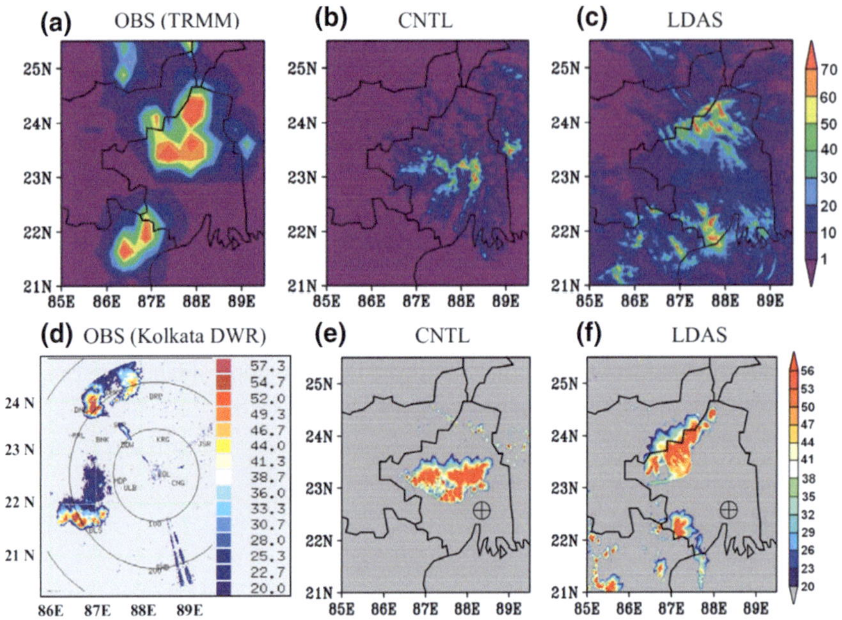

Fig. 10 Thunderstorm observations (left panels) and simulations using WRF as control/default soil moisture/temperature fields and those from the Land Data Assimilation System (LDAS). Results illustrate a marked improvement in the spatial structure of the storms when the gradients in the surface fields are better represented in the model (*Source* Osuri et al. 2017)

moisture fields could help simulate the timing, location, as well as, structure of the thunderstorms well showing the potential for such inputs to the predictions in the monsoon region (Fig. 10).

Considering the case of Monsoon Depressions (MDs), another series of studies have been conducted, where the hypothesis tested was whether the results improve with an enhanced land surface model used in the simulations. Consistent with the results found for the Atlantic Basin (e.g. Bozeman et al. 2012), for the Bay of Bengal region also, improved land representation helped simulate an improved track and rainfall from the MDs. An interesting finding related to the landstate and monsoon depressions inland track was also evident (Kishtawal et al. 2012; Chang et al. 2009). There was a higher probability of MDs dissipating within a short distance post landfall and cause heavy local rain if the land was anomalous dry. On the other hand, the MD had a higher potential for inland penetration and more inland rains if the land was anomalous wet. The period considered before landfall was about 1–2 weeks in terms of determining whether the land was wet or dry. In other words, the antecedent land state could provide guidance regarding the sustenance or dissipation of the MD post-landfall. This feature of the so-called 'Brown Ocean', i.e. land providing the latent heat for the storm to sustain itself post-landfall even after making landfall is evident over the IMR. This highlights the need for having a good land modelling system that is capable of simulating the soil moisture state (and not consider it to be a constant) for simulating storm systems such as tropical storms and MDs.

On a much broader scale, Krishnamurti et al. (2012) looked into the issue whether the land state could provide clues for the post monsoon—onset progress of the monsoon trough. Looking at the isochrones that link the points of monsoon progression, improved simulation of the progress was found when an improved land model was included in the simulations.

At a climatic scale for monsoon predictions, Niyogi et al. (2010) and Lee et al. (2009) studied the causal relation between land state and monsoon rainfall amounts. They found that pre-monsoon (February, March) greenness over IMR could have a negative causal link with the July monsoon rains. In other words, greener landscape prior to the monsoon onset could lead to a feedback, which may be linked to the higher latent heat flux and surface cooling, reduced pre-monsoon low pressure due to cooler land, and therefore a weaker circulation cell that could impact the monsoon trough progress. Halder et al. (2016) have pursued RegCM-based simulations and found that this mechanism can indeed manifest in terms of the land–atmosphere—monsoon predictions. On a much longer timescale, work by Roxy et al. (2015) has shown that the land–sea temperature contrast can govern the circulation strength on climatic scales. Therefore, the pre-monsoon land state becomes an important forcing for the monsoon rains, especially over the monsoon core domain.

7 Conclusions

Enhancements in the land representation are important for simulation of the monsoon processes. There is irrefutable evidence that at micro-, meso- and regional-, as well as, climatic scales, land surface representation has a detectable and important influence on the mean state (progress of monsoon), as well as, high-impact monsoon weather events.

There are at least three challenges that need to be addressed in near future to continue advancing the benefits that can accrue by effectively using the knowledge gained by the process-scale understanding and the model developments. The three areas are: (i) more realistic inclusion of land processes and model parameterizations that account for the land–atmosphere feedbacks; (ii) development of datasets including data fusion products that can be used for input conditions including initialization of the land models as well as validation of newer schemes; and (iii) more coordinated model calibration/validation efforts with focus on the monsoon region by community of users (such as Indian land model intercomparison project—using both offline and coupled studies). Example of the process-scale, realistic land representations include more datasets and modules, as well as, studies over human-managed landscapes such as urban areas and agricultural croplands. For these land surfaces, the morphology in terms of the impervious area coverage, associated characteristics such as building layout, average height, albedo values, anthropogenic heating rates and other human influences need to be developed. The challenge is to have this information available

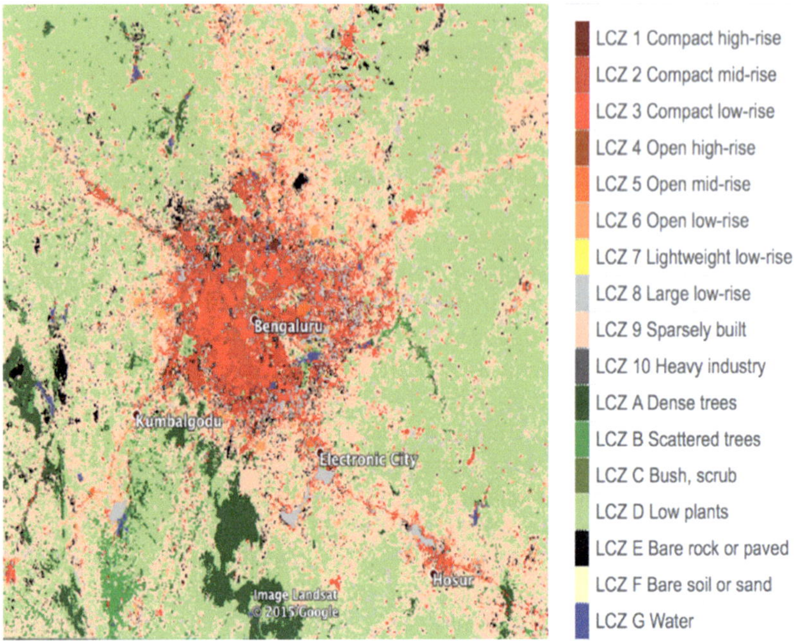

Fig. 11 World Urban Data Analysis and Portal Tool (WUDAPT) based Land Cover Zones (LCZs) mapped for Bangalore, India

in a multi-scale format that is suitable for models that are run from sub-km to climatic scales. Accordingly, community efforts such as WUDAPT (World Urban Data Analysis and Portal Tool, Ching et al. 2018) are underway and dedicated efforts for the IMR are needed for cities over India. Figure 11 shows an example of a WUDAPT map for Bangalore, India developed as part of a National Monsoon Mission project. This dataset is currently being compiled for other Indian cities as well and will likely aid the urban meteorology forecasting efforts over Indian cities.

Similarly, for agriculture, the information on crop biophysical characteristics and input that is required for developing dynamic crop growth are needed to be compiled and made available (see e.g. Liu et al. 2016 for Noah-MP-Crop). That is the community efforts need to be aligned for both the biophysical aspects of the land, as well as the human-management aspects. From the Indian modelling community perspective, there is also a benefit to creating a developmental testbed which allows for testing of land models and land–atmosphere parameterizations for select cases for which observations are available; and for model configurations (related to resolution, model physics) that is used by the operational community. This will likely speed the translation of efforts from research to operations and the success noted in published studies will likely be replicated when using the operational model configurations. The success in the offline models' performance seen by the hydrologic community will not directly translate to the atmospheric modelling efforts as the model performance is intimately linked to the boundary layer, cumulus parameterization, microphysics and related numerical options. Therefore, more case-specific efforts and community-based assessments are critically needed in developing this framework for improved land–atmospheric coupling terms rather than land models and boundary layer models that are being developed and tested by different groups. As monsoon forecasting continues to embark into an era of finer resolution, higher computational power and data-driven NWP, the fine-scale land information and data products integration have to become a priority to take monsoon–rainfall forecasts especially for extreme events to the next-level of predictability.

Acknowledgements The work in this chapter benefitted from a National Monsoon Mission project Earth System Science Organization, Ministry of Earth Sciences, Government of India (Grant no./Project no MM/SERP/CNRS/2013/INT-10/002), the U.S. National Science Foundation (NSF CAREER grant AGS-0847472 (Ming Cai), NSF CDSE-01250232, NSF AGS-1522492 (Chungu Lu), and USDA Hatch grant 1007699.

References

Barlage, M., M. Tewari, F. Chen, G. Miguez-Macho, Z.L. Yang, and G.Y. Niu. 2015. The effect of groundwater interaction in North American regional climate simulations with WRF/Noah-MP. *Climatic Change* 129 (3–4): 485–498.

Bozeman, Monica Laureano, Dev Niyogi, S. Gopalakrishnan, Frank D. Marks, Xuejin Zhang, and Vijay Tallapragada. 2012. An HWRF-based ensemble assessment of the land surface feedback on the post-landfall intensification of Tropical Storm Fay (2008). *Natural Hazards* 63 (3): 1543–1571.

Chang, Hsin-I., Dev Niyogi, Anil Kumar, C.M. Kishtawal, Jimy Dudhia, Fei Chen, U.C. Mohanty, and Marshall Shepherd. 2009. Possible relation between land surface feedback and the post-landfall structure of monsoon depressions. *Geophysical Research Letters* 36 (15).

Changnon Jr., Stanley A. 1980. More on the La Porte anomaly: A review. *Bulletin of the American Meteorological Society* 61 (7): 702–711.

Changnon Jr., Stanley A., Floyd A. Huff, and Richard G. Semonin. 1971. METROMEX: An investigation of inadvertent weather modification. *Bulletin of the American Meteorological Society* 52 (10): 958–968.

Ching, Jason, G. Mills, B. Bechtel, L. See, J. Feddema, X. Wang, C. Ren, et al. 2018. World urban database and access portal tools (WUDAPT), an urban weather, climate and environmental modeling infrastructure for the anthropocene. *Bulletin of the American Meteorological Society* 2018.

Collatz, G. James, J. Timothy Ball, Cyril Grivet, and Joseph A. Berry. 1991. Physiological and environmental regulation of stomatal conductance, photosynthesis and transpiration: A model that includes a laminar boundary layer. *Agricultural and Forest Meteorology* 54 (2–4): 107–136.

Fall, Souleymane, Dev Niyogi, Alexander Gluhovsky, Roger A. Pielke, Eugenia Kalnay, and Gilbert Rochon. 2010. Impacts of land use land cover on temperature trends over the continental United States: assessment using the North American Regional Reanalysis. *International Journal of Climatology* 30 (13): 1980–1993.

Gadgil, Sulochana. 2003. The Indian monsoon and its variability. *Annual Review of Earth and Planetary Sciences* 31 (1): 429–467.

Goswami, Bhupendra Nath, V. Venugopal, D. Sengupta, M.S. Madhusoodanan, and Prince K. Xavier. 2006. Increasing trend of extreme rain events over India in a warming environment. *Science* 314 (5804): 1442–1445.

Guo, Zhichang, Paul A. Dirmeyer, Randal D. Koster, Y.C. Sud, Gordon Bonan, Keith W. Oleson, Edmond Chan, et al. 2006. GLACE: The global land–atmosphere coupling experiment. Part II: Analysis. *Journal of Hydrometeorology* 7 (4): 611–625.

Halder, Subhadeep, Subodh K. Saha, Paul A. Dirmeyer, Thomas N. Chase, and Bhupendra Nath Goswami. 2016. Investigating the impact of land-use land-cover change on Indian summer monsoon daily rainfall and temperature during 1951–2005 using a regional climate model. *Hydrology and Earth System Sciences* 20: 1765–1784.

Holt, Teddy R., Dev Niyogi, Fei Chen, Kevin Manning, Margaret A. LeMone, and Aneela Qureshi. 2006. Effect of land–atmosphere interactions on the IHOP 24–25 May 2002 convection case. *Monthly Weather Review* 134 (1): 113–133.

Kellner, Olivia, Dev Niyogi, Ming Lei, and Anil Kumar. 2012. The role of anomalous soil moisture on the inland reintensification of Tropical Storm Erin (2007). *Natural Hazards* 63 (3): 1573–1600.

Kishtawal, Chandra M., Dev Niyogi, Mukul Tewari, Roger A. Pielke, and J. Marshall Shepherd. 2010. Urbanization signature in the observed heavy rainfall climatology over India. *International Journal of Climatology* 30 (13): 1908–1916.

Kishtawal, Chandra M., Dev Niyogi, Anil Kumar, Monica Laureano Bozeman, and Olivia Kellner. 2012. Sensitivity of inland decay of North Atlantic tropical cyclones to soil parameters. *Natural Hazards* 63 (3): 1527–1542.

Koster, Randal D., Paul A. Dirmeyer, Zhichang Guo, Gordon Bonan, Edmond Chan, Peter Cox, C.T. Gordon, et al. 2004. Regions of strong coupling between soil moisture and precipitation. *Science* 305 (5687): 1138–1140.

Krishnamurti, T.N., Anu Simon, Aype Thomas, Akhilesh Mishra, Dev Sikka, Dev Niyogi, Arindam Chakraborty, and Li Li. 2012. Modeling of forecast sensitivity on the march of monsoon isochrones from Kerala to New Delhi: the first 25 days. *Journal of the Atmospheric Sciences* 69 (8): 2465–2487.

Kumar, Anil, Fei Chen, Michael Barlage, Michael B. Ek, and Dev Niyogi. 2014. Assessing impacts of integrating MODIS vegetation data in the weather research and forecasting (WRF) model coupled to two different canopy-resistance approaches. *Journal of Applied Meteorology and Climatology* 53 (6): 1362–1380.

Lawrence, David M., Peter E. Thornton, Keith W. Oleson, and Gordon B. Bonan. 2007. The partitioning of evapotranspiration into transpiration, soil evaporation, and canopy evaporation in a GCM: Impacts on land–atmosphere interaction. *Journal of Hydrometeorology* 8 (4): 862–880.

Lee, Eungul, Thomas N. Chase, Balaji Rajagopalan, Roger G. Barry, Trent W. Biggs, and Peter J. Lawrence. 2009. Effects of irrigation and vegetation activity on early Indian summer monsoon variability. *International Journal of Climatology* 29 (4): 573–581.

Liu, Yaling, Zhihua Pan, Qianlai Zhuang, Diego G. Miralles, Adriaan J. Teuling, Tonglin Zhang, Pingli An, et al. 2015. Agriculture intensifies soil moisture decline in Northern China. *Scientific Reports* 5: 11261.

Liu, Xing, Fei Chen, Michael Barlage, Guangsheng Zhou, and Dev Niyogi. 2016. Noah-MP-Crop: Introducing dynamic crop growth in the Noah-MP land surface model. *Journal of Geophysical Research: Atmospheres* 121 (23).

Nayak, H.P., K.K. Osuri, P. Sinha, R. Nadimpalli, U.C. Mohanty, F. Chen, M. Rajeevan, and D. Niyogi. 2018. High-resolution gridded soil moisture and soil temperature datasets for the Indian monsoon region. *Scientific Data* 5: 180264.

Niu, Guo-Yue, Zong-Liang Yang, Kenneth E. Mitchell, Fei Chen, Michael B. Ek, Michael Barlage, Anil Kumar, et al. 2011. The community Noah land surface model with multiparameterization options (Noah-MP): 1. Model description and evaluation with local-scale measurements. *Journal of Geophysical Research: Atmospheres* 116 (D12).

Niyogi, Dev, Kiran Alapaty, Sethu Raman, and Fei Chen. 2009. Development and evaluation of a coupled photosynthesis-based gas exchange evapotranspiration model (GEM) for mesoscale weather forecasting applications. *Journal of Applied Meteorology and Climatology* 48 (2): 349–368.

Niyogi, Dev, Chandra Kishtawal, Shivam Tripathi, and Rao S. Govindaraju. 2010. Observational evidence that agricultural intensification and land use change may be reducing the Indian summer monsoon rainfall. *Water Resources Research* 46 (3).

Niyogi, Dev, Patrick Pyle, Ming Lei, S. Pal Arya, Chandra M. Kishtawal, Marshall Shepherd, Fei Chen, and Brian Wolfe. 2011. Urban modification of thunderstorms: An observational storm climatology and model case study for the Indianapolis urban region. *Journal of Applied Meteorology and Climatology* 50 (5): 1129–1144.

Niyogi, Dev, Subashini Subramanian, U.C. Mohanty, C.M. Kishtawal, Subimal Ghosh, U.S. Nair, M. Ek, and M. Rajeevan. 2018. The impact of land cover and land use change on the Indian monsoon region hydroclimate. In *Land-atmospheric research applications in South and Southeast Asia*, 553–575. Cham: Springer.

Noilhan, J., and S. Planton. 1989. A simple parameterization of land surface processes for meteorological models. *Monthly Weather Review* 117 (3): 536–549.

Oke, T.R. 1973. City size and the urban heat island. *Atmospheric Environment* 7 (8): 769–779 (1967).

Osuri, Krishna K., U.C. Mohanty, A. Routray, and M. Mohapatra. 2012. The impact of satellite-derived wind data assimilation on track, intensity and structure of tropical cyclones over the North Indian Ocean. *International Journal of Remote Sensing* 33 (5): 1627–1652.

Osuri, K.K., R. Nadimpalli, U.C. Mohanty, F. Chen, M. Rajeevan, and Dev Niyogi. 2017. Improved prediction of severe thunderstorms over the Indian Monsoon region using high-resolution soil moisture and temperature initialization. *Scientific Reports* 7: 41377.

Pielke, Roger A., et al. 2002. The influence of land-use change and landscape dynamics on the climate system: Relevance to climate-change policy beyond the radiative effect of greenhouse gases. *Philosophical Transactions of the Royal Society of London A: Mathematical, Physical and Engineering Sciences* 360 (1797): 1705–1719.

Pielke, Roger A., Andy Pitman, Dev Niyogi, Rezaul Mahmood, Clive McAlpine, Faisal Hossain, Kees Klein Goldewijk, et al. 2011. Land use/land cover changes and climate: Modeling analysis and observational evidence. *Wiley Interdisciplinary Reviews: Climate Change* 2 (6): 828–850.

Pitman, A.J. 2003. The evolution of, and revolution in, land surface schemes designed for climate models. *International Journal of Climatology* 23 (5): 479–510.

Rodell, Matthew, Isabella Velicogna, and James S. Famiglietti. 2009. Satellite-based estimates of groundwater depletion in India. *Nature* 460 (7258): 999.

Rodell, M., J.S. Famiglietti, D.N. Wiese, J.T. Reager, H.K. Beaudoing, F.W. Landerer, and M.-H. Lo. 2018. Emerging trends in global freshwater availability. *Nature* 1.

Roxy, Mathew Koll, Kapoor Ritika, Pascal Terray, Raghu Murtugudde, Karumuri Ashok, and B.N. Goswami. 2015. Drying of Indian subcontinent by rapid Indian Ocean warming and a weakening land-sea thermal gradient. *Nature Communications* 6: 7423.

Roy, Shouraseni Sen, Rezaul Mahmood, Dev Niyogi, Ming Lei, Stuart A. Foster, Kenneth G. Hubbard, Ellen Douglas, and Roger Pielke. 2007. Impacts of the agricultural Green Revolution-induced land use changes on air temperatures in India. *Journal of Geophysical Research: Atmospheres* 112 (D21).

Saha, Subodh K., Subhadeep Halder, K. Krishna Kumar, and B.N. Goswami. 2011. Pre-onset land surface processes and 'internal' interannual variabilities of the Indian summer monsoon. *Climate Dynamics* 36 (11–12): 2077–2089.

Santamouris, M. 2015. Analyzing the heat island magnitude and characteristics in one hundred Asian and Australian cities and regions. *Science of the Total Environment* 512: 582–598.

Schmid, Paul E., and Dev Niyogi. 2013. Impact of city size on precipitation-modifying potential. *Geophysical Research Letters* 40 (19): 5263–5267.

Schmid, Paul E., and Dev Niyogi. 2017. Modeling urban precipitation modification by spatially heterogeneous aerosols. *Journal of Applied Meteorology and Climatology* 56 (8): 2141–2153.

Sellers, Piers J., Compton J. Tucker, G. James Collatz, Sietse O. Los, Christopher O. Justice, Donald A. Dazlich, and David A. Randall. 1996. A revised land surface parameterization (SiB2) for atmospheric GCMs. Part II: The generation of global fields of terrestrial biophysical parameters from satellite data. *Journal of Climate* 9 (4): 706–737.

Shastri, Hiteshri, Supantha Paul, Subimal Ghosh, and Subhankar Karmakar. 2015. Impacts of urbanization on Indian summer monsoon rainfall extremes. *Journal of Geophysical Research: Atmospheres* 120 (2): 496–516.

Unnikrishnan, C.K., M. Rajeevan, S. Vijaya Bhaskara Rao, and Manoj Kumar. 2013. Development of a high resolution land surface dataset for the South Asian monsoon region. *Current Science* 105 (9): 1235–1246.

Webster, Peter J., Vo Oo Magana, T.N. Palmer, J. Shukla, R.A. Tomas, M.U. Yanai, and T. Yasunari. 1998. Monsoons: Processes, predictability, and the prospects for prediction. *Journal of Geophysical Research: Oceans* 103 (C7): 14451–14510.

Yang, Long, Fuqiang Tian, James A. Smith, and Hu Heping. 2014. Urban signatures in the spatial clustering of summer heavy rainfall events over the Beijing metropolitan region. *Journal of Geophysical Research: Atmospheres* 119 (3): 1203–1217.

Zhou, Liming, Robert E. Dickinson, Yuhong Tian, Jingyun Fang, Qingxiang Li, Robert K. Kaufmann, Compton J. Tucker, and Ranga B. Myneni. 2004. Evidence for a significant urbanization effect on climate in China. *Proceedings of the National Academy of Sciences of the United States of America* 101 (26): 9540–9544.

Author Index

D. A. Randall et al. (eds.), *Current Trends in the Representation of Physical Processes in Weather and Climate Models*, Springer Atmospheric Sciences,
https://doi.org/10.1007/978-981-13-3396-5

MIX
Papier aus verantwortungsvollen Quellen
Paper from responsible sources
FSC® C105338

If you have any concerns about our products,
you can contact us on
ProductSafety@springernature.com

In case Publisher is established outside the EU,
the EU authorized representative is:
Springer Nature Customer Service Center GmbH
Europaplatz 3, 69115 Heidelberg, Germany

Printed by Libri Plureos GmbH
in Hamburg, Germany